'Finished Labour of a Thousand Hands'

The archaeology of the Combe Down Stone Mines, Bath, Somerset

by Lynn Willies, Neville Redvers-Higgins, and Ianto Wain

with contributions by

Lisa Brown, John Cotter, Damian Goodburn, Jill Hind, Marek Lewcun, Chris Salter, Ian Scott, Ruth Shaffrey and Lena Strid

Illustrations by

Markus Dylewski, Magdalena Wachnik, Georgina Slater and Anne Kilgour Cooper

Reconstruction drawings by

Mark Gridley

Oxford Archaeology Monograph No. 13

2011

The publication of this volume has been generously funded by the Combe Down Stone Mines Project which was managed by Bath & North East Somerset Council with funding from the Homes and Communities Agency's Land Stabilisation Programme

Published for Oxford Archaeology as part of the Oxford Archaeology Monograph series

Designed by Oxford Archaeology Graphics Office

Edited by Lisa Brown and Alan Hardy

Front cover *Prior Park: The Seat of Ralph Allen* engraving by Anthony Walker *c* 1750 (courtesy of The Building of Bath Museum)

Oxford Archaeology Monograph No 13

ISBN: 978-0-904220-60-5

Typeset by Production Line, Oxford
Printed in Great Britain by Information Press, Eynsham, Oxon

Contents

List of Figures

List of Tables

Summary

Between November 2000 and August 2008 Oxford Archaeology carried out a programme of underground monitoring and detailed recording of a section of the underground stone quarries at Combe Down, Bath. The works were carried out prior to and during the stabilisation of the quarries, a process that was carried out by concrete infilling. The quarries (legally mines during the Stabilisation Scheme), were situated below the central Conservation Area of Combe Down, which lies between 1.5 and 2 km south of the historic centre of Bath. A large proportion of the parish was subject to either quarrying underground or at surface over a long period, but mainly between 1730 and 1860, when the area provided the main building material for the 'golden age' of Bath. Quarrying in the outer areas of the parish continued, but the main supply of Bath Stone thereafter came from the much larger, rail-served Wiltshire quarries.

The recording works were carried out as a long-running Watching Brief, during the construction of a network of supported engineering roadways designed to initially provide safe access to all areas of the mine and latterly to facilitate the filling of the mines with foam concrete. The recording was primarily by drawn and photographic methods, although a range of other techniques such as laser scanning and video photography were employed to augment the traditional techniques.

The survey, augmented by programmes of documentary and archival research, has enabled a detailed chronology and technological analysis of the development of the underground workings to be established. These have demonstrated that although stone quarrying was certainly being carried out in the early 18th century and may have been undertaken as far back as the Roman period, the first large-scale exploitation of the quarries occurred during the middle years of the 18th century (*c* 1730-1764) under the ownership of the entrepreneur Ralph Allen, who introduced innovative techniques and industrial organisation to the process. Extensive quarrying continued after Allen's death in 1764 although the central organising influence was ended and the work was conducted by a number of independent quarrymasters or freemasons. Underground quarrying continued in the central Combe Down area until the 1860s and both underground and surface quarrying continued on the fringes of the area until the 1930s.

The Combe Down Stone Mines Project is the first time in the UK that the full resources of a professional archaeological body, Oxford Archaeology, have been used to carry out a large-scale archaeological examination of underground quarries. The project successfully demonstrated that such works are compatible with ongoing mining operations where old workings are to be disturbed by development. The strategy integrated thorough archaeological investigation with an extensive examination of documentary and published resources, and the results are presented in this volume.

Zusammenfassung

Zwischen November 2000 und August 2008 unternahm Oxford Archaeology Beobachtungs- und detaillierte archäologische Aufzeichnungsarbeiten in einem untertägigen Teilbereich der Combe Down Steinbrüche in Bath. Die Arbeiten wurden im Vorfeld, sowie während des Stabilisierungsprozesses der Stollen durchgeführt, bei welchem diese mit Gasbeton verfüllt wurden. Die Steinbrüche, welche während des Stabilisierungsprozesses den rechtlichen Status von Minen innehatten, lagen zentral unterhalb des Flächendenkmalbereichs von Combe Down, welcher 1,5 bis 2 km südlich des historischen Stadkerns von Bath liegt.

In einem großen Bereich des Stadtteils Combe Down wurde über einen langen Zeitraum hinweg unter- oder obertage Stein abgebaut, wobei die Blütezeit des Abbaus zwischen den Jahren 1730 und 1860 lag, in welchen Combe Down die Materialien für das "goldene Zeitalter" Baths zur Verfügung stellte. Steinbrüche in den Außenbezirken der Gemeinde wurden auch im Anschluss noch weiter betrieben. Der Hauptlieferant des "Bathstone" wurden jedoch die sehr viel größeren, an das Schienennetz angebundenen, Steinbrüche in Wiltshire.

Die archäologischen Aufzeichnungsarbeiten wurde während der anhaltenden Baubegleitung für die Konstruktion des unterirdischen Straßennetzes angefertigt, welches anfangs sicheren Zugang zu allen Teilbereichen der Mine garantieren sollte und später dabei half, die Stollen mit Gasbeton zu füllen. Aufgezeichnet wurde in erster Linie durch Zeichnungen und Photographien, wenngleich andere Techniken, wie z.B. Laserscanner und Videotechnik eingesetzt wurden um die traditionellen Methoden zu unterstützen und zu erweitern.

Die Expertise, angereichert durch Ergebnisse der Quellen- und Archivarbeit, ermöglichte eine detaillierte Chronologie und eine technische Analyse der Entwicklung der untertägigen Grubenfelder anzufertigen. Es wurde aufgezeigt, dass –obwohl Arbeit im Steinbruch sicherlich im frühen 18. Jahrhundert und vielleicht sogar schon in der Römerzeit stattgefunden haben – die ersten groß angelegten Abbauprojekte im Steinbruch in der Mitte des 18. Jahrhunderts (ca. 1730-1764), in welchem der Unternehmer Ralph Allen Eigentümer des Steinbruchs war und innovative Techniken und eine industrielle Organisation einführte, ihren Anfang nahmen. Eine umfangreiche Förderung wurde auch nach Allens Tod im Jahr 1764 fortgeführt, jedoch endete die zentrale Organisation des Abbaus und die Arbeiten wurden von unabhängigen Steinbruchmeistern und Freimaurern durchgeführt. Der untertägige Abbau im zentralen Gebiet von Combe Down dauerte noch bis in die 1860ger Jahre an, in den Außenbezirken von Combe Down wurde sogar noch bis in die 1930er Jahre unter- und obertägig abgebaut.

Das "Combe Down Stone Mines Project", ist das erste Projekt seiner Art in Großbritannien, welches mit Hilfe einer professionellen archäologischen Institution, Oxford Archaeology, eine groß angelegte archäologische Untersuchung eines untertägigen Steinbruchs durchfgeführt hat. Es hat sich dabei herausgestellt, dass archäologische Arbeiten auch bei gleichzeitigem Betrieb der Mine durchführbar sind, wenn alte Grubenfelder durch Bauvorhaben gefährdet werden. Die Ergebnisse der ganzheitlichen, archäologische Forschungsstrategie, in Zusammenschluss mit einer umfangreichen Auswertung von Dokumenten und publizierten Quellen, ist in diesem Band wiedergegeben.

Résumé

Entre novembre 2000 et août 2008, Oxford Archaeology a exécuté un programme de suivi souterrain et de relevé détaillé d'une section des carrières de pierre souterraines à Combe Down, Bath. Les travaux ont été effectués préalablement et pendant la stabilisation des carrières, un procédé opéré grâce au comblement en ciment. Les carrières (légalement des mines durant le Projet de Stabilisation) étaient situées sous la Zone de Conservation centrale de Combe Down, localisée 1.5 km à 2 km au sud du centre historique de Bath. Une large proportion de la Paroisse de Combe Down a fait l'objet soit d'une extraction souterraine ou en surface sur une longue période mais surtout entre 1730 et 1860, lorsque la région produisait le matériau de construction principal de "l'âge d'or" de Bath. L'exploitation dans les environs de la paroisse a continué, toutefois la pierre de Bath provient ensuite principalement des carrières bien plus importantes de la région du Wiltshire desservies par voie ferroviaire.

Les travaux de relevé ont été réalisés sous forme de surveillance de travaux à long terme, durant la construction d'un réseau d'ingénierie de voies souterraines conçu initialement pour fournir un accès sécurisé à toutes les parties de la mine et dernièrement pour faciliter le remplissage de la mine de béton allégé. Le relevé a d'abord été effectué par des méthodes graphiques et photographiques, bien qu'une gamme d'autres techniques telles que le balayage au laser et la photographie vidéo aient été employées pour ajouter aux techniques traditionnelles.

L'étude, complétée par des programmes de recherche documentaire et archivistique a permis une chronologie détaillée et une analyse technologique permettant d'établir l'évolution des carrières souterraines. Elles ont démontré que, bien que les extractions de pierre aient certainement eu lieu au début du XVIIIe siècle et aient peut-être continué jusqu'à l'époque romaine, la première exploitation des carrières s'est produite au milieu du XVIIIe siècle (*c.* 1730-1764), alors propriété de l'entrepreneur Ralph Allen, qui a introduit des techniques innovantes et une organisation industrielle au processus. Une exploitation extensive a continué après la mort d'Allen en 1764 mais l'influence de l'organisation centralisée s'est achevée et le travail a ensuite été conduit par plusieurs maîtres de carrières ou francs-maçons indépendants. L'extraction souterraine s'est poursuivie dans la zone centrale de Combe Down jusqu'aux années 1860 et l'extraction souterraine et en surface a perduré aux abords du site jusqu'aux années 1930.

Le Projet des Mines de Pierre de Combe Down représente le premier projet au Royaume-Uni ayant occupé l'ensemble des ressources d'une organisation archéologique professionnelle, Oxford Archaeology, pour effectuer l'examen archéologique à grande échelle de carrières souterraines. Ce projet a démontré avec succès que de tels travaux sont compatibles avec la poursuite des opérations d'exploitation minières lorsque des chantiers existants sont perturbés par le développement en cours. La stratégie a intégré une enquête archéologique minutieuse à l'examen approfondi des ressources documentaires et publiées. Les résultats sont présentés dans ce volume.

Acknowledgements

In nearly nine years of archaeological investigation under and over Combe Down, and more than a further year of post-excavation work, the debt incurred to others by both Oxford Archaeology and the authors has been immense. Our primary debt of gratitude must go to Bath and North-East Somerset Council and to the Homes and Communities Agency (formerly English Partnerships) for their support and generous funding of this project over the course of the fieldwork and post-excavation phases.

It is from the recognition of the quarries' (Stone Mines) national importance by the Bath and North East Somerset Council in conjunction with English Heritage (regional inspectors Amanda Chadburn and Phil McMahon), that the impetus for the 'preservation by recording' programme originated.

Throughout the project the authors and Oxford Archaeology benefited from very fruitful co-operation with Bath and North East Somerset Council staff, notably the Combe Down Stone Mines Team and its leader Mary Sabina Stacey and the Council's Archaeological Officers Bob Sydes and Richard Sermon. We have been ably assisted by a number of project co-ordinators over the years including Hannah Atkinson, Jen Albano, Rachael Thomas and Jessica Turner (née Bowden). We are grateful to Susan Fox of the Roman Baths Museum for her continuing assistance with the finds.

We have worked closely with the engineering and contracting companies, including Parsons Brinckerhoff with Ross Singleton as their Technical Director, who carried out the initial engineering assessment, and Scott Wilson (Mining) and its Site Engineer Dr Colin Harris who took over for the main project. We benefited from the guidance of Annette Roe, the Scott Wilson Archaeological Monitor. The main contractors throughout were Hydrock Contracting with its Managing Director, Dr Brian McConnell and on-site Director Dr Rob Narbett. These, with the mine manager, John Harvey and the substitute mine manager John Lister, and with the determined influence of Her Majesty's Inspector of Mines, Mr A Forster, were collectively responsible for our safety and for ensuring we could meet our recording objectives. Our day-to-day fieldwork was considerably assisted by the diligence of Katie Alexander, Adrian Brooks and Phil Brown of Davis Langdon and the course of the project smoothed by Paul Wells and Neill Dowsett of Provelio.

On a more personal level, the authors would like to thank those who worked directly with us underground for varying periods: Pete Burge, Nathan Chichen, Tom Davis, Sarah Lane, Mike Simms, Phil Stastney and Kim Watkins. We cannot individually thank all the OA staff who contributed to this project during its eleven-year lifespan but special mention should go to the team of dedicated GIS staff who spent so many hours diligently digitising, analysing and shaping the plans that form the backbone of many of the illustrations in this volume. Particular thanks therefore to Anne Kilgour Cooper, Evangelia Kappa, Robin Latour, and Mercedes Planas. The project was initially managed by Robert Kinchin-Smith, a role then taken over by Ianto Wain. Bob Williams provided overall management and guidance throughout the project. The post-excavation programme benefited from the management of Lisa Brown. The initial draft of the report was read by, and benefited from, Dr John Barnatt's diligence.

We enjoyed a very good relationship with the Combe Down Heritage Society and Combe Down residents, and particularly wish to thank Professor Richard Irving, who was stimulated to write a book on the Byfield Mine (2005) and who also provided us with important documentary evidence, as also did Rosemary Simmons and Malcolm Aylett. Thanks also go to Roger and Julia Phillips whose house had its own underground workings accessed through their garage floor, and who donated a frig bob (stone saw) which we later used to saw out graffiti slabs. The landlord of the Hadley Arms, who also had his own access to below, via a well in his rear bar-room, allowed us access to his property deeds (Hadley Arms Papers) which filled several gaps in our knowledge.

Information came from a wide range of sources, both individual and institutional. Prior knowledge of the underground workings, freely shared with us, principally came from David Pollard, but also from Dr Brian Hawkins, Derek Hawkins, Philip Wooster and Paul De'Ath; on the local historical maps and plans, Mike Chapman, and on stone quarrying generally, Paul Sowan. Peter Davenport, formerly of the Bath Archaeological Trust, advised on the amount of stone used in buildings, and Mike Chapman and Brenda Buchanan did the same on Bath history. For help with documentary searches we thank the staff of the Bath Record Office (BRO), The Bath Central Library (BCL) and the Taunton, Somerset (SRO), and Gloucester Record Offices (GRO), and for access to the Hadley Estate papers (discovered there by David Pollard), the Lewisham Library (LL), and for detail on Bath generally, the Bath at Work and the Building of Bath Museums. We were helped in the supply of illustrations by

Dan Brown of www.bathintime.co.uk. and the Victoria Art Gallery, Bath, the Victoria and Albert Museum, London and the Bath Preservation Trust. Philip Wooster, Paul Deakin, David Pollard and Mike Chapman all supplied images used, but also many more for the research collection.

External expertise was brought in for laser scanning (APR Ltd) and for video recording, initially Peter Egglestone of IA Recording and, later, Rob Franklin. Assistance with the preservation and removal of graffiti and general conservation works for the finds were supplied by Dana Goodburn-Brown. Thanks must also go to Gerald Melksham for his assistance in the physical removal of some of the more significant graffiti pieces.

On a workaday basis, our thanks go to our mining colleagues. Initially this was a dozen or so people whom we were able to get to know very well indeed: most were former coal miners and some owners (in a small way) from South Wales – the exception came from the Forest of Dean. As the project developed, first to around 50 people, then to around 250 it sometimes became faces rather than names (apologies), but all the miners and other underground and surface workers appear to have appreciated what was being seen and found, and ensured it was brought to our attention. We are, certainly unfairly to the others, naming only two mining colleagues here, Elwyn Davis who became the underground deputy and Steve Dix the mine surveyor, both of whom saved hundreds of hours by their prompt actions in the allocation of work and in ensuring our maps were up to date. The rest of the miners who helped us on a day-to-day and week-to-week basis cannot all be listed here by name, but their help and co-operation was invaluable.

Thank you all.

Lynn Willies
Neville Redvers-Higgins
Ianto Wain

Extracts from *A Description of Bath* by Mary Chandler (1687–1745)

...View the brown Shadows of yon *pathless Wood;*
And craggy Hills, irregular and rude!

Where Nature sports romantic: Hence is seen
The *new made Road,* and wonderful Machine,
Self-moving downward from the Mountain's Height,
A *Rock* its Burden of a *Mountain's* Weight.
HAIL, mighty *Genius*! Born for *Great Designs,*
T'*adorn* your *Country,* and to *mend* the *Times;*

...PROPHETIC here, the Muse shall build thy Seat,
Great like thy *Soul,* in ev'ry Part complete:

....On this fair Eminence the fabric stands,
The finish'd Labour of a thousand Hands;

...Thy Taste refin'd appears in yonder Wood,
Not *Nature* tortur'd, but by *Art* improv'd:
Where *cover'd Walks* with *open Vista's* meet,
An *Area* here, and there a *shady Seat....*

...How from the Mountain's rocky Sides he drew
A thousand shining Palaces to view:
Temples, and *Hospitals* in ev'ry Land,
From Age to age, his Monuments shall stand.
Envy itself shall die, and fickle *Fame,*
When he is dead, do Justice to his Name.

Mary Chandler (1687-1745) was a poet and an early champion of women's independence, and became famous for this poem, which glorifies the city and environs of Bath, and extols the achievements of Ralph Allen.

Chapter 1

The stabilisation of Combe Down and its archaeological recording

Introduction

In the mid 18th century Combe Down saw the development of the first large-scale underground stone workings in England, which were fundamental to the creation of Georgian Bath.

In about 1730 Ralph Allen, with a fortune still being made from his famous reforms of the English postal system, bought the Down with its beds of freestone a mile south of the City as part of his growing landed estate. He introduced new techniques and integrated business methods to the quarrying industry. These increased the level of production and reduced the price of stone so effectively, that it permitted the vision of the architects John Wood and his son (also John), to be realised, removing the temptation to use materials other than the local stone, which has been so important in creating the cohesiveness of Bath's urban landscape (Fig. 1.1).

A century before the generally better known underground quarries at Box and Corsham in Wiltshire repeated Allen's success on an even larger scale, he systematised the methods of underground working and at surface introduced the use of cranes, a railway and roll-on-roll-off river transport, supplying both Bath, and via the Avon Navigation and coastal shipping, also more distant projects such as the St Bartholomew's Hospital in London.

This and later working at Combe Down left some 18 hectares of underground workings under the quarrying village Allen had established, which had gradually grown into the modern suburb on the southern outskirts of Bath. Located just a few metres below some hundreds of houses in Combe Down's central conservation area, the workings had become dangerously unstable and were considered to be liable to significant collapse within a decade. Between 2000 and 2008 they were stabilised by being infilled with foamed concrete.

Acknowledged as archaeologically of National Importance and of being Internationally Significant and part of the World Heritage Site of Bath, their effective destruction was therefore a major loss. Initial archaeological assessments had established the quality and variability of the extant underground features but also that they were poorly understood. Bath & North East Somerset Council in conjunction with English Heritage determined that a detailed archaeological examination to the highest possible standard would be the major part of the mitigation.

There are no precedents in the UK for such extensive and detailed archaeological work underground and the project offered a unique opportunity to study working techniques within a well-defined underground quarrying landscape and, in conjunction with documentary and literary research, to assess the economic and social impact on the village and on Bath. In particular, there was potential to examine in detail the technical equivalent underground, of Ralph Allen's well-known use of cranes and railways at surface. In the event, it was also possible to combine archaeological with historical investigation to assess the success of his stone business, which, though beneficial to the development of Bath, was probably less so to Allen himself.

The problems of archaeological investigation were complex, both in terms of health and safety and the archaeological methodology. The instability prevented access to many parts and close-up viewing of others and conventional archaeological excavation was not possible. It was thus necessary to devise suitable methods of recording and interpretation for these circumstances. It has also been necessary to determine a suitable terminology to describe and explain an underground landscape of which few people have any knowledge. Where possible we have chosen terms which are compatible with those used in mining and quarrying reports for the English Heritage Monuments Protection Programme: (Cranstone 1992; Willies 1993; Ashbee 1996; Chitty 1996). This report and the associated web site: (www.combedownstoneminesproject.com), and a proposed exhibition of the findings are the results which are an important contribution to the history of Bath and of significance for national and international research agendas for the post-medieval mining and quarrying industries.

This introductory chapter outlines the project history and administrative aspects of the report, along with the aims and objectives and the methodologies used to achieve them. A short note examines its place in the study of underground archaeology nationally and internationally.

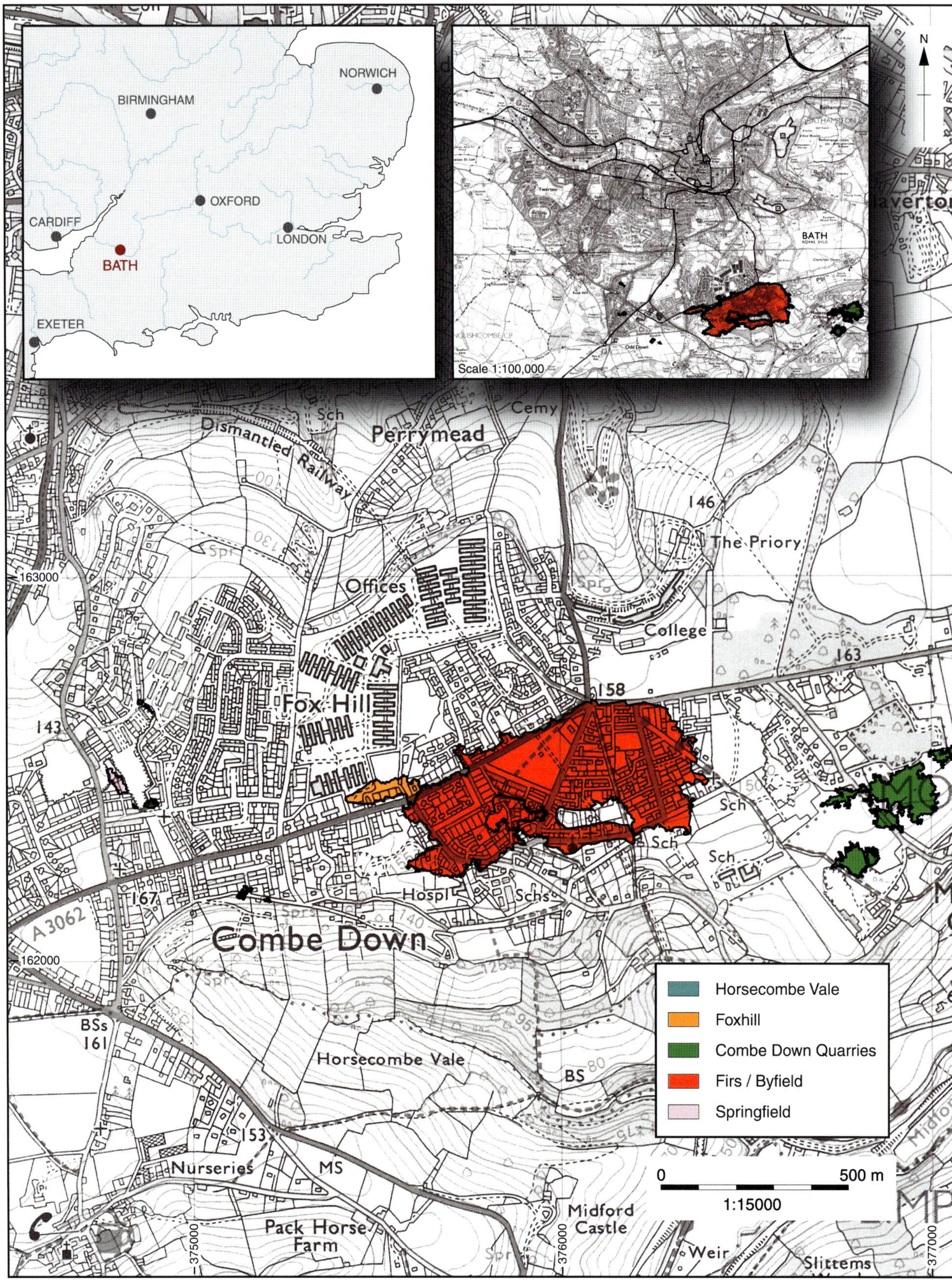

Fig. 1.1 Combe Down with the Combe Down Stone Mines site and other infilled underground quarries within the surrounding area. The centre of the main site lies at NGR ST 7662

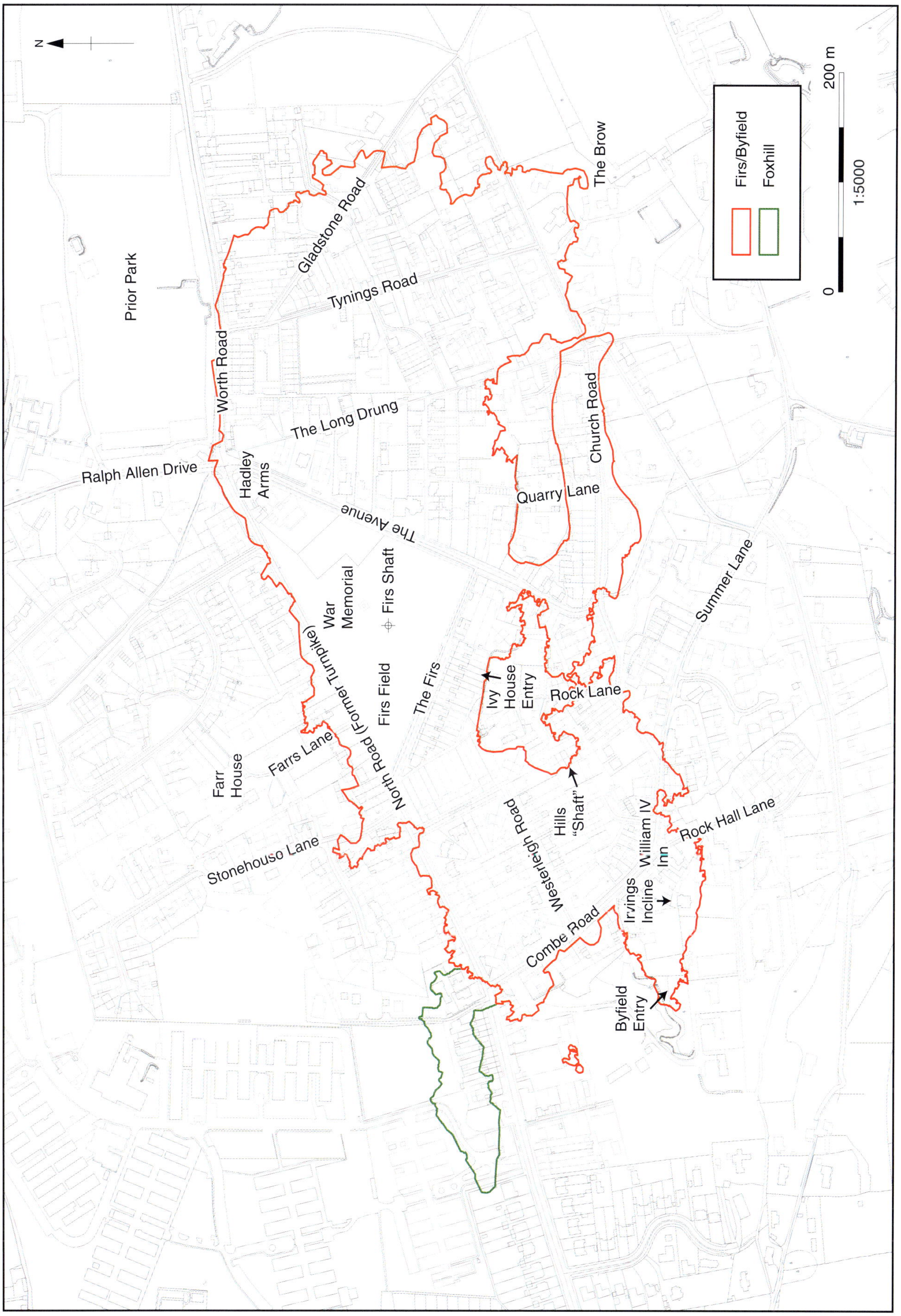

Fig. 1.2 Central Combe Down with named roads and paths and the Firs-Byfield underground quarries outline (this plan is also reproduced inside the front cover)

The Combe Down Stone Mines Project

The term Combe Down Stone Mines has been used to describe the 18 or so hectares of underground workings covered by the Stabilisation Scheme of the areas that had become known as Firs and Byfield 'Mines' in recent years, along with other outlying quarries in the local area, including those at Foxhill, Horsecombe Vale, Mount Pleasant, Springfield and Entry Hill, and Shaft Road, and The Lawns, St Winifreds and Freylings Quarries. (Fig. 1.2).

The Combe Down Stone Mines were in fact underground quarries – places where squared, dimensional, building stone was dug. Underground quarries have been legally classified as mines since the Metalliferous Mines Act of 1872.

Bath Stone, a similar stone to Combe Down stone, is still quarried underground today, and so is both a living industry and an ancient tradition which still uses the term quarrying to describe its activities. For the purposes of this report the traditional name is maintained and they are referred to throughout as quarries, as has been the local tradition (Pollard pers. comm.)

The structure of the report

The report is split into groups of chapters in four major sections:

The first section (Chapters 1-6) is essentially based on documentary and literary sources providing the geological and historical background of Combe Down and Bath, and the main features of the surface landscape including changes from surface quarrying. The concept of an underground quarrying landscape introduces some of the necessary terminology to understand the key chapter on Ralph Allen and his quarrying business and the following chapters.

The second section (Chapter 7-10) develops the results of the underground recording. Four chapters define and describe the broad principles and then the technical aspects of underground quarrying and its development over time in terms of methods of working, transport, and roof support and spoil management.

The third section (Chapters 11-12), following a summary of the specialist reports on finds, describes the geography and phasing of the quarries. The great area involved and the limitations of access into many parts of the quarries has led to the presentation, alongside the phasing, of a series of case studies representing each chronological phase of activity for areas in which a reasonably comprehensive picture could be reconstructed.

Finally, Chapter 13 presents a concluding discussion on the development of the stone quarrying in Combe Down, drawing together the archaeological and historical evidence, and seeks to evaluate the success of the project.

The archive

The project archive is currently with Oxford Archaeology, and will be deposited with Roman Baths Museum, Bath, under the Accession No. BATRM:2001.63. The digital archive will be available to view through the Archaeological Data Service (ADS – www.ahds.ads.ac.uk). Full artefactual specialist reports, plans and other contextual records of features etc, and a library of drawn and photographic images will also be available on the project website www.combedownstonemines.com.

A note on terminology

To aid the non-specialist reader Table 1.1 below lists a summary explanation of frequently used quarrying terms. More detailed explanations are provided at suitable points in the text.

Phasing of the quarry developments

The phasing of the quarries is discussed in detail in Chapter 12, but in order to refer to developments chronologically as they occur, the phases have been used throughout the text.

Phase I.	Pre-Ralph Allen – prior to his *c* 1730 developments
Phase II.	Allen period – *c* 1730 to his death in 1764
Phase III	Allen Estate period 1764 to *c* 1803 and the estate break-up
Phase IV	Early 19th century 1803-33
Phase V	Mid 19th century 1833-67
Phase VI	Late 19th and early 20th century 1867-1938

The phasing plan shown in Figure 1.3 is derived from a combination of the historical documentation and the archaeological evidence produced from this and earlier investigative work, and shows the interpretative sequence of underground quarrying by Quarry Area.

History of the Combe Down Stone Mines Project

The Combe Down underground quarries (legally mines during the Stabilisation Scheme), were situated below the central Conservation Area of the Combe Down Ward which lies between 1.5 and 2 km south of the historic centre of Bath. A large proportion of the Ward was subject to either quarrying underground or at surface over a long period, but mainly between 1730 and 1860, when the area provided the main building material for the 'golden age' of Bath. Quarrying in the outer areas of the parish continued, but the main supply of Bath Stone thereafter came from the much larger, rail-served Wiltshire quarries.

The principal area examined, known today as Firs and Byfield Mines, was centred more or less

Table 1.1 Glossary of frequently used terms

Term		Definition
barrow-way		narrow pathway for a single wheeled barrow
cartway		cleared way for a two-wheeled cart or four-wheeled wagon
chog hole		square-sided hole in the roof to allow the top end of a crane post to be inserted
cockles		local term for vugs (see below)
cogging		modern term (derived from coal mining) in which layers of timber logs with a hollow centre are stacked up from the floor to support the roof
crane	mast crane	the mast (or post) was kept in position by stays or guys, and the jib angled out from the base of the post, secured by an iron bar at the top
	post crane	similar to the mast crane, but the top of the post was secured in a chog hole in the roof
crib		box formed with walls of rubble (or timber) and filled with rubble to support the roof
cross-bedding		geological feature formed in beds with a shallow sea origin formed by wave or tidal movements moving sandy particles on the sea bed. The sinusoidal fronts of the beds which cross the main beds are often made obvious by weathering.
crown hole		small hole at surface opening below to a much larger upwards collapsing fall of roof below ground
fines		spoil – waste rock – composed of fragments below 5mm diameter
gull		natural joint or vertical crack in the freestone widened by landslipping, sometimes filled with washed-in clay
gullet		gap or trench between the working face and the waste rock or spoil stacked a short distance behind the face
heading		the face or driving end of an underground roadway
jad		iron bar with a sharp wedge-end, used with a pick to cut and trim jad slots and cuts
jad cut		vertical slot made using a pick, often to cut out and define a face of a pillar
jad slot	.	narrow and deep slot cut horizontally by jadding or picking one or more beds out from below the roof to allow the beds above and/or below to be broken out using a wedge
Lewis		tool formed of wedge-like iron pieces connected by a bar and used for lifting. The pieces were placed in a dovetailed 'Lewis hole' in the stone to be lifted
Long Room system		method of working in which the face principally advances forwards in one room between two lines of pillars
Long Wall, gullet and pillar system		method of working in which the face advances across a number of rooms in a line
Open or Wide Room system		method of working in which the spoil is removed further away to facilitate use of mechanical haulage equipment, (cranes and winches)
pillar	long	rectangular in plan – the length 3x the width
	direct	has near-vertical faces from floor to roof
	corbelled	the top bed(s) of the pillar are left horizontally protruding as 'ears'
	apophygate	the top curves out, like an inverted bell just under the roof (as an apophyge on an architectural pillar)
	sailed	the vertical plane is inclined and the ends are much narrower than the centre. Usually the result of faulting
ridding		unwanted beds or material above the desired beds
saw		the stone cutting saw was usually a thicker metal version of the timber saw
	frig bob	the equivalent of the cross-cut timber saw, but usually with a handle at one end only
	razzer	this was usually a worn down frig bob, used to start a saw cut where space was too small for the frig bob
scappling		breaking off the surface of the rock using a heavy form of axe
shear legs		pair of timbers erected in an inverted 'V' over a weight to be lifted by some form of winch
sprag (or scorter)		angled timber projecting out from a ledge or hole near the top of a pillar into the roof to increase the support
step(ped)-aside beds		beds displaced sideways because of faulting or local movement due to instability
stone pack		wall formed of rubble to either support a bank of spoil or to support the roof
tipping front		last phase of tipping from a barrow over, and forming the edge of, the spoil heap
vugs		natural mineralised cavities in the freestone
wedge slot		a triangular slot cut by a pick in the rock into which either a wood plank (for standing on, or as an attachment) can be inserted, or into which a wedge can be driven to break the rock out
wedge-and-chip		combination of a wedge driven between two triangular iron plates (chips), placed in the wedge slot to ensure a clean break in the soft freestone
whim		a form of capstan-winch with a vertical post or barrel, usually turned by horse-power
wrist stone		the first stone block broken out (wrested) from a bed exposed at the top of a face to allow the others to be wedged out whole

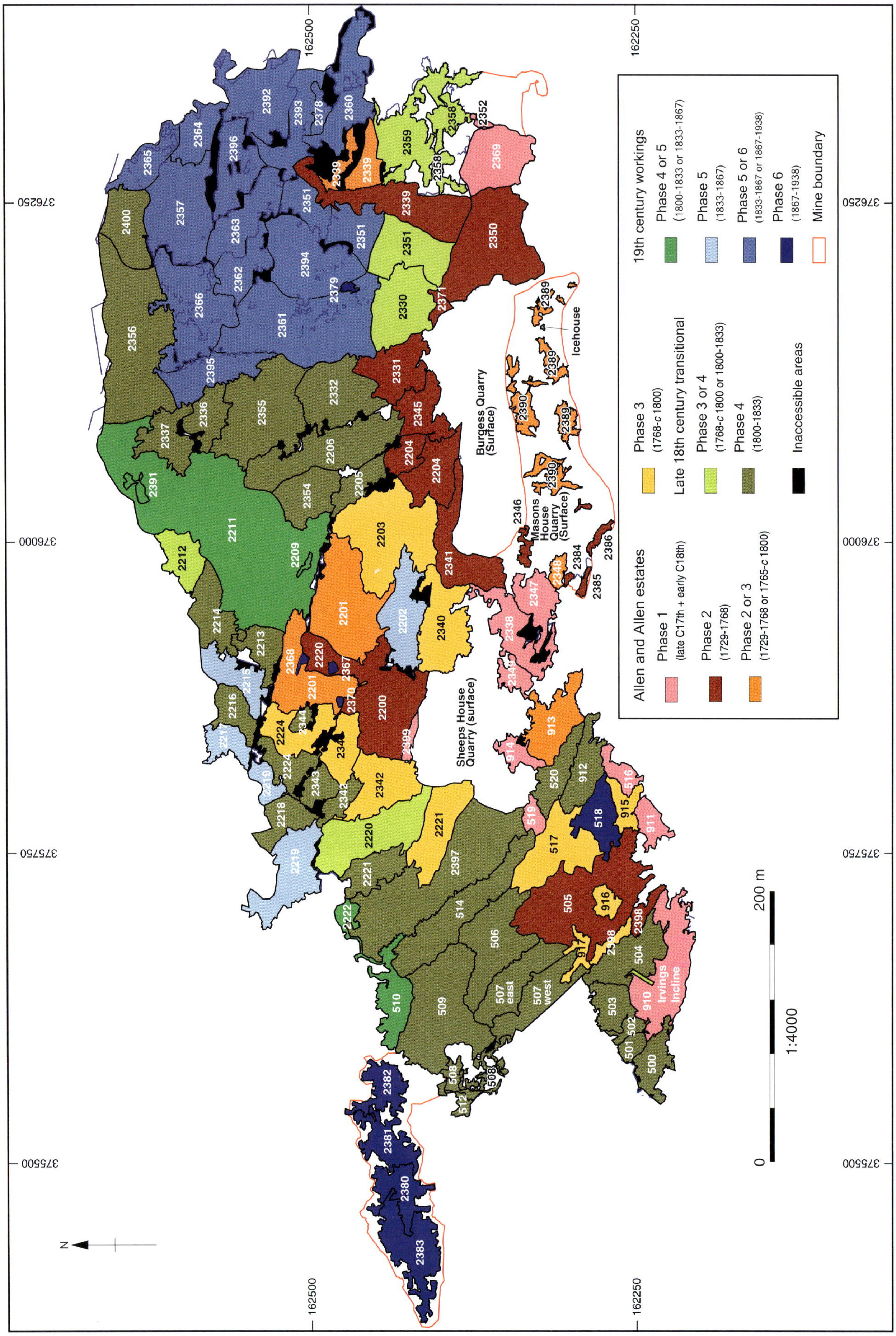

Fig. 1.3 Phase plan of the Quarry Areas within the Combe Down Stone Mines site (this plan is also reproduced inside the back cover)

around the reopened main access shaft (Firs Shaft) on Firs Field – at which point the surface infrastructure of the Stabilisation Scheme was established (see Figure 1.4) – and from the western side of Combe Road through a former surface quarry through the main Byfield Entrance. Their location in relation to the surface is most easily understood by their intrinsic relationship with the modern road system. Roads and paths formed the original boundaries for quarry leases or were developed either to serve the needs of the quarries or followed their working out and subsequent use of the areas for housing. The main underground quarry complex occupies the area south of the line of the North and Bradford Roads, roughly as far as Church Road. The western limit is a short distance west of Combe Road (the old parish boundary), and the so-called Byfield Mine there extends east roughly to Westerleigh Road. It is separated from what has become known as the Firs Mine by a largely infilled quarry (2397), with spoil which was previously inaccessible. Firs Mine extends east from west of Rock Lane across Firs Field (with the Firs Shaft), The Avenue, the Long Drung and over Tyning Road and Gladstone Road until the older area of houses ends.

Concerns about the stability of the mines were initiated by underground explorers in the 1980s. The quarry workings were normally within about 5 to 7 m of the surface and, in small areas, often less. Instability resulted from deteriorating pillar support and delamination of beds. Small collapses had taken place and engineering studies by Halcrow (*c* 1990) for the then Bath City Council, stressed the likelihood of major collapses within a ten-year time-scale. Planning permission was refused for the then proposed methods for stabilisation, but further work was carried out by Hawkins (1994) whose plans also showed floor and roof heights and high hazard areas, together with a report on stability. At the same time a report on the archaeology by Pollard (1994) established the presence of a large range of both typical and unusual features of underground quarrying, which, because of the dynamism of Ralph Allen, who entered the 'stone business' from about 1729 onwards with a wide range of innovations, was potentially of national significance.

Subsequently, Bath & North East Somerset Council (B&NES – the succeeding local government body) gained national government assistance for further investigations. This stage was carried out by engineering consultants Parsons Brinckerhoff and included access works from supported roadways, required by the Mines Inspectorate for safety, to assess the underground conditions. As well as some two kilometres of supported roadways constructed,

Fig. 1.4 Firs Field during the stabilisation project, view east. Firs Shaft is covered by the green circular roof in the centre foreground, and the Materials Shaft (the former Arched Shaft) is at the right of the long green-roofed building in the background (Image courtesy of Bath & North East Somerset Council)

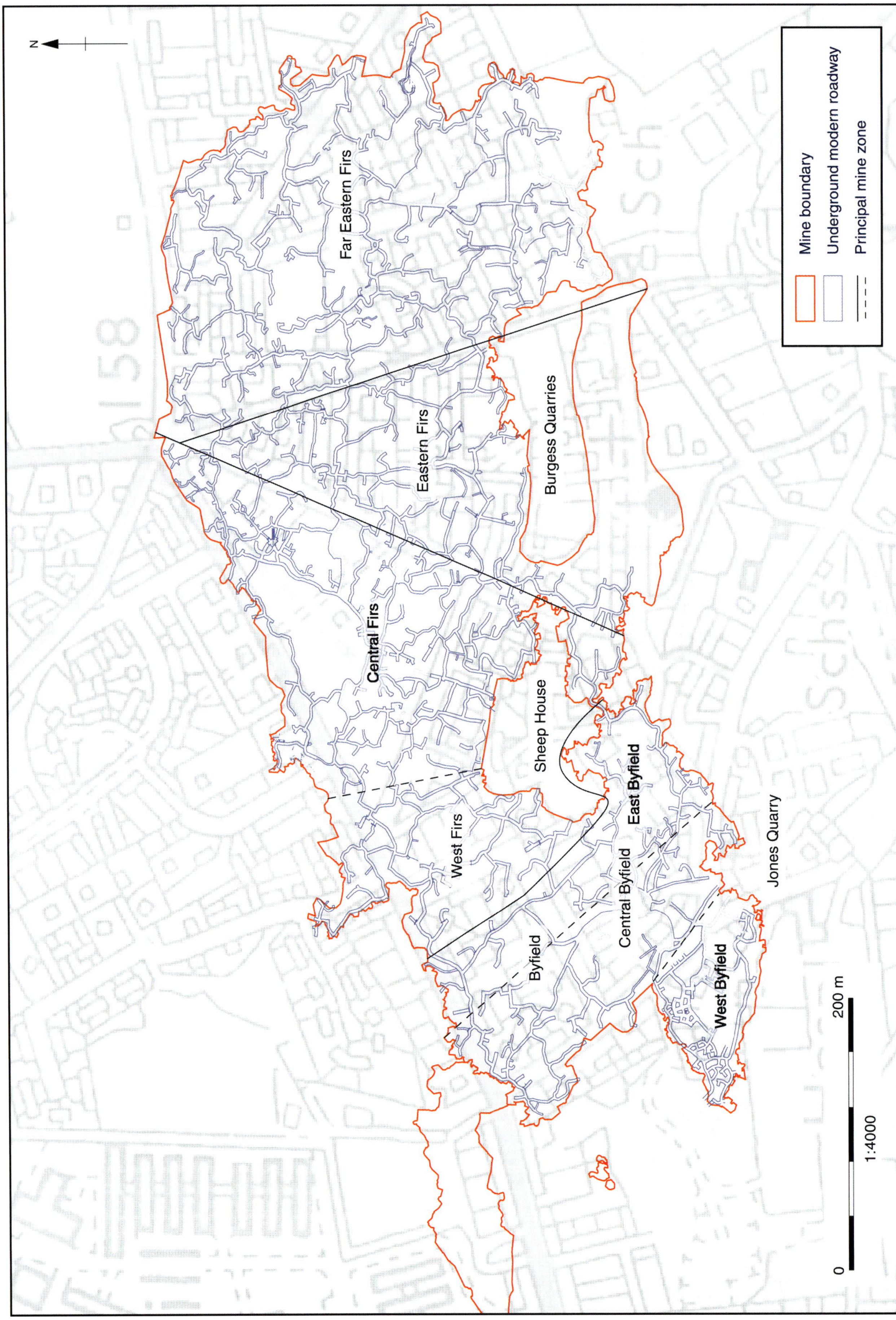

Fig. 1.5 Modern underground road network. Sheeps House, Jones and Burgess quarries were surface quarrying areas

further emergency works were carried out to remedy particularly dangerous situations and to support North Road. As part of this phase, in November 2000 Oxford Archaeology was commissioned to carry out the Environmental Impact Assessment for the proposed stabilisation works. The construction of the supported roadways inevitably resulted in damage to the below-ground archaeological resource and soon after Oxford Archaeology were asked to carry out an archaeological monitoring action or watching brief intended to monitor this process and record affected sites in advance of their destruction.

In 2003 a submission for planning consent for the stabilisation of the workings was granted. The work was put out to tender, with Scott Wilson appointed as the site engineers, and Hydrock, who had carried out the previous works, as the successful contractors. A number of archaeological conditions were attached to the planning consent in order to ensure that the mines were appropriately recorded as archaeological and historical assets, to allow preservation of features or areas of archaeological interest and to allow some aspects of the site to be displayed and presented to the public when physical access was no longer possible. It was agreed that analysis and publication of the results of the investigation and recording programme would contribute not only to the study of Bath and its post-medieval development, but also to national and international research agendas in mining and industrial archaeology.

During the main scheme stabilisation works, it became evident that the original archaeological mitigation proposals for preserving *in situ* some of the more significant areas of the mine were not feasible and the recording strategy for these areas was revised and a strategy focused on archaeological rather than engineering imperatives was implemented with time to allow more detailed examination. Five such High Grade Areas, accepted as being of national importance with features not available elsewhere in England, were recorded in as much detail as possible in late 2007 and early 2008. Additional engineered steel roadways were specifically driven to access these areas, were archaeologically led, and were used as recording platforms for the enhanced survey. Figure 1.9 shows these High Grade Areas and associated areas of enhanced recording where laser scanning and video filming were carried out along the archaeologically led roadways.

Monitoring of the main scheme works continued until August 2008. Work at three additional areas at Church Road, Horsecombe Vale and Foxhill was carried out between November 2008 and June 2009.

Aims of the archaeological project

The principal aim of the archaeological work was to preserve the archaeological aspects of the underground landscape by drawn, written, digital or photographic record, including the then visible remains and any which might be revealed in the course of the Stabilisation Scheme. The results would be presented in written and digital form, comprising the written and drawn record, along with finds archives, which would be suitable for presentation to a wide range of audiences including academia and the general public. In addition there would be a digital record of the underground archaeological landscape of a standard suitable to facilitate and create a 'fly-through' model and interpretation of the workings, including digitised photographic and video coverage which with advice and selection and interpretation could be presented in a visitor interpretation centre. It was also initially hoped to provide an underground visitor viewing facility, although this proved, for geological and engineering reasons, impracticable.

A further part of the aims was the provision of archaeological and historical advice during and after the duration of the project, and to assist in the preparation of the proposed display.

More specific objectives were developed both before and during the project, as possibilities unfolded. These included the location and planning of the full range of features, and would seek to understand the methods of working, the transport and handling systems including in relation to toolmarkings and artefacts found, and the methods of roof support and any constructions required, along with the disposal of spoil. It was clear that remains of underground working of wrought stone may survive, and that personal items would be found, relating to the lives of the quarrymen. It was also considered possible that the quarries may have had secondary uses and that evidence of these should also be recorded.

The completion of the physical recording aspects of the project allowed a reassessment, and in particular it was decided the provision of a Heritage Interpretation Centre programme should be separate to the output of the OA post-excavation and analysis programme, though material should be prepared with this original objective in mind.

It was also considered there was material evidence of quarry owners, quarrymen and children which should be presented, and that as the material culture of residents of Combe Down village was also represented in deposits of pottery, glass, bone and items such as clay pipes (domestic rubbish) in shafts and other entries and in graffiti, it too should be analysed and reported. It appeared also that the types of stone and architectural worked pieces produced by individual quarries could be analysed, partly in its relationships with the building of Bath, and where possible the wood resources and landscape of the area should be examined through species identification of wood used in the workings.

This present technical report has been produced with these aims and objectives in mind and is aimed at a wider audience rather than the very limited number of archaeologists and other specialists

Fig. 1.6 Timbered support on modern roadway 2 near Firs Shaft. The blue pipes delivered the concrete used for infilling

Fig. 1.7 Steel supported section of Roadway 2 in east Firs, after concrete had been emplaced

involved in mining and quarrying history. The main drawn, textual and image records which underpin it are digitally preserved and available on the web site for more detailed study.

The Stabilisation Scheme methods and archaeological recording

The former underground quarries had to be accessed through either timbered or steel-supported roadways. These and other working features are shown on Figs 1.5 to 1.8. Initially, the predominantly Welsh miners followed the familiar coal-mining practice of using timbered roadways about 2 m wide and 2 m high with side timbers slightly inclined inwards, with a roof timber located by notches on to the wedge-tops of the uprights. The sides were lagged as necessary with corrugated steel sheet, and the space under the roof made up by short timbers secured by wedges. Spoil was usually heaped outside the uprights to prevent spreading, and it or bought-in aggregate was laid to sufficient depth to bring the floor up to a level, and on which trackless vehicles could run. Steel supports used in the main scheme produced a 2.4 m by 2.4 m roadway, using 100 mm rolled steel joists sitting on a steel soleplate, and with a steel roof-joist chocked to the roof by timber. Sides were braced and lagged as necessary with the floor brought to level as above.

It will be appreciated that this roadway system, of which some 14 km were constructed, was most favourable with low roof heights, using cut-and-fill to bring to level where possible. Higher areas were best avoided. The exception to this was the area of Byfield designated for bats in which more complex high steel supports were used, concrete coated later. Thus most of the roadways cut through the original quarryman's backfilled spoil (which occupied about 50% of the original space) and for large distances the main visible archaeology was the excavation at the heading through the spoil. Since original working quarry heights were some 6 – 7 m, views of the lower third and even two thirds of pillars and of working faces, and along original cartways were very rare and often then difficult to get visual access to. Except for one instance no archaeological excavation was permitted. The roadways were pursued to, and often along the boundaries to ensure the limit of the workings had been reached, and boreholes were used to probe beyond for cavities. These and other surface and underground boreholes also provided useful data for the archaeologists. In the latter stages of the Stabilisation Scheme the boreholes were used as access for the laser scanning of inaccessible quarries at Church Road, a process which located several early eighteenth century quarries.

A team of miners advanced their roadway heading from 2-4 m daily unless the ground or roof was bad, either by hand methods or by use of a tracked excavator and belt-dumping into small tracked diesel dumpers. As the system lengthened, then branches were made and several headings, up to 23 at one time, were advanced. Surplus spoil was taken to the nearest open area suitable for dumping (ie preferably below the modern roadway level) where temporary timber support was provided. This gave access to more areas, and co-operation from the miners sometimes gave access for archaeological examination to a much wider area than the plan of supported roadways might suggest. Sometimes the number of headings meant that it was impossible to visit and record all headings and dumping areas daily, but again co-operation from the miners and engineers enabled significant finds or changes to be notified to the archaeologists. In particular cases work was held up to allow photography on the once-a-week visits by the consultant (LW). There was, of course, a corresponding need for the archaeologists to be as expeditious as possible, timing difficult recording to meal breaks and early evening or when otherwise work stopped. Frequently at later stages, the supported roadways, designed to encircle areas for infilling, approached an area recorded from a different direction, enabling the rear of pillars, for example to be completed on plans partly prepared two or more years earlier. But many areas, some fairly large, could not be approached at all.

Once the areas were accessed by the roadways, the sides of the roadways were boarded-up, a process which was not done, or completed until all visible archaeology was recorded and 'signed-off'. Concrete produced at an on-site surface plant was pumped along flexible pipes and, foamed at the nozzles, poured into the enclosed areas, raising the level in stages to allow setting and cooling. Pipes into the roof enabled the concrete to be pumped to the highest places, usually a collapse progressing upwards to the surface. Drilling of boreholes from underground or surface was used to check that these pockets were successfully filled. Finally the roadways and all shafts were filled, except, again, in the bat area where the much modified workings and two entries were left open. Effectively, therefore, the whole of the workings have been filled and destroyed, though theoretically it would be possible to excavate below the concrete filling in the spoil. This task was completed in November 2008 with some 550,000 cu m of concrete emplaced.

In terms of the archaeological monitoring and recording, the bulk of the work is best viewed as a form of rescue archaeology where the work process cannot stop except for very short periods, with the obvious limitations this imposes. It worked remarkably well in practice, thanks to both miners and management and the insistence perhaps, by B&NES that archaeology was a vital part of the operation. In the High Grade Areas the priorities were reversed, with mining purposely carried out so as to facilitate archaeology by providing specified access and a reasonable time for full recording,

Fig. 1.8 Modern roadway heading with tracked mini-excavator – East Firs

within the strenuous health and safety limits. This included use of laser scanning and video recording as well as more complete planning of the whole areas.

Subdivision of the underground workings for the archaeological survey

The archaeological plan

The mapping of the Combe Down Mines prior to the Stabilisation Scheme was devised and carried out by Brian Hawkins, Professor of Civil Engineering at Bristol University, who reported the results in 1994 (Fig. 1.9). His team had full access underground restricted only by time and by the limitations of the quarrying environment. Not all parts of the quarry were either accessible or recorded, especially where deposits of waste stone had been previously placed to the roof level or in areas of unstable roof. Limitations of time and money resulted in some areas, away from housing above, not being recorded. During 1994 the survey team began to map the external limits of the accessible quarries. Their recording, by electronic distance measurer (EDM) established a network of fixed points and the locations and extents of stone pillars which remained in the quarries to support the roof. It also included a number of other features including the location and extents of the roof-supporting stone packs and areas where quarrying waste had been stacked to the roof level. The whole was related to the National Grid and formed the basis for the Mine Survey during operations.

Location – in general and by Archaeological Areas and Quarry Areas

The large area involved required some form of division to aid location of features, both in general and in specific detail. The most basic division used was into the two main mines, Firs and Byfield, and the smaller dispersed underground workings (Fig. 1.2). To denote general locations, the terms West, Central and East Byfield were used, with whether it was north or south. Similarly Firs was West, Central, East (of the Avenue) or Far East (of the Long Drung). The numbers of modern roadways and 'stubs' were also used, notably on photographs where recording sometimes was irrelevant to or was ahead of planning and contexting activity. Hawkins had divided his surveyed areas into sectors designated by letters, whose relic is seen in the underground modern road numbering (eg E4 stub).

The underground quarrying landscape was subsequently subdivided into more manageable

Fig. 1.9 (facing page) Pillars and Sectors (after Hawkins 1994), with divisions used to communicate general locations within Firs and Byfield Quarries. The shading shows the High Grade Areas where laser scanning was carried out

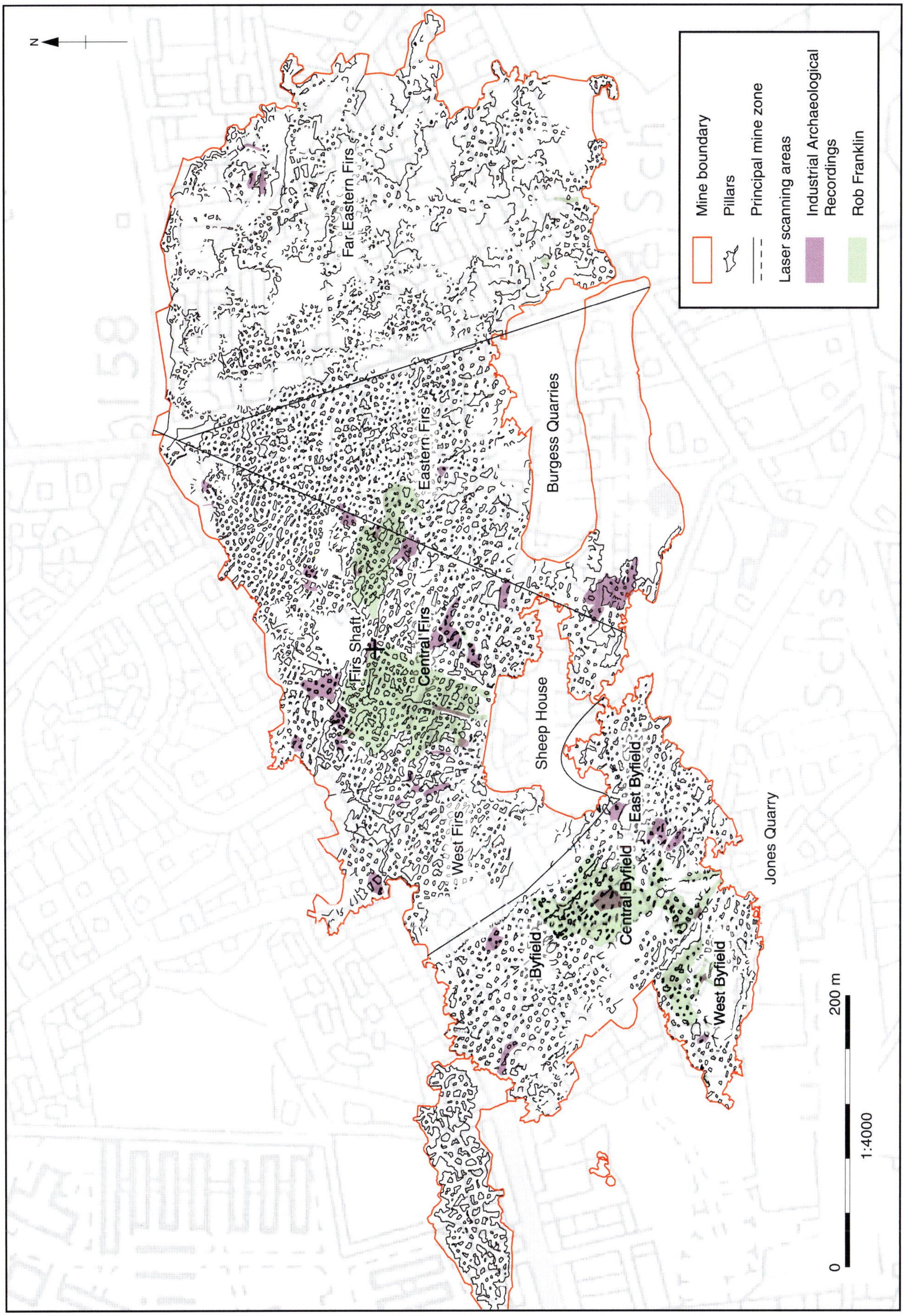
N
Far Eastern Firs
Eastern Firs
Burgess Quarries
Firs Shaft
Central Firs
Sheep House
West Firs
East Byfield
Central Byfield
Byfield
West Byfield
Jones Quarry
Mine boundary
Pillars
Principal mine zone
Laser scanning areas
Industrial Archaeological Recordings
Rob Franklin
0
200 m
1:4000

and describable areas, termed Archaeological Areas. These are areas which appeared to have common characteristics or were conveniently-sized areas on which to report the early interpretive findings. This division was devised so that artefacts, recording and photographs could be sorted and catalogued by area. This system had substantial limitations, including the fact that few complete quarrying boundaries were accessible and the ability to define features was extremely limited before more quarry complexes were seen and experience gained. Allocation of the Archaeological Areas from 2002 onwards were assigned to anticipate the individual Quarry Areas and so were generally larger in size than previous areas and followed more closely the actual quarrying practices. The recording remained predominantly by Archaeological Area.

What emerged was that many of the Archaeological Areas that had been assigned prior to the 2002 and within the 2005 recording period were indeed arbitrary and in many cases several archaeological areas could be incorporated into a single Quarry Area, ie areas likely to have had a single operator uninterruptedly within the same chronological phase. The Archaeological Areas and Quarry Areas are shown in map form and tabulated with their characteristic features and components on the website (www.combedownstoneminesproject.com) Quarry Areas are widely used in the chapters below. In the text they are identified by their assigned number, ie Quarry 501.

Individual features were given a context number, which was recorded on the drawn plans, and/or sections and recorded on a context sheet. The context and Archaeological Area numbers are essential in retrieving data from the project archive but are not used in this report.

Monitoring methodology: standard and adaptive methods

The recording methodology

Photographic recording began in February 2001 at intervals of two weeks. Systematic archaeological recording was initially and largely carried out by the site officer (NRH) from September 2001 in conjunction with a mining archaeology consultant (LW) who attended on one or two days a week. Other site staff and specialists were brought in as the work demanded.

The initial work focused on recording individual quarrying features or components and localised aspects of the underground quarrying landscape which were being immediately impacted by the Stabilisation Scheme. An adaptive methodology was quickly developed to use the standardised OA system to appropriately record areas as they became available, including surviving parts alongside earlier modern roadways not previously archaeologically fully recorded. Most conventional archaeological planning involves the establishment of a planning grid system and requires full access to the archaeological deposits and/or structures. At Combe Down access was restricted to a network of engineered roadways and stub roads, a problem overcome by using the techniques discussed below. Throughout the decade of recording, the principles remained the same, though increased time and much better access meant that High Grade Areas were more completely covered. Laser scanning and video considerably enhanced the record in these areas.

The principle problems affecting recording were that of visibility and distance because of health and safety access restrictions within a mining environment. In two instances, access was delayed due to a different contractor or contract where the mitigation for the main scheme did not apply and it took time to negotiate entry on a voluntary basis (Grey Gables and Foxhill sites). Different methods were tried, subsequently improved, adopted or abandoned throughout the period of archaeological recording. One of the most significant was the use of 110v mains lighting rather than just headlamps or hand battery lamps: this was sometimes available on road heading machinery and from fixed power points on the roadways, but was far from universal. All High Grade Areas were so lit, which was a requirement for video filming.

Recording was carried out mostly through a programme of monitoring the stabilisation roadways in advance of any impact on the elements of the quarrying landscape. The recording involved planning in front of each of the working headings, capturing features using still photography, recording of any vertical data using conventional section drawings and/or 1:1 scale tracings of features, and in some instances laser scanning was employed with scaled hand-drawings produced from that data. In areas that had already been impacted by the enabling works a programme of reactive recording was initiated to capture quarrying areas where roadways had already been driven. This involved context numbering, planning and photographing features seen ahead and on both edges of the roadways.

Recording the quarrying components

Recording methods differed depending on the nature of the underground quarrying landscape and the type of historic activities that had taken place. For example, rooms that had been backfilled from floor to near or at the roof were impossible to record using the detailed planning methods employed where a greater vista of open underground landscape was visible.

Quarrying toolmarks survived on pillars, roof and other worked faces throughout the mines. These illustrate the various quarrying methods used. The types of tools varied by period and individual quarrying enterprises and so provide an indication of the date and of working techniques within each quarry area. Most of the toolmarks were located and described in note form on the drawn plans. Marks that were unusual or that specifically indicated the working technique were allocated individual context numbers so that they could be recorded in more detail or linked to photographs and other records. Significant toolmark profiles, for example, were drawn at a scale of 1:1 when possible and a proportionate number were routinely drawn in any given quarrying enterprise. Tool profiles were also routinely recorded at scales of 1:20 and 1:50 in other drawings to show their distribution within the quarrying area.

Graffiti

Other features, such as graffiti on pillar faces, were also recorded. Section drawings complemented the photographic and/or written record and were generally drawn where the graffito was too faint to photograph or was of such significance within the general quarrying context that it required further analysis. Some graffiti were recorded remotely by video photography and laser scanning where preserved on otherwise inaccessible pillar faces at ranges beyond the usual photographic range.

Quarrying features preserved between stone pillars included cartways, barrow-ways, and deposits produced during working. Lifting and haulage features included the former location of cranes determined by the locations of Lewis or wedge slots and square chog-hole recesses in the roof. Stone packs, used to support the roof and waste stone were a common feature of the quarries. The relative size of spoil, that is, fines and rubble-stone and blocks, including their range and distribution were recorded spatially on the plans. This information was used to ascertain the methods employed during the historic working and in the management of spoil. Some of the information gathered was difficult to transfer on to digitised plan format and is thus best examined by direct reference to the hand-drawn field plans held in the archive.

Context records

The standard pro-forma context sheets were used during the early stages of recording. These include the standard provision for recording and describing archaeological deposits, cuts and structures. It became apparent that the conventional sheets were inadequate for capturing the unique underground quarry landscape and, given the high humidity environment underground, were prone to deterioration. The recording practice was therefore adapted so that the descriptions were either recorded in surveyors' waterproof books or were detailed on the Permatrace plans and later transferred to context sheets on the surface. Several new pro-formae were designed to incorporate, in addition to the standard data check lists, feature types and descriptive terms needed to describe the quarrying landscape.

Planning and the drawn record

The Hawkins pillar plans were supplied as required by the Mine Surveyor Dr R. Narbett, of Hydrock Contracting, and 2005 onwards from the mine plan by the mine surveyor Steve Dix. The plan was based on a series of fixed points with their accuracy confirmed by use of closed traverses. The 1:100 scale plans were copied at that scale on to Permatrace of a suitable size for underground use, so that a baseline of pillars was available to map the quarrying environment in the visible area just ahead of the construction of the engineered roadways. Since most of the underground quarrying landscape and features were dictated by the position of the pillars they proved detailed baselines on to which the archaeological inferences could be mapped.

Quarry feature outlines were plotted by combining a series of offsets from the engineered roadways and reference points on to which other measured and sketched details were added. The process was designed to capture the characteristics of individual features and relate them to stone pillars and other features within the quarries. The main survey apparatus used was a Leica Disto distance measurer used to plot archaeological features that lay between the pillars. This was achieved by lining up the instrument between two known points on the Hawkins plan and measuring the distance from the points, often corners of a pillar, to the feature being surveyed.

The main limitations of this process, used on the bulk of the investigation, for the archaeological plans resulted from the method of recording pillars used by Hawkins, who naturally had other priorities than archaeological recording. Pillar survey had been done (and continued to be done on the post 2005 Mine Plan) by a series of measured off sets using an electronic distance measurer (EDM) from a network of established survey stations. Pillars were recorded at their narrowest (ie weakest) points or at waist level, so that the individual pillar plan is an approximation of the whole, due to irregularities in their vertical form. As in many areas the archaeo logical survey had to use the pillars as a basis for location, the archaeological survey will have small errors as a result, which with other difficulties of access and viewpoints and working may mean errors of a few tens of centimetres in the field plans produced. Because the field plans were relocated to the mine plan at short intervals, such errors will be non- accumulative, and do not materially affect the final representation and interpretation of the quarries.

Where the Hawkins plans were inadequate to provide baseline pillar plans the current mine plans were extended by the mine surveyor using a chain survey. This involved the use of a series of survey stations set up to plot the engineered roadways. Temporary baselines and survey stations established in the roof were related to the mine survey so that the mine surveyor and/or the archaeological survey team in conjunction could produce accurate site pillar plans through measured offsets on to which the archaeological features were drawn. Data on the smaller sheets of permatrace were transferred to larger A1 size sheets when the survey station data became available. This process was laterally accurate for capturing the pillars on either side of a roadway but had more limited accuracy forward of any of the roadways due to the increased distance; 'best-fit' judgements had to be applied during transfer. Where rooms (between pillars) were not filled with floor-to-roof rubble, an advanced primary measured sketch was made forward of the roadway advancement that could later be correlated to the baseline survey.

As archaeological recording progressed, a dedicated full time mine surveyor and survey team were employed by Hydrock to survey the existing roadways. This included weekly surveys of the advancing headings, including the pillars ahead. This was especially important in the far eastern edge of the Firs Mine where access to the Hawkins team had been restricted by floor-to-roof infilling and restricted head room within the quarries generally. This meant that the steel supports within each of the roadways were individually surveyed by the Mine Survey team, and these provided good locations to laterally survey from the roadways, if necessary in advance of their survey. Where the pillars were not picked up by the mine surveyor (due to filling of the sides with spoil before he could attend), the archaeological results from off-sets from the roadway steel supports were provided to update his comprehensive pillar plan.

Sections, elevations and the drawn record

Section drawings were produced to capture vertical features of the quarries such as floor-to-roof waste rubble deposits, barrow-ways and stone packs. Conventional recording methods through measured off-sets were used when structures and deposits were accessible from the roadways, drawn at a scale of 1:20 or 1:50. Features such as toolmarks, which required more detailed recording, were drawn at 1:10. Graffiti on faces were drawn at 1:1 to supplement the photographic record, especially in the case of faintly preserved graffiti.

An adaptive use of the Leica Disto measuring instrument attached to an inclinometer facilitated recording of structures, deposits and floor to roof profiles located beyond the supported roadways. The quarry feature outlines were plotted using the distance and inclinometer angles to capture the limits of contexts, on to which additional sketched details and interpretation were added. An electronic distance measurer (EDM) was sometimes employed to record profiles beyond the roadways and produced similar results.

Artefact recording and recovery (including recovery of graffiti)

A wide variety of artefacts were located underground. Some related to the actual quarrying activity, including tools and personal equipment, others to items brought in by later explorers, In addition, a very substantial number of items had been dumped into shafts or other entries or, in lesser quantity, stored within the workings.

Artefacts were recorded on the standard OA small finds pro-forma, although graffiti was considered a feature and recorded on the context sheets.

In the case of obvious dumped, lost or stored material of a post-working type, samples were taken and recorded, for instance of bottles and pottery, whilst others, car body parts for example, were noted and left in situ. Where there was possibly a relationship with working, eg the material brought up from the well and discarded in the E4 stub, which yielded a range of early and mid 18th-century pottery, then it was fully recorded and recovered.

Quarrying-related material was all recorded and photographed *in situ* if possible. By the nature of their activities, many artefacts were first discovered by the mining teams. They soon appreciated the need for leaving *in situ* where possible, for instance, pipes left on ledges, and were careful to note the situation of any material they were forced to move. Such artefacts were placed at an agreed point or into a finds tray, and recorded on the next visit. Smaller finds could usually be conveyed out, suitably protected, on the archaeologists' equipment-wheelbarrow, while larger items, such as banker mason work was brought out by the tracked dumpers on pallets when convenient. A temporary store was set up at the buildings at Ralph Allen Yard (a former malthouse) and used until items could be conveyed back to the Oxford store. Some items were drawn on-site, others while in store.

Graffiti was a common feature, especially near entries, and some had useful content either for its direct relationship with quarrying, or for its contemporary social information which had potential for museum or interpretative centre display, so was physically recovered. This was done in one of two ways.

Most graffiti was drawn using a coal-like material, together with coloured earths, on the flat sawn faces of pillars. Some it was possible to recover by processes developed by a specialist consultant, Dr Dana Challiner. The surface was sprayed with a liquid silicone, then silicone rubber (similar to latex) was applied in a thick layer using a spatula. The following morning the silicone rubber could be peeled off, bringing the surface layer, including the graffiti, with it as a negative.

Subsequently this was applied to a sheet of long-life resin sheet, coloured appropriately in advance by ground rock. The result was the original image on a surface suitable for exhibition. The process worked well on a number of examples.

In other cases the graffiti was either inscribed, or was covered with a thin layer of calcite deposited as flowstone from seepages. Initially assistance was elicited from a local mason, Gerry Melksham and such examples were cut around until it was possible to saw the slab, 100-150 mm thick, off of the stone face. The largest of these pieces was some 1.8 m x 1.0 m.

The feature was first protected. White spirit was sprayed on and a wax composition (cyclododocane) was sprayed and then painted on in a thick layer. The wax would eventually sublimate over months when exposed to air, leaving no trace. This surface was then covered with a thick layer of the silicone rubber noted above. A large angle grinder with a 12-inch diamond cutting blade (110v or compressed air as available) was then used to cut a neat perimeter around the feature. At the base two horizontal cuts were inserted about 100 mm apart and the stone between was removed with a chisel. The feature was then covered by a thick plywood board, fitted with a wooden section which fitted into the slot created at the bottom which was wedged tightly into position. The feature was thus doubly protected. The three other sides were then chiselled back (using hand or powered tools, the latter if possible) until a frig bob saw could be inserted. This was the traditional one-man, cross-cut-type saw used for this work, and it worked well with a drip of water from a can to clear the teeth (and even better with the hose pipe we were able to use on some occasions). It is hard work to operate and the sawing was done in relays. It was a useful form of experimental archaeology. Once the saw was deep enough, ratchet straps were inserted into the top of the cut to bind the board tight to the slab being cut. Once cut through, the slab was tilted on to a prepared pallet and slid over wooden beams on to the bed of a tracked dumper and conveyed to surface, then with a fork lift truck to the store. We were grateful for abundant advice from passing miners.

After the graffito is re-exposed by removing the protective board and silicone rubber, the wax disappears (by sublimation) over a period of some months, leaving the stone slab with the graffito intact.

Artefacts and graffiti are considered in a series of detailed reports on the website, which are also summarised in Chapter 11 of this monograph.

Adapted methodologies

At the outset of the Stabilisation Scheme it was assumed that some archaeologically important areas of the mine would be preserved. These were to be supported and made accessible from the network of inspection and survey tunnels after the stabilisation works were completed. The OA archaeological team identified areas for preservation on the basis of their importance and visual interest in consultation with the project engineers. The Archaeological Preservation Areas were located beneath the Firs Field Recreation Ground, where it would have been possible to carry out the large scale engineering works required for insertion of a sub-surface and surface support framework to ensure long term stability. These areas were largely but not entirely coterminous with the High Grade Areas, which are shown on Figure 1.5.

The areas selected as Archaeological Preservation Areas were:

- Stub E4: an area of early 18th century, probably pre-Allen, quarrying.
- Two lengths of Allen-period (18th-century) cartway.
- An area of 19th-century deep quarrying
- The Grand Canyon: an area of late 19th-century deep quarrying, monumental in size (hence the name) and (unusually for the quarry complex) not backfilled with later quarry waste

It eventually transpired that the inherent engineering difficulties associated with stabilising these areas were insurmountable and it was reluctantly decided by English Heritage in consultation with B&NES to carry out a programme of enhanced recording of these areas and allow them to be backfilled along with the rest of the underground quarries. Enhanced recording was carried out using a mix of traditional archaeological techniques and other recording media such as laser scanning and video recording. Whereas, during the main scheme, roadways were driven for engineering purposes and archaeology recorded as it was encountered, for the Archaeological Preservation Areas archaeologically led roadways were driven to allow access to the core elements of the Areas and to facilitate the laser scanning and video recording.

Foxhill and Church Road: an alternative methodology

The Foxhill quarry complex, consisting of four small adjacent underground quarries, was archaeologically recorded using a variant of the standard archaeological method devised for Combe Down. Interpretation at the reporting stage utilised photographic images, provided by others (mainly engineering staff) of features impacted prior to the start of archaeological survey. Numerous artefacts were recovered from the site agent, which were related where possible to archaeological contexts or to a general location. Approximately a hundred artefacts were photographed *in situ*. The impacted quarrying features were assigned additional contextual information derived from photographs and their locations established in conference with the site agent. The photographic data was correlated with the archaeological survey data and with historic references derived from mines inspectors' records and those of others, to assist in dating the four quarrying areas.

Church Road voids

The Church Road voids located to the south and north of the road were not all accessible. Quarry Areas to the north of the road were partially surveyed where access permitted and were included in the Quarry Area descriptions. In order to record the voids located south of the Church Road a laser scanning machine was lowered down boreholes to record the limits of the quarrying, enhanced by video survey. No standard archaeological surveys were undertaken in these locations but the scans were interpreted to establish the character, dimensions and the direction of later surface-derived quarrying waste and so identify the former quarry entries.

Enhanced recording

Photography – still images

A combination of colour transparency, colour print and black and white negative photographs was used in recording, enhanced by digital photography from 2007 onwards. The latter was particularly effective in the recording of graffiti images and tools. Over 7000 colour images, 1000 digital images, and over 3000 black and white photographs were taken to depict the extensive and complex nature of the underground and related surface features. These totals included production of duplicated or multiple images to ensure satisfactory results were obtained by varying the lighting, viewpoints etc. Transparency and print films were digitised for archive deposition, in accordance with all other report data.

Video photography and the remote recording

In non-professional mining archaeology, video-imaging is in widespread use underground, though there is still a compatibility problem, largely because of different recording standards and the change from analogue to digital. Nevertheless, it has many advantages, especially for exhibition use. Video recording provides different and sometimes more information than equivalent still images, which it complements rather than replaces. It can, for instance, place a small feature in its larger context in a clear and obvious way, and can produce panoramic views, allowing full examination of an underground landscape with ease. Its ability to use very low light levels means simple hand lamps or 110v industrial electrical supplies can be used for lighting, though the latter is far preferable and was used in all High Grade Areas. Its zoom capacities combined with low light capability make features visible at ranges impossible to the naked eye or still camera, rivalling the data only otherwise obtained by laser scanning. Distant items of graffiti are particularly easily recorded and, if necessary can be converted to a still image.

At Combe Down video recording was initially used in 2003 and 2006, employing IA Recordings, a small, semi-professional unit, particularly skilled in underground video-imaging. They used a broadcast standard camera and the output is available on videotape and DVD. A second phase of video photographic survey, which concentrated on the High Grade Areas, commenced in March 2007. This was carried out by independent video film-maker, Rob Franklin. Filming was carried out in advance of the archaeologically-led engineered roadways and from the sides as access became available, usually as the next row of pillars was passed.

As well as using traditional tripod mounts, restricted access in an area known as the Grand Canyon necessitated the use of a large camera boom for surveying the floor and other areas inaccessible using the conventional tripod. An aerial cable was erected close to the roof along the eastern end of the Canyon to record aspects not visible from any other location.

Laser scanning and remote recording

Another key recording strategy was the use of laser scanning. This technique had a dual function, being used for both high quality recording of High Grade Areas and also acted as a the key element in the production of a 3-D digital model of the workings. The subsequent creation of this digital model represents a major output of the recording strategy, forming an accessible record of the results of the recording action (as a GIS) and is the basis of the 'fly-through model' intended for the proposed interpretation centre.

APR Services carried out a pilot laser scan survey in May 2003. The preliminary scans, coloured from still photographs produced almost film-like representations which could be viewed and interrogated. With a range of up to 100 m and a variable spacing of the point cloud down to about 1 cm, the equipment has the capacity to produce a high definition, three dimensional survey of areas either too far distant for other methods, or which would normally require a remote access device for entry. The survey instrument was set up on a tripod stand at intervals along the length of the roadway so that each scan could be referenced to the previous scan location and to a network of survey stations located using EDM by the mine surveyor. Scanning locations were also set up outside the roadway supports within a discretionary distance of 1m, so that the roadway structure did not impede the scanner's line of survey.

The introduction of a more extensive laser scanning programme in the High Grade Archaeological Areas in Byfield and the Firs Mines (Fig. 1.9) from February 2007 proved essential for the permanent 3-D underground landscape survey and allowed a more comprehensive and accurate survey of the pillar plan in areas surveyed from Roads 1, 2, 3 and 4, near the Firs Shaft, originally using conventional drawn plans in 2001 and 2002. Laser scanning, because each point of the 'cloud' was fixed in space and because of the long range feasibility, enabled pillars previously visually inaccessible from the roadways to be plotted from the survey data, and then provided 1:100 scale drawings that could be combined with the drawn record. This recording was carried out with the conventional tripod based laser scanner machine, used elsewhere during the project, and with the aid of the borehole laser scanner from several horizontal pipe locations projected over the Grand Canyon floor, that were inaccessible to the conventional laser scanner. This was particularly important in the Grand Canyon where the 6-8 m high quarry could be captured and depicted at the lower quarry floor level from a viewing platform positioned over 6 m above. Detailed plans were drawn from this pillar plan data on to which features between the pillars could be sketched without having to cope with the challenges of altered perspective from planning from a great height. The survey was also an important tool for capturing vertical elements of the landscape such as stone packs, inaccessible by traditional recording techniques.

Post-excavation methodology

Ongoing analysis and coloured plans

Post-excavation analysis began during the later stages of recording. During the recording survey the plans of the underground quarrying landscape were drawn at a scale of 1:100, then reduced to a scale of 1:200 to be used as synthesised interpretative drawings during on-site analysis. As more of the quarrying landscape was accessed the plans were extended and up-dated. In some quarrying areas the final plans were finalised several years after the original work due to piecemeal access. The synthesised plans were updated during final analysis to indicate Quarry Areas which are used to locate items mentioned in the text herein.

The features recorded on these plans included the transport routes (cartways, railways and the barrow-ways), and where possible the direction of quarry workings and spoil tipping. The locations of tipping episodes associated with barrow-ways, tipping platforms and tipping fronts, where they were tipped over earlier quarrying features enabled the stratigraphic sequencing of quarrying areas and phasing of some quarrying. The plans also indicated the location of boundary pillars and larger groups of linear spaced pillars. These were often located on the limits between quarry areas and better defined their boundaries. The plans allowed analysis of the extents and the locations of quarrying areas that could be linked, in some cases, to documentary sources.

Digitisation of records (CAD, GIS)

The site plans were scanned by the OA Graphics Department and saved in .tif (no-loss) format. Scans of the plans were inserted into an AutoDesk Map CAD drawing and transformed to the correct location, using points on the pillars and roadways. Each archaeological layer was traced in CAD to produce a closed polygon. Archaeological layers, such as spoil tips, cartways and surfaces had closed polygons, whereas features such as sawn faces, jad slots and roof joints were unclosed polygons, and small find locations, Lewis slots and chog-holes were represented by a point. Context, small find, section numbers and other data were attached to the polygons and points. Once the plans were drawn, scanned and digitised the boundary pillars were used to ascertain historical quarry areas. These areas were altered to closed polygons with the quarry number attached. On completion of the CAD work the polygons and points were exported as shape files with their data attached. The shape files were imported into ESRI ArcGIS to allow analysis of the data.

Analysis of GIS and records

The GIS data was used in a number of ways. Cartway and barrow-way routes were correlated with the primary underground survey data and their limits plotted. Plots included, for example, cartway ruts, their orientation and extent. Where ruts were obscured their extent was extrapolated using synthesised drawings, pillar patterns and locations of vertical winding shafts and/or boundary pillars. Barrow-way routes were better defined and the data has allowed an understanding of the total lengths of the routes and their orientations.

GIS data was also used to establish the spatial distribution of quarrying features including sawn faces, wedge pits on worked faces and the roof, crane anchorage wedges and Lewis slots. The plots were used to examine similarities and abnormalities of distribution of quarrying features. The limits of each Quarry Area was calculated and the location of roof supporting pillars established through GIS data. The plots excluded pillars inaccessible to primary survey that could not be closed into polygons. The area of the pillars was deducted from the total area of the established quarry limits to suggest the total area of stone quarried. This total did not record the actual cubic volumes of extracted stone, as the depths of the quarries could not be established due to spoil deposition within them, but it did allow rough estimates to be made.

Analysis of other recording media: laser scans and video, planar viewing and measurement

Non-standard archaeological recording methods used during the underground survey facilitated analysis of features beyond the engineered roadways and which were otherwise inaccessible to the survey team.

Laser scanning

By means of fly-through laser scanning the overall areas of quarrying could be examined to determine three-dimensional aspects of the landscape not achievable through standard archaeological data capture, and to inform general trends in understanding specific quarrying areas.

Planar viewing and measurement

Laser scan data was used to measure the heights of the roof in inaccessible areas, and along cartway and barrow-way routes, and also to inform analysis of the quarrying operations that had been undertaken. Features analysed included working heights below the roof negotiated by quarrymen during working at the quarry face, and spoil deposition. Laser scanning was also employed to record the inaccessible northern side of the Grand Canyon Quarry Area to show the height and position of the full height pillars, and to record the stone packs, barrow-ways and spoil deposits. The scans also provided the inclination, the height from the floor to the roof and the extent of an important inclined barrow-way. The data also provided the extents of the vertical winding shafts, including the scaled images of shaft capping, and the construction details in some of the wells.

Recorded evidence of the inaccessible historical graffiti and some of the worked faces was also analysed. This enabled scaled recording of sawn faces with graffiti, in some cases establishing the names of the individual quarrymen that were faint and illegible when viewed from the stabilisation roadways. Scale-drawings of the worked pillar faces were also prepared so that synthesised drawn sections could be completed and reproduced in the report.

Video

The video was used in correlation with the Planar viewing and measurement (see above) to record the sawn faces with quarrymen graffiti and obtain images of features which lay further from the stabilisation roadways and were therefore not achievable by conventional still photography, or that would have been impacted before good access could be achieved.

Video photography was used to record the areas of the underground quarrying landscape which were inaccessible to the standard recording methods in the former High Grade Archaeological Areas prior to the impact by the stabilisation scheme roadways. This data was analysed during the interpretation and reporting periods to better understand the quarrying environment.

Structural, stratigraphic and artefact analysis

The analysis of the artefacts (and their stratigraphic context) continued work begun during the on-site recording. This included the lost or abandoned quarrying tools, the quarrymen's personal effects and the discarded or buried stone artefacts.

Analysis of both archaeological and documentary sources attempted to establish the actual individual quarrying areas and their phase of activity. Each of the artefacts was assessed on whether they had been originally recovered from a closed context, for example, from a barrow-way surface, a platform or tipping front make-up; or whether they had been recovered from a pillar ledge or from within a niche on a pack wall. Those artefacts that were not attributable to a closed context may have been abandoned or lost during any of the quarrying periods, and also may have been moved from their primary to secondary contexts during either a later phase of working, possibly for re-use, or in the post-quarrying periods. Some of these artefact types were attributable to surface-derived material deposited through either vertical winding shafts or other entries including from the spoil from surface infilled quarries.

The possible redeposition of artefacts had to be considered. Even those artefacts that were recovered from secure contexts, for example the tipping fronts, had to be assessed to establish from which quarry area the tipping front had originated, regardless of which adjacent quarrying area happened to be nearest.

The small finds were added to the GIS which facilitated analysis of their distribution, their material and their type. This element of the GIS was extensively used by the finds specialists in order to inform their discussion of the distribution of the artefacts and to further refine analysis of the quarry dating. These reports are discussed in a later chapter, with the full report available on the website.

Combe Down and underground archaeology

The study of underground mining and quarrying is a relatively unusual branch of professional archaeology in Britain, though it has been established for several decades in Germany, based on the Deutches Berbau Museum at Bochum, usually published in the journal *der Anschnitt*. Probably the first substantial underground excavation in England was at Grimes Graves, in Norfolk (Sieveking 1979). Since the lack of recent work undertaken in the United Kingdom is due to increasing health and safety restrictions, a situation which has favoured work by amateur groups (who nevertheless often include professional archaeologists and others with the requisite skills). These groups have fairly frequently gained financial help via grants.

Carrying out such archaeological work during active mining operations, professionally or as amateurs, is also unusual, though less so at surface than underground (see for instance, Hartley's investigations of Tudor mining in a Leicestershire opencast mine (1994)). Combe Down is probably the first substantial such underground investigation in the UK. However one of the present writers (LW) has carried out (with colleagues) similar work during mining operations fairly extensively in India and at Rio Tinto in Spain (see articles under Willies on www.pdmhs.com). Underground archaeology, or projects with a substantial archaeological content, in the UK has been fairly extensively carried out by Timberlake and Barnatt, separately and independently, and Willies, under the auspices of the Early Mines Research Group or Peak District Mines Historical Society. Arthur J. Price of the Cotteswold Naturalists Field Club has recently published an account of Cheltenham's Whittington Quarries which has the results of a fairly substantial underground archaeological work (Price 2007).

Continental interest in underground archaeology has also been increasing in recent years and there have been or are ongoing several projects of comparable significance to Combe Down. The Bergbau Museum under its archaeologist the late Professor Gerd Weisgerber and now under Dr Thomas Stollen have many reports to their credit (see for instance Wagner and Weisgerber 1986 and Weisgerber and Goldenberg 2004) and Weisgerber in particular deserves the credit for establishing the discipline. In Romania the Department of Archaeology of Toulouse University is investigating the Roman gold mines at Monte Rosa.

At least two teams, made up of professional archaeologists combining with skilled amateurs in France have engaged with underground archaeology, including current investigations at the Laurium Silver Mines, Greece. The recently formed *Institute Europa Subterranea* has held three symposia on archaeological mining history, the most recent reported in Silvertant and Mhairtin (2008). This publication has a range of material from detailed archaeological investigation of mines in south-west Germany (Strassburger 2008) to more general articles with a strong emphasis on survey and photographic recording rather than full archaeology. Much the same is happening in Britain which has a strong tradition of such work, much of it unpublished: including, for instance photographic collections by Paul Deakin (of PDMHS) of underground mining and quarrying nationally, and, in underground stone quarries with strong links to Combe Down, Derrick Hawkins for photographs (NAMHO 2010 Conference at Coalpit Heath near Bristol) and for surveys, the Box Freestone Mine Survey by members of the Shepton Mallet Caving Club (*c* 1998). Video filming with strong archaeological content has been carried out by IA Recording (who also contributed at Combe Down) and by David Webb of PDMHS.

The Combe Down Stone Mines Project has, for the very first time in England and the UK, placed the full resources of a professional archaeological body, Oxford Archaeology, into a large scale archaeological examination of underground quarries, with all the back-up for specialist expertise and finds examination that that implies. It has shown that such work is compatible with ongoing mining operations within old workings (which has implications for mitigation strategies where old workings are to be disturbed by development), and that an integrated strategy of archaeological investigation combined with an extensive examination of documentary and published resources can yield a coherent result, which would not have been available by any other means for describing this fundamental part of the Building of Bath.

Chapter 2
The landscape of Combe Down

Topography

The landscape of Combe Down is based entirely on Jurassic rocks, including the area northwards to Bath and south to Tuckingmill (Fig. 2.1), which are easily but differentially eroded and have resulted in the dramatic topography of the valley of the River Avon, with steep slopes (between 1:5 and 1:10 are common) on either side of the river rising to limestone plateaux to the north (Lansdown) and south. Combe Down occupies part of the down on the south. The slopes are modified, by headward erosion by springs, into combes. Between Bath and Combe Down the steep valley of Lyncombe is directly opposite the city centre (Bath Abbey), with springs emerging just below Prior Park House. Widcombe is to the west, and was followed by the Roman Fosse Way through what is now the built-up area known as Odd Down. Horsecombe is over the hill to the south of Combe Down and descends to the east from the west boundary of Combe Down near the Cross Keys Inn, to Tuckingmill. Strong springs emerge on the north side of Horsecombe near the west end below Shepherds Walk. A further branch combe is developed out of the lower part of Horsecombe, the top of which has a very strong series of springs known as Whittaker Springs (near and just above the de Montalt Mill). Further springs emerge just above Tuckingmill and these and the Whittaker Springs are used for water supply. In the east, Brassknocker Hill forms the modern day boundary where the gradient is somewhat more gentle.

Both the Avon Gorge and Horsecombe are over-deepened, that is, the deepening has been faster, due to the vulnerability of the rock, than the adjustment of the slope angles to a stable condition, so both slopes are unstable and have fairly widespread evidence of landslip and what is known as cambering of the superposed limestone strata (below). The limestone beds, known as the Combe Down Oolite with the Twinhoe Beds above, form an almost level plateau running, from west to east, through Odd Down, Combe Down, Claverton and Bathampton Downs, for some 5 km, with a width of around 1 km. It forms an elevated tract rising to about 200 m OD, about 150 m above the River Avon. This is the tract which has the freestone quarries. The river partially encircles the upland tract which was important when the adjacent canal was built, in giving several points of access from Combe Down. Meanders in the river have provided a site for the historic core of Bath.

On all but the west side, the limestone edge forms an escarpment with either cliffs or very steep slopes down to the softer beds beneath. Combe Down has only a north and south 'rim', marked on the south side by Shepherds Walk from the Cross Keys Inn to the village, and, eastwards, under the cliff, Summer Lane, formerly Combe Lane angling down the hillside to Combe, known today as Monkton Combe. It is aided in this route by crossing the rim where a fault has disrupted the strata. On the north side of the down the rim has a walk above the steep slope of the 'Hanginglands'. This was formerly known as Popes Walk, though this name where it passes through a modern housing estate, has now been adopted for part of the Bath – Combe Down – Tuckingmill footpath, known also as the Long Drung, which crosses over the whole of the hill, using Horsecombe and the west side of Lyncombe.

The down land was, in former times, mostly sheep pasture and is naturally dry underfoot. What was formerly a track called the Claverton Road, which is now the line of the Bradford and North Roads crosses from east to west over it, bounding the top of Prior Park. On the north – the Bath side – roads use the combes to access the downland. However the beds immediately below the rim, mostly clays, locate the emergence of springs, and in winter especially the clays can prove difficult to traverse, so that routes over them, until improved in modern times, proved difficult, and were treacherously soft underfoot on steep slopes. The problem is probably reflected in the name 'Summer Lane' on the southern slopes. Other roads reflect either the former needs of quarries or result from subdivision for housing.

Most of the houses in the modern village of Combe Down, with the exception of some on Summer Lane, and two or three in the upper part of Horsecombe, have chosen to be on the higher, drier ground. Prior Park, just outside the parish of Combe Down but very close by, was built on the limestone close to the clay margin, the springs being used for water gardens and a lake.

Geology

The following account is based mainly on the six-inch geological maps, and commentaries by Forster *et al.* (1985) and Macdonald *et al.* (2000). Other sources, whether historical or from detailed observation within the quarries are cited as appropriate. These include observations by William Smith, the 'Father of Stratigraphy', whose early studies took

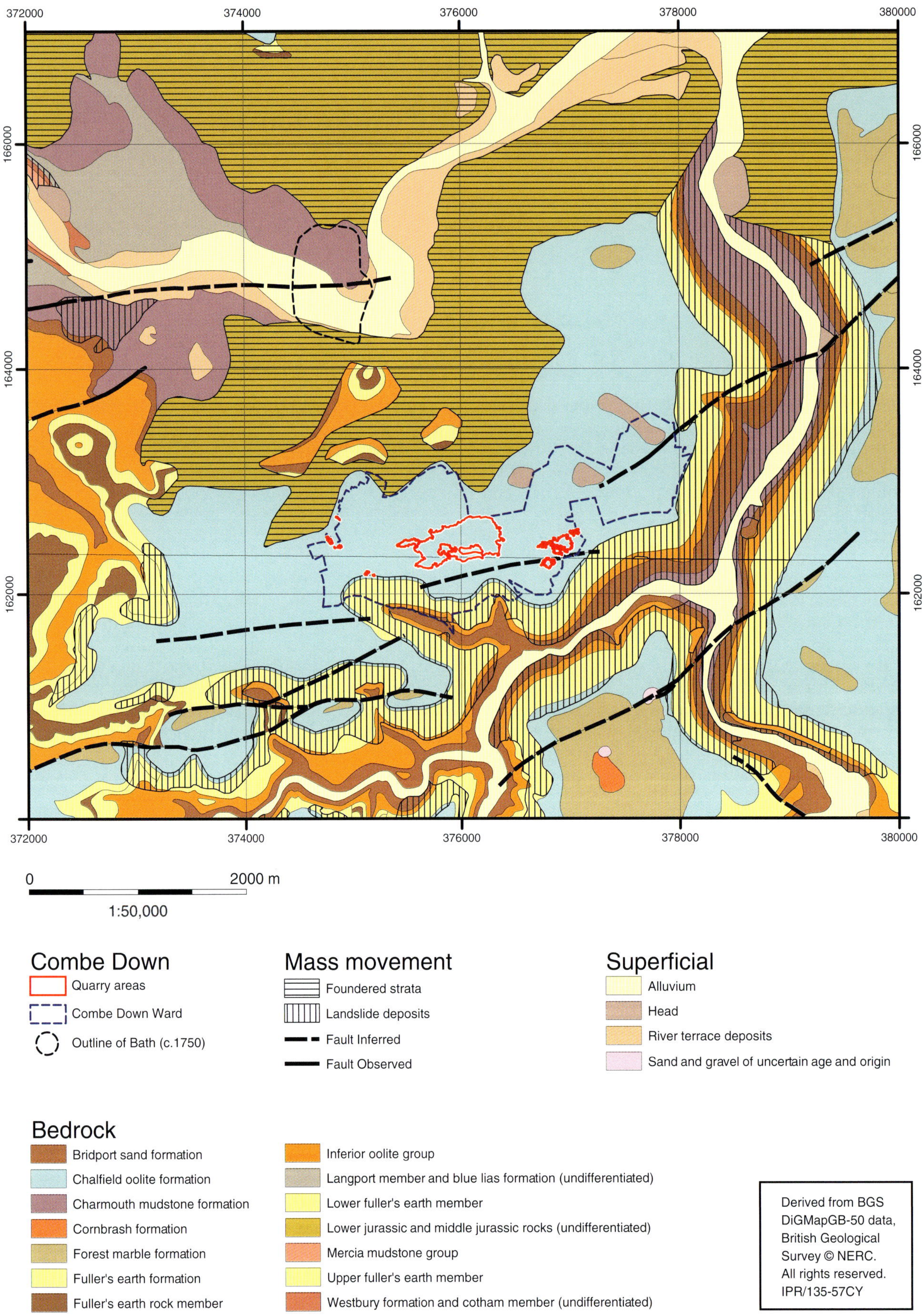

Fig. 2.1 The geology of Combe Down and the surrounding area, showing the parish boundary and selected place-names

place in this and adjacent areas, and who lived at Tuckingmill.

All the outcropping rocks (solid geology) in the immediate area of Combe Down, as noted above, belong to the Jurassic period. Figure 2.2 schematically shows the local stratigraphic sequence. Bath and the River Avon rest on Liassic clay and limestone rocks (Blue Lias), which underlie but are not exposed in the Combe Down area. Liassic limestone was quarried west of Bath and its clays were probably used for brickmaking. An area known as Brick-kiln Fields lies to the north of Widcombe village (Chapman 1996, 5). Overlying the Lias are the Midford Sands. These are silty sands, sometimes with hard bands of calcareous concretions that thin laterally southwards to nothing around Wellow. These were probably the source of sand in a sandpit in Monkton Combe (perhaps at Tuckingmill) and the sands also outcrop on the lower slopes of Widcombe where they were worked as foundry moulding sands (1:10000 Geological Survey). Within Combe Down, they outcrop in the floor of Horsecombe Vale. A depth of up to 24.5 m has been recorded, but this substantial thickness may have been the result of a slip. The Inferior Oolite above the Midford Sands is made up of fossiliferous limestones, including the Trigonia Grit, the Upper Coral Bed and an oolitic limestone, in all about 11-12 m thick. These were worked locally for a monumental limestone, the Yeovil Marble and, from 'everywhere' at its base a friable 'calcareous sand' was 'used by cooks to sand their kitchens' (Townshend 1813, 105). The outcrop is marked by copious issues of springs and is found in Horsecombe and just below the railway cutting near Tuckingmill.

The Fullers Earth Clays (now termed the Bathonian Formation) are the next higher in the sequence. The once commercially important true Fullers Earth is a montmorillonite clay (60-70% of the rock) which occurs in only a thin stratum, up to 3 m of the *c* 54 m total thickness of the formation. The montmorillonite clay was derived from volcanic dust from a relatively close source which settled in a marine environment and was reworked by tidal currents and concentrated into thin, discontinuous, layers. The bulk of the sequence, however, is made up of the more common illitic clays and mudstones, together with thin bands of limestone. A thicker limestone band, the Fullers Rock, separates the Lower and Upper Fullers Earths, and the commercial bed lies from 3 to 10 m below the top of the Upper Fullers Earth. The commercial interest in the deposit led to a borehole (Horsecombe No. 18) being sunk in 1971 at the former quarry west of Combe Road, finding the top of the sequence some 7.62 m below surface. Borehole No. 9, at Shepherds Walk at Horsecombe Head, passed through 11.4 m of limestone before intersecting the clay (6 inch Geol. Survey; Penn *et al.* 1979).

The original use for Fullers Earth was for degreasing woollen cloth during the fulling process, a property which depended on its power to absorb oils on to its crystal surfaces. Fullers Earth, at least in recent centuries, was not worked in or under Combe Down, although mines operated early last century on the South Stoke side of Horsecombe Vale, within the study area, with other sites near Odd Down. A small boom in Fullers Earth production emerged about 1880 with demand from the nascent north American oil industry (Parfitt 2001,

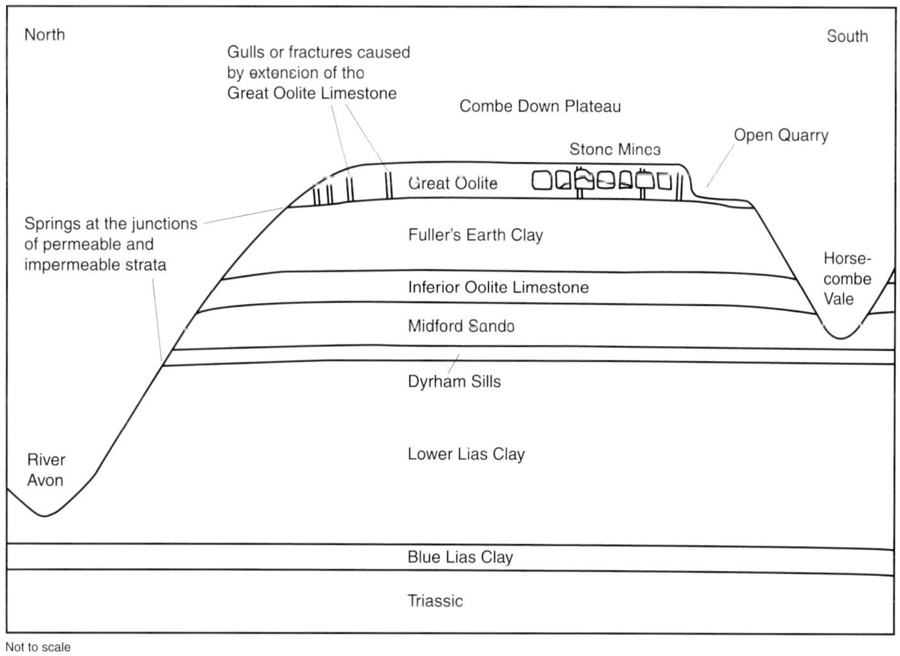

Fig. 2.2 The strata under Combe Down (after Hawkins 1994)

136. See also a recent published account by Macmillan 2009), and small works developed to process it, including at South Stoke and Tuckingmill. A gravity mineral tramway at South Stoke took the raw material, from several sources, down to the boundary stream with Combe Down where it was agitated with water, probably by a paddle wheel which survives in the stream bed, and the resultant slurry was piped about a kilometre to the Tuckingmill clay works. A section of this clay pipeline survives buried within Combe Down at the southern boundary below Beechwood. A Fullers Earth Union set up in 1890 to combine the various businesses was later taken over by Laporte, an industrial minerals company, who ran the Tuckingmill site until about 1970.

The clay-dominated sequence has been of immense importance locally in another way. Wells were sunk through the overlying oolitic limestone to reach it and use was made of springs issuing at the margin. The most notable of these was and is the Whitaker Spring next to the de Montalt Mill. There the water is delivered underground by a fault, which was tapped by tunnels bringing it to a central point. This, and water from springs at the same outcrop at the top of Horsecombe Vale, merge at Tuckingmill and provide much of the water pumped from there into local supplies. In former times these also provided power for the de Montalt Mill, Tuckingmill and, with additional water from the Wellow Brook, Monkton Combe. It was also used for growing of irrigated crops such as water-cress. Water in Widcombe's Prior Park also issues at this horizon and feeds the ornamental features and pond lower down.

Less positively, the clay, particularly in the Upper Fullers Earth, forms a weak substratum to the overlying rock and squeezes out, sometimes forming an almost level terrace just below the limestone rim on the otherwise steep valley slopes and, commonly, hummocky ground with the arcuate features of a typical landslip. It is also an extremely difficult material over which to build stable roads and tracks and created considerable problems for all those who descended from the plateau.

The Combe Down Oolite

The limestone plateau on which Combe Down is sited is formed from the lowest limestones in the Great Oolite sequence, the Combe Down Oolite and the overlying, and less useful, Twinhoe Beds. The Oolite incorporates the freestone which has been quarried over a large extent of the present territory and, especially, adjacent Odd Down. The oolite was deposited as a lime sand (medium grain-size, most grains between 0.1 and 1.0 mm in diameter) on a shallow, current-agitated sea-floor that deepened towards the south, during the Bathonian stage in the Middle Jurassic period of geological time, *c* 166 million years ago.

The Oolite as a whole is composed of generally fine ooliths of limestone calcretions around minute shell fragments created in shallow seas, but also has shelly and thin marly (limey-clay) horizons within it. More technically, the oolite can be described as an oolitic calcarenite or oobiosparite.

Not all the sequence can properly be described as freestone, which only occurs (reliably) in the upper few metres of the sequence, ie the part that is quarried. According to Palmer (2009) based on drill-core samples, the top is marked by a hardened, iron-stained layer that is perforated by contemporaneous mollusc and worm borings. This, termed a hard-ground, is due to a thin layer of cement coating adjacent grains, and opaque fine-grained sediment filling the remainder of the intergranular porosity. Below the worked freestone the beds are increasingly impure and clayey limestones become more common and are unsuitable for building or carving.

In samples taken by Palmer from stone artefacts found in the quarries, the ooliths dominated and were ovoid to spherical, depending on the shapes of their nuclei. The original sediment was well washed and winnowed so no fine grained material occupies the pore spaces and in this preferred material, the ooliths are formed round fragments too small to identify. The grains are un-compacted, just touching at contact points. They are cemented with a clear calcite cement which seems to grow out of the grain edges. Though the rock is still porous, and the individual grains are mechanically weak, it is this cement which provides the overall strength.

The porous nature of the ooliths makes them soft and crumbly on freshly-exposed surfaces, covering it with a chalky dust. This soon washes off to leave a surface on which the former positions of the ooliths are marked only by their impressions in the enclosing cement.

Slightly larger shelly fragments are also found which form a subsidiary grain type in its own right. These are from echinoderms, oysters (and maybe scallops), and rare brachiopods and bryozoans. Shelly fragments replaced by aragonite also found were probably derived from molluscs (mostly gastropods and bivalves), and maybe corals. During the burial process the original aragonite material formed a micrite envelope around the grains. This was less useful material and any horizons of coral debris or burrows seem to have been deliberately avoided of rejected by the quarry workers.

One sample examined was exceptional. It was strongly compacted and the already cemented grains forced into one another, becoming fractured and finally re-cemented. This is a condition where the rock has been subjected to tectonic movement nearby.

Palmer also notes that the material quarried from the same horizon in the Corsham area of Wiltshire, used widely in the 19th century under the name of Box Ground Stone (and other names), seems to have more, and larger, recognisable shell fragments. This apparent difference deserves a closer study, but it does suggest one reason why Box and Corsham found it difficult to directly compete with Combe Down stone and instead used freestone from higher in the Great Oolite sequence.

Unfortunately the term 'freestone', referring to stone easily workable in all directions, has often been applied to the whole sequence, up to 18 m thick. In fact, the quarryman (at least in honest moments) referred only to the very fine-grained oolite as freestone. Few historic descriptions appear to suggest more than about 7 m of true freestone was worked. Other horizons were also quarried – harder material made hard-wearing courses and floors, or roadstone, and the stronger shelly limestones were used for structures such as bridges and also to create decorative features.

The detailed geology affecting quarrying and use of the Combe Down beds

At the marginal areas of the Oolite, the combes dominate all but the plateau area and result from headward sapping at the springs with erosion by fast-running flows of water down the already steep gradients due to the deep dissection, overdeepening to the base level of the River Avon. The removal of the underlying clay by extrusion or other erosion causes the limestone margin to progressively collapse, resulting in slopes (some of periglacial, soliflucted origin) to be littered with limestone boulders and sometimes large masses (head deposits) which can obscure the underlying solid geology.

The combination of weak (Bathonian Formation) clays lubricated and weakened by water draining through the permeable limestone, and a thick overlying Combe Down limestone with a strong joint pattern, has led to very extensive and continuing land-slips, and to cambering or 'draping' of the limestone (referred to as 'foundered strata') at the marginal areas. With respect to landslip, the Lyncombe and Widcombe slopes are perhaps more mature and less unstable with a slightly lower slope gradient than Horsecombe Vale, which has minor active land-slippage in many places that has largely deterred house-building there.

These processes have had an important influence on the surface and underground quarries, since the strongly jointed limestone opens at the joints to form gulls – open or soil and debris-filled fissures up to a metre and more wide. Such fissures can be seen, for example, at the north-east face of Springfield Quarry at Entry Hill. Open joints and gulls allowed the quarryman to remove blocks of the stone with little effort, simply using a bar to pry them out, rather than having to cut back in the stone using pick and jad or stone saws. Such openness of joints, however, had the penalty of the roof to the workings being unstable. It produced a particular form of underground quarry landscape which is rare if not unknown elsewhere in England.

Hawkins (1994) investigated the rocks of the plateau area in some detail (Fig. 2.3), to which it is possible to add observations made during the archaeological survey. There is a regional dip of 2-5° towards the east, although local flexures, such as

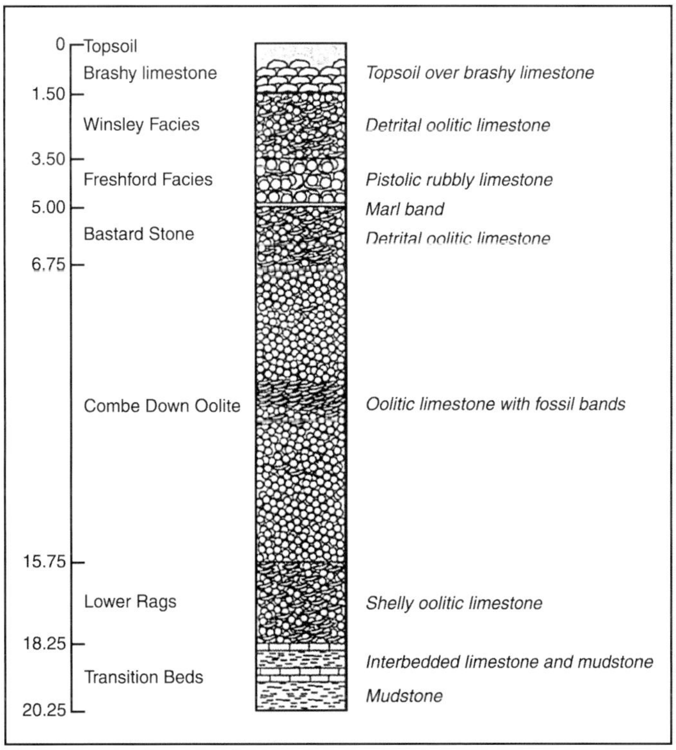

Fig. 2.3 The beds under Combe Down (after Hawkins 1994)

those running north-south through the Hadley Arms area, and variations in bed thickness can locally offset this.

The Combe Down Oolite is at its thickest around Combe Down, merging laterally into the Frome Clay. It is at a lower horizon than the Great Oolite 'Bath Stone' freestone horizons exploited near Corsham in Wiltshire. However a small exposure of this higher horizon occurs at Combe Down next to Summer Lane at Beechwood Road (Vinegar Down Quarry) due to graben-faulting (faults between which the strata have been let down). In the east part of Combe Down at Lodge Style, where current freestone working at Hancock's Upper Lawn Quarry allows easy observation, the Combe Down Oolite is overlain with up to two metres of the Twinhoe Beds, a detrital limestone (eroded from an earlier rock and redeposited) with shells and 'ironshot' grains, overlying up to two metres of a rubbly, shelly, friable limestone with a sandy-silty clay component. Neither of these two horizons is very competent. They sit unconformably on the eroded surface of the Combe Down beds, which has mollusc boreholes and is encrusted with oyster shells. The lower of these two horizons, up to two metres thick, is exposed at Springfield Quarry, forming a weak top to the cliff. Near Combe Down village itself both these horizons are largely absent, partly due to erosion, and partly, perhaps, because the horizons are discontinuous. Instead the Combe Down Oolite is found immediately below a metre or so of a regolith composed of broken rock fragments and soily material. In total, over the mined area these horizons result in up to seven metres of overburden which is usually very weak. If the top of the Oolite yields, there is a strong tendency for progressive upwards collapse to surface, leaving a subsidence or crown-hole. Particularly notable examples of progressive upwards collapse were found during the Stabilisation Scheme under the right-angle corner on Westmoreland Road, and of crown holing in the car park of Barclays Bank at the junction of Combe Road with North Road. Others, probably relating to surface collapses or subsidence were found under North Road, though of course these have now been stabilised.

At its base the Combe Down Oolite lies on an erosion unconformity over the Fullers Earth. The bottom two metres or so is a clayey, variable shelly limestone-clay, transitional sequence. Above this are two to three metres of thin-bedded, shelly, crystalline limestone, called the ragg or ragstone by the quarrymen, the Lower Ragstone today.

The worked horizons underground are up to some 9 m thick (the largest cavity height observed was 8.4 m. but this was a little above the true floor), but a metre or so less was probably more usual in the higher workings and about 5-6 m worked was probably closer to the average. It is clear that the lowest workings are still well above the ragstone, either being considered unsuitable due to greater hardness, or to caution about the stability if the underground workings were worked to greater heights. They have possibly discontinuous shelly horizons and, in the middle or top part of the sequence, a coralline bed. The beds are very variable and cavities, marl, shells, re-crystallization in vugs (with scalenohedral dogtooth calcite crystals known as 'cockles') were found fairly commonly. A known stratum with cockles lies under the north-east corner of Byfield (see below), but this seems to occur over a very restricted area. These and other features such as clay content either rendered the stone useless as a freestone, or led to much increased waste.

At the top of the worked sequence were the 'picking beds', apparently softer, of poor weathering quality, which were extracted to allow access

Fig. 2.4 The 'Bastard Bed' seen in the background behind H Pillar, East Byfield

for working the beds beneath. Above this there is up to 2 m of a shelly, pisolitic (coarser grained) more thinly-bedded and harder limestone unusable as freestone, called a 'bastard stone' by the quarrymen, which was left as a roof where it occurred under the regolith. Hawkins used this term specifically for the hard roof or ceiling stone, but Pollard (1994, 28) warns it was a generic term used regionally by the quarrymen for any such horizon within the worked beds.

According to Hawkins, the 'Bastard Bed' has a different fracture-pattern to the beds below and, in section can appear remarkably like a coursed rubble wall, as at near the margin of James Riddle's quarry (518) at H pillar in Byfield Quarry (Fig.2.4). Fallen blocks of Bastard Stone 'ring' when drilled, in contrast to the softer freestone beds. The Bastard Bed was not as distinctive, if present at all, in the eastern parts of the workings as in the centre and west of the underground quarries, and may have phased out laterally.

The freestone beds which were actually worked – below the Bastard bed and picking beds, were generally 5 to 8 m thick overall but possibly less in some eastern parts of the mines. Only rarely in the investigations was the full sequence revealed because of huge quantities of waste stone, occupying perhaps 50% of the original excavated space. Due to health and safety access requirements it was not possible during the archaeological investigation to make any meaningful geological close-up examination of a full-height pillar, so we were reliant on historical observations and a more recent measurement by David Pollard. The earliest description of a section through the Combe Down beds seems to have been by Richard Pocock in 1750 (Hemmings 1983, 46). This records that:

> *Under a foot of earth was*
> *4 feet limestone, full of small shells*
> *2 feet "strigery" stone mixed with spar*
> *pitching stone (used for paving) with spar and pea-sized*
> *4 feet ragstone, similar but with more spar* [used for] *paving*
> *5 feet picking bed, similar but less spar, softer than rag, poor weathering*
> *they at present work about 12 feet below the picking bed, in all about 30 feet. They say there is good stone 30 feet deeper than they work.*

This suggests a total of 48 feet (12.8 m) of freestone, which is still substantially more than seems to have been worked.

Geologist William Smith (1767-1839) briefly became a quarry operator at Jackdaw and Kingham Quarry, just south east of the modern village but his venture ironically was 'finished utterly by the unexpected deficiency of the stone on whose good quality the whole enterprise depended' (Phillips 1844, 78). The likely cause was a fault running through or close to his workings, presumably leading to fractured stone. Debts from this and other causes led eventually to his imprisonment. This is, however, no reason to doubt the acuity of his observations at Combe Down (Pollard 1994, 27-28):

> Letter from Bath, 21st April 1911
> *This stone on the north side of the Down* [roughly north of North Road today] *has but a few inches of rag lying over it but on the south side* [ie the mined area] *thick and strong beds of it which form a good firm ceiling. The beds of stone are thin on the north side* [of] *the Down* [and] *become much thicker on the south side.*
> *The stone from the quarries on the north side cuts good ashlar but cannot be procured in large blocks but blocks of large dimensions may* [be] *got on the other side – one 4 feet thick $4^1/_2$* [ft] *wide & 17 feet long was procured.* [306 cubic feet]
> *The stone is very thick on the north side* [of] *the hill. The two top beds worked will stand any weather. The bottom beds are the hardest and on that account the men were paid 6d more per score for getting and working the ashlar. The quarrymen have names for the different beds – one which is very hard and bad to work 6 or 8 feet from the top is called the Iron Bed.*

In a section from the north side of the down Smith noted:

> *A hard bed, a part would do for the road*
> *Rubble Stone*
> *about a foot of hard stone with Screws, very sharp*
> *Loose rubble*
> *Rag with Oyster shells (and holes on the top)* [presumably unconformity]
> *Picking stone*
> *2 Beds which will stand any weather*
> *Iron Stone*
> *Part quarried out for Freestone*
> *Hard beds at bottom*

and, in a letter of 1st November 1812 from Tuckingmill

> *The direction of the joints (many of which are faults which elevate or depress the stratum a little) is so prevalent that the principal part of the long joints seem to take this course. The gullys seem to have more of an east and west direction. These intersections & also with other joints more in a north and south direction makes the blocks and pillars of stone in a rhomboidal shape and sometimes triangular.*

Smith also commented on a 'cockly bed' below the picking stone, which Pollard (1994, 28) interprets as three beds, the lower two of which were rich in cockles left *in situ* in a small area in north-west Byfield, with freestone removed above and below. This is the one instance found in the mines of an upper- and lower-level working system. The cockles would render the stone useless for fine building purposes, blunting the saws. A few cockles were found nearby in full-height working, but otherwise were not noted elsewhere. Stabilisation Scheme boreholes made to investigate the cavities between the upper and lower cavities showed the three beds varied laterally over some 20 m with a total thickness from about 1.5 m down to about a metre.

Large blocks were certainly restricted in availability and that mentioned by Smith was very large. The largest blocks in the early 19th century were found in John Pierce's quarry east of Combe Road with others in the quarries on the west side of the road. The survey noted particularly large beds, somewhat over 2 m below the roof, in West Firs and East Byfield just south of North Road. Two very large blocks (about 12 x 3 x 3 feet in contemporary measurements) form the gate posts to Allen's Home Farm (now Priory Close), opposite the lower gates to Prior Park (Fig. 2.5). These have well-developed current or cross-bedding exposed by weathering. Pollard relates this feature to the beds under the picking and cockly beds which were particularly used in exposed situations for plinths and string courses. It was, perhaps, this hard-to-get bed which earned the workmen sixpence a score (20 cubic feet was a nominal ton) more pay. Hawkins (1994, 222) refers to cross-bedding at Springfield Quarry (a worked-out open quarry near the Combe Down boundary with Odd Down) but specifically notes the phenomenon was not found in the mines. The archaeological survey also failed to note any underground example of cross-bedding and it seems likely it is only easily noticeable in weathered sections.

Fig. 2.5 One of the two Priory Close gateposts, nominally 3.7 m (above ground) x 0.9 m sq. The pair are the largest extant blocks of Combe Down freestone

A sequence recorded in 1894 by George Harris (Pollard 1994, 26) in an open quarry north of North Road, (probably at that date almost opposite the Hadley Arms) of 14.94 m (49 feet) had 4.27 m of 'building stone best in quarry', ie freestone (1994, 27-8), but they also used the overlying beds for various purposes and were unlucky to have this cockly bed or bastard stone.

12 feet	*Tile like rubble*	(3.66 m)
4 feet	*White stone, road metal*	(1.22 m)
5 feet	*Ragstone, pavements*	(1.52 m)
6 feet	*"Picking" bed, ashlar*	(1.83 m)
8 feet	*3 beds bastard stone with cockles*	(2.44 m)
14 feet	*Building stone, best in quarry*	(4.27 m)
Total 49 feet		(14.94 m)

None of the sections described show the full sequence in detail, and health and safety considerations prevented direct examination close up during the Stabilisation Scheme. However, laser scanning permitted the thickness of some 5 or 6 m of the beds exhibited on pillars to be recorded in the Grand Canyon (Fig. 2.6). There is no data on the beds here actually worked below this, which, so far as could be seen elsewhere were usually not worked at all.

The above sections and descriptions make clear that the beds of stone were very variable due to lateral variation during deposition, erosion and re-deposition caused by sea floor currents and tidal movements. The thickest worked beds and sequences were probably in Central and West Firs and Byfield or the open quarries west of Combe Road, which was perhaps why they were worked first on a large scale. In the north-east the beds were generally thinner and the very thickest beds probably absent. The usefulness of the stone varied and a degree of pragmatism was required by the quarrymen in deciding which beds to work for what purposes. Certain of the Twinhoe Beds in surface quarries could be used for specific purposes, especially for rubble wall building, which were entirely neglected in the underground mined quarries. The occurrence of thicker beds in the south of the area, noted by Smith, may be the reason why early 18th-century quarrying started at the far, southern rim rather than the north rim closer to Bath, although Allen's plans for the creation of his grand Prior Park House may also have influenced this.

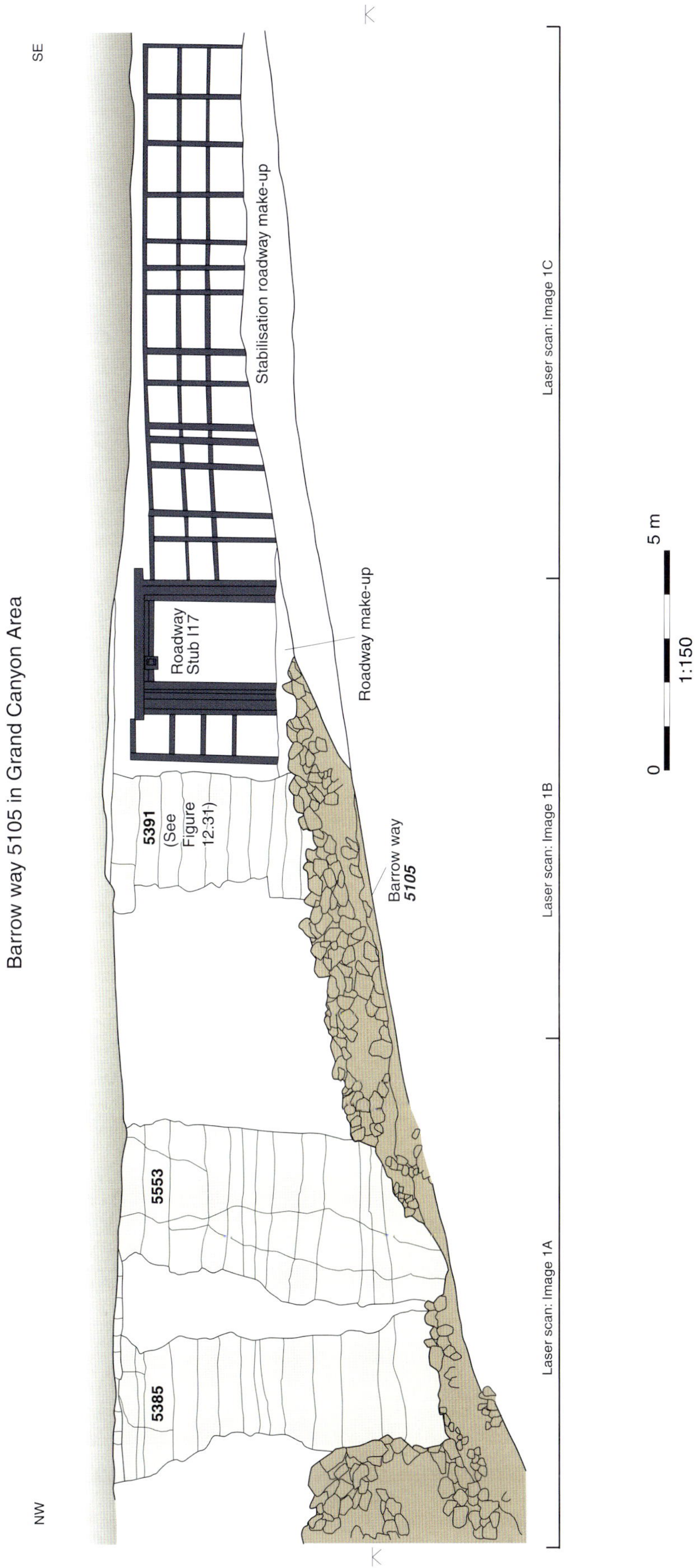

Fig. 2.6 The Grand Canyon inclined barrow-way and pillars with sedimentary beds, derived from a laser scan image

The decision of Allen and others to work underground may also have been an economic decision if the Twinhoe Beds were not then easily marketable. In later years Allen opened a quarry 'open to the day', the refuse of which 'was of such use' that 'part was burnt into lime, part was used for building into common walls, some cut into slabs for paving and the rest was converted into steps and other things as was proper to be made from the harder stone' (Wood 1765, 425). This clearly made surface quarrying a more favourable option. Underground working, however, continued until the mid 20th century, perhaps because it did not disturb the surface and conflict with other land-use requirements. Thinner beds were not necessarily a major disadvantage, since they were more easily worked and handled, and there was a large demand in buildings for blocks of modest thickness and the major demand by far was for good-weathering rubble or 'walling stone' though this was rarely mentioned. However, for the finest grand work, thicker beds were advantageous, and their relative lack must have been a major disadvantage to Combe Down when Wiltshire Bath Stone (Great Oolite freestone) became easily available in the mid 19th century due to railway development, which occurs and can be worked in large blocks.

Historically the Combe Down freestone as a building stone has been frequently described in a complimentary way. By freestone, it is meant that it can be carved without fractures generating in any particular direction. When freshly dug, 'it is soft, moist, yellowish, and almost crumbly, and it seems very little more than congealed sand and that not well concreted together' (Owen 1754, in Woodward 1876, 129). Opponents of the stone described it as like 'Cheshire cheese, liable to breed maggots that would soon devour it'. Wood, however, at the occasion of the competition for the Greenwich Hospital contract in 1730, provided a sample which the opponents' architect actually mistook for Portland Stone (Wood 1765, 426-7). It is easily wrought or sawn and on exposure it hardens and becomes whitish, probably due to migration of lime-containing-moisture to the outer faces where re-crystallisation would take place, 'becoming an excellent stone', but careful selection and honest dealing were required for the best stone. According to an article in *The Builder* (12 April and 28 May 1894), following a description of microscopic examination, the better class of stone had granules which were semi-crystalline and as lasting as the (calcitic cement) matrix with fewer open cleavage planes and with the constituent particles interlocking with each other. This is a briefer but fundamentally similar description to that put forward by Palmer, above.

The quality marketed was certainly mixed. Woodward (1876, 128) wrote that Lodge Hill Stone had been used in the Henry VII Chapel at Windsor (he may have meant St Georges Chapel in which Henry VIII was buried) but 'It has not proved very durable', but he also suggested the Combe Down stone was the best for weathering, though there was a finer quality for interior work at Farley Down, and Boxhill, Wiltshire's Great Oolite freestone had a superior fineness. It is probably the more crystalline character and lower porosity of the better Combe

Fig. 2.7 Sailed pillars – the result of a fault running through north Firs

Down stone which allowed it to be used immediately, rather than, as with the Corsham, Wiltshire stone, needing a period of weathering or seasoning after extraction to let it dry out. Nonetheless it probably also benefited from such seasoning, if only because it exposed flaws before the stone was actually used. Stone from nearby Hampton Down had significant failures in the locks of the Kennet and Avon Canal around 1810, exfoliating after frost, but the use of a porous stone in conditions of frequent soaking and drying is probably an unsuitable one. Less scrupulous, market-orientated entrepreneurs were apt to adapt any of the local stone to the term freestone, or, as one said, 'we would call it Portland if that was what the customer required' (ibid).

Both joints and faults have a major impact on the stone. The major fault affecting Combe Down is a normal fault, running north of Beechwood Road, letting down the rock to the south, displacing, for example, Vinegar Down Quarry from Kingham and Jackdaw on the north side, by 20m or more (Macdonald *et al*. 2000, 3). It is possible there are several small faults, as intimated by Smith, with individual small movements leading to minor fault scarps. It is difficult to explain different quarry horizons otherwise.

Another fault runs through Horsecombe Vale and down-faults to the north, this and the Beechwood faults producing a graben, or lowered block between them. These faults in conjunction with flexures in the bedding seem to be responsible for the substantial flows of water at Whitaker's Spring and possibly those in the upper part of Horsecombe Vale. What may be a small fault runs roughly east-west just north of the Firs Shaft,

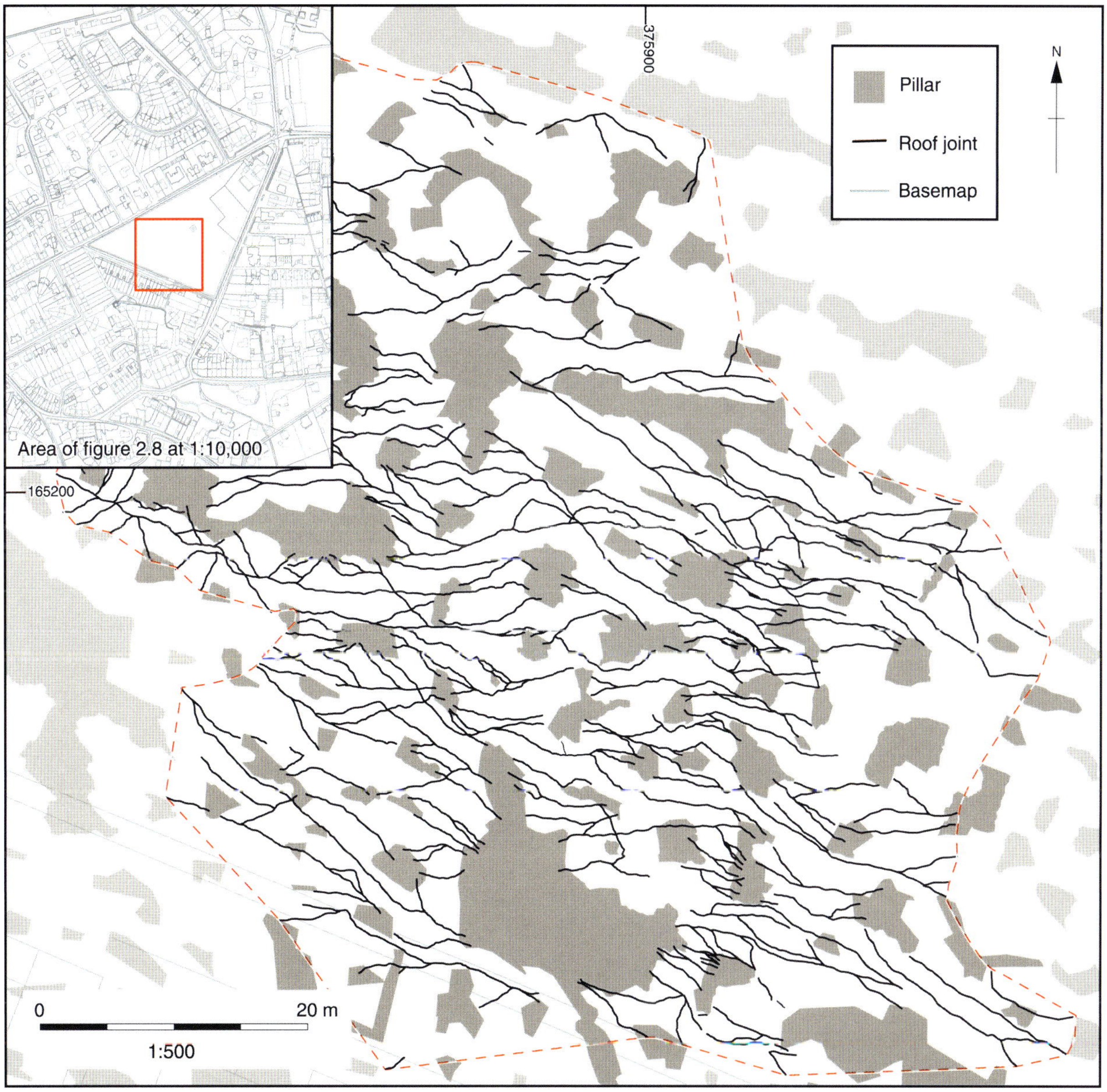

Fig. 2.8 Plan of roof showing joint system and pillar spacing in the Allen and Allen Estate workings in Central Firs

producing a zone of narrowly-spaced fractures with a strong inclination to the south, resulting in what we termed 'sailed pillars'. These are narrow with an elongated elliptical horizontal section (Fig.2.7). This may reflect the westward projection of a fault shown on the 1:10000 geological map which Hawkins (1994, 23) suggested may cut the northern boundary of Firs Quarry.

There is a strong system of fairly close joints or fractures through Combe Down roughly trending 70° and 140° with an average spacing of 0.8 m. According to Hawkins, the strongest and most continuous joints tend to follow the 070° direction, which is also roughly parallel to the general trend of faults and the southern outcrop. However this did not always appear to be so, and in Central Firs for instance the predominant joints appear to be on the 140° trend (Figure 2.8). The joints were often slightly open, but many joints in both pillars and roof opened as a result of post-working tensional stresses, so are anthropogenic rather than natural. The effect is two-fold, first that it assists removal of blocks by quarrymen, especially where the beds were also affected by foundering over the Fullers Earth at the margins, and second that it also limits the size of blocks available both by the fractures themselves and also, since they are trapezohedral rather than rectangular, by the need for considerable scalping to produce regular blocks. Faulting also, as well as displacing beds, also increases the degree of fracturing nearby and, as Smith found, could be disastrous for the business of quarrying.

This had implications for the placing of pillars, which tend to have their long direction aligned with the strongest set of joints. In some cases lines of pillars were left either using a single strong joint alongside a run of pillars, or systematically sidestepped joints from pillar to pillar. The former was seen in James Riddle's quarry (518) in Byfield where pillars had an alignment along a joint open some 75 mm wide. This was clearly a dubious practice but was not apparently punished by collapse. West of Firs Shaft and in Byfield, sidestepping of pillars to avoid exposing the same side of successive pillars to the same joints appears to have been used. However, many arrangements of pillars seem to have little logic in position or joint relationships. The general close spacing and openness of the joints which led to easier working is shown on pillar surfaces which are mainly natural discontinuities rather than split or broken faces. Hawkins suggests only 5% of pillar surfaces showed evidence of working (1994, 45).

The available stone at Combe Down of almost any type and quality was much more efficiently utilised during the 19th and 20th centuries. Some quarries only worked walling stone, perhaps scavenging from older workings, and it is possible that some of the 'honeycombed' and weathered stone found in many gardens in the area was deliberately worked for sale for this purpose. The architect Wood used some 800 tons of shelly-ragstone for the foundations of Allen's Prior Park House. This he referred to as usually left as the roof-stone (Hemmings 1983, 41). In Bath's Newbridge over the Avon, rebuilt after 1774 and enlarged *c* 1830, the voussoirs of the arch are shelly limestone, probably better able to stand the compressive forces than the softer, more friable freestone. Although it may have only accounted for relatively small quantities, stone was also burnt for lime, Ralph Allen adding this to his business probably around 1740, and there were still limekilns at Priory Farm and north of Brassknocker Hill in the late 19th century (1884 1:2500 OS.). Local mortars were made of crushed stone-dust and lime, the former perhaps made by the masons and builders from waste stone from the banker masons.

Chapter 3
Development of the 18th-century landscape

This chapter first provides an overall account of the background history, in summary form, for the history of Bath, and in more detail, for Combe Down. The landscape changes which developments in quarrying after *c* 1700 introduced at Combe Down, including development of the village are considered. More detailed accounts for both underground and surface quarrying will be the subject of later chapters and the documentary and archaeological sources for information used in describing the landscape will be found therein unless otherwise stated.

Bath and its rebuilding: a brief history

Bath's World Heritage Site status, which extends over Combe Down, has resulted from its development within the 'Valley of Pleasure' as it was called in the 18th century. Its site is on the north side of a loop in the River Avon rising to Lansdown in the north and to Odd Down, Combe Down, and Claverton and Bathampton Downs to the south. Its fame was established by the time of the Romans who renamed a former Celtic shrine as the Temple of Sulis Minerva and the settlement as Aquae Sulis (the Waters of Sulis) from its thermal springs. It was established close to the river crossing of what was initially a military road (Cunliffe 1986). The Romans built their baths, with lead linings, and bath houses in stone, and although no precise location for their quarries has been found – it might be Lansdown (north of the City) or at the southern Downs – there was almost certainly minor quarrying at Combe Down where a Roman villa and stone coffins have been found. The town and springs, though decayed in their bathing function for a long period after Roman society collapsed in the 5th century, remained important for those wishing for a cure of diseases and for those using the baths and the surroundings for pleasure, in a similar way as happens at the current time.

In medieval and early post-medieval times, however, it seems likely that use of stone was limited to the more prestigious buildings, such as the new Cathedral commenced in about 1091, of which the present Abbey is largely its nave (Davenport 1988), or to the city walls. Most houses then seem to have used cheaper timber framing methods of construction, and re-used stone, sometimes from the walls. In the late medieval period Bath had become a regionally important cloth town, but by the mid 16th century was in a state of decay, despite expansion elsewhere of population and the woollen industry. This decay was despite (or perhaps because of) the advantages of the hot baths which were pointed out by Dr William Turner in 1562. They were then described as a danger to body and health with a lack of drainage and with indiscriminate mixed bathing. However a visit by Elizabeth I in 1574 appears to have stimulated civic pride, and the 'rebuilding of Bath', including a separate Women's Bath dates from about that time (though mixed bathing was again condemned in 1625).

Soon after the streets were described as full of 'swaggering swordsmen' and the days of 'ancient and honest trades of merchanting and cloths were over'. Fear of fire and civic pride led to the widespread use of stone, and in the mid 17th century it was described, not quite accurately, as being a city built entirely of stone. It catered for medical patients and those with fat purses in 'petty palaces' rather than common lodgings (Manco 1992).

By the beginning of the 18th century and the onset of the Golden Age it was thus already the resort of the noble or rich and the city had been described as the 'prettiest in the kingdom'. Bath's accommodation was in handsome but vernacular building style, mostly within the old walls but without any of the formal planned layouts and classical styles of later. The earliest phase of improvement had used recovered stone, but by then the main source of the new stone used was from around present day Odd Down alongside the old Roman route from the south known in the 18th century as the Fosse, which led, via the Holloway and over Bath Bridge, into the city.

The golden metamorphosis of the City was associated especially with three men. Richard or Beau Nash (1674-1762), who exhibited 'a veneer of gentility ... personal dignity and a willingness to do anything for money' who, arriving early in the 19th century as a mere gamester, transformed the social aspects of Bath as Master of Ceremonies (Eglin 2005), away from that of 'swaggering swordsmen' to something closer to his own persona. Ralph Allen (1693-1764), a Cornishman who arrived as a thirteen year-old boy about 1710, provided entrepreneurial and business skills and the stone by which the new parts of the city were built, though his extreme wealth came from his Post Office contracts. He will be considered in much more detail later. The third was John Wood (1704-1754), originally a joiner who had already developed into an architect and building developer when he arrived about 1726. He championed high-quality building design in the classical mould. He and his son, also John (1726-1782), were responsible for many of the exquisite buildings so well revered in Bath today. These men, of course, were the most notable of those who developed Bath; there were also the rich landlords, such as Mr Pulteney who is remembered for his shop-lined bridge, and others who backed them financially. There were other architects and skilled masons too, less fashionable but busy and competent, like John Strahan whose accomplishments included Avon Street, the City's red-light district. This combination of quality and sleaze, as in 20th century Las Vegas, proved irresistible, and though its zenith of fame and quality came by about 1790, Bath's fame and attractiveness have lived on to produce the modern city.

Combe Down prior to 1700

Prehistoric

Perhaps because Combe Down was subsequently so extensively quarried and built upon, few pre-Roman remains and no certain Iron Age artefacts have been recorded. Early occupancy of the wider area including Combe Down is demonstrated by a flint-working site, which also produced Neolithic and Bronze Age artefacts, in the grounds of a school south of Claverton Road (SMR/NMR number ST 76 SE 5), and there is another site with numerous flints on Claverton Down (ST 76 SE 28). A round barrow (ST76 SE7) seen on Claverton Moor in the 1920s is now ploughed out, and a late Bronze Age spear-head was found at Midford Brook in the southern part of the area (ST76 SE 6).

There is evidence within modern Combe Down of a later prehistoric field system at Claverton Down Hospital, which may have been part of a wider field system at Claverton Down and Rainbow Wood (ST76 SE 3). Aerial photographs of Bathampton Down *c* 1.5 km to the north show a large area covered by field boundaries and it is possible that Combe Down originally had similar features, which may have survived in use into the Roman period.

Roman

The proximity to Roman Bath is reflected in a much greater assemblage of finds in both Combe Down and the wider area than for prehistory. It is reasonable to assume that quarrying was carried out in Combe Down to provide for local building and for stone coffins found there. However, for the principal buildings in Bath it is more likely that stone was carried down the old Fosse Way or Holloway route from quarries at Odd Down (via Lyncombe), the then major route from the south into the city (Peach 1895a, 11), or from Lansdown on the north side of the city, both closer to the centre of Bath than Combe Down.

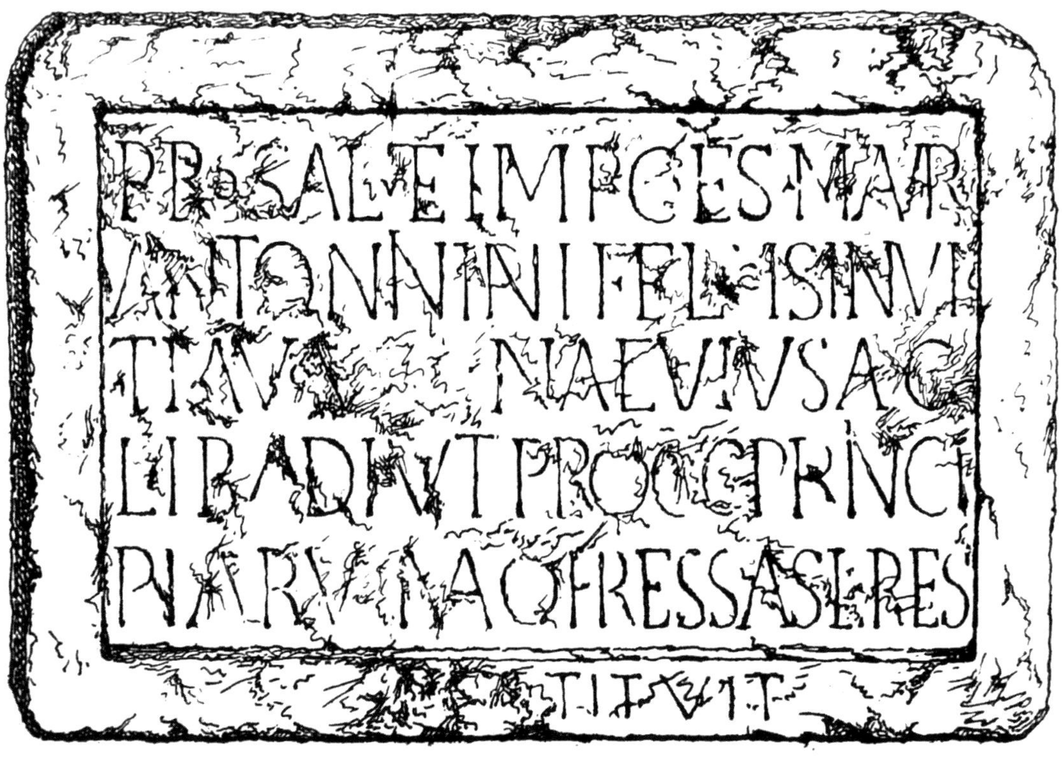

Fig. 3.1 Inscription found on a Roman stone coffin-lid at Combe Down. Illustration by C S Beckett, in Scarth 1864

A courtyard-style villa was found about 1860 at Belmont whose stone Addison (1998, 14) suggests came from Vinegar Down Quarry. A more likely source may have been just uphill at the main outcrop into which it is built. Stone coffins were located nearby. Pitcairn and Richardson (1924, 47) list these as found in 1822 in the field adjoining Summer Lane, below Grosvenor Lodge. In 1844 more were found in the foundation works for Belmont House, with a coin of Licinius. In 1854 two coffins and a skeleton were found in levelling mounds between Belmont and the Upper Schools and below the vicarage garden (Scarth 1864, 75). A large slab over a coffin was inscribed (Fig. 3.1). A hoard of coins of the later emperors were also found (see also Page 1906, 309-11). Construction of the Foxhill housing estate in the 1950s produced pottery, broken querns and animal bone, and a number of shallow pits yielded more pottery, building material, a bronze bracelet and a child's skull (Cunliffe 1979, 131). Another stone-coffined burial and some Roman pottery were found at Combe Down Quarry in the 1930s when a gallery collapsed. There was another cemetery about half a kilometre north of Combe Down at Perrymead where three stone coffins were found in the 19th century and a further one in 1952 (Cunliffe 1979, 132).

The field systems noted above on Claverton Down and Rainbow Farm may well still have been used in Romano-British times with pottery beads and building material recovered from the ditch fills suggesting a Roman occupation site not far from the round barrow. It is possible that, as well as engaging in agriculture, with perhaps some quarrying, some Roman-British people used its fine outlooks and bracing climate to site villas outside the city, much as happened there after 1800.

Anglo-Saxon

Little archaeological material of the early part of this period has been found, the main exception being a spear from the site of the Roman villa excavated in 1860 (BN1814 ST76 SE15). There is, however, the major feature of the Wansdyke, which crosses the area from west to east, en route from Portishead near Bristol to near Andover. This is a 6th- and 7th-century fortified embankment and ditch probably marking the boundary between the kingdoms of Wessex and Mercia. It is visible as it enters the area near Odd Down, and appears to terminate at the head of Horsecombe Vale. It seems likely it then followed the natural north side escarpment of the combe (Shepherds Walk), leaving it to cross Firs Field and the south end of Prior Park and over Bathampton Down (Major and Burrows 1926, 88). An aerial photograph shows a soil mark in the Park on the suggested line. This and another feature appeared on earlier Ordnance Survey maps as crossing Prior Park School sports ground, but are not shown on the modern maps. The possible former route, based on topographic considerations, was shown as a footpath or ride across from Shepherds Walk and Firs Field to the Prior Park gates (opposite the Hadley Arms Public House today) on the post-1764 Allen Estate Map (Chapman 1996), though the route hereabouts is now lost. However, a later reassessment (Fox and Fox 1960, 36) suggested that the soil marks described above were possibly earlier field systems or natural landslip features. Features of possible medieval origin north of Combe Down towards Perrymead are now considered natural landslips partially modified for agriculture. This is perhaps more credible on the latter site, which is on or just above the underlying clay slip-strata, than at the former, which are on the top of the Down.

Medieval

A manor of Lyncombe was held by the church from the 7th century. The name Widcombe first appeared in the Domesday survey of 1086. That year the Bishop of Wells bought Bath, including Widcombe and Lyncombe, from William Rufus (Peach 1895a, 6 and 23), and in 1091 the building of a cathedral was begun (its nave is the larger part of Bath Abbey today). Subsequently the transfer of the See from Wells to Bath led to the Bishop becoming Abbot of the older Bath Abbey, and the Prior becoming subordinate to him. Disputes between Wells and Bath led eventually to the emergence of a Bishop of Bath and Wells and the Cathedral was designated a Cathedral Priory. (Davenport 1988, 7-8).

Under the Bishops the slopes and top of the whole area must have been used for grazing the bishop's sheep, perhaps giving the name Shepherds Walk to the track along the northern edge of Horsecombe Vale. Lyncombe and Widcombe was not sold after the Dissolution, but leased. By the early 17th century it was held by Hugh Sexey whose executors established Bruton's Hospital with the estate as a trust (Chapman *et al.* 1998, 4).

A priory and deer park had been early established at Combe Down, the boundary of which followed Claverton Down Road to the south and east, ran west just north of Rainbow Wood and came up Hanginglands Lane (Popes Walk today). The outline is shown on the mid 18th-century map of Ralph Allen's estates. King John probably hunted in this park in 1212-16 when he visited Bath. In 1223 part of the Bishop's holdings, including the western half of the deer park, were transferred to the Prior of Bath Abbey (Chapman *et al.* 1998, 2). Thus there was subsequently both a Bishop's Park and a Prior's Park and to the present day a wall runs along the boundary between the two. The Prior's Park had apparently been detached in the 16th century (Peach 1895a, 40) and parts were sold off. When Leland visited Bath around 1540 he described the walls as ruinous and the parks without deer (Toulmin Smith 1910, 98). The walls have been repaired at intervals, but some medieval stretches appear to have survived (Chapman 1996, 17). By the early years of the 17th century the Park was owned by the Poole family of Monkton Combe.

Most of the land around Combe Down beyond the parks was also in church hands during this period. Monkton Combe, then called *Cume,* was also owned by Bath Abbey until the Dissolution. After the Dissolution Combe was granted by the Crown to Humphrey Coles, but it soon reverted and in 1564 was given to John Robinson of Gravesend, whose descendants held it until 1706 when John Robinson sold it to Thomas Poole. At some stage in late medieval times the waters available at Tuckingmill and lower downstream in Monkton Combe village must have become utilised for the mills there, probably also using the Fullers Earth from a nearby outcrop higher up Horsecombe. By the early 18th century the Tuckingmill was in ruins (BRO. 339/4/10). The various bishops also held a 'Liberty of Claverton and Bathampton' from 1250 (Chapman 1996, 19). By the early 18th century Claverton was held by the Rector and Bathampton by the Holder family, into which Ralph Allen later married. Small-scale quarrying may have taken place during this period, but agriculture would still have been very important despite Bath's decline as a centre for the cloth industry from at least the mid-16th century.

Quarrying certainly was taking place in the area around the Combe Down in the mid 16th century, most likely at the present-day Odd Down and probably at Claverton and Bathampton, supplying the two villages and the city. Monkton Combe must also have had quarries, and field observation suggests these were probably at a site immediately above the village around the present Mount Pleasant and Shaft Road with the old route from there down to the village a shorter one than Summer Lane.

Leland described his journey in *c* 1540 from Midford Brook to Bath through 'mountayne ground and quarre' (Toulmin Smith 1907, 139), suggesting quarrying was a significant feature of the then landscape. In 1663 Oluf Borch visited Bath and recorded the first mention of underground working. The site was identified by John Wood as that bearing the name Horsecombe Quarry, no longer identifiable but most likely near the head of the combe just east of the present day Cross Keys Inn (Addison 1998, 15). Probably the main area of quarrying then was at Pitts Quarr, working the escarpment near Barricks Farm, the most convenient of all sites by distance and road to Bath.

In the late 17th century and the first decade of the next, the area now known as Combe Down was, however, still predominantly agricultural and its 'sleights hills and mountains' in Monkton Combe and the area to the west in Lyncombe and Widcombe as far as Entry Hill (known as Combe Down or Greendown) were still largely regarded as sheep grazing. Some enclosure had taken place, on such as the large Widcombe Field, which was at least partially arable, and there were many small parcels of enclosed land on the slopes of Horsecombe, as well as Greendown and Claverton Down, Bathampton Down and an area (sheep pasture) east of Greendown and north alongside Prior Park known as The Lawns (Chapman 1996). Areas marginal to the Down, closer to settlements or on gentler slopes below the escarpment were mostly suited to arable, for which enclosure by walls was considered necessary.

Historical overview of the area after 1700

The term Combe Down originally referred to that part of the downland most remote from the parent parish of Monkton Combe, when it was usually also referred to as Greendown which included the adjacent part of neighbouring Lyncombe and Widcombe. After Ralph Allen built his small settlement there for his workers in *c* 1729-30, the name transferred to the settlement and adjacent quarries on both sides of the parish boundary. The present Combe Down parish and ward of the City of Bath originated in 1854 when the growing village formally took in the outlying uplands from the three parishes of Monkton Combe, Lyncombe and Widcombe, and Claverton. This study extends outside the parish to the contiguous part of Prior Park, the home of the principal landowner(s) and via linkages, mostly transport, to the Avon Navigation and later canals and railways and, in a very limited way, with Bath itself. The boundary, with the known former quarries at surface and underground is shown in Figure 3.2. This demonstrates that a very large portion of area has been quarried, though, as the key shows, much smaller portions were affected by active quarrying at any one time and the land thus as a whole had substantial, and often more important alternative uses, as it does today.

In the 18th century the main impact came from features related to the underground quarries, and the impact of surface quarrying in that century was still subdued, though surface quarrying was growing in importance much faster than underground in the late 18th century. The middle years of the century were dominated by Ralph Allen's quarries, but only the surface expression of these is considered here.

The end of the 18th century ended badly for the Combe Down Quarries with extremely difficult economic conditions in the late 1790s leading to very low output indeed but, after the 1801 Peace of Amiens, there seems to have been a considerable degree of economic buoyancy in Combe Down, presumably reflecting that in the City of Bath and despite some very bad years of depression around 1825 and 1830, quarrying seems to have recovered rapidly.

In the 19th century surface quarrying dominated, with the core areas of underground mining almost exhausted by the mid century, though they staged a

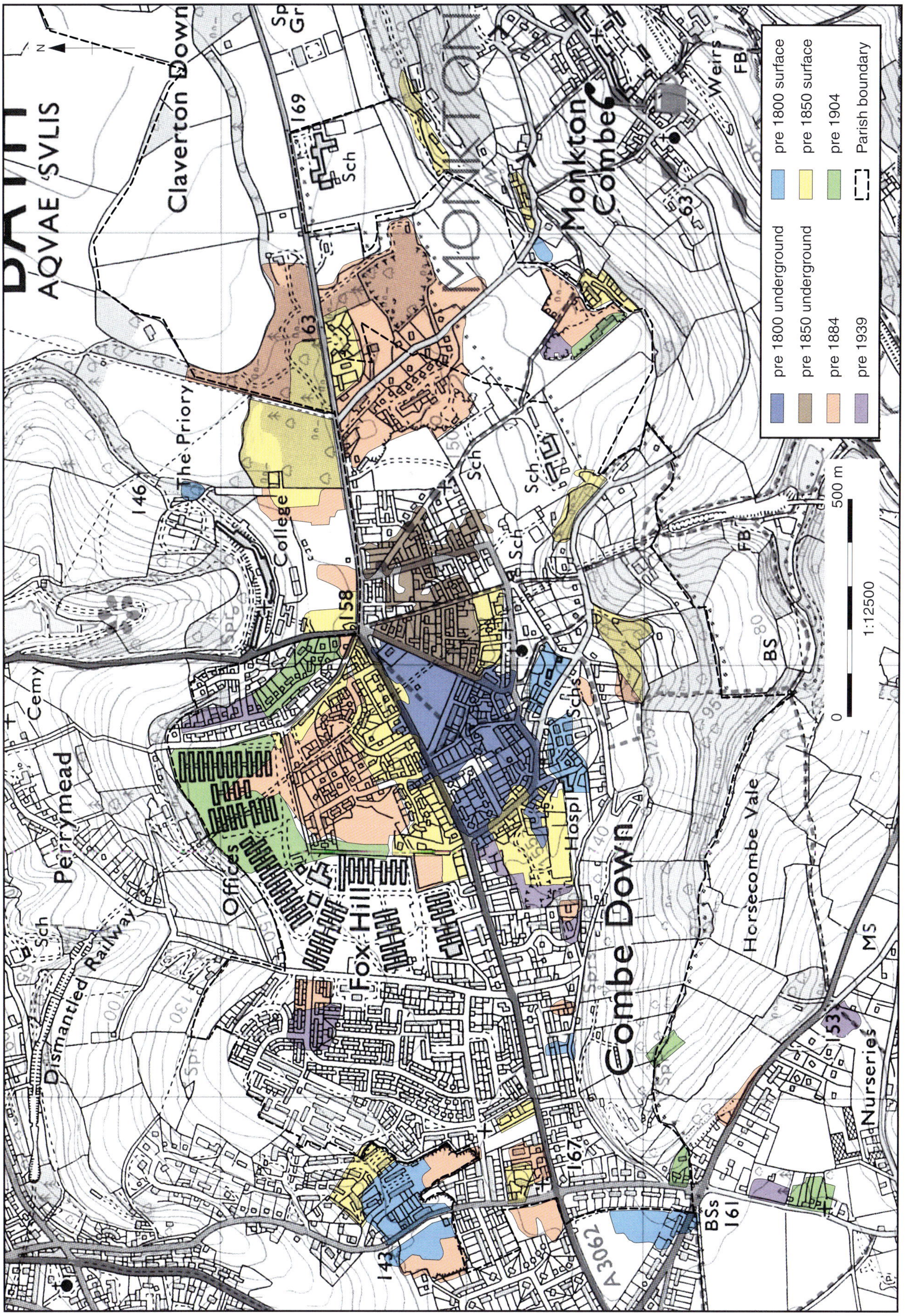

Fig. 3.2 Map of the modern ward of Combe Down, showing surface and underground quarries, updated from the 2000-2001 field archaeological survey

late revival near the century's end. As active quarrying retreated, so housing development advanced, beginning in the late 18th century, with planning for a settlement around 1800, and developed enough to form the nucleus of a new parish by the mid century. The modern core village, still dominated by quarrying, was largely developed by the 1860s, though there has been much subsequent infilling and the village then would have had a much more open aspect than it does today. The extensive building outside the core area seen today is very much a 20th-century phenomenon, but in many cases has also taken advantage of land released as quarrying ended. There are wide areas of quarrying dereliction (or land regenerating spontaneously) but many old quarries were filled after the Second World War by building debris from Bath and can be identified today as 'greened' recreational areas such as football and rugby grounds.

The pre-quarrying landscape

The landscape of Combe Down around 1700 was dominantly agricultural, though quarrying must have taken place for local purposes. Most of the area was rough grazing for sheep, the 'sleights and mountains' of Green Down or Combe Down as they were described in documents, which seem, partly to have been stinted pastures. The name 'The Lawns', a substantial area just east of the modern core of the village and extending down the east side of Prior Park, gives us some idea of the likely appearance of the down itself, which had only slight undulations and would have been close-cropped by sheep. The north-western area, formerly part of Lyncombe and Widcombe, was known as Widcombe Field and parts of it were certainly used for arable, probably the parts closest to Entry Hill at the west end of the area. A switch to arable from pasture seems to be reflected in documents, but this has not been quantified and it was probably overwhelmed by the impact of quarrying anyway. Parts of the steep slopes beyond the south rim down to the Horsecombe Brook were also arable and enclosed, mainly by hedges. Part of this area was known as the Vineyards, (now Vinegar Down) though this use was long abandoned. Parts of the steeper slopes on the south margins were wooded, above Tuckingmill and south and west of de Montalt Mill, including the Beechwood. Prior Park on the north side in 1700 was a decayed medieval deer park with low remains of walls and probably buildings, which had originally taken advantage of the valley overlooking Bath for hunting (Chapman 1996, 16).

The earliest proper pictorial indication from which details of the pre-quarrying landscape can be inferred comes from the *Survey of Five Miles around Bath* (Fig. 3.3) by Thomas Thorpe published in 1742, though by that date it naturally showed Mr Allen's freestone quarries so his house and park and quarry features must be ignored for present purposes. Routeways were few. The Down and head of Prior Park was crossed by the Claverton Road – a rough track which ran east-west, and which was later turnpiked (*c* 1770). It was joined by Summer Lane, which slanted up the south side slope from Monkton Combe and before or soon after 1700 it had a small farm at the junction, probably giving its name, Stonehouse, to the site. These were probably the only significant ways for wheeled traffic, leading across to the old Fosse route to the west and down to Bath. There were two through footpaths: Shepherds Walk came along from the top end and north edge of Horsecombe and possibly marks a former boundary, now a linear bank, of Wessex and Mercia, extending from the Wansdyke, to the west. It possibly continued across by a path over modern Firs Field and the top of Prior Park. The second, known sometimes as the Long Drung, came over the Down from Monkton Combe via Tuckingmill, and went on down to Bath. Though a farm cart might use either of these paths on the Down, they would have been often impassable on the steep slopes. There were probably another two footpaths from the western side down to Bath, both also very difficult for wheeled traffic.

There were a couple of small farms or similar east of the modern core area of the village by 1742 at least. By or soon after 1700 there was the Cross Keys Inn at the important crossing point at the head of Horsecombe, and, within the Combe, there was Horsecombe House. Both were possibly built with stone from a small quarry nearby within Horsecombe, but there was little other settlement until later and then mainly below Entry Hill. To the south-east, Tuckingmill was derelict.

Some areas were made into smaller, individually occupied enclosures. Hedges and perhaps fences were used on the slopes of Horsecombe which was divided into long and narrow enclosures, and stone walls were probable or on one side of the Claverton Road or on the Widcombe Field, It was a partially enclosed open field, though internally it was still partly divided into strips. Probably some small surface quarries had developed to supply material for walls, on the escarpment rims or perhaps at the Sheeps House Quarry on Combe Down, using the poorer shallower stone of the Twinhoe Beds.

The impact of early quarrying pre-1729

The archaeological evidence for the location of pre-Allen underground quarries is somewhat negative, in that some underground workings seem to have early characteristics, but were apparently abandoned in the area he took over. These and comparative features within other quarries not then affected by his ownership, point towards some quarrying on a scale greater than purely local use on Combe Down even before the documentary record indicates, that is before about 1715. Later quarrying near the south rim below the modern village may also have removed evidence of any early activity there.

Fig. 3.3 A segment of Thomas Thorpe's Survey Five Miles Around Bath (1742). It shows Summer Lane in its earlier position extending to buildings where now Stonehouse Lane meets the Claverton Road. The Long Drung is only sketchily shown. Note the Cross Keys Inn, and the road there and another to the west going by the 'Old Stone Quarries' at Barrocks (Barrick's farm), associated with the old Roman route down to Bath. (Image courtesy of Bath Record Office)

What is likely to have been called Sheeps House Quarry, now known as Rock Lane and Quarrymans Court, had demonstrable pre-Ralph Allen workings with one underground area having a dated inscription of 1725 when it was clearly well established. The Sheeps House Quarry had several other small underground workings from entries on its sides and there may have been similar early underground working on the east and west sides of the Long Drung, perhaps especially near what is today The Brow.

Underground quarrying of a similar type also took place north of Shepherds Walk at both sides of the former Lyncombe and Widcombe/Monkton Combe boundary either side of Rock Hall Lane. Possible such workings were destroyed by modern construction of Byfield Place. Competition from Allen's quarries would have made it hard for others to continue in business for long in the area.

Underground quarries such as these would have little impact on the landscape. They would have small scars of tipped spoil in front of the underground entries and perhaps a small cliff over the level or sloped entry, since the freestone would have been several metres below ground level. Underground workings might have utilized older surface quarries to make entry underground easier. A network of tracks would have led from the quarries on to the main track over the Down or back to Monkton Combe.

Further east, above Monkton Combe village, around what is today Shaft Road and Mount Pleasant, field examination suggests early quarrying probably took place at the surface. This was the most convenient point for stone used in building the village. No other site has suitable stone and ease of access. Remains of possible but undated quarries can be seen along the edge of the Shaft Road and along the edge near Combe Manor and extending towards Brassknocker Hill.

It seems clear, therefore, that the core area of Combe Down and some parts outside it, were already involved in quarrying, even if the larger part was hidden underground. In the core area which is now the village, the absence of any substantial group of contemporary buildings built of local stone implies that stone from there was conveyed a considerable distance from its point of production rather than being simply a resource for parish use, as the workings above Monkton Combe and at Horsecombe probably mainly were. What there may have been in this overwhelmingly agricultural landscape was overspill activity from the long-established quarrying area further to the west, around Barrick Farm, with the old Fosse Way to Bath. These were old and apparently substantial quarries in 1742 and at Combe Down we may be seeing competing small specialist masons opening their own small quarries, or already at work anxious to secure control of their own supplies as Bath began to expand.

Expansion of quarrying 1729 to 1764

Ralph Allen's substantial investment into quarrying at Combe Down from about 1729 had an immediate and increasing impact on both the surface landscape and underground. He seems to have allocated a specific area of his eventually much larger estate to quarrying from the beginning, concentrating his activities there and keeping both this and other areas minimally affected, a policy apparently continued by his estate after his death in 1764. Most of his investment went into an improved stone handling and transport system between the quarries on Combe Down and his wharf, stoneyard and building sites in Bath, a distance of some two miles, down a fairly steep slope. These will only be outlined here, as they affected the landscape, with the development of Allen's business examined (and sourced) in Chapter 5.

In quarries without a crane or railway, typical of Allen's competitors, the stone had to be loaded on to horse-drawn wagons outside the underground entrance and was carted away directly to the stone-yard or building site. But the generally already-poor roads were damaged further by this, and it was a laborious, slow and hazardous process.

Each of Allen's underground quarries had a similar and relatively efficient arrangement (Fig. 3.4). The entrance was placed at the base of a small quarry dug into the rock some 10 m below the general ground surface and reached by a sloping track from the top of the Down. Above the entrance was a vertical cliff. The entry or entrance went almost level into the freestone beds that were to be worked, and worked stone was dragged to the outside on barrows, carts or wagons.

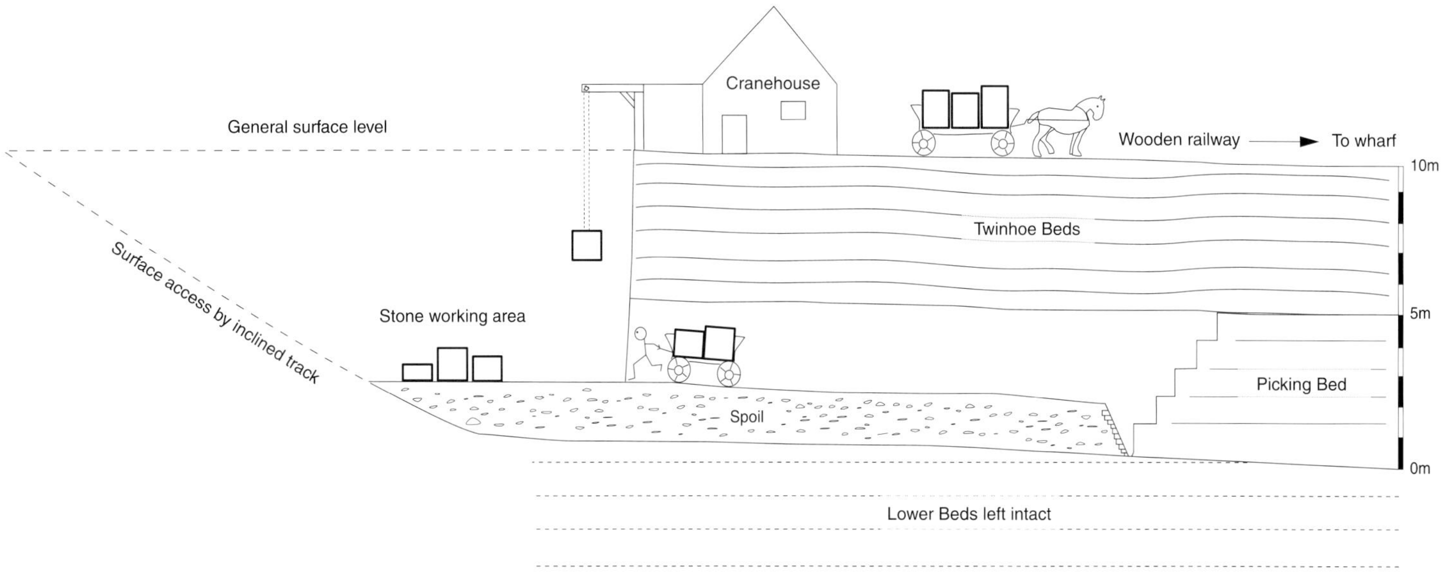

Fig. 3.4 Schematic representation of typical surface and underground arrangements at the entries of Allen's underground quarries

At the top of the cliff, at the normal ground surface level, a crane was mounted with its jib projecting over the edge to haul up stone from below. At the top the stone would be lowered dirctly onto a carriage on the wooden railway, from where it was dragged by horses to the top of the hill and down to the wharf.

At the wharf the stone could be slid off the waggon, or unloaded by crane at the stone yard, or loaded by crane on to a boat (to Bristol and

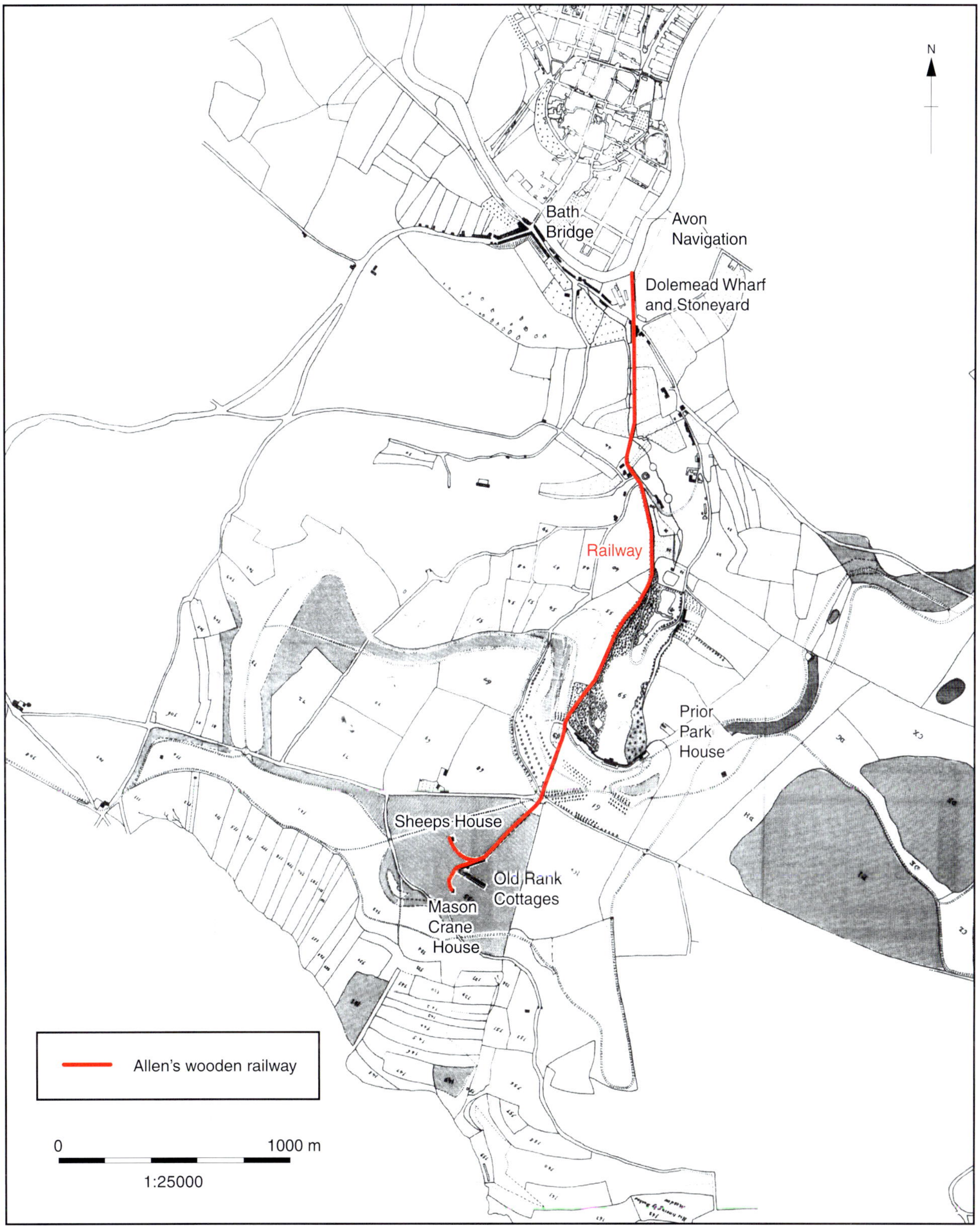

Fig. 3.5 Route of Allen's wooden railway from the Combe Down Quarries to Dolemead Wharf at the Avon Navigation, Bath (after Chapman 1996)

onwards). Alternatively the loaded wagon could be run on to a barge and moved by river (and another wooden railway if necessary) to the building site.

The area involved in Allen's quarries was south of the Claverton Road (North Road today), east of today's Combe Road (the then parish boundary), and west of the Long Drung, extending from here to the southern escarpment or rim. Allen also reduced the impact of quarrying by concentrating all the underground entries and associated surface areas on the southern side of the area, furthest from his intended house. This is the area within the Monkton Combe part of his initial purchases, though the remaining areas bordering it came into his ownership soon after. Previous quarrying within this area, and sometimes outside it by others seems to have stopped, but some underground quarrying seems to have continued in others' ownership even after he acquired the surface, perhaps because it was underground and barely visible.

The main surface impact away from the southern area came from his one and a half mile (3 km) wooden railway which connected both surface and underground quarries to the River Avon Navigation. It followed his own (and friendly neighbour's) carriage drive at the side of Prior Park (Fig. 3.5). Its novelty, at a time of considerable interest in industry and science, perhaps encouraged this unusual dual usage, and certainly it created great interest among visitors to Bath (Fig. 3.6). The railway led to Allen's Stoneyard on the south bank of the Avon Navigation, from where it was loaded onto barges or river boats for transport to site (Fig. 3.7).

At both the Combe Down and the Navigation ends of the railway Allen erected a small group of houses and an inn (Figs. 3.8 and 3.9). The houses were designed by John Wood as a classical style terrace, with a pediment for the central house at Combe Down. This was occupied by the Clerk in charge of his stone business, Richard Jones. Each house had a small garden and the row was central to and overlooked the nearby quarrying activities. Other houses and the Carriage Inn alongside the railway were erected and probably designed by Jones, who was a very competent draftsman as well as mason. It was not until late in the century that more houses were added to the proto-village. This early settlement and route of the railway is shown on the 1742 Thorpe map, though at a very small scale, and a better representation is given by an estate map of about the time of Allen's death (1764) but missing some of the earlier elements. Figure 3.10 is a reconstruction derived from these maps and from data found underground, showing the probable four crane sites and related entries and a surface quarry and stoneyard and crane as they probably were sited about the mid 18th century.

Fig. 3.6 Allen's wooden railway, at the lower end of Prior Park about 1750. The nearer carriage contains block stone, the other seems to contain walling stone. From an engraving by Anthony Walker Prior Park: The Seat of Ralph Allen (courtesy of The Building of Bath Museum)

Fig. 3.7 Ralph Allen's stone yard and wharf at Dolemead, alongside the Avon Navigation in about 1734. The nearby large buildings are a brewery. Detail from The South East Prospect of Bath, an engraving by Samuel and Nathaniel Buck (image courtesy of The Building of Bath Museum)

Fig. 3.8 The Old Rank of the eleven cottages at Combe Down, designed by Wood for Allen's workmen 1729-30

The cranes and house and railway must initially have been very visible across the down, but Allen soon began planting rapidly-growing fir trees (see Figure 3.10). These initially defined the carriageway (as the railway was known) and other roads and tracks, but later infilled the whole quarrying area and other adjacent areas. On the *c* 1764 Allen Estate map, the quarrying area was defined by the trees. He also moved the end of the old Summer Lane from Stonehouse to the present location of Combe Road. His industrial quarried landscape was thus part of a fairly typical estate landscape with estate activities concealed within woodland. The railway also functioned for building his house, probably using a small branch, and an early self-acting branch, where the weight of one wagon going down drew the empty back up, was certainly used by Jones to build the Palladian Bridge in the grounds.

Sometime around the mid-century, by which time Allen's trees must have provided useful concealment, he also began to quarry substantially at surface at or near the southern escarpment. The largest quarry ran along the southern edge, almost gradually linking along the whole length of the modern Quarrybottom, extending from Byfield across to Belmont. The top of the quarry, *c* 1759 on Thomas Robins sketch is shown on Fig. 3.11 in front of the row of cottages. Earlier, as shown on the 1742 Thorpe Map the crane had been sited further to the west, but expansion of the Masons Crane House Quarry had probably led to this new location, with a substantial area below occupied by a stoneyard with the usual piles and heaps of stone and spoil and small sheds for the masons. There may have been a limekiln too, since he also began lime-burning, but he may have preferred to place this nearer the navigation and cheaper coal supplies, as well as closer to his market.

After Allen's death 1764 – 1803

After Allen's death in 1764 the railway was abandoned and transport reverted to the use of horse and carts. It was perhaps about this time that a second surface quarry was begun behind the row of cottages with its west end near the Carriage Inn and extending east to what became Quarry Lane, adjacent to Claremont on Church Road. The west end was worked out by 1800 and the east end, behind Hopecote Lodge, almost so. There was almost certainly surface quarrying too at the Sheeps House Quarry and possibly, for a short time, at Jones Quarry, at the old parish boundary. The open quarries growing in extent must have had a very significant landscape impact, with men very visible at work.

The increased use of horses and carts meant inevitable damage to tracks and to the newly developed (by 1770) turnpike. This followed the old Claverton Road, across the down, either west to an improved Entry Hill route to the city, or east either to the city by a longer, gentler route via Bathampton, or down Brassknocker Hill. It is doubtful too that the individual quarry-masters who took over

Fig. 3.9 The cottages built for the stoneyard workers at Dolemead, at the lower end of the railway

Fig. 3.10 (facing page) The railway and probable quarry entries at Combe Down, derived from the Allen Estate and Thorpe maps, and archaeological evidence

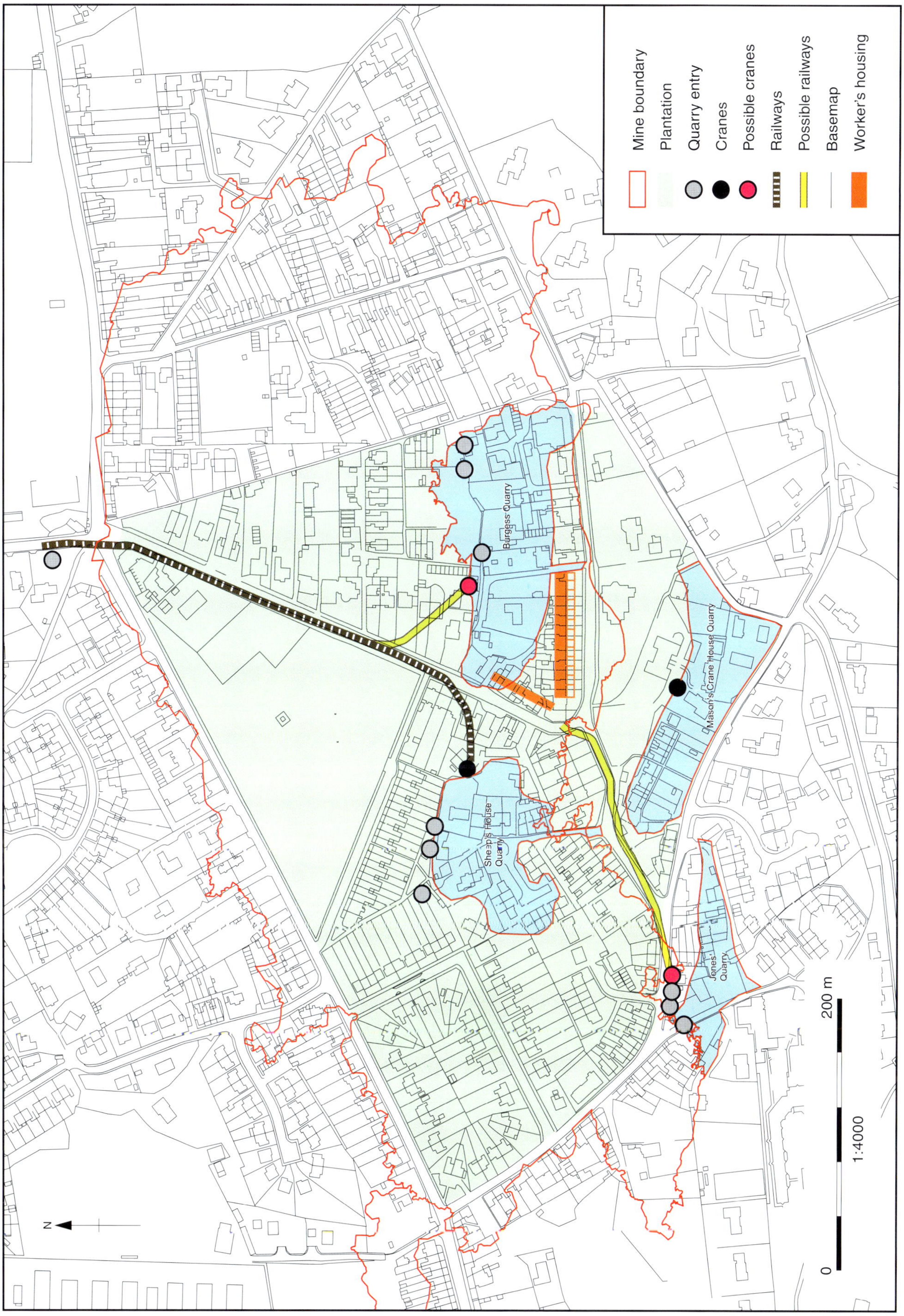
Mine boundary
Plantation
Quarry entry
Cranes
Possible cranes
Railways
Possible railways
Basemap
Worker's housing
Burgess Quarry
Mason's Crane House Quarry
Sheap's House Quarry
Jones' Quarry
N
0
200 m
1:4000

Fig. 3.11 Ralph Allen's Freestone Quarry at Combe Down showing the Masons Crane House Quarry and crane, with the Old Rank, including Dial House (where Richard Jones lived) in the background. Sketch by Thomas Robins the Elder, probably around 1750 (by permission of the Victoria and Albert Museum, London)

Fig. 3.12 Freestone Quarries near Bath by J. Hassall, about 1800. The site may be near Entry Hill. Note the use of bow-sawing and the wagon awaiting loading. (copyright Victoria Art Gallery, Bath & North East Somerset Council)

were as concerned to protect the area as Allen had been. Prior Park itself was isolated from quarrying impact by the maturing fir trees and by banning of wheeled vehicles from the Prior Park Drive, other than by specific consent: a small illustration (BRO. 313) shows a fine ornamental double gate built between two pillars blocking access at the top of the drive.

Underground the quarries were also advancing to the north, concealed under the surface. However, as they advanced, the difficulties of drawing stone out of the levels southwards and, after closure of the railway, having to taking it back northwards across the surface to the turnpike must have been increasingly troublesome. The need to ventilate the workings had led to the use of shafts. These at first were very small, a metre or so across, and thus were easily concealed in the trees. Some time in the second half of the 18th century, however, it became normal to use what became known somewhat misleadingly as 'light shafts', which in fact were used for winding stone. Each was some 4 m or more across, needing a fence or wall and a lifting mechanism, store yard and track to service it. With perhaps half a dozen of these in use by the end of the century, when the quarries were leased out to independent owners, they would become obvious landscape features to people walking or riding within the trees.

The Masons Crane House Quarry was the largest of those at work and exploited the southern cliff-like edge north of Summer Lane, and seems to have developed from the western end eastwards. It would have had a face some 10-12 m high, of which the top 5 or 6 m would be exposed as rather 'brashy' stone or 'ridding' – soft, broken, thin bedded and clayey – which would form a slightly sloping cliff. Some could be used but much would simply be dumped. Small spoil heaps seen on the 1759 Robins' sketch of the site are possibly the same heaps visible today in the churchyard. Much of the spoil from these and stone-working must have been dumped on the quarry floor to form a level working surface, or used to improve the tracks and roads around the site. The freestone beds would be more carefully worked, probably in steps or benches. A contemporary picture of a 'Quarry near Bath', which may have been at Entry Hill (Fig. 3.12), shows large blocks being levered off the cliff to the floor, with a man below using a large bow saw to cut 'wrought'

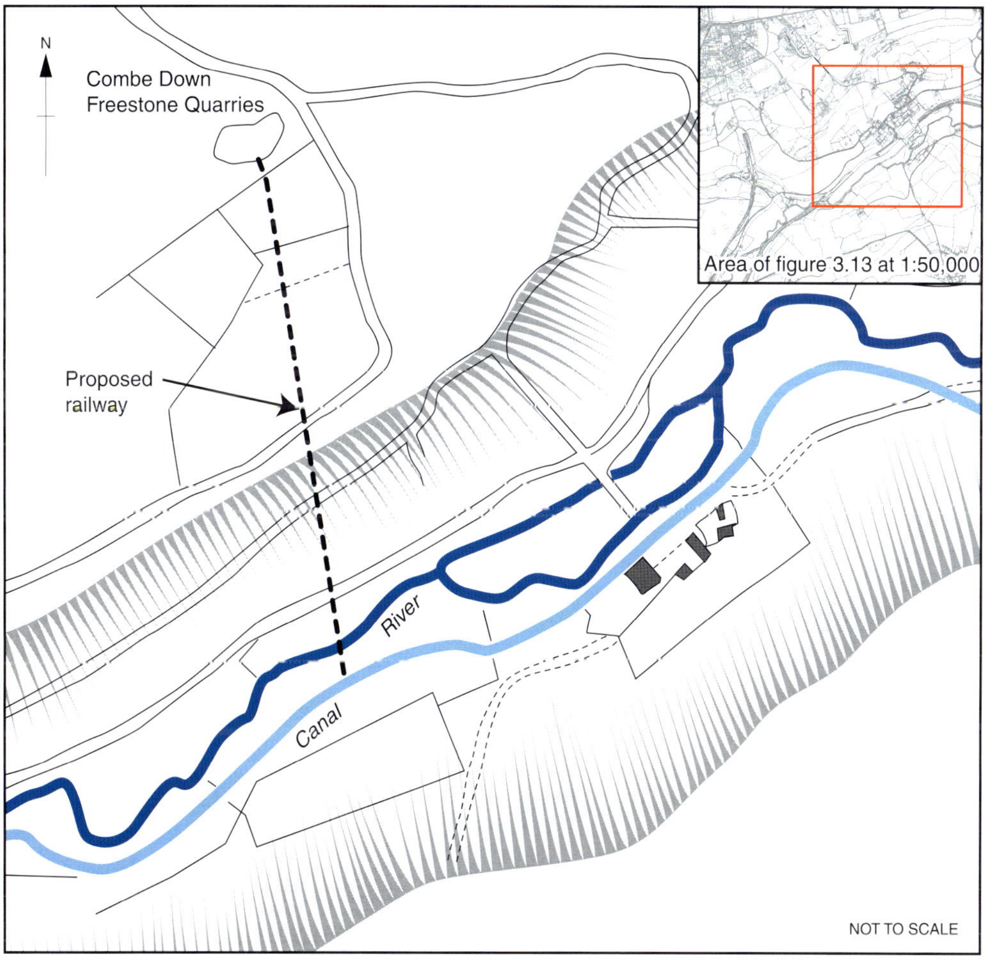

Fig. 3.13 A proposed railway from Mount Pleasant Quarry (Shaft Road) down to the intended Somersetshire Coal Canal. There is no evidence it was ever built. Detail from plan supplied by David Pollard

slabs or ashlar off a rough block. A horse and cart awaits loading nearby. The large quarry at Combe Down would have had similar scenes. The picture shows substantial vegetation, indicating a cessation, if temporary, of working in some parts of the quarry, but most of the working areas and faces would be raw, whitish rock with the quarry floor away from the banker masons workplaces, either dry, dusty and rutted, or wet, muddy and rutted. There would be a number of small simple stone or wood and, possibly, thatched sheds for the banker masons, and a map of *c* 1803 shows a stable. There may have been a pump over a well and horse trough, as there certainly were at the Carriage Inn. As well as the crane, it is uncertain whether other lifting devices such as shear legs were used, but it seems likely such simple tools of the mason would be there.

The second open quarry behind the Old Rank must have been similar, though this was a 'hole quarry' reached down the slope of Quarry Lane next to Claremont and presumably worked east and west from its bottom. This lane also appears to have been made to serve the underground entries which apparently preceded the surface quarry. This quarry also facilitated access to 'undermined' workings under the back yards of the row of houses erected by Allen for his men.

There may have also been small areas of surface working associated with the Rock Hall Lane and Rock Lane (Sheeps House Quarry), both of which appear to be much larger than would have initially have been required for the underground entries. At the Sheeps House Quarry, the original crane site as shown on the *c* 1764 Allen Estate Map was quarried away, almost certainly by this time, and quarrying may have been developed to the west along the same line of frontage to allow new entries and cartways to have been driven in from there.

There are several references in documents to unidentified quarries, possibly associated with Combe Grove, then belonging to the Vaughan family. These may have included what became the Mount Pleasant (surface) Quarry (2373), at the south end of Shaft Road, from which a tramway was planned about 1795 (Fig. 3.13), leading down towards what was to become the Somerset Coal Canal branch of the Kennet and Avon. However, the tramway was probably never built (Pollard pers. comm.) and there is certainly no remaining trace on the ground. The probable quarry there is today occupied by the row of houses built for later quarry workers while their gardens (often car parking spaces now) appear to be on the waste tips of the quarry. Surface quarrying had also begun along both sides of Shaft Road near the main Turnpike but in 1800 this still had little overall impact on the landscape (see Fig. 3.2) for a plan of these sites in their developed form). There may also have been a small development just east of Prior Park, possibly for estate purposes.

The largest new developments were related to the improvement of the Entry Hill route into Bath (Fig. 3.14). They are shown on the 1799 Charlton map of Lyncombe and Widcombe (SRO DD/583 c/212: copy in BRO.). Quarries are shown on the west side of the Entry Hill road across to Crossway on land owned by 'The Parsonage', one at Crossway, the other at the steep section near the top of Entry Hill itself, later used as a corporation depot. On the east side at Entry Hill, the land was owned by the Magdalen Hospital, and was part of the Widcombe Field. Two quarries are shown within what became Combe Down (two more are shown in the adjacent part of what is now Odd Down). One was the roadside part of the later Springfield Quarry, and was partly an underground quarry which was filled in as part of the recent Stabilisation Scheme (Quarries 2374 and 2375). The other borders the Bradford Road Turnpike. No other quarries are shown on the Charlton map including the area of Green Down or Quarr Down, 'Late Collibees', west of the Combe Road boundary with Monkton Combe. These became an important area of quarries after 1800. These Entry Hill sites would have particular advantages with respect to

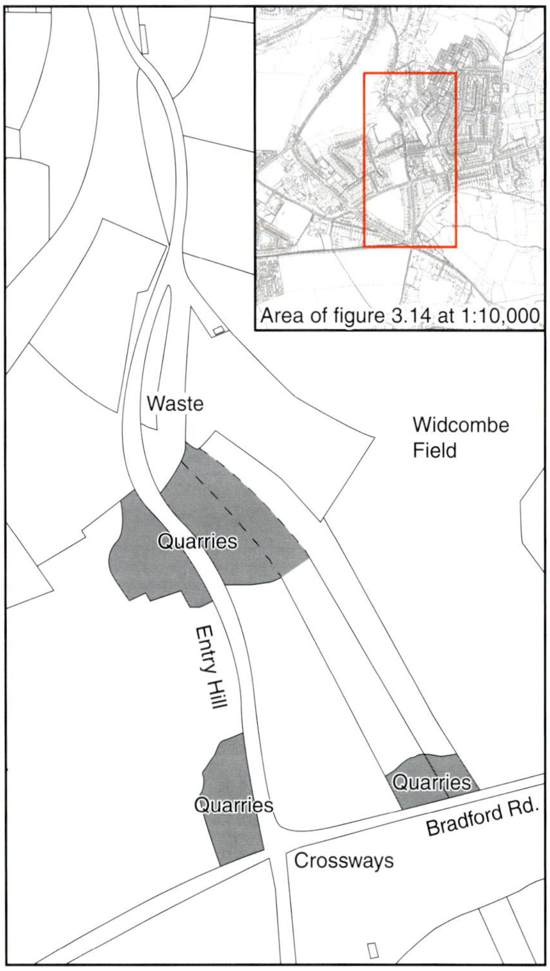

Fig. 3.14 Quarries at Entry Hill in 1799. Those on the east side were within Combe Down. Detail from the 1799 Charlton Map of Lyncombe and Widcombe

their distance from Bath but those on the east side of the wider area were also closer than the core area of Combe Down, via the Bathampton route, to the City. In about 1800 the level of economic activity was still very low, but normally the turnpike roads would have had scores of two-wheeled carts or four-wheeled wagons on them, the wheels retarded on the slopes by scotches to prevent runaways.

Open quarries, unless very small, are nearly always very conspicuous with high impacts on the surrounding environment, and the commencement of surface quarrying indicates that the original discrete estate aims of Allen were being abandoned even in his lifetime in response to commercial pressures, and to a growing extent thereafter. However, there were still extensive tracts of grass-land remaining between these sites, though this appears to have been in a process of private and piecemeal enclosure, as were the turnpike road bounds. This may have just been fashionable among landowners, but fences and walls protected animals from the nearby quarries and may have been a response to the needs of the growing city not far distant.

Chapter 4

19th- and 20th-century developments: the spread of surface quarrying and the growth of the quarrying village

Introduction

The first years of the new century saw the break up of the old Allen Estate structure at Combe Down. New landlords emerged and new entrants came into the freestone quarrying business, both in the old core area and the peripheral areas of the future Combe Down parish. This chapter concerns the developments at surface and the story is continued, briefly, to the end of substantial quarrying in the 20th century.

Two new major transport developments affected Combe Down in the first 40 years of the century, though they hardly affected the landscape of the Down itself. First was the opening of the Kennet and Avon Canal and its branch the Somersetshire Coal Canal in 1810. This led to wharves being built in Bath near the old Dolemead site, another near the Dundass Aqueduct at the bottom of Brassknocker Hill and others at Tuckingmill and near Midford, which opened up markets to the east. The Nowell family opened a wharf in London at Pimlico (Irving 2005, 68). This was to be a golden age for Combe Down, with perhaps its highest levels of production, probably for underground quarries, but certainly for the outputs of numerous surface quarries which came to dominate. The canal also linked quarries in Wiltshire to Bath, a competitive threat more fully realized in about 1840 when Brunel's Great Western Railway came via Box Tunnel, actually cutting through the Wiltshire freestones on route to Bath and Bristol. The huge reserves there of thick, massive and easily worked freestone became marketed as 'Bath Stone'. Its production on a much larger scale opened a new era of fairly cheap and abundant stone for the building industry, and it thus became a massive and ultimately overwhelming competiing source. Combe Down went quickly into relative decline and, after about 1860, absolute decline. Coinciding with exhaustion of the older quarries, although Combe Down probably was still the preferred choice of the Bath market, the future was becoming clearer and by the early 20th century, Combe Down was largely a relic quarrying landscape, and a growing suburban settlement.

Break-up of the Allen Estate

In the latter part of the 18th century, the still largely intact Allen Estate came into the hands of Sir Cornwallis Maude who had married Mary Bennett, Allen's niece. He later became 1st Earl de Montalt, and in 1793 became the first Lord Hawarden. He died in 1803 and his son, the second Lord died in 1807 (see Chapter 6 for details). Under the first Lord, the estate seems to have envisaged the end of surface quarrying in what is now the core area of Combe Down, the southern end of the area denoted by Allen's Fir plantation and centre of the modern conservation area. They clearly also envisaged the continuation of underground quarrying, including from the Sheeps House Quarry and the Jones Quarry which were slightly away from the centre.

The growth of the settlement of Combe Down

Lord Hawarden seems to have initiated the housing development process, improving the Old Rank, subsequently de Montalt Row, about 1795. The re-fronting, with a 'polite façade', of Isabella Place (named after his wife) a row of some slightly older cottages, added a degree of architectural as well as marital balance to the settlement until after 1805 (Fig. 4.1). This suggests Hawarden probably envisaged a settlement of mainly middle-class housing. These first houses seem to have been marketed for 'invalids', requiring somewhat cheaper accommodation than was on offer in the City of Bath, in a bracing and pleasant, wooded and elevated environment. In the next thirty years or so several substantial houses were built, such as Mrs Cruikshank's Belmont House, the Brow, and Claremont and Hopecote houses, in addition to a chapel, church and later vicarage and schools.

However, the sale of the estate (in part after the first Hawarden death in 1803, and the second in 1807), led, over three decades, to a substantial area coming into the hands of speculators who then divided the land piecemeal into leasehold or rented properties for building, gradually developing into the core community of Combe Down. The quarries in or close to the core area were somewhat similarly divided into fairly small, one- or two-acre plots, perhaps to ensure fairly rapid working across the area and the freeing-up of more land for building. Land well outside this central area was probably considered to have less building potential and seems generally to have been divided into larger plots, as around Shaft Road and Entry Hill, for example. The land east of the Long Drung, affected by later speculation (around 1830), was sold partly on the basis that housing and quarrying could co-exist, if the latter was on a small scale.

When de Montalt purchased Prior Park and the Estate in 1788 he immediately mortgaged it. In the downturn of the 1790s the interest, even at 4% must have been crippling, as receipts probably fell dramatically, not least from the quarries. It seems likely he used Benjamin Wingrove to advise him; he was a surveyor and conveyancer, and later (1817) the Bath Turnpike Trust's first professional surveyor. Probably between them, they conceived a new plan for Combe Down, with wide roads, 50, 40 and 30 feet wide. Perhaps, as for the Turnpike Trust later, Wingrove was an expensive man to appoint, probably demanding needlessly high standards of construction (Buchanan 1986, 242). There seem few signs that Hawarden prospered under the arrangements, though Wingrove and his fellow speculator Harry Salmon almost certainly did.

Wingrove was considered trustworthy enough, however, for Hawarden to sell him some of his land, reducing the mortgage on the estate by a few hundred pounds, and to be promised a further £1000 against a payment made, but without any grant or conveyance being completed. At this point Hawarden died, in 1803, leaving heavy mortgages on his Combe Down property and obligations by the sale of it and other properties to pay the £5000 legacy of his daughter Sophia (by Mary, née Allen). His estate trustees struggled and one resigned, but came to agreement with Wingrove to transfer yet more property, effectively the heart of the settlement of Combe Down as known today, to Wingrove on payment of a further £1000 (LL. A91/18/5/16). On the 4th of January 1805 the deal was done, but on reading it one can almost feel the expensive lawyerly disapproval in the huge deed which sorted out the complicated sequence of events since 1723. Wingrove now owned a large part of the area south of the Bradford Turnpike (North Road), and east of Combe Road to the Long Drung (the area

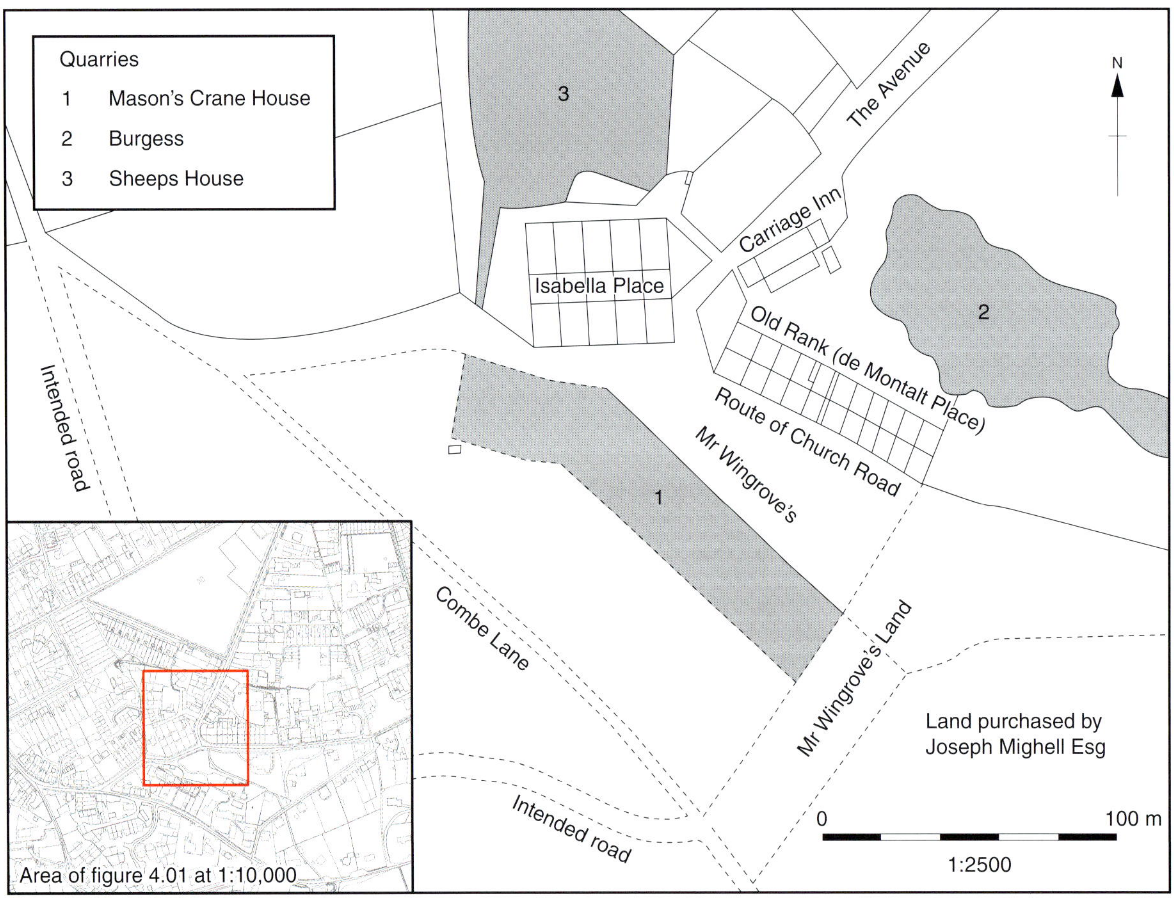

Fig. 4.1 Land at the centre of Combe Down sold by Benjmain Wingrove to Edward Layton in 1805. The three sets of buildings, from right to left, are the Old Rank which became de Montalt Place, the Carriage Inn and Isabella Place. Three surface quarries are shown. The rectangular area is Masons Crane House, Burgess' is behind the de Montalt Place and Sheeps House behind Isabella Place. Note the stable at the south-west corner of Masons Crane House Quarry. (Source LL. A/91/18/5/17)

between The Avenue and the Long Drung became known as Combe Park) and south to what is now Summer Lane.

Much the same was happening west of Combe Road, where Harry Salmon, another land speculator, was the purchaser of the 'Late Collibees Greendown' from Hawarden in 1803, one of the sales 'made or intended to be made by Lord Hawarden' before his death. He was also to have purchased the adjacent land east of Combe Road and he was also to widen roads (Irving 2005, 28) which included Combe Road from its junction with the Turnpike. This had a new wall built down the east side, and perhaps due to location errors found by the recent Lyncombe and Widcombe Survey of 1799, it left a narrow strip owned by Salmon on the east side of the wall, which he agreed to transfer to Wingrove to consolidate the holding on the Monkton Combe side. In the event the land east of the road, including the narrow strip, seems to have passed directly to Edward Layton (ibid).

East of the Long Drung, the Poor Tyning, once part of the Combe Farm, was less clear in its dealings. A small part of the north-west corner was conveyed to Wingrove by a Mrs Sarah Drinkwater in 1809 and a month or two later, what seemed to be this part of it passed to Andrew Greenaway, with rights to quarry etc. Greenaway a quarryman and builder, built two houses and sold them leasehold, and then sold his rights to John Burgess, quarrymaster about 1810 (Pollard 1994, 18). Before 1809 the quarry area to the north of The Brow belonged to the Houlton family, a builder, who sold some to another builder and the larger part to John Ovens Thomas (see below) of Prior Park. He had John Cotterell, a local surveyor and estate agent lay out the land and the Tyning and Gladstone Roads for dwelling plots: the whole passed leasehold for 1000 years to Cotterell in 1830 (Pollard 1994,16-21).

John Ovens Thomas was a wealthy Quaker, a Bristol businessman who accepted an offer to sell Prior Park to him for £28,000 from the Trustees of the second Lord Hawarden in 1808. He was a wealthy shareholder in the Kennet and Avon Canal and saw himself, perhaps, as repeating Allen's success in using the canal to help exploit his estate. The area involved included the land west of the Prior Park Estate in Monkton Combe, sometimes previously known as Colthurst Down, north of the Bradford Turnpike. This was let to several quarrymen and was an active quarrying area until the end of the century, continuing to pay royalties to a family trust after Thomas' demise in 1828.

To the south of Combe Down John Connolly of Midford Castle was the purchaser of at least some former Allen Estate lands, including Kingham Fields and what is now Vinegar Down. Both were worked from about 1810, stimulated by being the nearest parts of the outcrop to the wharf of the Somersetshire Coal Canal at Tuckingmill.

The Hadley Estate

Just before he died in 1803, the first Lord Hawarden, in conjunction with Harry Salmon and Benjamin Wingrove, had planned the core area as the nucleus of a much larger village. They modified the older tracks and laid out new roads to their specified widths and dedicated them to public use. These formed a loop with the turnpike at the north or top end, though with restricted and gated access via Prior Park Carriageway to Bath. This included Combe Road, Rock Hall Lane, Beechwood Road, Summer Lane, The Avenue and the path south of the (later) churchyard. Thus the modern road layout of the core area became a permanent feature of the landscape.

The land east of Combe Road and between the Turrnpike and Combe Road as it turns east into Church Road was sold (by release of mortgage) directly by the Hawarden Trustees in July of 1805 to Edward Layton, described as then as 'of Cornwall' (LL.A91/18/5/19). Significantly it was witnessed by Edward Wingrove and it is likely also it had been agreed in Hawarden's lifetime to have been sold to Harry Salmon so this must have followed a series of negotiations between the parties. The land extended east to a straight line boundary from the centre of Stonehouse Lane at its junction with the Turnpike in the north (which had been the Road to Combe on the Thorpe Map of 1742 and was thus probably a significant ancient boundary line on Greendown), to a point on the west side of Rock Lane where it meets Church Road today. A small strip of land was, as Salmon had agreed with Wingrove, added to it next to the new wall on the east side of Combe Road. Layton, by then of Bath, then began to lease the ground to various quarrymen in a series of strips ranging north and south. He also seems to have acquired, then or later, the ground east of the Stonehouse – Rock Lane line, across to The Avenue. He also bought the village area settled around the junction of The Avenue and Church Road (Fig. 4.1), and with it the quarry at the Sheeps House entries (2312), of which only the late 19th-century western entry (close to the steps today up to the houses on the Firs) was shown, already developed for some 120 feet during the Allen Estate period. The Sheeps House quarry boundary was virtually as it is today and by 1811 included John Davidge's cottage and brewhouse at the bottom, which gave its name to the location. These buildings seem to have replaced the wooden huts formerly used by the quarrymen. Nevertheless, there is evidence underground and in documents of the use of the cartways there long after these changes.

The area of the Masons Crane House Quarry was also shown and included in the village purchase, but symbolically as a long rectangle. From their design it was not long before it was built upon to become the site of the Quarry Vale Cottages. The quarry behind the Carriage Inn and de Montalt Row, west of the Lane to the Quarry (adjacent to

Claremont today) was also shown as bounded pretty much as it is now. All these older village areas can thus be considered worked out by about 1800, though work continued on the surface quarry on the east behind Claremont and Hopecote Lodge and possibly to where Belmont Road and the old vicarage is today, for some years. John Burgess was the probable quarryman of the Hopecote end of the quarry behind it by about 1810. The owner of the land to the front of the quarry is not known, but may also have been him. He was probably facing the end of his stone reserves there, since construction of new houses like Claremont and the Unitarian Chapel of 1815 were creeping eastwards, built in the quarry or on its waste stone.

Layton died about 1808 and his estate was left to Nathaniel Hadley of Lewisham in Kent, whose wife Ann was Layton's daughter. There was no rapid build-up of houses, but more a steady series of purchases of leaseholds by, usually, men of modest means, such as John Davidge, quarrymaster as well as brewer and inn-holder whose house was sited at the bottom of Rock Lane. This sale of leaseholds yielded a steady and increasing income for the Hadley Estate for the remainder of the century. The Hadleys were absentee landlords with their home near Lewisham in south London, where their papers survive in the Library. Their estate was sold in about 1901 (Hadley Arms Papers, private collection, Combe Down).

The Layton/Hadley estate also leased-out the underground quarries between Combe Road and the Avenue and just beyond, to various quarrymasters, some of whom probably previously held leases from the Allen Estate. They worked through the mid century, and on a lesser scale to beyond the end of the century. The extent of their leases is shown on the map (Fig. 4.2) for east of Combe Road, with the similar disposals at very near the same time by Harry Salmon on the west side. These leases, all for underground quarrying, will be considered in detail in Chapter 6.

Harry Salmon's purchase of Quarr Down west of Combe Road in Lyncombe and Widcombe

Harry Salmon bought the area west of Combe Road. He promptly sold it leasehold to several quarrymasters who worked there for the next three decades, mainly at surface. Three at least of these built houses on their plots, Abraham Sumsion, Jonathon Rudman and Samuel Nowell, but the development there, for some three or four decades remained dominated by surface quarrying.

The development west of Combe road has been recently described by Irving (2005), and is largely summarised here including indicating the boundaries of the actual quarries on a modern map (Fig. 4.2) using a 'best fit' approach from the two incomplete source plans (BRO.MH/7/14 and Irving 2005, fig. 13). Irving has also researched the individual quarrymasters involved (including some on the Hadley properties), which, for this reason is not repeated in any detail here except where it directly relates to the quarries.

Harry Salmon purchased this part of Late Collibees Greendown from Lord Hawarden in 1802, probably in anticipation of the sale of leaseholds to several local quarrymen which quickly followed. The tracing of the boundary on the ground is somewhat complicated so is described in detail here. Its south boundary followed Shepherds Walk along the edge of Horsecombe Vale, south of Rockhall gateway from where it projected across to the back of what is now The Paddocks. From there it followed a curving path to what is now Beechwood Road. The actual Rockhall Lane as is seen today did not in fact follow the original intended road, the projected new roads set up by Wingrove and Hawarden, which followed the Lyncombe and Widcombe, and Monkton Combe boundary. This was probably along an old trackway from the Shephards Walk path. Instead, from a point near the junction of Rockhall Lane and Combe Road, it diverted to the west, partly for the convenience of the quarry operators, but also, with future property sales in mind, to serve properties on either side of the road within the same parish – a common ploy when formerly unenclosed land was laid out. This south-east extension beyond the Lane included Byfield Place, where underground workings were reported during recent building operations (Pollard pers. comm.). The date of these is not known, but it is a likely site for early quarrying (Phases I and II) rather than quarrying currently being considered (Phase IV).

The east boundary thus followed behind the Paddocks and Rock Hall Cottages, probably rejoining the line of Rockhall Lane near Ralph Allen Yard (where a stone today marks the old parish boundary), following close to the course of Rockhall Lane and then Combe Road to the Bradford Turnpike then following the southern side of the Turnpike to just east of the Foresters' Arms. From there the west boundary was almost straight down to the Shepherds Walk corner and was delineated as a 15-feet-wide service road, intersecting with the 12-feet-wide road discussed below. Generally the boundaries fit very well to the present boundaries where this is relevant, though the link which seems originally proposed for Rockhall Lane to Beechwood road seems to have been lost, possibly due to quarrying across the line of the road.

This area was divided into two parts, to be separated by an east-west 12-feet-wide service road for the quarries from the modified route of Rockhall Lane. On the south, the land was divided for quarrying into three roughly equal portions, all fanning out from a single position from a previous entry at the west side just below Jones Quarry (now Ralph Allen Yard), thus giving an underground access available to each of the quarries which ranged from there south-east to north-west. These are the three shown in the plan of the sale deed illustrated by Irving (2005, fig. 7).

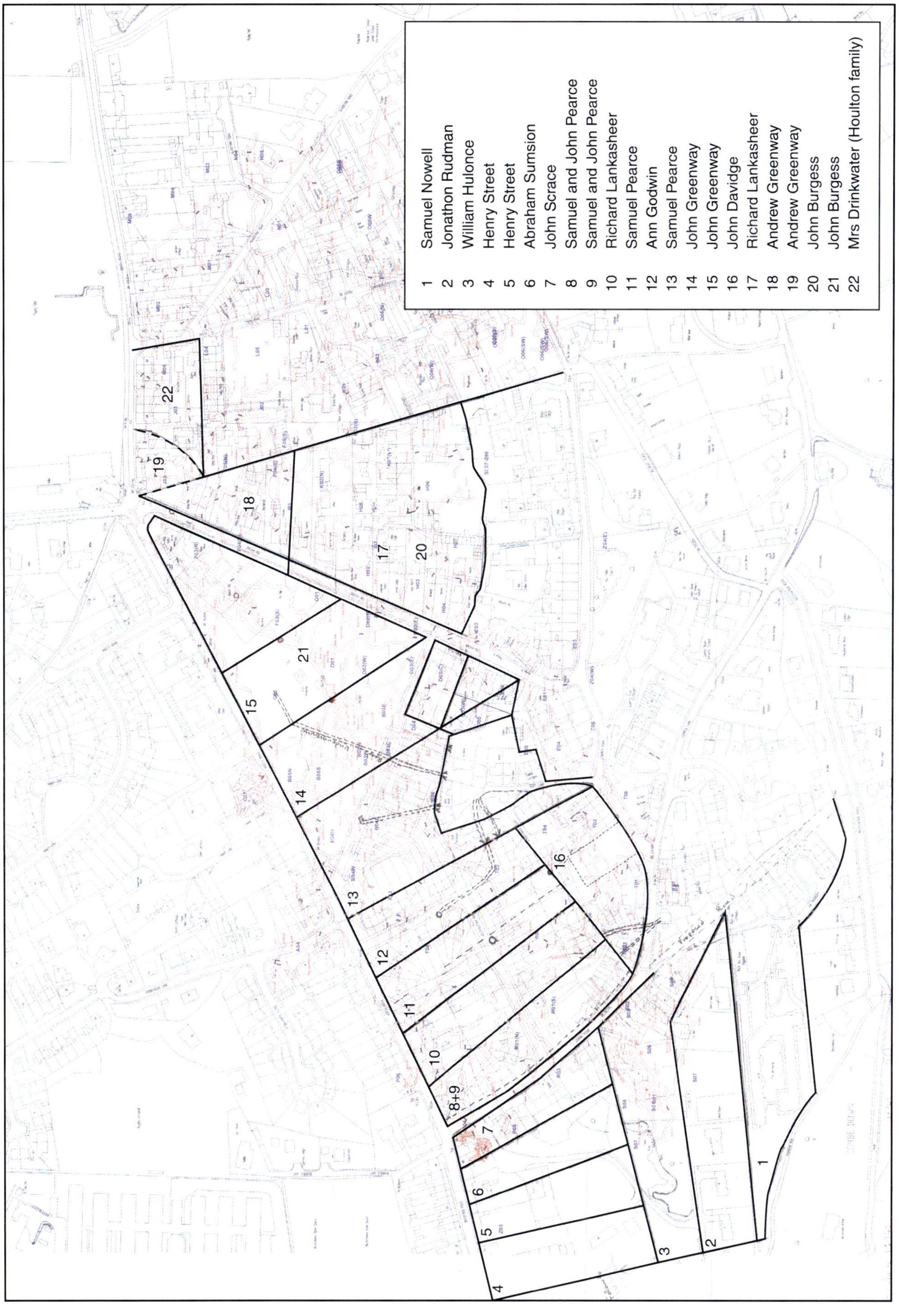

Fig. 4.2 Plan of Hadley and Salmon quarry leases (derived from LL./A91/18, and Irving 2005 and pers. comm.)

The southernmost of these quarries was sold by Salmon to Samuel Nowell, and was operated as a surface quarry. It was later sold to Phillip Nowell, whose family house, Rock Hall, was built on this plot. Nowell was to become one of Combe Down's most important masons, working on such projects as Windsor Castle. The central of the southern plots was actually bought by Nowell and Hulonce jointly, with Jonathon Rudman first renting it for £20 annually, but having an option to purchase it, an option which he apparently later took up. This plot extends over what is now Cherry Lodge and Tor View, generally following the present boundary walls (the boundary here is the least satisfactory fit and is only an approximation). It too was largely surface quarried, though there is a small, Phase I or II underground section under Tor View, previously unsuspected (Quarry 910), which has now been stabilised. The third plot was initially operated by William Hulonce who worked the eastern part underground (Quarry 503 and possibly 504) until about 1816, with Patrick Byrne apparently responsible for its underground western end (Quarry 500, 501, and 502 respectively). The western half of the plot, worked by surface quarrying was sold to Patrick Byne in 1805, and sold again to Henry Street (see below) in 1841.

On the north side of the area, fronting the Turnpike on the north side for which no original plan is extant, it seems likely the division was into four properties, with the length running north to south, and originally three ownerships. Only a small portion in the north-east corner was worked underground, by John Scrace up to about 1810 (Quarries 508 and 512), the remainder being surface-quarried. Adjacent to this area of land, where the Foresters Arms now stand, was a quarry in Widcombe Field opened by Henry Street in 1795 (Irving 2005, fig. 6) and this made expansion eastwards especially useful to him. The western plot was bounded on the east by more land belonging to Henry Street, with the 1836 Jewish Burial Ground at the top Irving (2005, 52). This is believed to have been after the establishment of a synagogue in Kingmead Street, Bath in the 1830s (www. Combe Down Heritage.org). This suggests that both the western plots were owned by Street, since the road was to his quarries and they are referred to as such by William Smith. Street sold the land for the Jewish burial ground in 1812, and the plan with the deed shows the second plot as Street's. He was also, in 1809, to acquire the next plot east, from Abraham Sumsion when the latter died, and John Scrace's corner plot fell to him a year later after he went bankrupt. The boundary shown on Fig. 4.2 is here a 'best fit' between the two easternmost plots at a small entry-road from the Turnpike, and shows equal division of the eastern plots. This fits the modern inside or west boundary of the eastern plot (originally John Scrace's) which follows the present day boundary of the garden of the houses along the west side of Combe Road.

It seems likely that exploitation of these quarries was first carried out near the Combe Road and Turnpike margins, since this allowed housing to be constructed at the earliest possible time, certainly at the junction, by about 1825 (Irving 2005, 49). Surface worked plots were restored as far as possible, and it seems likely these earliest workings have probably generally been brought up to the original level (i.e. the adjacent road level), while the later certainly remained sunken. Some of the houses on Combe Road however, on the southern part of Scrace's ground, are partly built on the quarry floor. In answer to a question posed some years ago to one of the writers, given the time gap between the availability of the land for quarrying and the building of the Jewish Burial Ground in the 1830s, it is likely that the burial ground was quarried too and restored. It would make the grave-digger's task that much easier. Much of the central and southern area of the quarry was shown as still being worked on maps in the mid and late 19th century, though this area is now obscured by infilling in recent years with spoil resulting from a new sewer laid along Summer Lane. What was perhaps the last ever working may be suggested by a remaining gullet (working trench) near the south-west corner of the quarry.

John Ovens Thomas and Colthurst Down

Colthurst Down ranged on the northside of the Turnpike to the northern escarpment, and east of the Combe Road junction as far as what is today called Popes Walk (formerly Hanginglands Lane, the footpath to Bath), which was all within Monkton Combe. The area east of that to the Prior Park Road (today Ralph Allen Drive) was in Lyncombe and Widcombe, as was Prior Park itself. These were all bought by John Ovens Thomas of Prior Park, sometime after 1807, and it seems likely the surface quarrying on Colthurst Down started soon after. At his death in 1828, the estate was left to his niece, Mrs Cruikshank, who lived at Belmont, and she put it all up for sale. The sale notice of 10 July 1828 included the quarries, and emphasised their eventual value for building. A proposed road (never built) was set out from some way down the Prior Park Carriageway ranging south west to the site, 'which might be a mile long'. The stone, it read, 'is of the very highest quality and might be worked in a very lucrative way being only a mile from Bath and the Kennet and Avon Canal, which stone might be conveyed easily to London where it has been used in many of the public and other buildings and is in great demand'.

It is not certain who operated the Colthurst Down Quarries until the 1851 Monkton Combe Tithe survey and plan (copy in BRO). They occupied enclosures at that time extending about half-way back to the escarpment. The Combe Road junction corner enclosure was shown as a wooded enclosure next to which quarrying by Isaac and Giles Sumsion was encroaching northwards, probably from the western end. The area eastwards was just beyond

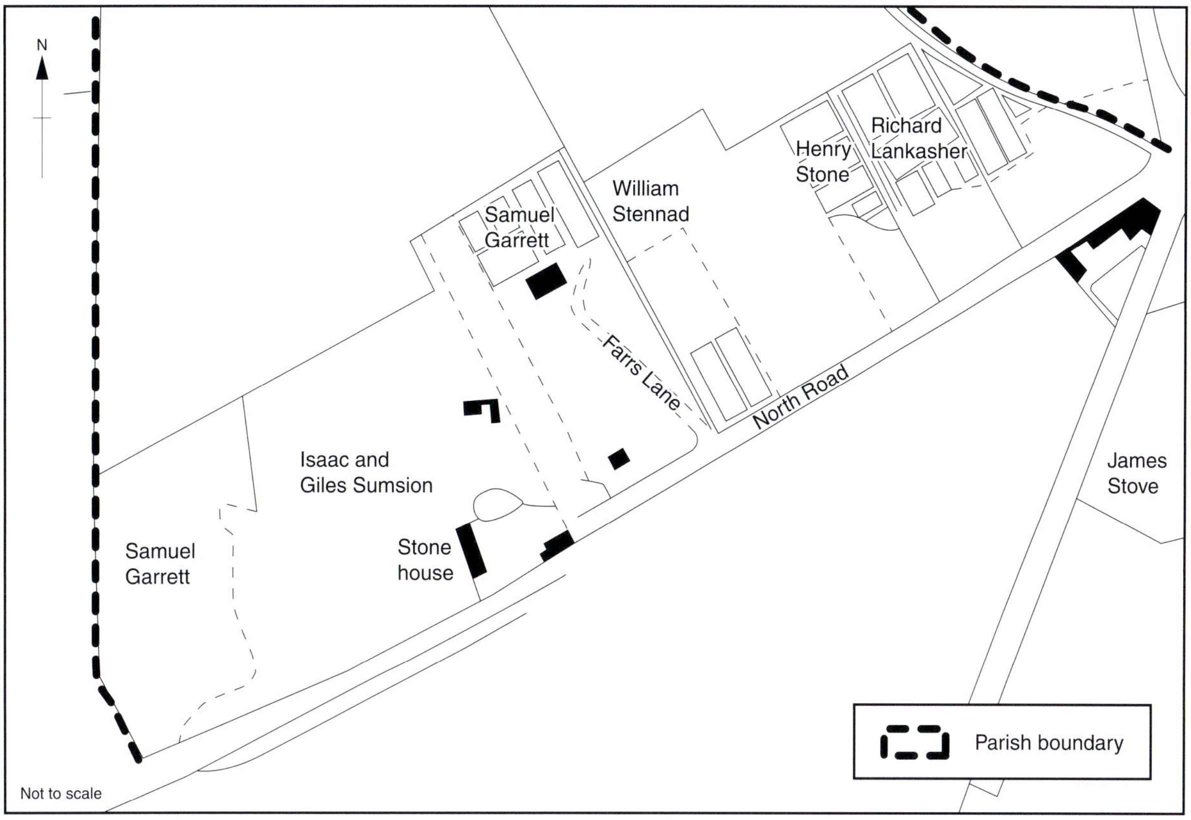

Fig. 4.3 Quarrying north of North Road (The Turnpike), based upon the Cotterell map of 1852 with data from the 1851 Tithe award

Fig. 4.4 The Farrs Lane Cottages built on the quarry dump. The second from the right was the first to be built

Fig. 4.5 Farrs House, built on the bottom of the quarry

Stonehouse, the houses there shown in a small enclosure almost as was on the 1742 Thorpe Map. The Sumsion site was the largest on the Colthurst Down with some 40% of the area. From the enclosure east of Farrs Lane the owners of the four quarries can also be determined from the tithe award schedule: From the west they were owned by William Stennard with the largest, Henry Stone, Richard Lankasheer and at the east, James Love. A map of 1852 (Fig. 4.3), possibly by the same surveyor, J.H. Cotterell (Addison 1998, 36), shows the last three quarries to have only worked a portion (the remainder is still gardens), amounting to about half of the enclosure within which they were sited.

Part of Stennards Quarry, next to the Turnpike, already had a west-facing house built upon it, which may have been have been of pre-quarrying date. The area west of Stennards Quarry with a lane running within it,was owned by Samuel Garret who may not have been a quarrymaster. It appears as though the quarry may have been worked out, at least to the rear of the then enclosure. The lane was still a curved quarry road, which, later straightened, became Farrs Lane. At the west side was the first of what became a row of cottages, (similar to earlier cottages at Mount Pleasant Quarry, so possibly for quarrymen too). These were built on the quarry tip (Fig. 4.4). Down the lane was what is now Farrs House, which site examination shows was built on the bottom of a former quarry (Fig. 4.5), with the quarry face a little beyond in 1851-52. The bottom floor, now the cellar, of this house appears to have sat on the quarry floor and it is possible it was associated with the working quarry, perhaps a count-house or similar. The possible quarryman's cottage (and other buildings erected behind it by the time of the 1884 OS 25 inch sheet), sit on a large spoil heap which extends back from the Turnpike Road (North Road today), with a 7-10 m high slope down to near Farrs House. It is evident this whole central area had been quarried by around 1880.

By 1884 and the Ist Edition OS 25 inch sheet (Fig. 4.6), quarrying was pushing northwards towards the northern rim or escarpment edge. The quarrying faces by 1883-84 are shown as gullets about 25 m wide, the actual working faces indicated by cranes, four in all and, from the spacing, probably one for each of the old quarries, though by then the ownership was down to two, Love and Stennard. The discontinuous faces of the quarry form a large arc extending almost to the Combe Road junction across to Hanginglands Lane (now Popes Walk). More houses had been built eastwards from Farrs Lane. An opening to surface at Stennards Quarry (where Selwyn Court is now) was still functioning, serving the last remaining underground working in the vicinity (Quarries 2214 or 2212). A fifth crane was shown on a new working east of Popes Walk, extending across to the Prior Park Road, known as Junction Quarry, This was later worked by the Bath Stone Firms Ltd until about 1925. Stonehouse Lane was still curved (with Quarries 2218 and 2219 below), and no houses had been started there.

William Stennard was succeeded by his son Henry in 1870 and the Misses Stennard sold their quarry in 1903. Henry Stone became involved with Fullers

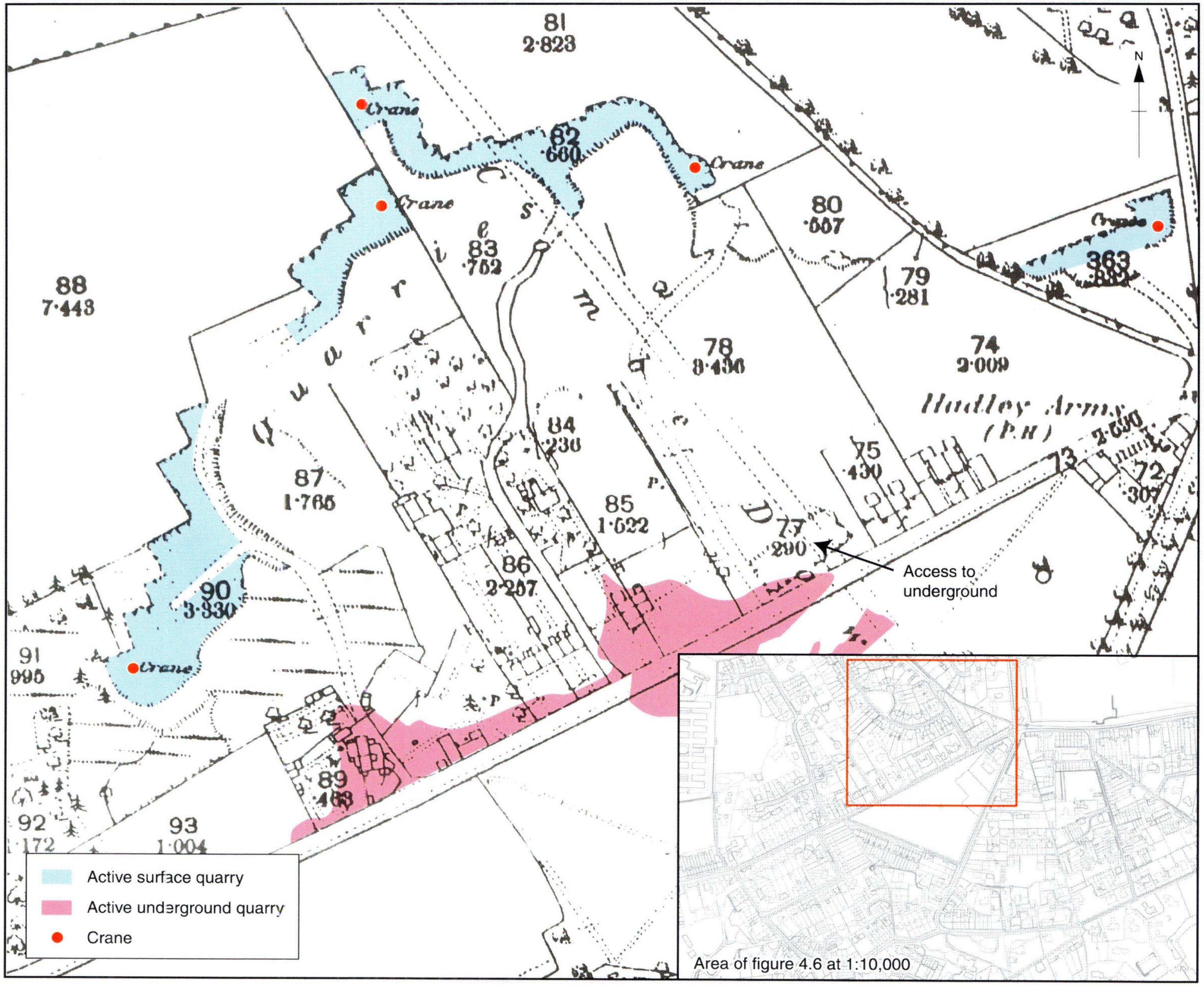

Fig. 4.6 Quarrying north of North Road c 1880 (1882-84 OS 25 inch map). Note use of cranes. Underground workings were accessed from the small area left open next to the Turnpike

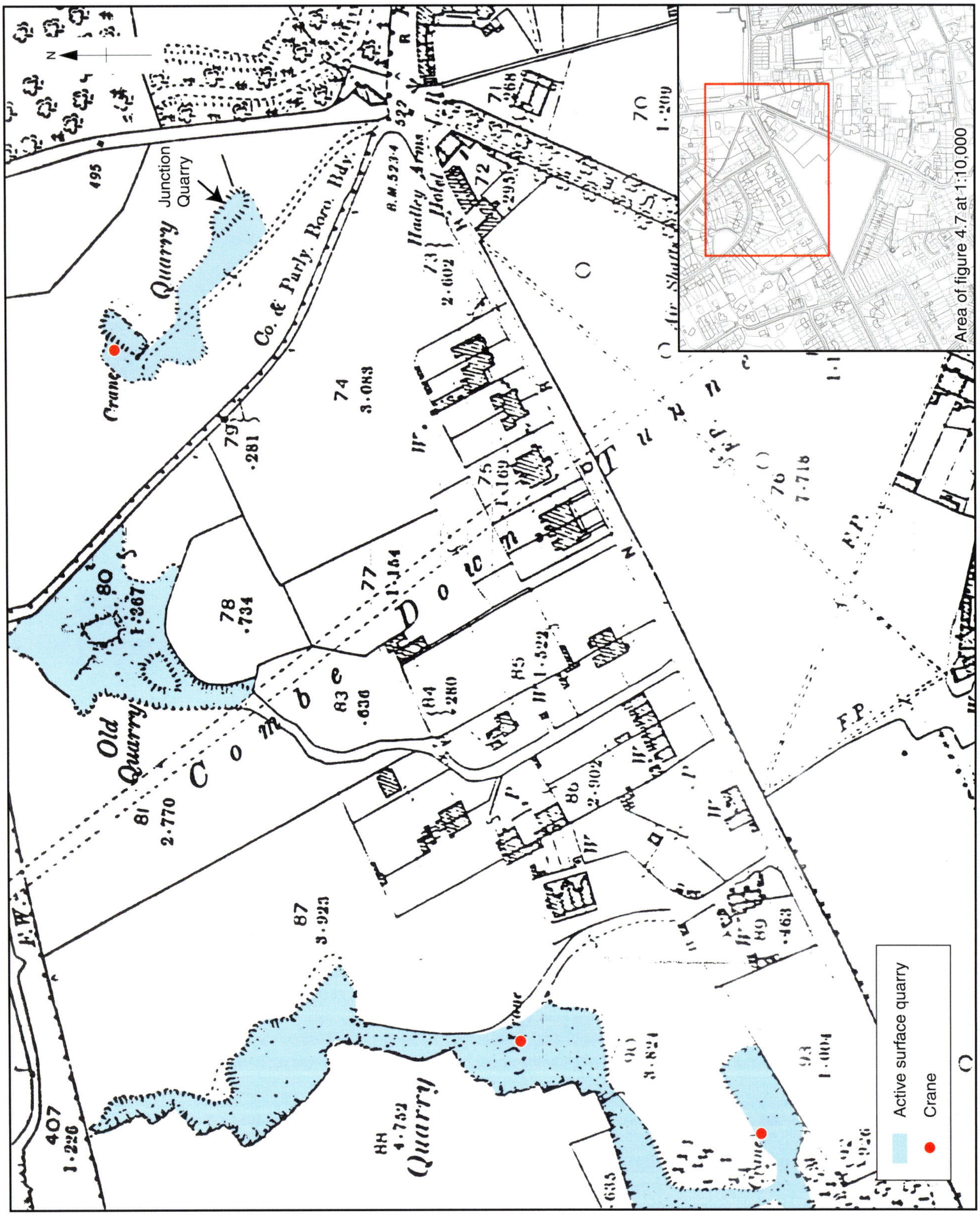

Fig. 4.7 Quarrying north of North Road c 1900 (1902 OS 25 inch map). Most of the area to the west has been quarried, with Junction Quarry still extending over what is Prior Park today

Earth and other materials and possibly sold out to Stennards. Richard Lancasheer (there were alternative spellings used) sold his rights, in 1865, presumably to James Love and a year later was listed in a local directory as of Claverton Street, Gentleman. James Love who also quarried at Odd Down, linked with his son in 1865 to form J. and E. (Edwin) Love and Sons until 1872 when only Edwin was listed. In 1899 Edwin's son joined him and the firm became Love and Son until 1927 (Hemmings 1983).

The leaseholds to the quarries were sold again in 1877, on behalf of the Thomas devisees, in two lots, one occupied by Messrs E. Love, the other by H.J. Stennard. In lot one (the easternmost), there were about 4 acres of stone out of the total 11 acres and 24 perches, and in the second some $7^1/_2$ acres of stone out of 15 acres, 2 roods and 37 perches. The extensive building frontage to both was emphasised. They appear to have worked until 1893 (Pollard 1994, 22). The quarry east of Popes Walk appears to have been in operation until 1925 (Hemming 1983), continuing through what is now the Priory Estate. Comparison of the 1882-84 map with the 1902 equivalent (Fig 4.7) shows almost the whole of the area back to the escarpment had by 1902 been excavated.

The quarry eventually became owned by the Bath Stone Firms Company. In 1872 the son of Henry Stone and Son and James Stone formed Stone Brothers. This joined, in 1887 with other local companies including the Isaac Sumsion quarries and some of the Wiltshire quarries to form the Bath Stone Firms Ltd. The first chairman seems to have been James Pictor, but a director was Isaac Sumsion. In 1898 Sumsion became Chairman and was succeeded by his son, also Isaac in 1910. In 1907 the title became The Bath and Portland Stone Firms and two years later the Firms were shipping Portland Stone to Glasgow (BRO. 388). They dominated production in the Combe Down area subsequently.

Prior Park

In the 1828 sale, Prior Park and House was sold to a 'Popish Prelate of Pious character', and was changed into a 'college for Popish youths' (Grenvill 1841, 442). It was possibly under the Bishop that the upper part of the Park became used as quarries. In 1839 the Lyncombe and Widcombe Tithe Award map has no indication of them, but on the 1883-84 OS 25 inch map (Fig 4.8), quarries are shown across

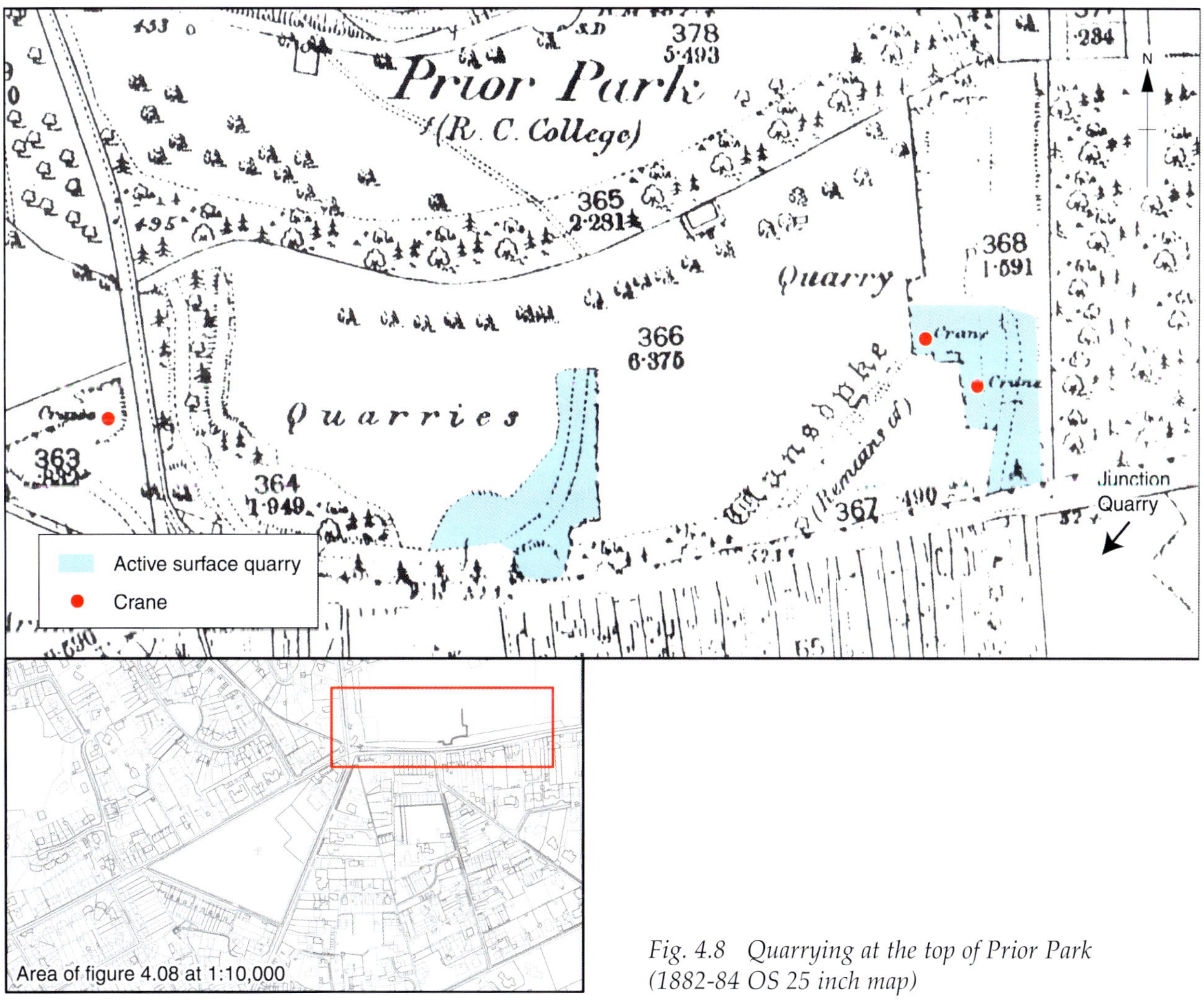

Fig. 4.8 Quarrying at the top of Prior Park (1882-84 OS 25 inch map)

the whole width adjacent to the Turnpike Road, except for a short section near the east end where the supposed Wansdyke crossed. It seems likely that the westernmost part of quarry also extended under the junction at the top of the Prior Park Drive, since a chamber under the road here was found driven from that direction in the underground archaeological survey.

Freefields and Rainbow Wood

This extensive area of open quarries, north of the (North Road) Turnpike and east of Prior Park was probably worked mainly in the mid and late 19th century, but was totally finished before 1900. The owners are not known but Wooster reports that the Hancock family began their family operations there in the mid 19th century (1978, 10). This may have been residual working of already worked-out quarries.

Combe Down Quarries

This is the name by which the large quarrying complex either side of Shaft road, extending east to St Winifreds near Combe Manor was known from the mid 19th century onwards, including the older Mount Pleasant Quarry. They were started as surface

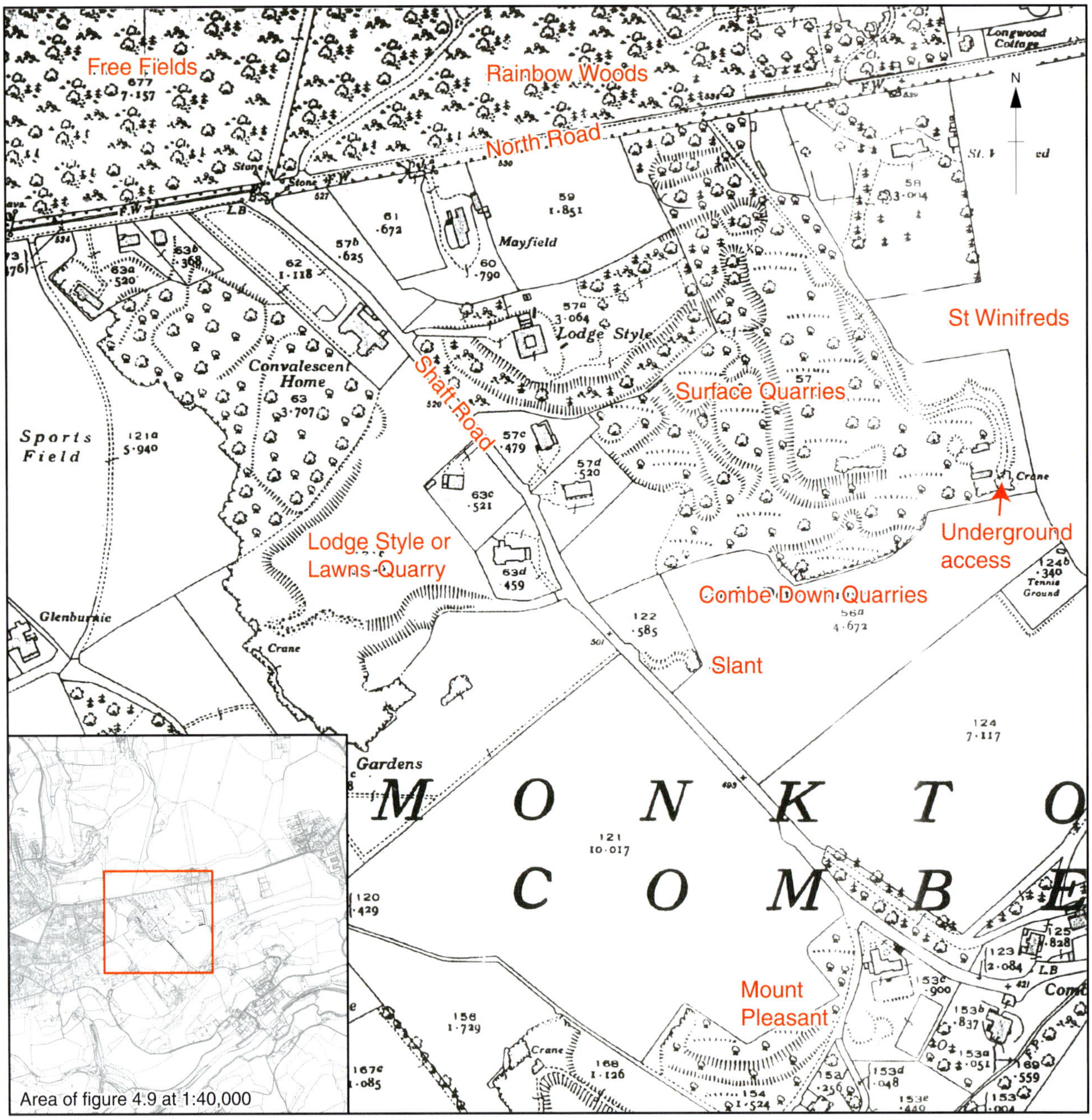

Fig. 4.9 Combe Down Quarries, near Shaft Road, about 1930. Lodge Stile Quarry is on the left of Shaft Road (OS 25 inch map)

quarries, becoming very extensive (Fig. 4.9) but by the century end also had underground quarries which continued working until about 1928 or after. It is possible these were operated by the Sumsion family (shown as by Isaac and Giles Sumsion in the 1851 tithe award) until they merged their interests into the Bath Stone Firms Ltd (see above).

A small quarry at Mount Pleasant is shown on a plan (see Figure 3.13) of the proposed Kennet and Avon Canal of 1795. By the middle of the 19th century it had grown and had cottages built on its oldest spoil tips. Part was said to have been owned by the Stone Brothers in the 1870s (Addison 1998, 51) who were party to the later Bath Stone Firms merger. Mount Pleasant Quarry itself was closed in 1914 by the Bath Stone Firms, but reopened, probably in the 1970-80s, by Bathite Ltd, who used crushed stone to produce a composite concrete block with a oolitic limestone concrete face. Unfortunately these building blocks, used on less expensive housing projects, were prone to splitting at the joint and the process was abandoned. A substantial quantity of discarded blocks were used to support the underground roof near the Hadley Arms. The derelict bases and buildings of the Bathite plant were still *in situ* in 2000.

The 1851 Tithe Map shows quarrying on the east side of Shaft road adjacent to the Turnpike, which

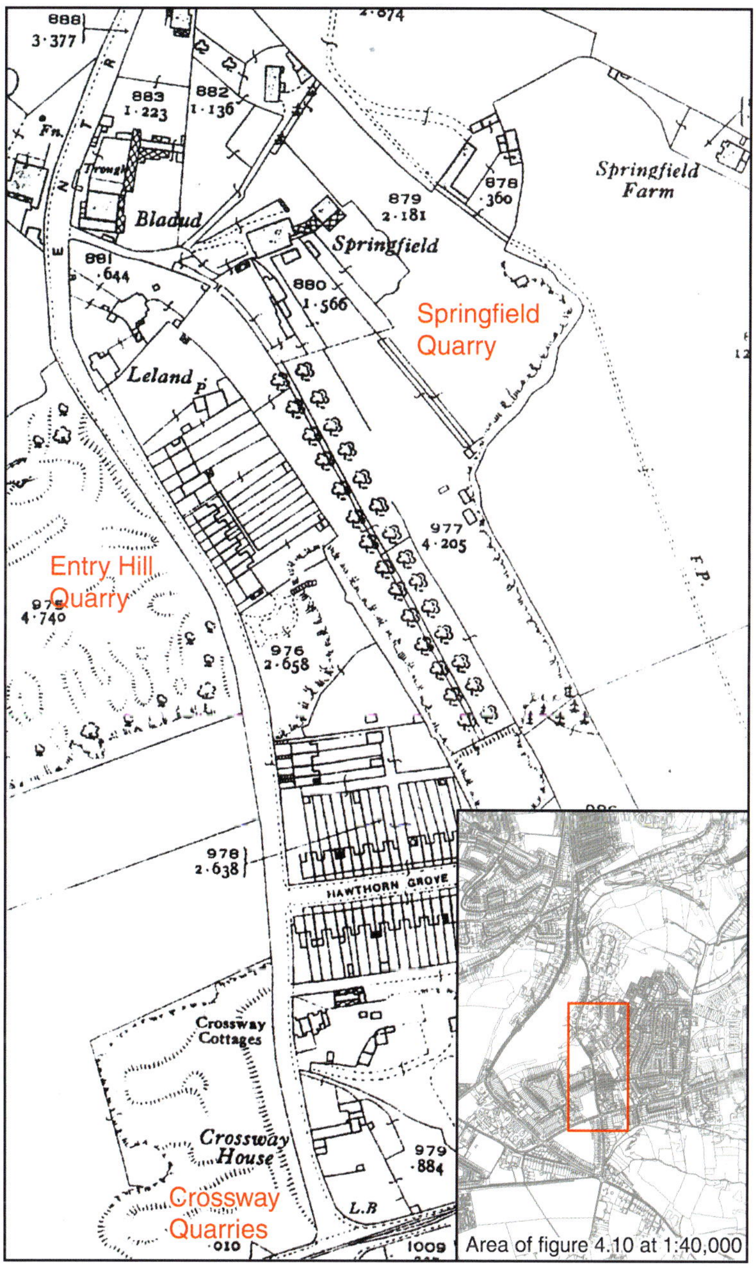

Fig. 4.10 Entry Hill and Crossway Quarries (1930 OS 25 inch map). Springfield Quarry had become the largest in the area about 1900

was carried out by Sumsions. At the roadside a house has been built on a site that seems to have been quarried, so stone was probably removed here long before, possibly in the early 19th century. By 1884 these workings were very extensive though as the 1930 OS map shows, they were not quite yet complete. The latter map also shows the slope entry to the underground quarrying near the escarpment on Shaft Road. The quarries at Lodge Style, west of Shaft Road, remain in production today (Fig 4.9).

Entry Hill and Crossway

Quarrying had begun here in the late 18th century on the westernmost area of Widcombe Field, then used as arable on land which probably belonged to the Magdalen Hospital (1799 Charlton map of Lyncombe and Widcombe (see Figure 3.14). By the mid 19th century this had spread south-east towards the Turnpike but probably terminated at Hawthorne Grove before 1880 (1882-3 OS 25 inch map). No ownership is known. However, after this time, what became known as Springfield Quarry developed to the south, becoming, according to Addison (1998, 40) in 1900, when worked by Armstrongs, the largest quarry in the Bath area, producing overall 2.5 m million cubic feet of stone. It certainly has some of the highest quarry faces in the area. A small area in the corner between these two was also worked underground, probably in the late 18th century. The underground elements, Quarries 2374 and 2375, were infilled during the Stabilisation Scheme.

Just to the south of the Turnpike was Crossway Quarry, begun before 1799, with another on the other side of the road later. It was worked by Philip Bennet in 1839 and both were sold 1863 to Thomas Shepherd. Building on the site was shown on the 1882-83 OS 25 inch map. The quarrying had ceased by 1930 (Fig. 4.10).

Kingham and Vinegar Down

Irving has recently investigated William Smith's quarrying and stone-sawing business in some detail, and the following is a summary of the main data from his account. South-east of the village, Kingham and Vinegar Down were still owned in 1799 by Lord Hawarden (SRO. SDH/6 Renewal of Mortgage) but in about 1808, after Hawarden's

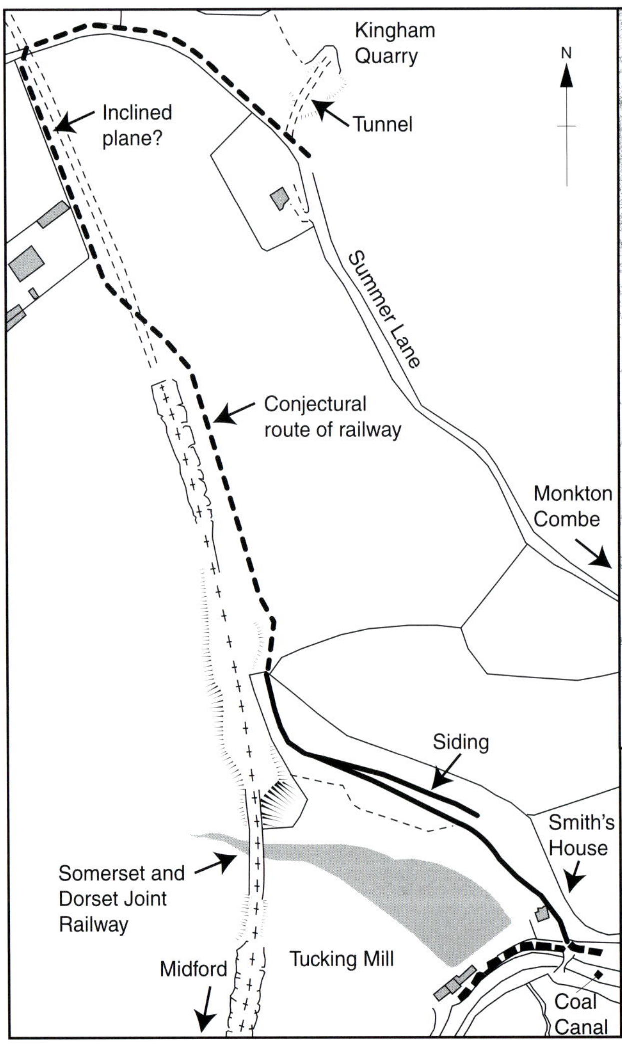

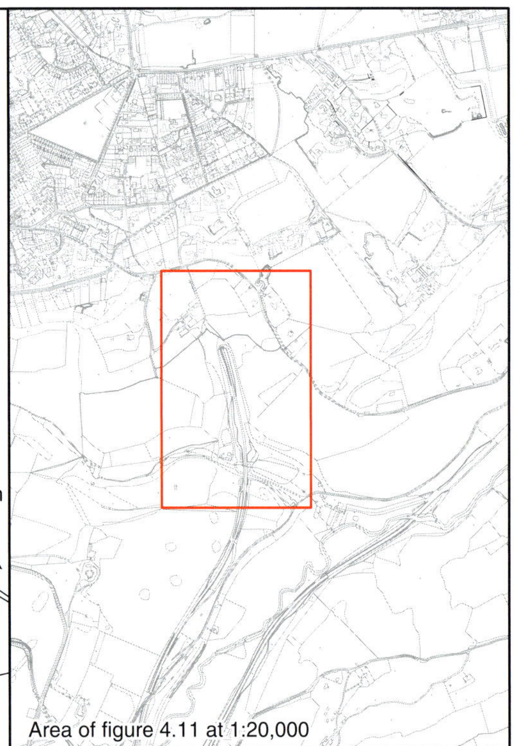

Fig. 4.11 Kingham Quarry Railway. Possible route of William Smith's railway to his stone-works at Tuckingmill next to the Somersetshire Coal Canal (after Pollard 1982)

death, they came into the hands of Charles Connolly of Midford Castle. Smith had bought the nearby Tuckingmill, becoming aware of it as the surveyor to the Somersetshire Coal Canal in 1798. He saw the opportunities the canal would offer to a stone business, so once open, he acquired it. Smith failed to sell his estate a few years later and decided, with the imminent opening of the canal, to raise capital and work the stone himself. For this purpose he arranged a mortgage with Connolly. There is little doubt that Smith saw himself as repeating Ralph Allen's success. He took a lease from Connolly on the Kingham Field Quarry about 1810 and made some form of arrangement with Connolly and with John O'Neal, the latter who leased the Vineyard or Vinegar Down Quarry from Connolly also. In a joint arrangement they began the development of a plate railway, using mainly second-hand rail from the canal construction (Fig 4.11). It linked the two quarries along Summer Lane, then followed the line of the Long Drung footpath down past de Montalt Mill in a self-acting section, then crossing the slope to go past Smith's mill to the canal wharf using horse traction.

At Tuckingmill Smith built an ingenious if not entirely original saw frame for cutting ashlar and devised the use of circular stone saws to cut the ashlar into thin plates (perhaps tiles or cladding) suitable for masonry repair work, a growing London requirement. All this was worked by water power, though at first this was not sufficiently powerful.

It all went wrong for Smith. The quarry had problems with cambering (landslipping) of the stone, with one gull following another in his arched entry. His railway section was neglected by him, to the ire of the others involved, and the sawmill suffered not unexpected initial operating difficulties. Money was very short, and Smith neglected the work at both a surface quarry, a short underground entry and the mill, preferring to stay either at his London house or go on one of his geological tours. Eventually Connolly lost patience, and sued Smith for debt, with the result he resided for a time in the Fleet Prison in London. It was the end of his venture. The course of the railway has been traced by Pollard (1982) and was recently confirmed by the Combe Down Heritage Society on Pollard's route.

The operations at Vinegar Down by O'Neale after 1810 do not appear to have continued much longer than the ill-fated ones of William Smith. However, the quarry appears to have been re-opened as in 1852 it was worked by Robert Russel (Hemming 1983 and 1851 Tithe Map) and in the 1870s was exploited for the building of the Somerset and Dorset Railway, which passes in a tunnel deep under Combe Down. A light railway was apparently built to convey stone to the works where they emerged from the tunnel and over the Tuckingmill viaduct. The route is unclear, but may have been similar to that of Smith, O'Neale and Connolly. There are substantial buttresses in the cutting leading to the tunnel built of the stone and more may have been used in the tunnel. The line was disused about 1880. About the end of the 19th century there were small quarrying operations there by a Herbert Jones, and a Frederick Davidge was killed underground in 1913, after which the mine closed (Addison 1998, 52-3). Today there are indications, in a collapsed entry and subsided lawn, of a small underground quarry extending from in front of Beechwood House on the south of the road, to a possible working on the north side.

Landscape and working of 19th-century quarries

Surface effects of underground workings

The surface aspects of underground quarrying changed relatively little. The older entries to the cartways largely remained in use at both the Rock Lane (Sheeps House Quarry, or later, Davidges Bottom) and Rock Hall Lane (Jones Quarry at Ralph Allen Yard today), but a new cartway was driven in on the west side of Sheeps House Quarry to work western parts of Firs (Quarries 2220 and 2221) and two from the west side to work northern and southern East Byfield areas (Quarries 514 and 520). The use of these was directed by the Layton/ Hadley Estate to cope with the needs of different lessees from Combe Road across to the Avenue. Apart from the entries below the face of what by then was effectively a worked-out surface quarry, there needed to be an area for storage and loading and for banker masons to work. Prior to the end of the 18th century, Sheeps House had small wooden huts for the bankers, but these were apparently cleared for houses in de Montalt's time, and certainly were when Davidge's brewhouse and other cottages in the bottom, were built by 1811.

More shafts were sunk and these probably handled the bulk of the stone quarried. The immediate areas around their tops must have been busy with the handling of stone, and the tracks from them to the turnpike, now totally disappeared, must have been conspicuous with stone-laden carts. The sinking of these shafts had to be specifically agreed with the Hadley Estate, although at first consent seemed grudging. The OS maps and sale deeds (Hadley Sale in 1901, Hadley Arms Deeds) show a line of five wide shafts ranging north from Jones Quarry, and another line of three ranging north and east from Firs Shaft on Firs Field, towards the Hadley Arms (Fig. 4.12). There were a further two west of Combe Road and several east of The Avenue and across to Far East Firs. The only formal requirement from the landlord was for such shafts to be fenced, or in practice, surrounded by a low wall, as for example that exposed in 2010 in an archaeological trench at the Chestnut Tree shaft on Firs Field (see Chapter 9). By 1830 when the Fir trees planted by Allen up to a century earlier had largely been felled, the lines of shafts and their mounds and haulage gear would have become very conspicuous in the northern part of the core area, with the perimeter gradually becoming occupied by houses.

What structures they required when working is not at all clear. The alternatives were, first, some form of crane, probably with a swinging jib, either similar to the 18th-century crane types introduced by Allen, or, more likely, a version of the jib crane with a hand-operated windlass mounted on the jib, similar to that used on surface quarries, described below. This arrangement was certainly used around 1870 at the small quarry at the north-east corner of the Combe Road/Bradford Road junction. Such structures would have been very noticeable, especially as between 1800 and 1830 the fir trees planted by Allen up to a century earlier were felled and apparently not replanted.

The second was a simpler arrangement, probably only really suited to smaller quarries not producing large blocks, such as those under the Tyning and Gladstone Road areas, where frequently a banker mason produced wrought stone underground. This would probably be a simple hand-windlass mounted on a timber frame, similar to a jack roll used in mining. They would hardly be more obvious than a well, though stone brought to surface may have needed a small storage area too.

The impact of surface quarrying

Surface quarries had a very much greater impact on the landscape than quarrying underground. First the useless ridding, or Twinhoe Beds had to be removed to expose the freestone below. In the first instance this required removal of spoil to some distance away, but when some quarrying had been done, there was space available for dumping behind the current working face area. The freestone below the ridding or overburden was worked in benches, of a few beds thick. In the early part of the century the working of the stone was done with little more than bars or a winch to move the blocks, but in the first decades of the century a simple form of stayed-derrick crane was developed. Surface working, even when carried out with the sequential restoration of the surface, left fairly large areas of bare ground and rock faces, with jibs of cranes projecting above the horizon, sufficient landmarks for the Ordnance Survey to mark them on their large-scale maps. There were clearly the usual problems of dust or mud, and damage to roads by heavily laden carts. Steam cranes and steam lorries also made a minor additional impact in the early 20th century.

An indenture of 12th August 1817 between Edward Salmon and Patrick Byrne makes clear how

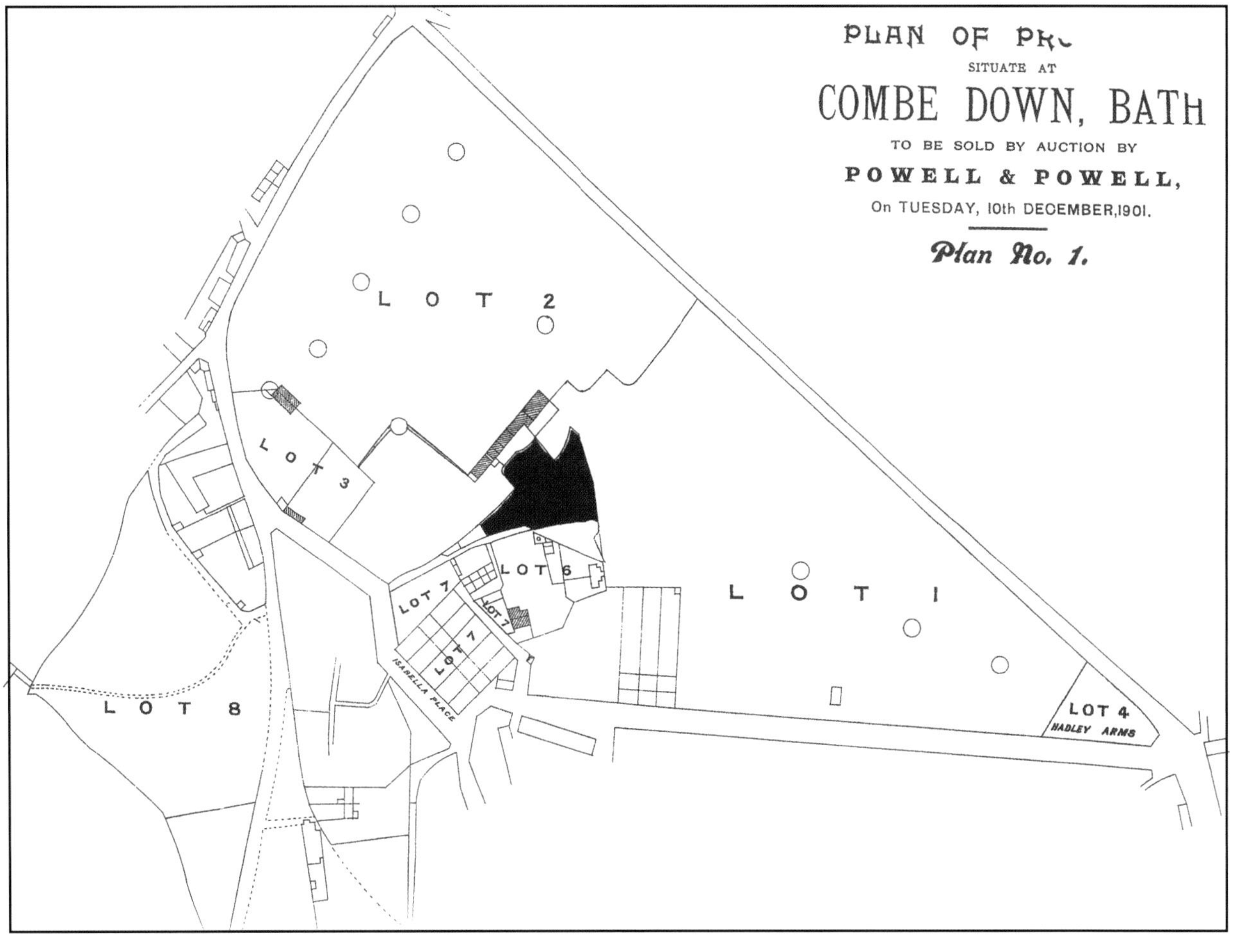

Fig. 4.12 Shafts left from quarrying under the Hadley leases by 1900, detail from auction plan (Hadley Arms Papers)

the underground and surface quarries were to be left, at least in this case where possession was by leasehold tenure, with an annual rent to the lease owner. It particularly emphasises the idea of quarries as potential house sites in the landlord's mind. Byrne had possession of the western acre of the (originally) Hulonce plot west of Combe Road. He was directed to quarry in a sequence of five stages, the first the underground section at the east of his plot (the Byfield bat entrance area today), then in four fairly equal stages west to the boundary – which were expected to be done, and were done by surface methods. Before the expiration of the lease, he was to:

> *. . . level the whole surface of the ground either by filling up or otherwise throwing arches over the same, and shall carry away or otherwise bury all the loose stone and rubbish . . . and well and effectually level the surface of the quarried part not more than five feet below or above the present level of the unquarried ground at the western boundary . . . and to set aside all the mould dug from the surface . . . and spread the same (or other good mould) to not less than three quarters of a foot depth to make the ground fit for agriculture or garden use.*

He was also to erect a house and to sink a well to supply a good quantity of water, walled from top to bottom. There would obviously be an advantage in sinking such a well in the bottom of the quarry, building the wall up around it before or during backfilling the site (copy of deed and plan supplied by Richard Irving).

There seem to have been two systems of stripping the overburden in use at Combe Down. The first, probably continuing a practice from the previous century, was a common method in stone quarrying and evidence of its use still, just, survives on the site west of Combe Road close to Shepherds Walk. In this a slit trench, just wide enough for a cart, was kept open to the full height of the quarry. At the commencement the overburden had to be removed before any work could start. Once this was done the removal of the freestone left space behind the working face, where further spoil could be stacked behind a series of advancing rubble walls built as the site progressed. The spoil was removed continuously just ahead of the freestone working using stout planks as a barrow-way. In the case of Sophia (widow of Jonathon) Rudman's quarry, this was the system in use, since a workman, in 1825, was killed falling with his barrow from the high overhead plank (Irving 2005, 40). Stone was either tumbled or slid, possibly using a winch, to a suitable position for dressing, or was cut or sawn almost *in situ*, into easily manageable sizes. When John Scrace went bankrupt in 1809, his quarry was advertised as having a share of a crane and two winches (Pollard 1994, 13), the share presumably had been with his neighbour, Abraham Sumsion and, later, Henry Street. Sharing a crane suggests use for loading at a central position common to the two quarries with the winches used to drag stone to that position.

The alternative method of surface quarrying, which had become usual by the late 19th century, relied on a fairly substantial crane of the Scotch derrick type using a central post and jib supported by (usually) five cables, attached to drilled or weighted anchor points around the quarry. The crane, with a hand-cranked cable drum mounted on the jib, used a friction brake and ratchet to control the load. It was used to lift or drag blocks from their bed, from a much wider radius than its jib might suggest, and to load them, after trimming, and ringing to detect cracks, on to carts. A preserved example of such a crane from Union Quarry at Odd Down was probably typical of the type (Fig. 4.13).

As well as handling quarried stone, the crane may also have sometimes been used to remove overburden by use of some form of tray or 'muck box', hand-loaded and swung over to the spoil heap some reasonable distance away, though use of a horse and cart for distances too long to plank and barrow over would be more usual. This may be the

Fig. 4.13 A quarry crane preserved at the former Union Quarry, Odd Down. The mast has been slightly shortened

origin of the spoil heap extending back from the North Road just west of Farrs Lane, to near Farrs House. By having a high heap, it released space to form a wide quarry floor for working, handling and storing stone, and for buildings, such as an office, perhaps, for example, Farrs House.

The advantage of the crane lay in the greater ease in moving stone, and in allowing the face to be worked vertically rather than in benches. The capital cost was only modestly higher. In substantial quarries, as those north of North Road were by the late 19th century, one or two cranes were used, as shown on the 25-inch maps, in each quarry, moved as necessary as the site and the working face advanced over the landscape. The typical large surface quarry had active operations over an acre or two, in some cases followed or accompanied by restoration to an acceptable degree for agricultural use or housing. This was especially the case in the inner parts of what in 1854 became the new parish, particularly along North Road.

The largest quarries around Shaft Road however, seem generally to have been restored to a much lesser degree, so that approaching a working section usually meant crossing a growing area of land characterised by a series of dumped heaps roughly parallel to the original quarry faces. After a few years the quarry face was a fairly insignificant feature compared to the much larger area of spoil dumps behind it, and it is perhaps not surprising that some landowners again began to prefer work to be done underground, as appears to be the case near Shaft Road.

The relict landscape of the 20th century

By the end of the 19th century the older quarrying areas had almost entirely ceased production, the main underground exception being James Riddle's Quarry (518), off Combe Road. A new small underground working of the late 19th and early 20th century took place north of North Road west of its junction with Combe Road and was marked by a surface crane and later a steam crane used for raising stone from below (see Chapter 12). Minor underground working was probably still spasmodically undertaken in the Gladstone Road area using shafts and some form of hoisting gear, but these would have had minimal impact on the surface. The last surface work in the central area was the development between Ralph Allen Drive and Pope's Walk, and at the top of Prior Park, which finished by around 1910. By then the quarrying area north of North Road, where not built upon along the frontage, had become partly derelict, partly returned to agricultural use.

Aside from the Hancock Upper Lawn Quarry, which continues today, the last surface and underground quarrying probably took place in the late 1920s or early 30s. Underground working continued longest. The Combe Down Quarries at Shaft Road, St Winfreds and Mount Pleasant, had underground working amounting to some eight acres and St Winifred's had about one acre. Both were accessed by slope (or inclined) entries with a surface yard and crane, but otherwise were relatively insignificant landscape features. In contrast most if not all the huge area between Shaft Road and St Winfred's lay derelict, shrubs and trees gradually taking over, as they also have north of the road, in the former Rainbow Wood and adjacent Claverton area of Combe Down. The dominant surface feature of the early 20th century was one of dereliction and/or natural regeneration of the areas amounting to possibly half of the total land area of Combe Down, but no longer directly impacting on the main settlement area.

Settlement, mainly in the form of recreational areas but also housing, gradually took over much of these areas. In the first housing phases old quarries were built in, as at Quarryvale, or over, as at Combe Park and Gladstone Road, more or less as soon as quarrying had completed. In the case of west of Combe Road and North of North Road, housing and quarrying lay alongside each other, sometimes with the new housing built on the former spoil heaps barely months after a retreat of the quarry to a new face and crane position. What can be considered modern development began after the First World War. West and north of the Combe Road intersection with North Road became Ralph Allen Park (*c* 1920) and houses were built along the frontage, only to be replaced during the Second World War by the Defence establishment. The Hadley family sold their holdings in 1905 and the building of The Firs followed. The 1930s saw much building near Entry Hill and much infilling of the Combe Down central area. After the Second World War there was almost total infilling from Entry Hill to Shaft Road, with the Foxhill Estate dominant. Many old quarries were filled with rubble following the rebuilding of war-damaged Bath and either used for housing or sports or recreational grounds. In a sense this landscape development was as much related to quarrying as it was to expanding needs of the overspill of the ever-growing city of Bath.

Chapter 5

Combe Down and quarrying in the time of Ralph Allen

Ralph Allen was one of the three men, along with Beau Nash and John Woods, who became best known for their influence in the rise of Bath and 'The Valley of Pleasure' in the early and mid 18th century. In this chapter the available and documentary sources concerning his rise to wealth and fame are reviewed, with particular reference to aspects of his stone business, in which he introduced new technical and organisational methods which for a third of a century gave him a virtual monopoly of the supply of stone in Bath at a crucial time in the City's growth. The archaeology of his mines will be discussed in a later chapter.

The early 18th century – setting the scene

Any quarrying activity at Combe Down/Green Down was certainly still very small-scale at the beginning of the 18th century. As described earlier, the Down had a track across the top joined by Combe Lane from Monkton Combe, a significant path in terms of land division led from Tuckingmill over the top to Bath and another crossed from the Cross keys and the Wansdyke via Shepherds Walk and the top of Prior Park (see Figure 3.3). It was still dominated by sheep grazing but there were only a handful of small farms actually on the Downs, with Stonehouse (part of which may survive today) the only one on Combe Down itself. A couple of other tracks down the slope to Bath were probably impassable to wheeled traffic once off the drier downland. Bath, still with much of its medieval walls, was emerging from its older role as a regional market for the woollen trade and beginning to embrace the idea of a future providing for the needs of more than just invalids and 'swaggering swordsmen'. It was indeed on the brink of commercialised tourism with all its regional spin-off and long-term importance. Combe Down had an important role to play in this.

The earliest underground quarrying seems to have begun along the southern edge of the limestone outcrop of Monkton Combe's Combe Down/Green Down, east of modern Combe Road. Archaeological evidence is slender (a set of initials and a date adjacent) but John Pitcher's Quarry (possibly 2347) may have been located (in today's topography) under the south end of the Avenue at the junction with Church Road with its entry probably from Sheeps House Quarry. There are several other small underground quarries with features which can be inferred as at least partially pre-dating Ralph Allen's works, in the area he denoted for his own quarrying activities. These quarries date to around 1715, at which time Richard Jones probably began his apprenticeship with Pitcher, though direct documentary evidence only starts in 1723 (see below). Archaeology also identified similar workings with no specific documentation for this time which extend across the boundary into Lyncombe and Widcombe at modern Rockhall Lane near Shepherds Walk, and there may also be one or perhaps two on the Long Drung's east side, north of 'The Brow'.

This mainly new quarrying area may reflect economic opportunity from the beginnings of Bath's growth, or perhaps the decline of opportunity in the Pitts Quarr and Barricks Farm areas at the northern edge of (present-day) Odd Down, or simply a landlord seeking new means of exploiting his estate. Combe Down's location was almost as far from Bath (by then available routes) as possible on the outcrop. The north and closer side of the Down was not exploited until substantially later. It may have been favoured partly by the proximity of Summer Lane, crossing the outcrop and leading to Stonehouse at the track now succeeded by North and Bradford Roads and the tracks from it down to Bath.

The ability of Ralph Allen to put together his Combe Down estate and stone quarrying business were made feasible by the break-up of the Monkton Combe estate of the Poole and Shute families, possibly because they had overextended themselves in the purchase of the estate from the Robinson family. The Robinsons had been given Monkton Combe by a royal grant of 1564 (Pitcairn and Richardson 1924). Thomas Poole, of a long-standing Monkton Combe family, purchased the estate and Lordship in 1706. However, other estates in the area – Bathampton, Claverton and in Lyncombe and Widcombe – also seemed to be in distress or at least willing to participate in a very active land market in the first third of the 18th century in which Allen was much involved.

The process by which Allen finally acquired the wider area of Combe Down more or less from The Lawns to Entry Hill was complex, beginning in 1723 but not completed until 1737 for even the central core area. Even apart from some missing links, the deeds involved intermediaries and mortgages and remortgages (Bath Record Office Accessions 339 and 446). The Monkton Combe lands were all east of present day Combe Road and should not be confused with those in the Lyncombe and Widcombe, which, however, came into his hands later. The whole area was known as Green Down, or as 'alias East Down', which west and north-west of Combe Road was also known as 'Late Collibees' (see also Irving 2002, and the 1799 Charlton Lyncombe and Widcombe Survey (BRO). The area involved is included in the *c* 1764 Allen Estate Plan, BRO), though it is not always possible to positively identify specific parcels of land mentioned in deeds (Fig 5.1).

The first extant mentions of the Combe Down area appear in 1706 (BRO.339/046, 339/050). The former refers to the common or common of pasture on the south-east of the Great Green Down called Combe Down while in the latter John Robinson of Combe borrowed £700 against a mortgage which included:

> *. . . the sleight, pasture and feeding of 360 sheep called wethers in the ground called Sleights Hills and Mounteynes late parcel of the possessions of the dissolved monastery of Bath.*

A sleight in this context appears thus to mean a stinted pasture, primarily for sheep. Subsequently the lands clearly passed to Poole and then to his son-in-law, John Shute. There was then no mention of quarries.

Quarrying at Combe Down had probably started by 1715 at the latest, since it was then that Richard

Fig. 5.1 Detail of the c1764 *Allen Estate Map showing the triangular, tree-planted quarrying area at Combe Down (BRO)*

Jones was apprenticed to John Pitcher the Elder, freemason at Combe Down (Jones, *Life of Richard Jones*, Bath Library), at which time Thomas Poole owned the land. In 1723 two parallel but separate lease-and-release transactions were made by which Thomas Poole transferred land, including Greendown/Combe Down, to two members of the Lane family, James Lane of Priston, a tailor, and John Lane of Yeaton, yeoman, in each case for a consideration of £990. This was the normal means of conveying property involving preparing a lease of the land for a nominal sum, which lease was then sold to the buyer a day later.

By the first deed from Poole, James Lane of Priston (BRO. 339/014) received (among other land, houses and a grist mill conveyed):

> *. . . a piece in Greendown of $31^1/_2$ acres bounded on the east by Thomas Poole's land, west with the lands of Thomas Bushel, north by lands of Thomas Colthurst and on the south by land belonging to Thomas Lane; and also a moiety of the half part of Green Down called Combe Down or the Sleights Hills and Mountains, together with a moiety of the freestone quarries in possession of John Pitcher the Elder and his son John Pitcher and William Biggs; and also the common or waste ground lying on the south and south east of Greendown . . . 34 acres . . . purchased by Lane which is intended and agreed to be and remain a sheep sleight in common between them.*

John Lane of Yeaton (BRO. 339/015) received:

> *. . . part of Greendown called Combe Down with mearstones (boundary stones) $31^1/_2$ acres and a moiety of the remaining part of Greendown called Combe Down on the Sleights Hills and Mountains with a moiety of the freestone quarries in possession of Pitcher and Pitcher and (in this deed) Joseph Biggs.*

In these and later deeds there is mention of moieties which were undivided half parts. There is a brief reference to these in a schedule of deeds in 1799 (SRO. DD/ SDH 6) relating to the then Lord Hawarden's ownership of the former Allen Estate. This refers to a moiety to the Sleights etc, formerly in the possession of John Pitcher the Elder, John Pitcher, James Allen, William Biggs, Thomas Greenway and John Pearce, and to the second moiety . . . 'planted in the time of Ralph Allen with Firs', fairly clearly identifying the triangular area involved in conjunction with the Allen Estate Map, and providing the fullest list in any of the documents of quarrying freemasons there around 1729-30. It suggests six quarries were in operation but there may have been at least one other at some time operated by Milo Smith (see below), possibly on the other moiety.

It is feasible that James and John each became the owner of a moiety in all the quarries and in the 'remaining part of the Down'. This was north of modern North Road, referred to as 'lately Milo Smith' on the Allen Estate Map schedule (BRO). The Down was apparently divided into two areas of $31^1/_2$ acres each, probably divided by the Long Drung, the footpath between Monkton Combe and Bath, which seems always to have been a fundamental divide. There was also an additional 34 acres to which Poole retained common grazing rights with Lane, south and south-east of the Down, which must be land extending above and beyond Kirkham and Jackdaw Quarries. After a long series of transactions the entire remaining interest of the Shute family in Combe Down was finally conveyed to Allen in 1737 and the actual divisions never become absolutely clear. John Lane's portion eventually seems to have passed to James.

All of the land included here seems to have been sheep pastures or sleights on the limestone plateau, though later deeds also refer to part being arable. The mention of Colthurst (Down) to the north and a mention, involving a further deed relating to the same areas, of the land being bounded on the north by the road leading to Claverton (North Road today) makes it almost certain the land was nearly all south of the (present) North Road. Colthurst Down is that part of Monkton Combe lying north of the North Road between the present Popes Walk on the Combe – Bath footpath and a line ranging north of Combe Road. In 1723 Poole still owned The Lawns, the sheep grazing between the old built-up area of modern Combe Down and Shaft Road, which is mentioned in a related deed of 1731. Essentially what the Lanes purchased was the core settlement area of modern Combe Down, east of Combe Road extending to the western boundary of the Lawns.

On 3/4th April 1729 the same land, moieties of land and the moieties of the quarries, was conveyed by James Lane alone, in two portions both expressed in broadly similar terms to the 1723 deeds. One was conveyed to Milo Smith and John Smith of Ingiscombe (today Englishcombe near Odd Down) and Thomas Whittington (BRO 446). This was probably the portion originally sold to John Lane and, from its (and the other deed's) relationships, was the land west of the Long Drung as well as the moieties. It involved a tripartite deed in which a Thomas Whittington paid Lane £500, with the two Smiths paying Whittington £525. The difference was presumably for Whittington's trouble or, more likely, for payment of a debt by Whittington who appears to have had no further interest. This portion was then almost certainly conveyed to Ralph Allen, although no deed survives. However, by March 1730 (old calendar) Allen had 'lately made' a wagon way or road to convey stone from his quarries (Peach 1895b, 78, citing an agreement between Allen and Philip Bennet). It is unlikely Allen would have done this without prior ownership of part or all the land and quarries involved.

The other agreement was between Lane and Milo

Smith only for a consideration of £540. 'A day and two later' (as stated in a deed of 1737: BRO. 446/5), the property involved, along with other properties acquired by Smith from others, were the subject of a further lease and re-lease to John Horton of Wolverton. This was simply a formula for a mortgage, since there was a reversion for £450 due the following October.

In April 1730, by agreement between Smith and Horton, the moieties within the last deed for the remaining part of Greendown and the quarries were transferred to Allen. This may have sustained Smith's undoubtedly hazardous financial position for a few more months, but in April 1731 Smith seems to have needed to lease and re-lease The Lawns (which had been purchased by Smith previously, probably from Poole) to Horton, again fairly certainly for a mortgage. On the same day the mortgage held by Horton, with the exception of the items now in Allen's hands, was transferred to Jerry Pierce, a Bath surgeon, with a redemption in the event of a payment of £400 plus 5% by October the same year from Smith to Pierce. It was not paid and the properties were forfeited to Pierce . In December 1734 Pierce then sold the mortgaged lands (back) to John Shute (son-in-law of Thomas Poole, the latter by then deceased) for £400 with a redemption of £450 plus 5% annually. The other property included 'four ox chutes and three swathes of grass in the Great Dolemead' which Smith had purchased from Thomas Colthurst of Lyncombe and Widcombe (deceased) with some four acres and premises, all in the occupation of Milo Smith, his undertenant. Thus Smith had land giving access to the Navigation, possibly predating that owned by Allen there.

This should have given Allen an opportunity to buy the lands or, at least, the mortgage, from Shute, as he had done at a slightly earlier date with other properties. However, once again circumstances benevolent to lawyers intervened and both Shute and Smith died. The former died intestate so a curator held his estate in trust, and in legal limbo, for his daughters who were both minors. It was necessary to take the position to the Master of the Rolls in order to allow disposal of their capital assets. Smith left his property, with instructions to pay off his debts to his wife, Joyce and eldest son, Richard. Richard immediately renounced his share and sold it to his mother for five shillings. In February 1737 (BRO. 446/5) the parties, the infants and Joyce Smith represented by their respective administrators being agreed, the remaining titles to Combe Down and other lands in Lyncombe and Widcombe were finally transferred to Allen for £509 1s $8^1/_4$d. He later paid a further £40 to Joyce Smith for ox-chutes in Dolemeads which had not been accounted for. This was probably all that Joyce did get and it may well have gone on legal expenses.

It is hard to know what Smith's motivation in all this was. Jones, in his *Life*, implies that after Allen started on the Down, only Greenway and Smith were left quarrying there, probably east of the Long Drung, possibly just north of the Brow where early workings were found underground (Quarries 2339, 2350 and 2360). It is possible that Smith, a shareholder in the new Avon Navigation and with land at Dolemeads for a wharf, had a similar vision to that of Allen's. However, he was clearly unable to compete successfully against Allen, though it seems likely it was he who reduced prices per ton of stone delivered in Bath by a further sixpence on top of Allen's reduction of two shillings and sixpence (see below). Perhaps he bought the Down to ensure that Allen could not force him out, but it seems certain that he over-extended himself from the very beginning and, without the financial backing as was available to Allen, paid the price. The old had been moved out, and the new, a virtual monopolist, was in possession.

Ralph Allen – early life and rise to prominence

The early 18th century saw remarkable changes in Bath. The most significant of these affecting nearby Combe Down took place in the 1720s and 1730s. The development of the Combe Down quarries revolved around the career of Ralph Allen and his wealth created from his postal business and later, his burgeoning estate and stone businesses. The basic history of Allen's life in this section rests primarily on three sources: Benjamin Boyce (1967) *The Benevolent Man*; R.E.M. Peach (1895b) *Life and Times of Ralph Allen*, which is occasionally somewhat unreliable; Richard Jones (late 18th century) *Life of Richard Jones* (Bath Library). A further useful source on his estate is Mike Chapman (1996) *A Guide to the Estates of Ralph Allen around Bath*. However, the conclusions drawn from the evidence presented by them are ours and are sometimes at variance with theirs. Other sources are indicated.

Ralph Allen was born in 1693 at St Columb Major in Cornwall and died in 1764. His father Philip Allen was an innkeeper and his mother's family, from St Austell, may have been gentry, which together may suggest that a relative degree of prosperity might have given him a good start to adult life. He married Elizabeth Buckeridge in 1721; it was described as a good but not spectacular marriage, no doubt in both a social and financial sense. She bore him a son in 1725, his only child of two marriages, but the child died three months later and she died in 1736. The second marriage was to Elizabeth Holder in 1737, daughter of a neighbour, Richard Holder of Hampton Manor, with whom Allen had social and business connections. She died two years after Allen. Figure 5.2 shows a mature Ralph Allen, seated, with Richard Jones, the clerk of all his stone works (anonymous artist, Building of Bath Museum).

Allen's life-long ability to take advantage of almost any business opportunity seemed to begin with his employment in 1708 at his probable grandmother Gertrude's post office in St Columb Major. She left the position shortly after, but it probably gave him the opportunity to move on, and by his own claim he became deputy postmaster in Bath in 1710 or by 1712 at the latest. However, it seems more likely this shift was via Exeter where he probably became assistant to Joseph Quash. Quash had a substantial contract from the Post Office, possibly as early as 1708. If so, Allen's move to Bath by 1712 was either astute or fortunate since Quash was dismissed and prosecuted soon after. Meanwhile Allen was in the ascendant and by age 19, about 1712, was certainly postmaster himself at Bath. His first task there 'was to deal with the disorders of the local office' (Peach 1895b, 61). In successfully doing this he clearly ensured his abilities were recognised by both local business and those in the mail or government service, such as General Wade, with whom he came into contact during the Jacobite disturbances of 1715. Allen, by good sense or fortune, backed the winning side and provided information for the Government forces. Wade, who became Member of Parliament for Bath, remained Allen's friend for the rest of his life and possibly became his bondsman when Allen gained his first government contract.

Fig. 5.2 Detail from Four Bath Worthies (artist unknown), showing Ralph Allen (seated) and Richard Jones c *1735. (Bath Preservation Trust, Building of Bath Collection)*

This first major contract from the government, which probably arose out of his experience with Quash a few years before, was to operate the bye-way and cross posts between Exeter, Oxford and Chester, an area which expanded later. This was a badly neglected service, which Allen reformed and extended. In the first quarter he expended £1500 but his receipts were £2946, which, given the high returns this meant for the government and for various sinecures, delighted the postmaster general. This situation did not last for at the end of the third year his business was £270 in debt so that he certainly required capital backing at this stage. However, he had by then been elected a member of the Bath Corporation, the body which effectively ruled Bath. In 1727 and 1734 his contracts with the Post Office were renewed, in the latter case for life. According to Richard Jones (*Life*, Mss Bath Central Library) his annual income from this alone amounted to some £16,000 a year, equal to that of a great ducal house and perhaps the spending equivalent of some £16,000,000 today.

By the mid-late 1720s, still a young man, he had secured himself a large and secure income for life. He had established himself politically locally and nationally, and as a member of Bath Corporation had access to influence and money from his fellows and from the wealthy classes who frequented the City. He must have felt it was time to consolidate his position in society by acquiring land and turning his skills to the more conventional gentleman's occupation of developing the resources of a landed estate. At the same time, however, he must presumably also have seen himself, in his friendship with John Wood the architect, as a patron of Wood's vision of a classical rebuilding of Bath. In typical Allen style Wood's ideas fused with his own in the formation of a 'stone business' on a scale perhaps only previously envisaged in Britain by the stone producers on the Isle of Portland in Dorset. These had, some decades earlier, included Sir Christopher Wren at the time of London's rebuilding after the Great Fire. Allen intended the stone not just to rebuild Bath, but also to export it on a substantial scale to other main centres, including London.

Allen must have seen the improvement of the River Avon, already navigable to a point some three miles away, as a keystone in his own and Bath's expansion. Extending the Avon Navigation into Bath seems to have become a pre-occupation about 1724. It was a long-desired development. A Bristol merchant, John Hobbs, put the scheme into motion, but it was very heavily backed by Allen. The Navigation Trust had 32 shares with no person permitted more than one. Allen bought one and so did his wife Elizabeth, his grandmother Gertrude and his father or brother Philip, his brother-in-law Anthony Buckeridge, his sister-in-law Sarah Hudson, all bought others. The share of Charles Holder, later to become his father-in law, went to one of the Philips, probably his brother (Buchanan 1996, 67). His friend General Wade bought another and so did Milo Smith, soon to become a competitor to Allen in his quarrying business.

It is uncertain how some of Allen's family, at least, were able to afford the heavy outlay, but some were presumably nominees and Allen the true owner. It would be consistent for him to use the eventual dividends on the shares as the basis of pensions for family members. By this means he would have had support and probably effective control of at least a quarter of the shares, so it is not surprising he was the most active of the three Navigation treasurers, entrusted with the newly purchased 'iron chest' which he kept either at the Post Office or his home (Buchanan 1996, 67). In December 1727 the first shipload arrived in Bath. Bath was then directly linked to the sea-going and river port of Bristol and thus to a viable bulk transport system inland to north-western England, and coastwise to London and beyond, to Ireland and the European continent. These were to become Allen's chosen markets.

In 1724 he also began addressing how he could best take business advantage of the Navigation's opening. About this time he bought some 18 acres of land at Great Dolemeads from William Chapman (Hemmings 1983, 22), which was to become his stone wharf (this is now the site of the Kennet and Avon Canal junction with the River Avon). The lease of Prior Park in 1728 was his first large known purchase from Thomas Poole, by then living in Lyncombe and Widcombe, with one acre of land at the water mill and 60 acres of park, for 1000 years at a rental of five shillings (BRO. Allen Estate Papers: Abstract of Title 1728-1783). As well as providing the land for his intended great house, this also provided the main part of the route for his railway, begun soon after. He was able to complete its route to Dolemead where he intended his wharf to be, by an exchange of way-leaves with his neighbour, Philip Bennet (Peach 1895b, 78-80), a happy circumstance perhaps connected with the marriage in 1732 of Allen's brother Philip to Bennet's sister Jane.

The opportunity of acquiring the whole of Combe Down or Greendown in either 1723 when James and John Lane became involved, or 1729 when it was bought by Milo Smith using a mortgage, seems to have passed Allen by. In the former case, it is possible the cost of expanding his postal business, then probably making a loss, and of anticipating the purchase of the navigation shares were more important to him. In 1729 he may have been over-extended by the Prior Park purchase and the anticipated expenditure on the railway and the other infrastructure required.

He was negotiating with Wood in London in 1728 to sell stone to the Greenwich Hospital Commissioners, and it seems feasible that by then he had arranged with the Holder family of Hampton Manor for a quarry on Hampton Down which would probably have been the supplier had he won the contract. This seems to have preceded the Combe Down adventure. A compromise might have been arranged with Milo Smith to limit Allen's involvement in 1729 to taking over just a moiety of the quarries and attached 'remaining part' of the land over the quarries on Combe Down. This must have included the route of the proposed railway there. There is no extant transfer to Allen but he certainly held both the moieties later, and must have had some security on the land before 1730 to enable him to risk the cost of building his railway. Whatsoever, he did acquire the moieties of the remaining part of the land and of the underground quarries in 1730 for £400, while the remainder of Greendown passed to him in 1737 for £510 (BRO. 446/5).

Allen also at some point acquired the adjacent part of Green Down in Lyncombe and Widcombe and eventually most of the land as far west as Entry Hill came into his hands. The land immediately adjacent to Combe Road on the west side had the term 'Late Collibees' (Edward Collibee, presumed to be the previous owner) or 'Quarr Down' attached to it and extended to 58 acres. It was apparently owned by the Duke of Kingston from the 1720s but was not sold to Allen until 1744 (Addison 1998, 44). Previously it was probably held in trust for the Bruton Hospital (Irving 2005). It was described in the 1720s as lately enclosed. Early quarrying observed during archaeological survey (Quarry 910), under an area just south of the present-day William IV Public House may have been carried out by independent quarrymen rather than under Allen since it came into his ownership somewhat later. In either case some leasing arrangement must have been made by the quarrymen (other than Allen) for this in the 1710-30s but no information has been found.

Allen's estate continued to grow throughout the 1740s and 50s, but as far as quarrying was concerned it seems likely that sufficient title to the areas affected by operations in Allen's time was accomplished by 1730. Both the extent of underground workings of his time revealed by the archaeology and on the surface by plantations shown on the Allen Estate Plan, are consistent with Allen locating his quarrying area in an area south of the track to Claverton at the top of Prior Park between Combe Road, the modern road at the parish boundary, and the Long Drung footpath, extending as far south as the southern escarpment.

The financing of these new activities, according to Boyce, provided proof of Allen making a profit in his postal business (1767, 23), which he eventually did, but it is difficult to see how he could at this early stage in his business have sustained his other activities from this source alone. For the time being the value of his very substantial cash flows from the Post Office would be that they would sustain any interest on loans with repayments as they became due. Such large and regular cash flows were invaluable for use as security against mortgages or other sources of finance to leverage and raise capital, enabling him generally to seize opportunities as they arose.

It was probably to make use of his cash flows, or at least, or at times, to ease cash flow difficulties that Allen acted in effect as an early form of banker. He seems to have both received and loaned sums on behalf of the Corporation (Boyce 1967, 45fn) and was the active treasurer of the Navigation Trust. He also seems to have lent by means of mortgages, either directly or by paying those of others. In 1735, for example, Holder family mortgages of £15000 were consolidated, on which he appears to have paid interest, with the property itself falling into his hands in 1742 (Peach 1895b, 72). So long as he provided the cash when required, used careful cash-flow management and maintained a reasonable cash reserve so as to maintain others' confidence in his abilities to settle accounts at some reasonable time, then he would have had effective short-term use of all money in his care available to his wider business activities. The story retold by Boyce (1967, 72fn) of the architect John Wood's experience when he doubted Allen's ability to pay for the grand concept of his intended Prior Park house is a classic banker's ploy to instil confidence. Allen took Wood into a room where he kept his money and opened chest after chest full of guineas. Wood was re-assured. To how many would Wood in some way have related this story? Whether true or not, the story certainly circulated and helped create a sense of invincibility around Allen's affairs which was invaluable at this stage.

Allen's stone business 1727-64

Early developments

In 1727 Allen's contract for the postal service was renewed, ensuring his cash flow from this source for at least the next seven years. The navigation was completed, turning three years of heavy outlay into a further cash flow as well as opening huge possibilities for expansion of business. His friend and most famous of Bath architects, John Wood, was engaged in redesigning Allen's town house (Fig. 5.3). Wood was also both attracting influential friends such as Lord Chandos and Dr Robert Gay and embarking on his first Bath building projects for them. At the same time he was architect for the proposed St Johns Hospital (for poor sick people, run as a charity) and, at the beginning of the following year, he began building Queens Square his first major project (Boyce 1967, 33-35). If Wood's actual and intended projects proved a success, the demand for stone would be enormous.

It is not clear when Allen actually entered into the stone business. Boyce suggests about 1730, but in 1728 Wood and Allen were in London trying to convince the Greenwich Hospital Commissioners to use Bath Stone in their hospital development. Their competitors were the Portland stone proprietors, whose stone Wren had made so much use of. The Portland interest derided the Bath Stone as like Cheshire cheese and there was joking about it being liable to breed maggots which would soon devour

Fig. 5.3 Allen's town house in Bath, one of John Wood's first commissions, seen from Lilliput Alley

it. However, Wood had prepared samples of stone which the Greenwich architect mistakenly chose as being Portland stone. The Portland proprietors, however, then reduced their charges by a full third and won the contract (Wood 1765, 426-7). It seems likely this had a profound effect on Allen, who subsequently set out to reduce costs of stone delivered at the Navigation at Bath with great determination. It perhaps also had a cautionary effect on Wood, who later built the walls at Prior Park with wrought stone on both faces (normally the inner side was of 'walling' or rubble stone) to equalise pressures and prevent buckling, a tacit admission of his doubts about its strength (1765, 430).

Where did Allen anticipate acquiring this stone in 1728? Not, it seems, at Combe Down, where the initiative seems to have been taken by Milo Smith, but a quarry on Hampton Down, which was nearer to Dolemead and his wharf. The first reference to this is with respect to a device for lowering stone down the steep hillside. Peach (1895b, 71) refers to 'a series of short trams laid on to a centre where was constructed a large drum worked by some kind of machinery, but we cannot quite tell what'. Elton (1962-63, 39) suggests it was an 'in-quarry system', with a tramway descending the steep slope below the quarry for a short distance to St Georges Hill on the modern Claverton Road. There the stone would have been loaded and sent down the lesser and regular gradient to Allen's stone-yard and wharf alongside the Navigation. It was almost certainly not a self-acting system using cable and pulleys arranged so loaded wagons pull empty back up, since Richard Jones claimed credit for introducing this system when he built the Palladian Bridge in Prior Park. The drum may have been some form of horse or hand-powered whim, with or without a brake, used to lower stone down the high gradient to where a cart could be used. Allen may have seen such devices on his postal travels to his Chester outpost. There were several that led down to the River Severn around Broseley, for example, where they had long been in use (Lewis 1970, 95-108). The distance from the Hampton quarry to Dolemead is slightly shorter than from Combe Down and there was an established track or road of a fairly regular gradient. It seems likely that stone for some projects was supplied from the Hampton site, even possibly after the Combe Down quarries started operation, but lack of information suggests this was not for long.

Allen was acquainted with the Holder family who lived at Bathampton Manor, gradually buying land or acquiring rights from them. His second marriage was with Elizabeth Holder in 1737 (Peach 1895b, 71: the marriage date of 1727 in Peach is incorrect: Boyce 1967, 71). A quarrying lease there would thus not have been difficult to arrange, but Bathampton (or Hampton) stone was sometimes considered inferior to Combe Down. Any difference in quality would certainly have become apparent within a year of Wood and Allen using it and may have given added attraction to the purchase of Prior Park with its direct connection between Combe Down and Dolemead Wharf. Most likely Allen had secured the first moiety of the land and quarries at Combe Down by 1728-29 by agreement with James Lane and / or Milo Smith at the time of their acquisitions. Smith was certainly strained financially and, as a fellow Bath Alderman, it would not be at all surprising to find them, in an age where oligopoly was the natural order, accommodating each other. Whether Smith was fully cognisant of Allen's plans would be another matter.

As noted earlier, Combe Down's quarries were then occupied by half a dozen freemasons. Their objective was to quarry the stone required by themselves largely to support their own local building activities, which resulted, as observed during archaeological survey, in quarrying on a very small-scale. 'It was at that time a very good trade and gave them a good living', said Richard Jones in his *Life*. They would have conveyed the stone directly from the quarry to building sites in the City. In the scale of the operations, the technology of transporting the stone and the organisation of the stone building business, Allen's methods were to change their trade forever.

Management

A key figure in the managerial development of Allen's stone business was Richard Jones, whose unpublished autobiography *Life of Richard Jones*

(Bath Library Mss), gives a valuable and very detailed eye-witness perspective on the character of Allen, and his motivation.

Jones dictated his autobiography in his old age, and only a handwritten copy from *c* 1858 is extant, in which an added preface notes his style as 'garrulous and rambling'. It is indeed so, and he was probably a little hazy over his dates, but this appears partly to be responses to a series of questions designed, perhaps, to bring him back on track. Boyce (1967, 101) refers to his 'senile ramblings' and him being 'seemingly important' and involved in an 'improbable amount of building and design' (1967, 116; 101) but in this Boyce probably underestimates him. It is doubtful if Allen would have kept him in such a responsible position had he not been extremely capable.

Though Jones said he had been born in 1703 to a Bath cordwainer, a later footnote observes that the parish record has it to be the 5th of November 1700, the son of Robert Jones. He was 'kept close to writing', ie at school or in an office (then a longer than usual time), until he was 15 years old, when he was apprenticed to John Pitcher the Elder, a freemason quarrying at Combe Down. After serving his time, in 1725 he went to Monmouth, presumably as a journeyman mason, where he worked on the County Hall. He returned to Bath when it was finished, at about which time he married the daughter of Charles Tyler, with whom he had eleven children.

In Bath he gained further experience working on the Council Room, the Old Town Hall and the Monument at Lansdown and seems both to have designed and built a small summer house in the garden for Squire Bennet, at the same time as Wood was rebuilding the Bennet house, at Widcombe. adjacent to Prior Park. Bennet's mother had been a Chapman, and Chamberlain Chapman had paid for the Town Hall work. The Bennet daughter was to marry Allen's brother Phillip a few years later (1732). Even without further connections it is not hard to see how Jones' capabilities might have come to Allen's notice (*Life of Richard Jones* and Boyce 1767).

Jones states that 1727 was the year Allen began his great works, perhaps at his Bathampton Quarry, which probably means he began to work for Allen about that time and, in 1731, he was made clerk to carry out all his stone work and building. These projects included the row of eleven houses at Combe Down and those at the bottom of the hill near Dolemeads for Allen's workmen, plus the White Hart opposite, both sets designed by Wood. Jones also built and probably designed those alongside the wooden railway at Combe Down (the Carriage Inn). At this time he 'taught himself to draw' (architectural drawing) which he 'did to his master's satisfaction'. He originally lived at Dial House in the row at Combe Down (the central house with a pediment), probably until Prior Park House was complete, after which he lived there, presumably to be close at hand.

For Allen he went on to build two houses in Abbey Green, a house in the Church Yard, and Shears Buildings and then began the houses in Wood Street, serving all the stonework on each side of the square. He built the 'obilisque' in the Square at the expense of Mr Nash. He enlarged the Pump Room and installed a balustrade by order of the mayor. He then began Allen's own stables and the Porter's Lodge 'up the hill' (Prior Park), and, among other buildings in Bath, the New (St John's) Hospital. In 1736 he built the long Malt House and Brewery at Dolemead and, at the same time, appears to have been responsible for the building of Prior Park House. The first storey he built under the direction of Wood, but the second storey of the house and roof were built perhaps using drawings completed by Jones himself, with the east extension also partly by Jones. He also had some involvement, perhaps the supply of stone at Bath Gardens, with the huge project of the building of the Parades.

In the 1750s onwards, possibly because Allen was in his sixties and enjoying himself entertaining his socially important friends and in the management of his large estate, Jones seems to have been involved in estate rather than major city projects. He designed and built the Palladian Bridge over the lake in Prior Park, Sham Castle, and houses, farm, church and school at the Claverton and Hampton estates after Allen bought them, and supervised the building of Allen's house at the fashionable seaside town of Weymouth. His public works, probably on his own behalf, included the design and building of the New (Newton) Bridge at Bath which, unfortunately, was washed away in a flood after his death.

He became embittered in the last decade of his life. In his last days Allen walked with him to inspect his (Allen's) proposed burial place, and had Jones design the vault and monument, which he did promptly and brought them to him. Allen never saw them built, as he had hoped; and after Allen's death in 1764 Jones found himself profoundly in disagreement with Allen's niece and heir Mrs Gertrude Warburton. She ordered the cessation of the stone business, had the equipment sold, putting the quarries out on a royalty basis. After having to finish a piggery for her before being allowed to erect Allen's monument, Jones was discharged. By this time about 67 years old, he had been promised he could spend the rest of his life living at Prior Park but received only a legacy of what was a surprisingly small annual salary of £45 (he had, however also had considerable parish business as well as that of Allen), and found himself having, for a time, to stay with his son in the city.

It seems likely others also felt Jones was treated badly. He was appointed City Surveyor of Bath by Mr Collibee the then Mayor, and the corporation, and was responsible for a reservoir under Beechen Cliff in 1767, bringing water via lead pipes into Bath, and with the beginning of another reservoir at Bathwick to bring water down from there over Mr Pultney's bridge. He replaced the old market and shambles and built a new prison, and was present when the then mayor, Alderman Chapman, (one of the two trustees to the Allen estate) laid the foundation to the new Town Hall in 1769. In his honour and no doubt to assist in his retirement, he was made Sarjeant to the Mayor though he also remained Surveyor until 1771. He died in 1773, the *Bath Chronicle* referring to him as 'Wood's foreman' (slip in Bath Library card index). No doubt this was an understandable error given his involvement for Allen on the old projects of John Wood the Elder.

Despite Boyce's doubts, Jones was clearly an outstanding building and managing mason but, perhaps, in comparison to John Wood, father and son, only a reasonably competent 'architect'. There seems little difficulty, with his continuity and close involvement in Allen's service, that he was trusted to carry out the most difficult of tasks, for which the marker might be the substantial part of Prior Park House he was responsible for. In his own words, he always had too much business, both for Mr Allen and other parish business. He is included in a portrait (see Fig. 5.2) with Allen (the original in The Building of Bath Museum also includes Robert Gay and John Wood) which perhaps better reflects the contemporary view of his abilities.

As for Jones' involvement with Allen and his business, some time after Jones returned to Bath from his time in Monmouth, about 1727, Allen bought all the quarries on Combe Down and from about 1727 (the date may be suspect, or related to the old style calendar), according to Jones, 'Mr Allen carried on great works'. The great works were those connected to the quarries and to building and the implication is that these gathered momentum from 1727 (or thereabouts). Jones took a definite part in several aspects of this, and his involvement with the business probably grew until he was given full responsibility in 1731 as Allen's Clerk for the stone business. Whatever the means of taking over the quarries at Combe Down, it would have been logical for Allen to have initially continued using the various freemasons already there, such as the Pitchers and Biggs and their workmen to work the quarries at Combe Down under contract for him. It is possible that Jones was employed to do similarly or replace one or other of the freemasons. One of the quarries at the Down was called Jones Quarry and the continuity of the name may reflect his appointment by Allen and the very beginning of the quarrying expansion. Biggs and his brother at least continued as Allen's foremen at various sites for what virtually was the rest of their lives.

Allen seems to have followed much the same business practice in both his postal and stone businesses. With his great variety of personal activities and the difficulty of communications whenever he or his foremen were distant, it had to be based on a mixture of trust, personal supervision and to some extent, reward. If it was reward, the only information available (in Ralph Allen's will) was Jones' annual salary in 1764, which was only £45. It is likely that Jones and possibly others had earning activities beyond their direct employment, which Allen tolerated as part of customary practice in his postal business as well and, in his case, Jones also received his keep as he was part of the Prior Park household. Allen gave very substantial authority to his trusted clerks (or foremen – then very senior positions). Of the two members of the Biggs family who were probably among the Combe Down freemasons before Allen took over, Richard Biggs became his London representative 'Manager of the Stone Works' responsible for, among other matters, the building of the four wings of St Bartholomew's Hospital until its completion about 1764. William Biggs seems to have been responsible for the shipping of stone at the Dolemead stoneyard, and, presumably, all other business carried on there. Later, in 1742-43, he was in charge of building the Corn Exchange in Bristol. Other members of the Biggs family were involved in his postal business.

The successful sale in bulk of stone to any who needed it, anywhere, was only possible if Allen could guarantee efficient production on time, with reliable transport, and ensuring its quality was maintained. It was probably to ensure a substantial part of the demand for stone was directly under his own control, or that of his foremen, that he employed skilled masons both at his own sites and at distant sites from Bath to control the downstream masonry-building business. These were the banker masons working at a bench in stoneyards or on site, or rough masons involved in the erection of the buildings. It is likely that he recruited the best of these on to his permanent staff. In the case of St Bartholomews' Hospital he initially intended to send twenty men, but problems over their freedom (which would normally result from serving apprenticeships actually in the City of London) led him to reduce their number to seven, whose liberty to work there cost Allen £30 each (Boyce 1967, 43).

In his third edition of the *Essay Towards a Description of Bath*, the younger Wood (1765, 426) provided a eulogy to his friend who had died the year previously. Allen, in entering the stone business, intended the 'exportation of block stone [as] the branch of trade in which he engaged himself'. Blockstone was stone only roughly squared into blocks at or in the quarry, ideally leaving the production of wrought stone to be done as close to the site as possible, a procedure which Wood, who railed at the damage done during transportation, clearly approved. In practice it was certainly blockstone that was exported but much of Allen's locally used stone seems to have been wrought or 'squared' either at the Masons Crane House Quarry at Combe Down or at the Dolemead stoneyard. However his transport methods were less damaging to the stone than the older use of carts, and his yards allowed him to capture and control the market and costs downstream as well as at the quarrying stage.

His immediate objective, wrote Wood (1765, 425), was reducing the price to increase the consumption of the stone. This may be a *post hoc* rationalisation since Allen, as a newcomer, presumably needed to undercut the existing suppliers unless the demand had risen from other causes. It is also possible that he did not intend quite such a substantial reduction as actually occurred. From the beginning at Combe Down, as seen in both documentary sources and in the archaeology recorded underground, methods changed in ways which would have reduced costs. It could be done by increasing the scale, permitting a lower profit per unit of production; by increasing the time efficiency of the workmen and by lowering their wages or, alternatively, their unit costs; by introducing better extraction methods and underground transport, and by improving surface handling and transport. The latter was the most obvious manifestation and is considered first.

Surface transportation of the stone

In the typical underground quarry of the time, stone had to be loaded on to horse-drawn carts or wagons outside the quarry and was carted to the stoneyard or building site on roads then notably poor and made worse by the stone trade. On the steep hills around Bath, it was a laborious, slow and hazardous process. Each of Allen's underground entries was sited in a small quarry some 10 m below the general ground surface which was reached by a sloping track from the top of the Down. The underground cartway went almost level into the freestone beds which were to be worked from the entrance and blocks of stone were dragged to the outside on barrows, carts or wagons. Above the entrance was a vertical cliff and at the top of the cliff, at the normal ground surface level, a crane was mounted with its jib projecting over the edge to haul up stone from below. It was slewed by the jib on to a carriage on the wooden railway, from where it was dragged by horses to the top of the hill down to the wharf. These carriages carried loads of 4-5 tons with relative ease. From the edge of the Down the carriages ran downslope on a wooden railway to the stone yard of wharf at the Avon Navigation. There they were unloaded in the stoneyard, loaded directly on to river boats, or the carriages of loaded stone were shipped on a barge on the river to the building site.

Stone in block form was transported in larger pieces than most of the finely wrought stone used in building, probably as large as trimming to roughly cubic or rectanguloid form could produce. This meant the smallest blocks would weigh around two hundredweights (0.1 ton) rising to as large as three or four tons (20 cubic feet were reckoned to the ton, but later practice reduced this to 16). The fractured nature of Combe Down stone beds meant a large proportion of smaller blocks were produced but the larger blocks also had to be handled. Prior to around 1730 and Allen's railway the contemporary horse cart with one or two horses, when road conditions were reasonably good, might reasonably take a ton or two, but on steep slopes or on deeply rutted roads over the Fullers Earth clays movement of such loads could be a huge problem. There is no evidence that sophisticated cranes were used at Bath before Allen. Rather the reverse was the case, with stones on building sites commonly slid manually up ladders. But where simple lifting by one or two men or sliding on timbers using levers was insufficient, timber shear legs operating with chains or rope and pulleys would have been available to masons. But this was all time- and labour-intensive.

According to Wood (1765, 425), Allen had gained the idea of the wooden railway – variously called a tramway, wagonway, even just road – during travels for his postal service near Newcastle in the Great Northern Coalfield: 'of the roads and carriages in the North the late Mr Hedworth sent proper models to Bath'. At Newcastle wagons filled with coal ran under gravity on wooden rails from the pitheads down to the staithes on the Tyne. Similar systems were used in Shropshire and it seems likely that John Padmore, a Bristol engineer with a wide practice around Bristol and South Wales, who had earlier been appointed as the engineer on the Navigation, was familiar with the Shropshire methods. He was engaged by Allen to design the system used at Combe Down and possibly had been for the Hampton Quarry system also. The detail of this and comparable systems has been discussed at some length by early railway specialists (Elton 1962-63; Lewis 1979; and Pollard 1994) so only the main technical outline gained from contemporary sources and reasonable inferences from them is provided here.

The top of the railway incline (at the current Ralph Allen Drive junction with North Road) at Combe Down lies at about 150 m (500 feet) altitude above the wharf at the River Avon. This was the section which operated by gravity and it is about a 1.6 km (one mile) from Dolemead Wharf. It is roughly level to the southern escarpment where the early quarries were located, and with the several branches the total length of track was about 3 km (two miles). It seems likely that the wooden railway to Combe Down was being built from about 1728. This and the stoneyard and wharf infrastructure at Dolemead, the cranes there and at Combe Down, the workmen's houses at Combe Down and Dolemead and the underground cartways were all developments which needed time, money and management. Jones claimed the total expenditure was some £10,000 which was an enormous investment at that time.

In an agreement of 20th March 1730 the 'wagonway for the carriage of stone from [Allen's] quarries' is referred to as 'recently completed'. Since the old calendar was still in use, this suggests a late 1730 completion. The deed is between Allen and Bennet, by then neighbouring landowners, which permitted the course of the railway to pass with wide and smooth curves and regular gradient over both their properties. Allen and Bennet were allowed to pass over each others' carriageways (Peach 1895b, 76-81). It is a reminder that affairs could be conducted by agreement, the lawyers' writings coming later, and may apply to other aspects of Allen's business which, above all, depended on trust.

The wooden railway was made to a 1.14 m (3 feet 9 inches) gauge of oak timbers, 6 inches deep and 5 inches wide (150 x 125 mm), mounted on stone sleepers and/or timber cross-ties. A rough calculation suggests the oak timber alone would weigh in excess of 150 tons. A form of level crossing seems to have been used where the line crossed a road just before the Wharf with the road there built up around the timber rails. On an initial almost level section from just across Dolemead Wharf at the Navigation side, the railway ran via a well-engineered gradient of about 1:10 up Bennet's and Allen's private carriageways past their houses (confusingly the wagons running on the railway were termed carriages too), over a small bridge to the brow at what is now North Road. From there it crossed the Down via the present Avenue, almost level to the top of the quarries. Its original route at the southern end was aligned with the Carriage Inn frontage rather than the present end of the Avenue. A branch ran to each of the quarries which were entered by levels (tunnels) underground some 10 m or so below the surface, or possibly initially to a wide shaft (which appears to be shown on one sketch) which would have entered the level below. A crane was used to raise stone from the level to the railway. Where the railway split at its various branches, some form of points would have been necessary but there is no detail of these. Simple moveable timber rails operated by hand or a lever would have sufficed.

Two maps show the crane positions. Thorpe's (1742) map *Bath and Five Miles Around* is very small-scale but shows the railway coming along the (later) Avenue, one branch appearing to go to above the Rock Lane entries (Sheeps House Quarry) and what is presumably a crane building and another branch to above those at the Rockhall Lane (Ralph Allen Yard) site. The main line continues in a direct line to a position which may have served entries below at an early quarry under the end of the Avenue. The position might be in or above what is today Quarrybottom. While the scale means the branches might be symbolised, the straight line relationship to the west of the cottages built by Allen for his workmen is fairly convincing in its placement (see map detail, Fig. 5.4). The Allen Estate *c* 1764 map shows two branches, one leading to above the two entries at the bottom of present day Rock Lane and another to the top of Masons Crane House Quarry just above what is today the Junior School and

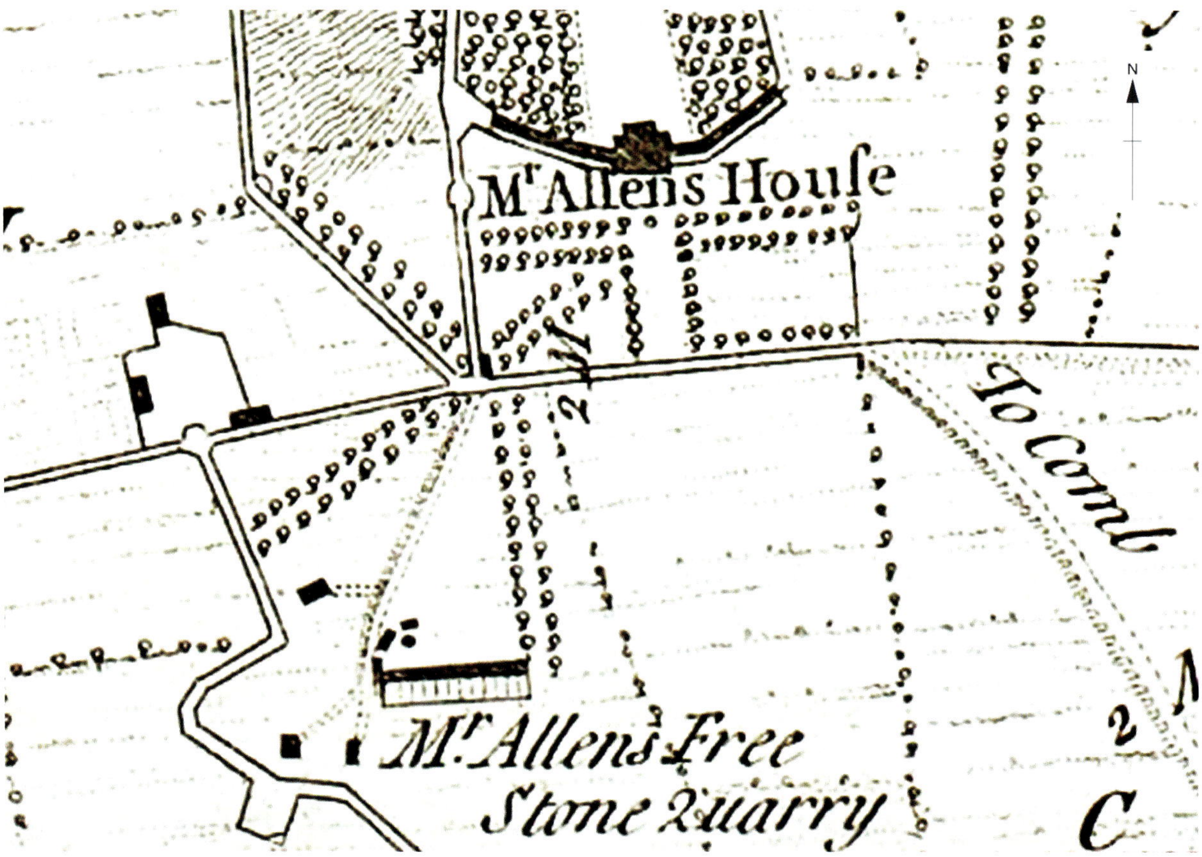

Fig. 5.4 Detail showing Allen's quarrying area on Thorpe's 1742 survey Five Miles Around Bath (BRO)

below the church (Fig. 5.5). This agrees approximately with the position of the crane showed on the 1759 sketch of the quarry, crane and Old Rank – workmen's houses. This may suggest that the crane was moved, either because the underground operations had ceased in the immediate area and the entries had relocated, or because surface working on the open quarry had progressed eastwards, or both. Though not shown on the later map, it is reasonably certain from cable grooves inside the underground quarry (505), that a crane at least served the two entries at what is now Ralph Allen Yard down Rockhall Lane, and that a railway probably followed what is now Church Road. The site is shown on the Allen Estate map as an area clear of trees, which may indicate abandonment of working at the site prior to 1764. This, from the archaeological survey and leases (see later) is not inconsistent with the position of quarrying restarting there after 1800.

Jones in his *Life* says there were 'four cranes on the hill' and another at Masons Crane House Quarry. Jones listed the material of cranes sold after Allen's death and this and the underground developments recorded by the archaeological survey suggest there were also two crane-operated sites east of the Avenue between there and the Long Drung which were not marked on either map. These underground quarry entries were behind Allen's row of workmens' cottages, including Dial House, and were accessed via Quarry Lane adjacent to Claremont off Church Road and, possibly behind Hopecote Lodge. Quarry Lane is analogous in position to the inclined surface track to the Sheeps House entries down Rock Lane. Unless one or both were accessed from the Masons Crane House surface quarry these would probably also be served by rail from the cliff top with a branch or branches somewhere north of the Carriage Inn buildings. These are shown on the map on Figure 3.10, a reconstruction of the possible railway branch system. Later quarrying has removed evidence that might have cleared this up.

Jones drew a little diagram (Fig. 5.6) in his *Life* to illustrate the position of the crane at the Masons Crane House Quarry. It was placed centrally between two parallel quarry roads that are labelled as such. This position is puzzling since there seems to be no related surface road location, but the explanation may be that the roads are the parallel underground levels or tunnels as seen elsewhere, as the parallel cartways running in from the Sheeps House

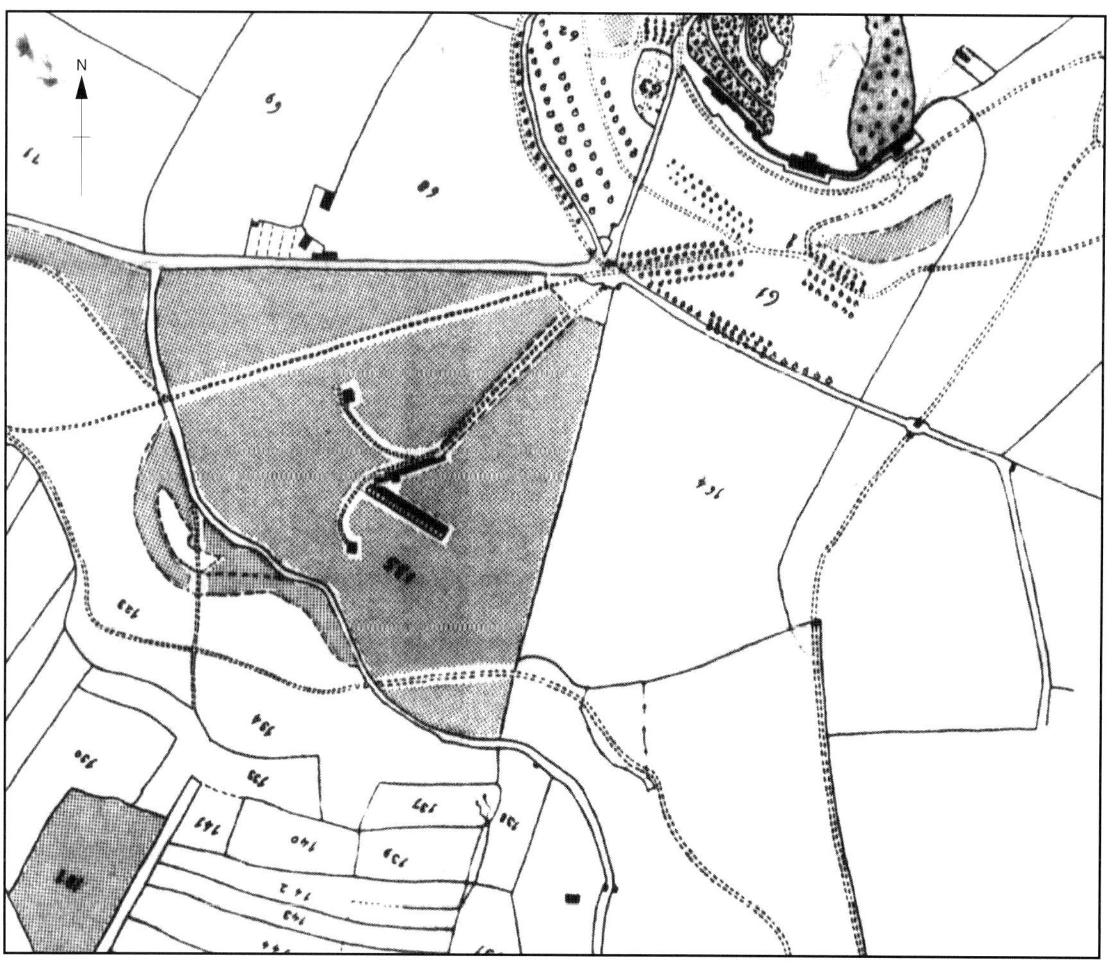

Fig. 5.5 The same area on the Allen Estate map, c 1764 (BRO)

Quarry site. If so it may be that initially some workings east of The Avenue were accessed from the Crane House Quarry area (and if so the little plan may suggest a shaft entry) and that the surface quarry there was developed later – which is consistent with comments that it was only later that Allen began surface quarrying (Wood 1765, 425).

The branches gathered together somewhere near what was the Carriage Inn (named after the carriages on the railway) on what is now The Avenue. The quarrying area, the land between the now Long Drung, North Road, and the then Combe Road / Summer Lane, is distinguished on the *c* 1764 map by the trees on it. Allen established this plantation with fir or larch (then an unusual species in southern England). Along what is today The Avenue, the railway runs centrally from the northern apex of the triangle before branching, neatly bisecting the area, while the sloping surface roads down to the quarry entries at Rock Lane and Quarry Lane have a pleasing symmetry, probably deliberate. The Hopecote and Rock Hall Lanes down to the entries, both now substantially modified, may have been to the same pattern. It was a carefully designed scheme for the exploitation of the estate with the intended production clearly on a very substantial scale.

Somewhat later, perhaps coinciding with the main planting of the trees, the old Summer Lane was diverted at its north-west end at modern Stonehouse Lane (1742 Thorp's *Five Miles around Bath* map) to its current position at Combe Road, running there along the parish boundary at the western side of the triangle. This is shown on the *c* 1764 Estate Map. The reason for the diversion is not obvious but Allen acquired the land in Lyncombe and Widcombe somewhat later and it might have been a tidying-up of the then estate, or possibly it was placed to serve the parishes on both sides of the boundary.

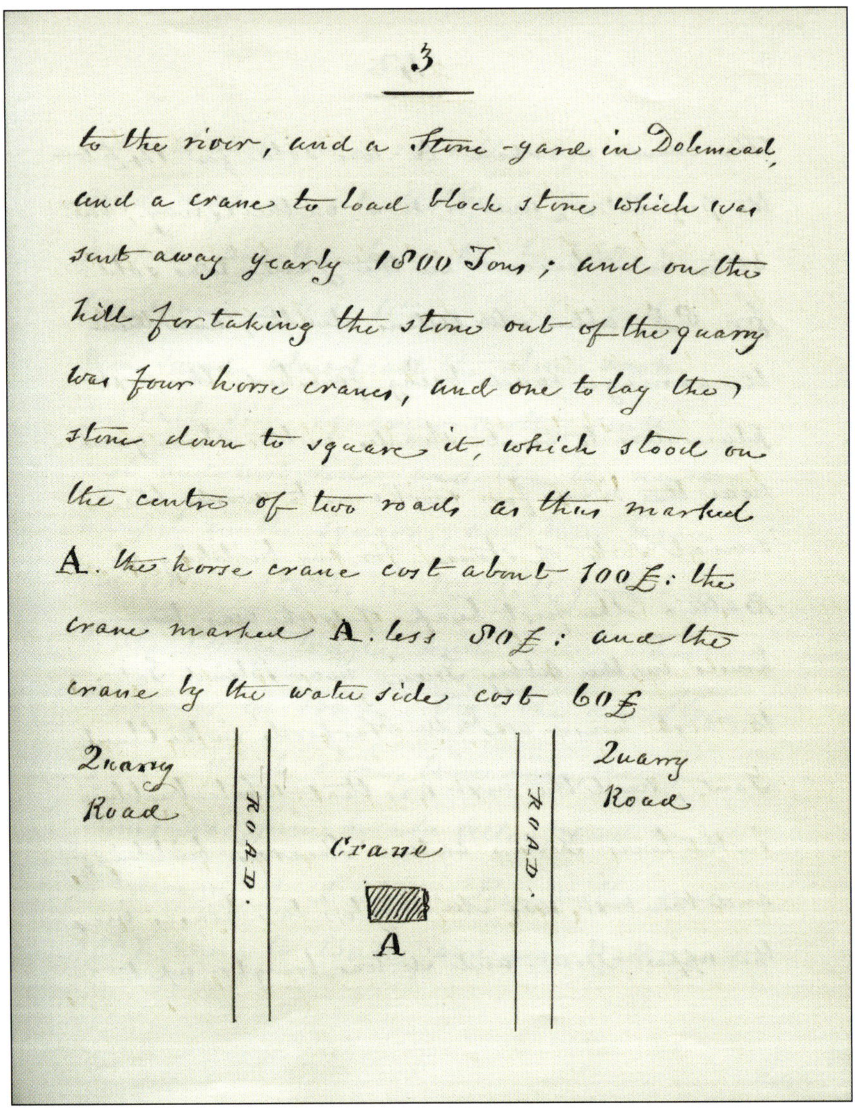

3

to the river, and a Stone-yard in Dolemead, and a crane to load block stone which was sent away yearly 1800 Tons; and on the hill for taking the stone out of the quarry was four horse cranes, and one to lay the stone down to square it, which stood on the centre of two roads as thus marked A. the horse crane cost about 100£: the crane marked A. less 80£: and the crane by the water side cost 60£

Fig. 5.6 Richard Jones' sketch, in his Life, of the position of the Masons Crane House Crane (image by permission of Bath Central Library)

It took two horses to drag loaded railway carriages with 'often upwards of four tons' (de Labelye in Desaguliers 1734, 276) or 'seldom less than five or six tons aboard' (marginal note by Wood cited in Elton 1962-63, 44) from the quarry cranes to the top of Allen's original Carriageway alongside Prior Park. Beyond here they descended under gravity. Contemporary illustrations show the carriages travelling singly with a driver following on foot in order to operate the brakes. With care, momentum would probably bring the carriages across the bottom, almost level, section to the Dolemead stoneyard and wharf crane. Empty carriages were hauled back uphill with two horses, which would have had to be available at the bottom as required. It is probable that they had other duties too.

The carriages were particularly well-engineered (Labelye in Desaguliers 1734, 274-9 and pl. 21) and strongly made of oak and iron plating, about three and a half feet wide and 13 feet long, with removable sides (Fig. 5.7). The carriage platform was heavily made, and wear was reduced and lifting of stone blocks made easier by timber crosspieces. Padmore provided two means of controlling them, a device which could lock any of the wheels through the spokes and a 'jigger-pole' to operate a friction brake (post-brake) acting on the wheel rims. This latter had a chain affixed to its rear end, which was wound around a roller fitted with a 'rochet and click' (ratchet and pawl) actuated by a bar. This enabled the brakeman who walked behind to secure the pole on the wheel, until the ratchet was released.

The gauge was around three and a half feet (in his diagrams de Labelye's measurements do not quite match the scales given and other sources suggest 3 feet and 9 inches). The axles were of three-inch-thick wrought iron, running in well-oiled brass bearings. Each had one square end attached to the wheel, the other round, on which the opposing wheel could freely rotate, thus lessening friction and wear on bends and making the brake system more versatile. The wheels of about 20 inches diameter were of cast iron with spokes. The flange or 'flaunch', in effect a giant washer placed on the inside of the wheel, was separate, that is, not cast with the wheel as in later practice, but with the same function of keeping the wheels on the rail. This axle arrangement and brakes gave the brakeman considerable flexibility in adjusting the braking to the load and to the conditions, but it seems usual to have locked one wheel for the decent. However, wet or icy conditions must have been especially difficult under full load and would require careful management by the driver. Allen told de Labelye that each carriage cost upwards of £30 (Jones says £40).

Jones added that when the river-crossing barge was in use, four carriages were going up the hill constant, four were across the water, four leading up the hill, two loading blocks and four spare in case of any misfortune. This meant 18 carriages in all, although those going up the hill and those leading up the hill may refer to the same four carriages. The river crossing system will be described below.

Two horses to drag the carriages from the cranes to the brow of the railway incline was the minimum required. At busy times this was probably not enough, since there were four carriages 'on the hill' and reasonable utilisation of the railway (unless there were bypass sections) suggests they moved in either twos or fours. One or more may also have been needed to move carriages during loading and even to go underground occasionally. Horses might have been hired and brought from a distance but, more likely some were owned by the business. This

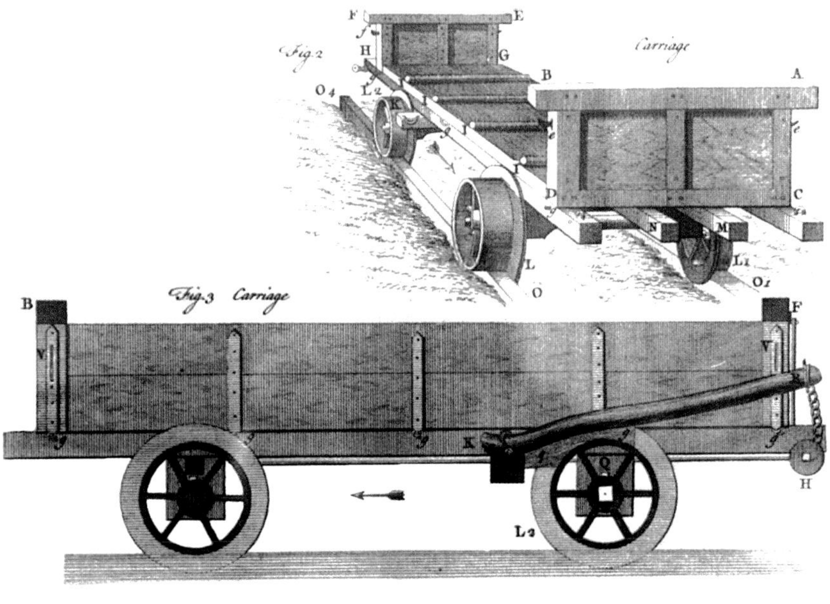

Fig. 5.7 The stone carriages used on the wooden railway (Desagulier 1734)

would require a stable, the most likely position for which would probably have been at the Carriage Inn buildings where a nearby well was located underground. Artefacts found by and from the well were contemporary. Alongside the original track of the railway the well would have served a trough for horses as well as local people.

The cranes, mainly constructed of wood, were placed at the top of the face or cliff above the mine entries. The cranes appear to have been sited above each of the four sets of quarry entries (the archaeology suggests all four areas were in use more or less at the same time), and at the Masons Crane House Quarry. These were all designed by John Padmore the Bristol engineer. The fundamental design of the winding mechanism resembled the common horse gin or whim used in mining, or indeed a ship's capstan, fitted with a rotating gallows-like jib. A 1731 illustration of one of these by William Halfpenny (Fig. 5.8) shows it was a man-powered capstan under a roof, sufficient perhaps because its use was probably discontinuous, but there are also references to horse cranes to which it could easily be adapted. The sketch by Robins in 1759 (see Figure 3.11) possibly shows the same crane, and a similar crane-house as a square building with a pyramidal roof. The view faces north towards Dial House, which identifies it as Masons' Crane House. According to Jones the cost of this crane was £80, and it and the associated

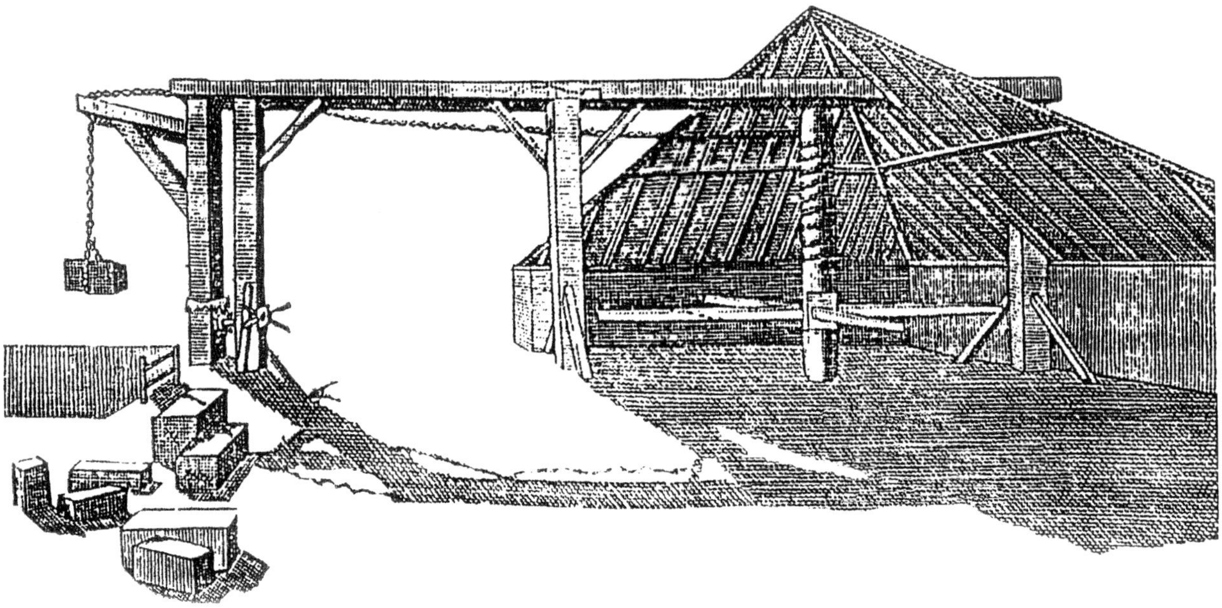

Fig. 5.8 Crane house, probably the Masons Crane House (William Halfpenny 1731)

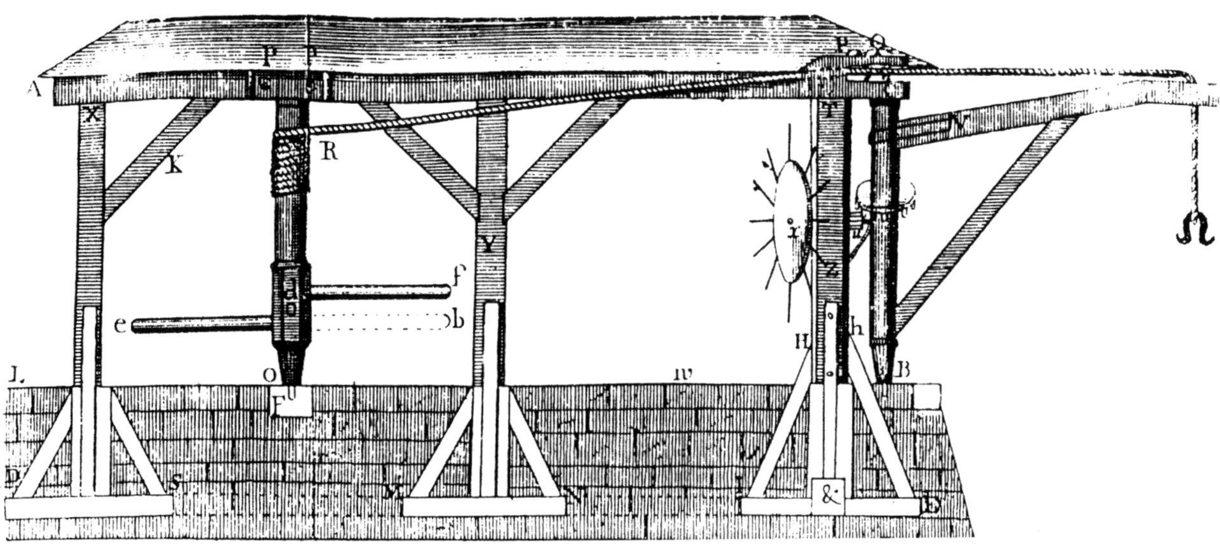

Fig. 5.9 Capstan crane (Desagulier 1734)

quarry was for the squaring-off of blocks. As shown by 1759 it was clearly a very substantial open quarry too but, unless the original site had the stone yard at the upper level, it is unlikely blocks of unfinished stone were brought here for dressing in the mason's yard from other crane sites as it would be counter-productive to double-handle them. It is not certain that all the other cranes were roofed over, though a horse crane noted by Jones cost £100 which is fairly expensive. At this cost it was unlikely to have been that illustrated by Desagulier (1734, 172, pl.12) which was a similar crane with a much simpler roof (Fig. 5.9). Allen was sympathetic to providing good shelter, believing it led to higher productivity.

These cranes were handling blocks of stone of up to several tons, susceptible to fracture if dropped and, of course potentially very dangerous to workmen around. The raising and lowering depended on the men on the capstan. To assist in handling Padmore mounted the jib (gibbet) on a post at the front of the house which could rotate through a semi-circle allowing the stone to be raised from a wide area at the bottom of the cliff, probably also by dragging from a much wider area than the jib radius covered. At the top the jib could be rotated to either side of the crane for loading stone on to the ground or a carriage. Unlike on later quarry cranes, which mounted the windlass so it rotated with the gibbet, the arrangement necessitated fitting a slewing mechanism powerful enough to overcome the weight of the load moving off-centre to the line of the chain. Padmore designed it with a wheel with large spokes with gears on its axle operating on a toothed ring on the jib post, which allowed 'one man standing out of harms way to move the axis of the gibbet with great ease' (Desagulier 1734, 179). The technology, even the design of the hand-wheel used to work the mechanism, may have been derived from ship's practice, notably for rotating the rudder. Padmore, based in Bristol was doubtless familiar with such mechanisms.

At the Dolemead Wharf Padmore installed his 'rat-tailed' crane (Desagulier 1734, 179-81; pl. 16), shown in detail in Fig. 5.10). Surprisingly, at £60 this cost less than the cranes at Combe Down, possibly because the jib arrangement was simpler. Here the lifting problems were different. The height of lift was very small, just enough to ease the block followed by lowering on or off a carriage or down into the boat, for which manpower was sufficient. The danger here was of a large block going through the bottom of the vessel being loaded if the load overcame the winding mechanism or the strength of the winding men.

The Dolemead crane rotated on a vertical post, rather like a post-mill, with a curved projecting jib formed of a single timber which could swing out over the vessel alongside the wharf. Inside the 'cabin', which rotated with the jib, thus avoiding slewing difficulties, two winding-handles permitted four men to raise very large weights through a simple geared winding-drum and a chain. There was a pawl and ratchet attached to the geared winding drum and a circumferential post-brake to control lowering. It was claimed this was the first time both devices had been used in conjunction with each other, which prevented 'the great mischief

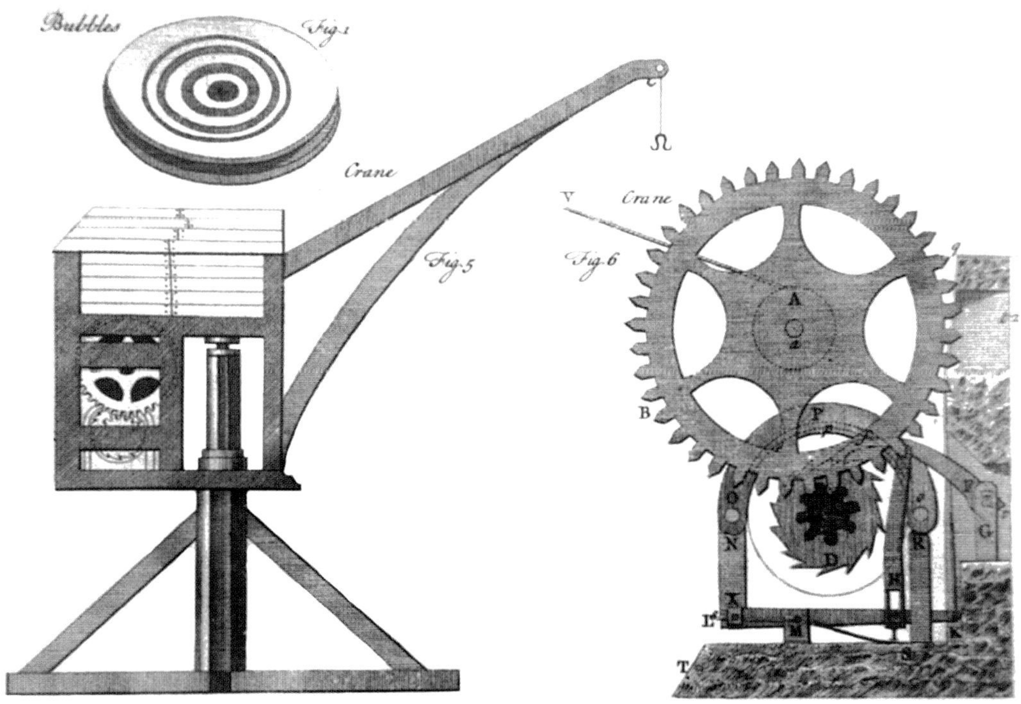

Fig. 5.10 The rat tail crane at the Dolemead Wharf (Desagulier 1734)

which often happens by the carelessness of the men employ'd to raise and let down the heavy burdens by use of a crane' (Desagulier 1734, 180).

Soon after the initial installation of the railway and cranes, a roll-on-roll-off system was used to take the stone-filled carriages over the river into Bath. Jones relates this was done, probably at first, to take stone to Monks Mill (near Pulteney Bridge) 'and so rolled up a road made up into the Grove for the purpose' (the road would be a length of wooden railway). This is slightly enigmatic, but he later more explicitly says that throughout the 'great work at Bath Gardens' which took place between 1739 and *c* 1746 , for the four piles of buildings called the Parades, four carriages at a time were taken over the water. These were brought upstream on two barges each with two carriages, four times a day, and hauled ashore by a capstan.

There is no detailed description of the 'roll-on–roll-off' mechanisms by which the barge was loaded, moved upstream, docked and unloaded in order to pass the carriages laden with stone across the river and into the City. It can perhaps be assumed the carriages were arranged in tandem rather than abreast, since this would provide more stability and a better loading angle along the length of the boat. There was, however, one possible and intriguing mechanism for drawing heavy carts or carriages of stone, which could well have been used for unloading the barges. The 'cratchet winch' was illustrated alongside drawings of the other Padmore machines at Combe Down, and published in a *Supplement to the New and Universal Dictionary* of *c* 1750 (Science Museum Library). The description does not include the place used or origin of the drawings but it is a machine ingenious enough to have come from the mind of Padmore and was certainly powerful enough for this purpose (Fig. 5.11).

The cratchet winch used a horizontal drum fitted with a ratchet wheel (a toothed rim) which was engaged by two extended pawls attached to either side of the pivot of, and operated by, an overhead rocking beam. The drum, when the ratchets were engaged, was drawn round but could move in only one direction. The whole was mounted on a trolley, which could be secured by ropes or chains to a post or pillar. The drum, using a rope or chain, gained further leverage by running round a pulley secured at the end of the loaded trolley, could then pull a very heavily laden carriage of stone. The beam appears to be some 5 m or more long, rocked up and down by four men. It would, with this number of men, based on their weights, be capable of dragging five to 10 tons, equivalent to one of the carriages or even one of the larger blocks, with relative ease. Much smaller devices working on the same principle are available today. It would have been possible, if this machine was not used, to design a simple four-man cranked, windlass mounted in a strong frame, with ropes over multiple pulleys, much as used to haul cannon on board ships. Such a machine would be unlikely to have any brake or ratchet, which were clearly still very unusual except on Padmore's machines, and thus might pose more hazard.

The Monks Mill and Grand Parades sites are only a few hundred metres from Dolemead, but it saved transhipment on to ordinary carts at the wharf and the considerably longer journey over Bath Bridge and through the difficult traffic of the City. It may also have enabled wrought stone to be offloaded at the building site prepared at either the Dolemead Stone Yard or Masons Crane House Quarry without the damage to the arrises John Wood considered one of the worst aspects of building stone in Bath. At upwards of four tons a carriage load, this implies some 60-80 tons of wrought stone a day were delivered to the site, which would otherwise have been very difficult to work on in the space there. Wood commented that the work at the Parades commenced on 10th March 1739-40 and that 'a five hundred feet length had been raised' (making it likely to have been the southern block, today's South Parade) ready for roofing by the following Michaelmas, that is, in six and a half months (Wood 1765; 1969, 248). Pre-cut (ie wrought) stone delivered to site would have been invaluable in aiding such progress and, from Wood's comment, it is clear that this was an exceptional work rate, demonstrating how well Allen's integrated system could work (Fig. 5.12).

The railway, allowing 'such heavy carriages to be drawn along it with such little strength', halved the cost of haulage to the Navigation side and enabled Allen to reduce the price of stone delivered at the

Fig. 5.11 The cratchet winch (Supplement to the New and Universal Dictionary 1751)

Fig. 5.12 South Parade, one of the Grand Parade developments c 1742, designed by Wood, with stone and some building works by Allen

Avon side from 10 shillings a ton of 20 cubic feet (a nominal measure) to seven shillings and sixpence. Such a huge reduction would undoubtedly increase consumption, as Wood claimed was its purpose but, even more, it placed immense pressure on his competitors. However, to stay in business they then reduced their prices to seven shillings, presumably at the expense of their workmen, the carters and their own profits. Allen had to reduce his prices also, and seven shillings a ton was still being charged late in the century.

Changes in quarrying underground affecting transport

Some of these aspects will be considered in detail in Chapters 8 to 10 which deal with the archaeology, so here only the linkages with Allen's general business methods are emphasised. In general the changes were not documented, though some inferences can be drawn from occasional comments.

The evidence presented above, in conjunction with archaeological evidence below ground suggests that Allen set up four (five if the Masons Crane House served the underground too) separate underground production units, all with cranes served by a railway branch. It is possible he did this to ensure that at times of high production haulage from underground did not delay the production process and perhaps so that collapses or other unforeseen circumstances such as bad ground would not have a similar effect. Such separate units might also increase the competitive element and the management would have useful comparisons of output between teams of workmen.

For each set of underground entries the surface track led down to some 3-4 m below the overburden of the Twinhoe Beds, forming a cliff 10-12 m high at the level entrance. Here a small area was formed to provide outside working space either for banking masons or for stacking space for the crane operation. These are shown on the *c* 1764 Allen Estate map (see Figure 5.1) as semi-circular areas clear of trees. The entries may have been at previous quarry working sites which Allen had taken over. From the entries, tunnels or levels (called cartways if evidence of use is extant) were driven into the freestone and the working places. In at least one case, pairs of levels were driven in nearly parallel, about 10 m apart, and in another, at Rockhall Lane, the two Allen levels diverged into the hill from what is now Ralph Allen Yard. Initially the levels seem to have been driven forwards for perhaps as much as 40-50 m. This opened up a large number of working places for small teams of men to operate, each of which was close to the main level along which barrows or small carts of cut stone (and initially, any spoil) could be moved directly to the outside.

The method of haulage into and along the main levels was the key to major productive improvement underground. The usual methods described included flat barrows, two-wheeled quarry carts and four-wheeled wagons. These can be hand-hauled over short distances, or pulled by horses.

Evidence of horse-use was restricted to a few fragments and a couple of more complete horse-shoes that were thought attributable to the 19th century, but few obvious traces of hoof marks which could not anyway be dated. In 1818 a visitor did comment on a horse and carriage (cart) being seen about the openings (Price 1984, 15), but this might have been for surface use. Their use remains debatable, and perhaps only ever occasional.

However, there is evidence that almost certainly by the late part of Allen's time of the introduction of cable haulage over short distances, either from the mouth of the level or from down shafts at others: in either case it is likely the surface crane was used as the hauling mechanism, in conjunction with a pulley at the entry or the base of the shaft to change the direction of pull into the cartway. Movement of materials underground was always more difficult than at surface but the methods used do seem to be an improvement on older ways, and there is evidence of modification to improve them. With the relative slowness in extracting and scappling stone blocks, and the short distances to surface, it is likely that the transport component in the list of underground tasks was small and could be managed by a small team either devoted only to transport, or by any or all members of the team in each work area, depending on the size of load.

Probably in the 1750s, Allen began quarrying from the surface. According to Wood (1765, 425) he did this to cut down on the number of accidents. A namesake of his, probably James Allen, an independent freemason (BRO.339/071), was killed in 1751 and in 1755 there were three more deaths in the area (Boyce 1967, 198; 213) so that it may have been about this time. It has been suggested by Boyce (1967, 213) that this comment indicates the adoption of the 'light shafts' for haulage, as Wood refers to it as 'quarrying from the top' rather than using the ever-lengthening cartway-levels. This certainly did take place, but it is probably not what Wood had in mind.

Wood in the same context also refers to the 'refuse of which was of such use for building that part was burnt for lime using coal from three or four miles away, harder beds were cut into slabs or used for road metal', so that he is clearly referring to the layers within the overlying Twinhoe Beds for which formerly no substantial market seems to have existed. The use of the 'refuse' was not entirely new – the hard beds just over the picking bed had been used in the foundations of Allen's Prior Park House, for example, but being able to use or sell the overburden would have enabled the exposure of the freestone below for working more cheaply than was possible by underground quarrying. It also avoided the loss of stone in pillars and would have facilitated improved handling. The 1759 Robins sketch illustrating the crane at Masons Crane House also shows the surface quarry there (Quarrybottom today) as well-developed, and it is to there perhaps that Wood was referring. The later widespread adoption of surface quarrying at Combe Down suggests that Allen's change was an important step in improving the economics of stone production, perhaps further reducing costs but also widening the market for additional quarry products.

Surface quarrying involved first the removal of the soil and subsoil which, would be placed aside for later use in restoration of the site, or barrowed directly on to worked-out, subsoil-filled areas as a continuous process but filled to a lower level than originally. It was probably a common way at the time for working minerals with only a shallow cover; the Twinhoe Beds would not have generally been difficult to remove, and useless material would be wheeled over stout planks to a worked-out section, the spoil bank separated from the working faces by little more than a wide trench or gullet, with quickly-erected rubble-stone walls holding back the waste. At a later date the spoil heaps were either returned to agriculture or built over, just behind the working area. Although it probably did not apply in Allen's time, the rehabilitation of old quarries must have been as useful to landlords as undisturbed land with underground working.

Lowering wages and costs

According to Wood (1765, 425) Allen embarked on 'reforming the domestic Masons' Trade in Bath and proposed to lower the Rates for all Manners of Workmanship'. He seems to have done this from the very beginning, believing it would not only be to his own and general benefit, but also to the benefit of the workmen by a '..saving to the Workmen of all the Time they lost in going between their habitations in or about the City and the quarries at Combe Down (and other sites), by finding them constant employ and paying them wages regularly each week'.

He achieved this by increasing the degree of specialisation, in which the blockstone underground at Combe Down was only extracted by the freemasons ('freemasons' at Combe Down became to refer only to the men at the quarries) but some of the smaller-scale operations carried out all aspects of the trade, thus allegedly, travelling and wasting time between operations. Banker masons worked at a permanent bench in the stoneyard to cut and carve the block into wrought stone, while 'rough masons' carried out the on-site erection. This was not necessarily entirely popular and even Wood was wary of separating the freemasons 'from being the same as carried out the erection of buildings'. They thus 'handled it carelessly, causing joints and moulding often to lack neatness and provision' (Boyce 1765, 136). Allen tackled this problem by his improved transportation system and by employing his own men at all stages (but in different specialisms) from quarry to building and, no doubt by rigorous management supervision. In this his stoneyards were the key feature, replacing work in the open air either at the quarry or the building site, with one at

Combe Down and another at Dolemead. He provided sheds (seen in illustrations) for the workmen, and the cost of workmanship by the piece and measure (piecework) was much reduced with payment by the week. Apart from the greater degree of supervision possible in the stoneyards to maintain quality and output, the sheds made work in hot or bad weather easier and undisrupted. By these means the cost of freestone work for local consumption was reduced by a tenth. The workmen probably accepted this system more easily during the building boom in Bath, which was under way in the late 1730s. Its success can be seen in Allen's ability to supply the demands of building at the Parades, which alone used up to 60-80 tons a day, while several other projects were building too, including his own Prior Park house.

Adequate supervision was the key element in almost every aspect of the changes Allen introduced. The number of well-built cottages at the townships at Combe Down and near Dolemead were far too small to house the general workmen, but would have accommodated some twenty or so key staff. The Combe Down houses seems to have been ready about 1730, those near Dolemead in 1734. Jones was the prime example of those he accommodated, and also of Allen's early recruitment methods. At least one of the Biggs was known to work there before Allen took over and one other of the family went to work for Allen as foremen. The houses may have been attractive inducements, as well perhaps, as promises of jobs or contracts on one of his building projects. Other masons may too have seen advantages in business specialisation, abandoning the less skilled (or less rewarded, once Allen had started) production of blockstone. There must have been resentment against change, though by recognising the inevitability such men probably gained substantially in the long run as their workmen and supply costs were reduced also. Possibly for those like Jones and the Biggs brothers, there were substantial gains in the short run too from more constant employment.

Increasing demand

The final option open to Allen to reduce his costs was to expand the market and gain by the economics of scale he had introduced, at the same time diminishing competition from those with less capital, business skills or no captured market. This could be both in the local market and one further afield. An important result from his cost reduction was the competition against alternative materials, notably brick and render methods of building. With coal only a few miles away, with fairly easy access by a short road to the Navigation, cheaper brick may have been a credible alternative to stone in Bath (and in Bristol), which would have had an incalculable impact on Bath's future. The successful alliance of Allen and Wood in persuading Bath to keep to high quality stone buildings was obviously much strengthened by the introduced economies in stone production and its building costs in the crucial early years of Bath's development.

The first documented act we have of Allen's involvement in the stone business was his and Wood's meeting with the Greenwich Hospital Commissioners in 1728. This and his more successful meeting with the Governors of the St Bartholomew's Hospital in 1730 illustrate his direct approach to winning large contracts in and outside Bath. The 1742-43 Bristol Corn Exchange was a later success. Establishing a large contract for a substantial period encouraged others in that area to follow the initial example and consider using Bath Stone. By establishing a flow of blockstone in regular shipments, the possibility of additional casual sales begins to grow. In London, as early as 1731 the sale of stone intended for the St Bartholomew's contract on the general market, which was generally desirable, became embarrassing to the performance of the contract and had to be temporarily forbidden for his London foreman, Richard Biggs (Boyce 1967, 44).

The St Bartholomew's Hospital was the first success, in 1730, in a quest for substantial stone business in London, which Allen clearly intended to be his major 'export' market. There he offered very generous terms which secured the contract for the first of the proposed four wings. Probably as a result of this, in some years some 1800 tons of blockstone was shipped annually from the Dolemead wharf (Jones, *Life*). As it transpired, the contract proved seriously unprofitable in itself, but it was a substantial success in introducing stone to the wider market, which was probably his main intention. It was a direct challenge to the much longer established Portland stone producers, who effectively controlled the London market for the highest quality stone, though it was probably not economic, due to the cost of transhipment at Bristol and the long coastwise journey around Lands End, for Allen to compete with direct loading from the quarry onboard at Portland and their much shorter journey. Successful competition was not established until the first London-Bath canal connection was opened about 1810. In the decades after 1840 and the Great Western Railway opening, Bath Stone – albeit mainly from the huge reserves at the Wiltshire quarries around Box and Corsham – probably became the dominant stone both in Bath and elsewhere, capitalising on the fairly similar Combe Down stone's reputation in Bath.

Whatever Allen's intentions, the Bath market must have been substantially larger for him, taking the poorer grade stone too for inner walls and backs, than more distant markets. This was because of Wood and other architects' grand projects there only in stone, compared with fierce competition from lower-transport-cost local producers in markets elsewhere. All sources acknowledge that Allen almost immediately dominated the provision of stone in Bath, though other producers from Combe Down can be identified in the mid century.

Among these was James Allen, in 1736 leasing a quarry in Combe Field (south-east of Combe Down (BRO 339/071), Thomas Hughes in 1747 at the Little Tyning (BRO. 339/072) and Thomas Greenway possibly continuing quarrying east of the Long Drung where the archaeology suggests the style of work was by someone other than Allen's men, possibly in Quarries 2339 or 2351. There were also complaints about road damage which suggest that the Odd Down producers were still reasonably active. It is possible these survived as when it came to work of specifically high quality, the cost of stone was less important and the level of production would likely not be high. Thomas Greenway in particular was well known as an ornamental sculptor, with the Garrick Head, Nash's famous house, perhaps the best example of his work.

Despite winning the St Bartholomew's contract, the defeat and reflections on the Combe Down stone's properties at the hands of the Portland stone supporters at Greenwich particularly irked both Allen and Wood. According to Wood this was part of the rationale behind Allen's decision to proceed with his house at Prior Park, to exhibit it 'in much greater variety of uses than it had ever appeared in any other structure' (1765, 427). Allen's ideas perhaps took ornamentation to excess, at least in Wood's view. He comments that 'the warmth of this resolution at last abating, an humble simplicity took place', the latter most probably to Wood's considerable relief, and with the principles of design thus decided, work started on building the great house in about 1735 '...for to see all Bath and for all Bath to see'.

Situated high up the Widcombe valley just below the Combe Down escarpment, the house, with a central colonnaded mansion and two long curving wings or pavilions, was magnificent and attracted enormous contemporary attention (see Figure 3.6). Wood wrote that the wings housed the doves and tamer poultry and horses in scarcely less magnificence than the mansion itself (1765, 429-31). It was probably the most spectacular site in Bath and permitted Allen to entertain some of the highest political and aristocratic figures in the land. Whether it was the social prestige or as an understated form of advertising which appealed most to Allen, it certainly had successes on both fronts, the Earl of Leicester, for instance, succumbing to the golden hue of the stone with an order for his Holkham house in Norfolk.

Locally his closeness to John Wood the architect, who had a five years agreement with Allen for building (Peach 1895b, 101) was probably as important to Allen as the lower prices and speed of delivery that he was able to provide. Nevertheless he was able to considerably increase his outputs by building other properties on his own account on his substantial and growing urban and rural estate and as a speculation in conjunction with other builders of projects such as the Grand Parade. He seems to have used Richard Jones as his foreman mason. In the late 1730s, while providing stone for the Parades, he had Jones also building the large brewery and malthouse in Widcombe at Dolemead (see Figure 3.7), as well as the great house at Prior Park. There were many other small buildings built for Allen and others and the flow of great buildings within Bath associated with Wood and his son would have been difficult for any other quarries to provide in competition with him. While his postal monopoly was by government fiat, it seems likely that there was a virtual monopoly in stone in Bath which was Allen's own intention and accomplishment.

Financial returns

Allen's annual income from the postal business was substantial. Jones (in his *Life*) claimed to have seen 'a stated account' in about 1738, showing it as £16,000 clear of all deductions, while Gregory (1886, 10) suggested it was about £10,000 annually, the two figures perhaps illustrating its range. From this, let alone other sources, Allen was clearly able to provide cash flow, which, on a temporary basis with leverage from mortgages on his property and other forms of loans, could obviously sustain huge outgoings. One insinuation was that Allen used the mines as a plausible pretext to conceal the exorbitant gains from his postal contracts (Thickness's 1788 comments, cited by Peach 1895b, 98). Even if this was untrue, the huge postal income and lack of surviving accounts makes financial analysis of the stone business difficult.

After 1730 the business must have gained enormous impetus. In 1731 Allen began supplying St Bartholomew's Hospital in London and Jones suggests some 1800 tons were shipped annually for such export contracts. He may have been able to supply Wood's Queen's Square, begun in 1728, first from his Bathampton Quarry and later Combe Down. By the late 1730s Jones was involved with the Dolemead brewery buildings as well as numerous houses at Combe Down and Dolemead. From about 1737 to 1742 Allen also supplied stone for the rebuilding of the Mineral Water Hospital (St Johns Hospital, John Wood's first major project) as a gift to the value of £1000, probably about 2750 tons at the prevailing price for blockstone. From about 1737 the building of Prior Park may have consumed some 30,000 tons of stone (Peach 1895b, 104). In 1739 building of the Parades began and there must also have been many other buildings. Neale's (1981) imperfect figures for the total houses in Bath suggest 500 houses built in the 35 years of Allen's control of his stone business. This was about 15 a year on average but this would vary as would their size, and does not indicate public or other large buildings, or those outside the City as in Widcombe which included Prior Park, which may have dominated demand at times.

This early phase of quarrying and building was probably extensive and innovative enough to be termed 'heroic', with buildings involving John

Wood the Elder (1706-64) as architect (and developer). Its scale may be reflected in the Avon tolls (Neale 1981, 385-6) charged on timber and other building materials brought in. These were at a fairly high level in 1730-33, falling slightly then rising 1734-37 with a new peak forming about 1742-43. After a decade of poorer years they rose again in the 1750s, increasing generally until well after Allen's death, the tolls being twice as high then as in the prior poorer decade and treble that of the early years. The rise in the 1750s and 1760s might be though of a second heroic phase involving the elder Wood 's son, John Wood the Younger (1728-82) and such key developments as his Kings Circus (1764-67) and the Royal Crescent (1767-74).

What these changes in economic activity meant for the quarrying business is hard to quantify, but some indicators are possible based on historical information. At the Parades from around 1739/40 to 1745, Wood tells us that some 500 feet of building were completed in about six months ready for the roofers. This suggests it was the south of the two piles which made up the Parade buildings in Bath Gardens, next to the (later) Grand Parade Bridge. The south building has three sides faced with ashlar etc. with the fourth, facing the river, of coursed rubble. The central part of the buildings face into a rectangular garden area so the inside faces are also rubble. The uniformity of style and repetition of elements in Wood's designs were ideally suited to the larger-scale specialisation provided in Allen's stone yards and possibly only his scale of operations was capable of coping with such a demand in so short a time. Very approximate calculations suggest that this first part of The Parade would have needed 8860 tons of stone of all types. This included ashlar, dressed mouldings and rubble stone (walling stone) for internal walls and backs, together with mortar. Jones description of about sixteen carriages a day (about 9600 tons in 150 days) being taken over the water roughly corresponds with this. Both accounts of quantities seem to suggest a very large building project could use up to 12,000 tons of stone a year, though substantially less was probably averaged over a few years of building.

With two or three large buildings and other lesser buildings often progressing, and with possible export shipments, peak demands of up to 20,000 tons a year were clearly possible at the best of times. The most likely time for this peak was when Prior Park House was under construction in roughly 1737-1745 (though the mansion was complete in 1742). However, it may not be accidental that the building period of Prior Park House coincided with the lows in Navigation Tolls reported by Neale (1981), so that Allen may have taken advantage of reduced demand elsewhere. Undoubtedly the railway, even at these high levels of output, was still far from being fully utilised and the option of additional capacity using horse and cart, especially for sites south of the river and away from Dolmead was always available and possibly preferable at times.

The seasonal and cyclical nature of the building industry often meant a level of production could be much less than its peak. If a low figure was 5000 tons when building was very slack, then some 12,500 tons per year overall may reflect an average level of production. This under-capacity would permit a degree of stockholding at all stages of production (quarry, entrance area, stoneyard, building site) to relieve strains at other times in what was notoriously a cyclical and weather-affected industry. Building up stocks might also offer some winter work underground, which could employ surface workers displaced by the weather, allowing Allen to achieve his aim of regular employment and keep his workforce intact and modestly paid.

The important output from the quarries was blockstone and it is likely it was the only material exported down river, certainly if it was to be transhipped and sent onwards from Bristol. Since we cannot determine relative outputs at the Combe Down and Dolemead stoneyards, it is uncertain what this meant in terms of the mix of block and wrought stone transported down the railway. In a modern mason's yard some 20% of large block is discarded in producing wrought (varying with the product), so perhaps 10% less in total weight transported was the result of dressing some stone at Combe Down. On a half-and-half basis of wrought and block, on the calculations which follow, some 618 tons would fall into these two categories needed and transported for the building of the southern pile of the Parade.

Since the Bath buildings used mortared rubble for non-visible portions, including the 'backs', it seems likely that much material of this type was supplied by the same quarries and sent, roughly dressed and crushed, for building. It would amount to some 8200 tons for the six months work on the Parade building (sides based on 0.45 m-thick walls five storeys high). Such stone resulted from splitting of bed-blocks to produce blockstone or from stone in the many thinner freestone beds not suitable for block, and would require only labourers to produce. It was a huge tonnage and, though no mention appears to have been made of it in contemporary sources, it was likely to have been despatched by the railway as by far the lower marginal cost form of transport. The carriages were equipped with removable sides so they were suitable for carrying rubble stone, which must have been by far the largest category carried. The lower of the two carriages shown in Figure 3.6 appears to carry block or ashlar, the second seems to have rubble. The cost of such roughly dressed stone was presumably markedly less than for blocks, but not necessarily less profitable since wages and prices had been squeezed by Allen and its extraction was a necessity so as to get at the better material so otherwise it would have gone to waste.

At Prior Park, Wood ensured ashlar was used inside and outside on all external faces 'to prevent buckling', with the implication that this was the case for at least some interior walls too, which remained unplastered for a time. (In buildings built at the end of the century partition walls were often very thin as ashlar had become more easily available and was used single-thickness (Peter Brown, pers. comm.). Thus the proportion of ashlar to rubble in walls needs calculation for the particular type as well as date of building. In calculating output and the area of quarry involved, this also makes it more difficult to approximate what proportions of rubble and block were produced at the quarry and what this meant in terms of overall stone produced in the quarry by the freemason. Currently this must remain unresolved and for the following estimates the two are regarded as the same.

About half the full excavated space in the underground quarries was left open, but due to bulking of broken stone, this implies some 60% of the excavated material was brought out, of which possibly half was blockstone (rubble was not exported and other suppliers probably produced walling stone too). To summarise the calculations made for the estimating purposes, every 1000 tons of useful stone brought to surface required about 7.4 cubic perches. There are 160 (square) perches of $5^1/_2$ yards (almost exactly 5 m square to the perch) to the acre and, as the average quarrying height was also equivalent to a linear perch, the postulated possible average output of 12,500 tons, had it all been worked underground, would require, allowing 10% space left for pillars, some 0.64 acres per year, equivalent to a total of about 21 acres over Allen's life. This is too high for the available areas underground.

The average of 12,500 tons may be too high and not all the stone was worked underground, but we do not know the full extent of his surface quarries at the date of his death. The Masons Crane House Quarry was certainly fairly extensive in 1759 judging by the Robins sketch (see Fig. 3.11), but whether the surface quarry behind the Old Rank was working, or the probable surface quarry down Rock Lane, is unknown. It is unlikely that underground provided less than half this level output, even allowing for less losses in surface working. Thus over Allen's Phase II period of working something like a minimum of 10.5 acres were possibly worked underground in Firs and Byfield, or, in metric terms, about 4 hectares out of the 18 hectares.

In daily terms, equalised over a working year of 300 days, at the higher level this would be between about 16.6 to 66.6 tons daily, divided between the possible five sets of entries and surface cranes, or later four underground sites and one surface quarry. Peak daily production, because of seasonal and building-specific demands, would undoubtedly be much higher than this (f only to supply the Parade in the six months described), though a substantial part of the extra could be met through stocking stone underground and alongside the surface crane in anticipation of demand. This would be easy and possibly better for the product as it gave time for drying-out (even if not necessary, as was claimed) and would allow production to remain constant through the winter, again helping achieve constant employment for his workmen. This does not appear an unreasonable level of output given the number of cartways and their working places and the straightforward movement of stone to the surface via Allen's cartways.

The profit and loss account – a reconsideration

There can be little doubt that Allen set up an efficient and effective business with quarrying at its core and that he was technologically and financially responsive to the almost visionary incentive of the building boom in Bath. However, Wood (1765, 426) in what amounted to a eulogy to Allen less than year after his death had this to say. 'His designs in the stone business have from time to time been very much defeated partly by his troubles and the disappointments inseparable from trade, partly from wanting freight for exporting blockstone, and partly by the chief master masons of the County reducing their prices still lower than he himself had abated them to obtain work and thereby keep as much of the domestic trade in their own hands as possible.'

It is worth examining this informed but sympathetic judgement further. We have already noted the sixpence per ton the master masons reduced their charges below the 7s.6d per ton charge at the wharf, which Allen proposed. Using the ranges of output used above it is possible to project what would have been Allen's revenues and possible profits had all gone to a reasonable plan at this lower price level. The basic information is that a ton of blockstone, defined as 20 cubic feet, was delivered at the stoneyard and Navigation side for 7 shillings but that Allen had initially proposed 7s.6d, a reduction of 2s.6d on previous prices, and that Allen halved the cost of transport. This is complicated by some output being roughly dressed rubble, but for comparative purposes here all the stone is valued at 7 shillings a ton, although this probably leads to slightly overstating of profit. We do not have carters' regular prices around Bath and although Pollard (pers. comm.) has supplied details of a range of contemporary haulage prices, which appear to vary wildly, a regular trade would probably produce a regular price somewhere around a shilling per ton-mile on good ground in summer and 1s.6d to 2 shillings a ton-mile on poor ground in winter. These suggest that 1s.6d a ton-mile (commonly cited in mid-18th century documents) would be a reasonable figure to use in aggregate calculation, since winter tonnages were almost certainly much less than summer. Thus a carter's charge would be about 2s.3d a ton from Combe Down to Dolemead, which is reasonably in

Table 5.1 The profitability of Allen's stone business

Production/sales	*5000 tons*	*12,000 tons*	*20,000 tons*
Revenue			
7 shillings (84d) per ton	£1,750	£4,375	£7,000
Costs			
Transport 13.5d per ton	£282	£703	£1,125
Labour etc 56.4d per ton	£1,175	£2,939	£4,700
Gross operating profit	£293	£733	£1,175
Capital opportunity cost	£995	£995	£995
Net profit	£-702	£-262	£180

line with Allen's proposed reduction in price he thought necessary to capture the trade. The extra sixpence by which his competitors cut the price would thus be very irksome.

According to Jones, the quarrying enterprise cost Allen some £10,000 to set up, though exactly what this expense included is uncertain. If it included property, for example, this would have a residual value, whereas worked-out quarries and cranes and railway might have had relatively little. A very crude calculation or business plan he might have made to see what kind of return he could make out of his proposed business (supposing he foresee the reduction in the full cost per ton) can be postulated.

The opportunity cost of the £10,000 capital of an investment is shown as a $3^1/_2$% compound return – available from, for instance, the Bath Turnpike Trust which raised capital at this level in 1757 though it rose subsequently (Buchanan 1986, 238). This is shown in equal payments of £995 per year over 20 years, like a modern house mortgage repayment scheme. Transport is shown at $13^1/_2$ (old) pence for the 1.5 miles per ton (half the probable cost of his competitors), and the operating costs of labour and materials is calculated after 20% is taken off the seven shillings (less transport cost) per ton price at the wharf, leavıng his operating margin.

The impact of that sixpence is that had Allen's planned reduction to 7s.6d held, then the sixpence would have meant his business would have broken even (using net profitability) on the above figures, at about 10,900 tons output (there would be a £40 profit at 12,500 tons). After the break-even point, any economies he was able to make, or any higher tonnage, would have yielded real profit. But, even if these projected figures are substantially wrong because of the original capital spend or by different costs and revenues when surface working was introduced, they do show that without large export markets the stone quarrying aspect of his business was probably always doomed to be at best marginally lucrative in real terms.

This, of course, is based on hindsight and informed guesswork, but it is accurate enough to see the impact of the capital expenditure on the enterprise. For double the transport charge, ie double the unit cost of the railway and cranes as being the alternative cost of using carts, he could have avoided most of the capital charge by loading on to a hired cart at the quarry. At 5000 tons output he would have been about £700 better off; at 12,500 tons, possibly an average production in a year, he would still be nearly £300 better off. In fact the railway, cranes and other technical aids do not appear financially worthwhile until output is around 17,000 tons per year. This seems a very high level of activity, probably infrequently reached.

However, the £10,000 was sunk capital and was likely not all irretrievably lost. Allen gained an operating profit at virtually any reasonable level of production, so it was worth continuing in business so long as he did not let his actual quarrying costs rise above those of his competitors who, unlike Allen, might also have to find rents. There were also other advantages. Stone for his own building could be obtained at cost on his own property and he might have only taken it when other demand had first been met. The same would apply when he gave away stone, for instance to the Mineral Water Hospital, or effectively for the St Bartholomew's contract, and even for the present of stone for the Mansion House in Truro made to his niece Elizabeth's new husband 'Guinea a minute' Thomas Daniels (Jennings, P. ND) when they married. It lessened the cost of his well-known benevolence.

It also allowed him to produce and market stone in quantities which the use of carts might not have effectively allowed, at least not without probably raising ton-per-mile charges and creating objections about the roads and congestion in the city. It disadvantaged his competitors downstream in the stone-yard and construction aspects of his overall business, almost to the point that he was effectively a monopolist, though this might be considered, in his case, to have been of direct advantage to Bath. After twenty years he could consider the capital burden as lifted but, presumably, repair costs had risen by then. The major problem would arise if very major repairs were required to the system.

Export markets

Allen's problems with shipping stone beyond Bath, which he considered to be his most desired market, began in 1730 when he won the contract to supply stone and supervising staff and masons to build the four wings of St Bartholomew's Hospital in London (Boyce 1967, 43). In 1730 William Biggs at Dolemeads began assembling a small shipment of stone to send to his brother Richard in charge in London.

It was blockstone rather than wrought stone and there were (unidentified) problems in assuring the supply. Allen had to order Biggs not to make sales to anyone else. By September, the best time for shipping, there was still not enough vessels available and by the following January work had to stop (Boyce 1967, 44). The difficulties were again overcome and prospects must have seemed bright when the Prince of Orange visited the quarries. This might have been connected to a purchase soon after of stone by the Earl of Leicester for Holkham House in Norfolk. But again in May 1734 he was again having trouble securing a ship for his stone to London (Boyce 1967, 54-55).

Matters went from bad to worse with the third wing at St Bartholomew's in 1744 when there was war with France, and Allen had to break his contract. Explaining to the hospital later that though it had obtained a claim for protecting the seamen from being pressed into the Royal Navy and had made use of their influence, the expense in first waiting for a convoy from the Downs to Portsmouth, staying there for a second to Plymouth, for a third from that port to Scilly, and a fourth from there to Milfordhaven, and the men 'running from the vessel' to privateers (vessels licensed by the Government to capture French ships – with large 'prizes' as the reward for success) at Bristol, the ship could now hardly make one voyage in a year, having at other times made three. He recommended that another stone be used for the contract and paid £200 compensation for the third wing and £400 for the fourth. He went on to say that Captain Wade would sail on to London and sell what stone he had elsewhere. However, two months later the ship was still in Bristol and the new crew had left for privateering for want of a convoy.

Originally this contract had been for £1700. Allen claimed he had lost £2500 by it. In 1757, his friend William Pitt was all for prosecuting the Seven Years War with France. Allen's feelings may have been more mixed. He was probably more successful with trade via the Irish sea, to Liverpool and Ireland, notably Belfast, Dublin and probably Londonderry but, for the London market, it was nearly 50 years before the inland link was provided, via the Kennet and Avon Canal, which was so vital in wartime. Though he had provided the essential local links, the inefficiencies of sea trade and the long journey around Lands End placed him at a cost disadvantage to the direct loading from quarry on board ships at Portland and also to the quarries in south-east England which also supplied London with its stone. The time for distant markets had not yet come.

It is clear from his text that Wood saw Milo Smith as the difficult master mason affecting Allen's business. Wood wrote (1767, 426) that Milo Smith, notwithstanding all the means taken to reduce the price of stone and workmanship'...soon became a competitor with Mr Allen . . . tho' he did not long continue so, and his works dropping, the quarries at last became the property of the person he opposed'. This is not wholly unbiased; his phrasing seems to suggest Milo Smith came to the Down after, rather than before Allen started work there. This does not appear to be so and it was Milo Smith, using mortgages, who purchased most of Monkton Combe Green Down, including, almost certainly, that with the quarries (or at least a moiety of them). It is certain he, as a shareholder too, saw the potential benefit of the Navigation (in which he also had a share) and he also seems to have had a wharf at Little Dolemead. Perhaps, from concerns over the decline of the older quarries in which he was involved at the family property near Barrick Farm (Odd Down today) it was he who first foresaw that Combe Down was in a prime position to replace them.

Milo Smith may have, like Allen, burdened himself with debt (without having Allen's deep pockets from other activities) and at the lower price of 7 shillings a ton in Bath would clearly also have felt the pinch. He may also have been distracted by his aldermanic duties which had him as mayor in 1732, and it is possible there was an underlying health problem, with his death in 1735. In any event, by 1736 his property, including the moiety in the quarries, had passed to Allen and by 1737 his other (mortgaged) assets followed. Odd Down probably still had fairly substantial quarrying potential and there was a major revival later in the century (perhaps connected with Allen's death) but in Allen's time the competition from there and other minor quarries was probably very muted.

Allen's last years and the aftermath

Allen died in 1764. If the *c* 1764 estate map is any guide, then either he, or Jones on his behalf, had either concentrated production of stone into two areas or the level of production had severely declined. As evidence against decline Neale's (1981) statistics suggest the 1760s were fairly good years for building in Bath, but the map shows only two cranes (Sheeps House and the Masons Crane House) with no indication of activity elsewhere and a definite indication, in absence of the crane and an area without trees, that former activity at the Rock Hall Lane site had ceased. There is no indication of quarrying behind the Old Rank, east of The Avenue, though there is plenty of evidence underground of earlier exploitation contemporary and comparable with the other sites. However, from considerations above, it is clear that the railway was substantially under-utilised, and that as the access levels got further under the Down, then workplaces would multiply. Concentration of the workforce and use of surface quarrying might well maintain and even increase both efficiencies of labour, and equipment such as cranes, for the same or even greater output. It might also of course reflect cannibalisation of railway and cranes as their condition deteriorated.

It is very likely that Allen was considering alternative means of transport. The early advantage of Odd Down (Barrick Farm), had been its transport links with Bath, with the Great Pitts, and the Upper and Lower Quarr Ground belonging to Milo Smith and an area once known as the Pitts but known today as the Tumps, belonging to Magdalen Hospital (SRO. DD/SAS/212). From the Pitts it is about a mile to Bath, down the Fosse and Holloway and over Bath Bridge. Allen's railway overcame the disadvantaged location of his Combe Down quarries away from a principle road transport link, and, as the building of the Parades shows, was particularly advantageous in building operations for the link to his Dolemead masons yard near the river. By the 1760s however, the direct and shortest route to building sites in the town began to be diverted away from the riverside, to sites uphill. These included the King's Circus and Royal Crescent ,which resulted from the younger Wood's architectural skills and promotional abilities, and were key sites in the building boom of 1762-71 (Neale 1981, 168). To these sites, stone dressed at the quarry could be brought from the quarries directly by cart and delivered over Bath Bridge to the sites without the need for transhipment. This was probably at least as convenient from the Entry Hill direction as from Combe Down on the railway.

The Entry Hill area (sited between modern Combe Down and Odd Down) was especially favourable, as the road downhill from there could serve the Upper and Lower Quarr ground. Certainly two substantial quarries adjacent to the hill on either side of the road were well-established by the end of the 18th century (Charlton Map of Lyncombe and Widecombe, 1799; Keevil 2000, 27-31) and may have begun operations around the time of Allen's death (a 1782 advertisement refers to building stone quarries there; Keevil 2000, 40). Revival of quarrying in Odd Down was probably due to road improvement by the Bath Turnpike Trust which underwent a major re-organisation in 1759. They had Parliamentary consent for a substantial capital programme (Buchanan 1986, 236 *et seq.*), and their activities in the late 1750s and 1760s concentrated improvements to the road up Rush Hill (from Bristol) to Odd Down and Combe Down to link to the old Wells Road and Entry Hill there.

Allen had a clear interest in those roads which were within a mile of the Combe Down quarries and his house. They were already linked by the old Claverton Road (Bradford Road and North Road today) across the Down, though clearly this was not in a satisfactory condition. Seizing an opportunity when the Trust wished to raise capital, he offered to invest £1000 with them, but on condition £700 of this was spent on improving the road over the Down from Entry Hill eastwards. That this was particularly important to him is indicated by his attendance at the relevant meeting of the Trust, on 7th of June 1759, where he took the chair, presumably having the most senior status of those present.

His attendance was certainly unusual and this seems to have been the sole occasion at about this time. It seems likely several others were there only because they supported his interest, including his brother Phillip, John Chapman (who became his executor later) and another mason with a Combe Down interest, Thomas Greenway. The decision on his offer was adjourned until the next meeting, and on the 27th July his proposal was agreed. However, at the 17th of August meeting the proposal was suspended and it was agreed to concentrate efforts on the new road to Rush Hill. In October a plan was proposed for easing Entry Hill. The following year petitioning took place for an Act to bring the road from Bath up Rush Hill which was allowed the following year, 1761. In March 1761 Allen renewed his offer of £1000 with the £700 petition but this was postponed until the road from Rush Hill to Crossland at the top of Entry Hill was completed (SRO/30/T/ba 6).

Given Allen's importance in Bath – he was satirised around this time (Peach 1895b, 180-81) as the 'One-headed Corporation' – it was a serious rebuff on the first occasion and that he repeated his offer after this is an indication probably, of its urgency for him. The most likely reason for this was his quarrying business and the state of his railway, the latter probably deteriorating badly by this date, and the threat posed by improvements at Entry Hill and the Old Wellsway facilitating direct access for actual and potential competitors to the main new city building sites. No-one would have been better able to foresee the huge demand for building stone that Wood the Younger's imminent major projects would provide. Allen might have taken the same course as a neighbouring entrepreneur, Jacob Mogg, who linked his colliery by a private road development to Rush Hill (Buchanan 1986, 232), but Allen had unhappy experience of the costs (in his own railway) that such private developments could incur. In fact the Bradford Road Turnpike was not completed until some five years after his death.

We have very little information about Allen's business including his quarries in the period between the mid 1740s and his death. Presumably, despite rather more difficult times, the work simply went forward under the capable attention of Richard Jones. Allen was, however more cautious when St Bartholomew Hospital proposed starting the fourth wing about 1756. He wrote that he had 'not the least expectation of being able to send any stone from Bristol to London during the present (Seven Years) War', but if they wished he would put aside the stone, although the responsibility for shipment would be theirs. They decided to wait. By 1763 and the peace, he and they relented and shipments resumed. Because of Allen's benevolence the original price he had given, though long unprofitable, was maintained (Boyce 1967, 235; 275). Less agreeably he also agreed to carry out repairs to stone damaged by the London smoke; possibly it was not so suitable as Portland after all! The wing was virtually completed just before his death.

Aged 70 by this time, Allen was clearly aware of impending death and ordered Jones to draw and construct his burial vault and memorial but, perhaps misjudging the imminence of the event, died before it was constructed. He more importantly misjudged his financial position at his death in 1764, probably because of the ending of his postal contract which took place almost immediately. His legacy was enormous, the estate by then some 3000 acres in extent, bringing in £5000 annually. His will (Peach 1895b, 226-241) took great care of his extended family and included dozens of friends and servants. Unfortunately the end of the postal contract meant that the huge cash flow it created ended very quickly and Philip, his brother, who attended to the day-to-day aspects of this business for him, thereafter effectively managed only the Bath Post Office. This left insufficient revenue to easily pay the legacies due in the next two years. The estate was left in trust, and an income and life interest in the house was left to his wife Elizabeth. She died two years later, and the whole of the estate passed to his niece Gertrude, but was still inadequate to cover the liabilities.

Gertrude Warburton, by then the wife of the Bishop of Gloucester, soon found herself faced with outlays from her expenses growing faster than the residual income after accounting for legacies and such like. She reduced the size of the household, and in 1769 sold off the moveable items in the house and shut it up, though she re-opened it some years later. So far as the quarries were concerned, she had little interest in them, and though the operating cash flow probably remained positive it was fairly insignificant and the cranes and railway system may have needed serious expenditure. These however, were becoming less important as the large buildings in progress were at least equally well served by horse and cart via improved roads and Bath Bridge. She no doubt took advice, and as one who, 'did not like being overlooked by anybody', and who had, to Jones horror, demolished a good house which had done just that, dismissed the main opposition, Richard Jones, and closed the railway, though only after he had sold off the cranes and other materials at both the quarries and stone yard. Probably by the end of the same year as Allen's death, she let the quarries out on a rent and royalty basis. Jones and his son became Gertrude's first quarry tenants. The sale of crane parts, chains and ironwork by Jones on her behalf fetched just over a hundred pounds, with a similar sum raised at auction on the cranes and other equipment.

Conclusions

In general these are the type of changes in scale, organisation, equipment and management more commonly found in industries (such as textiles) at the time of the industrial revolution, combining a traditional squirarchical paternalism with a ruthless drive for efficiency. Allen was an early exponent of this and developed the first industrialised and integrated quarrying and building organisation.

That his system was successful is probably beyond doubt, but as with so many innovatory schemes, the financial return was less dramatic. As a Post Office contractor he was financially wildly successful, but in quarrying, the cynical Philip Thickness, who had dined at his table but was no friend to Allen, though cognisant of his abilities, succinctly characterised his relative achievements 'While he was supposed to be gaining a Princely Fortune by Digging Stones from the Bowells of the Earth, he actually picked it off the surface by traversing the whole Kingdom with his Post-Horses.' (Peach 1895b, 98)

After Allen's death (see below) the system, except for some cranes, was abandoned and transport reverted to horse and cart. Retrogade as this may appear to have been, just as today, a freight railway functions best in continuous point to point delivery and where trans-shipment is necessary, this is a serious disadvantage, especially over short distances. Simpler systems such as horse and cart have greater versatility in their destinations and, using the road as a common facility, do not have to assume the total capital cost, if any. Despite its efficiency, it was undoubtedly the high capital cost of the railway-based system which was so damaging to the quarrying aspect of his financial success, though his overall position was much underpinned by extraordinary profits from the mails and latterly by his estate income.

Milo Smith may not have been the only quarrying freemason to find his business and profits lost or diminished, though the building boom may have ameliorated this for others. Clearly only a few of the more skilled workmen could be accommodated in the houses Allen provided and the reductions in wages necessitated by the price reduction he forced through probably fell mainly on carters (typically local farmers) and on the unskilled or semi-skilled workmen. Against this the lower prices, by over 25% for stone and 10% for skilled labour, and the efficiencies in building his business fostered were among the factors which allowed Bath to grow quickly and gave more employment. With a coalfield just a few miles away, Bath might otherwise have been built of brick or brick and stucco, and been just another pleasant market town. Instead his achievement and his support for the Woods and their achievements helped accommodate and amuse, in some splendour, the wealthy visitors who flocked to the 'Valley of Pleasure'. The Woods, at least had no doubt about the benefits that Allen had wrought.

Chapter 6
Underground quarrying after Ralph Allen

After Allen's death, the documentary and literary supporting material to complement the archaeological investigations is much less, and it has thus been gathered under a single chapter. The data is especially weak for the remaining part of the 18th century, though it has been possible to describe the reasons for the break-up of the estate which do not appear otherwise adequately described, and there is a brief glimpse of the quarrymasters and their production for some three years at the end of the century.

In the 19th century, the underground quarries, like their surface counterparts described in Chapter 4, entered into their major phase of output. The fortunate survival of a series of quarry leases dating to the beginning of the century allow the ownership and working areas to be projected forwards and, to a certain extent, backwards in time. This illuminates, in conjunction with archaeological findings, the problem of chronologically phasing the underground workings. The limited data on late 19th- and early 20th-century underground working is also summarised.

Quarrying under the Allen Estate 1764 to *c* 1805

This chapter follows on from the previous one , outlining the business of underground quarrying and the areas affected. There is emphasis, as the documents allow, on leasing arrangements under which most quarrying was carried out thereafter. The exception to this was for small parts of the far eastern part of the underground quarries where owners of the land above seem sometimes to have exploited the freestone below their own small bounds.

Allen's will and the subsequent owners of Combe Down

The new ownership of the land and quarries underneath, and the eventual break-up of the Allen Estate at Combe Down were the most significant factors in the continuing working of the quarries and the changes which began to affect the Combe Down settlement as the older workings near the southern escarpment began to be worked out. The timing of change was markedly affected by the only partially anticipated effects of Allen's will, the relevant parts of which, as a result are considered first.

Under Ralph Allen's complex will (see Peach 1895b Appendix 1 for the full text of the will), Prior Park and the House, and other property were all placed in a trust controlled by John Chapman, Alderman, and James Sparrow, Clerk, both of Bath in order 'to support and preserve the . . . estates . . . from being defeated or destroyed' . The main legatee was Gertrude Warburton, but she was not to be fully in control, that is – able to make disposals of the estate other than those specified, during the life of Mary Allen. Mary was Allen's niece, the only daughter of Allen's brother Philip and Jane, née Bennet. She married Sir Cornwallis Maude in 1767, who later was to become the first Earl de Montalt and the first Lord Warburton. She was the residual legatee should Gertrude (and others named) die with no issue. Gertrude Warburton, née Tucker and her younger brother William had been orphaned children of Allen's sister Elizabeth of St Blazey in Cornwall. The two children came to live with Allen when Gertrude was eight years old (Boyce 1967, 62).

Allen's widow, Elizabeth née Holder, was entitled to enjoy the use of the Park House and other property for her life and £1300 of the rent and profits arising from freehold property as an annuity. There were also many listed payments to legatees to be paid out of the freehold estate. Sufficient could be sold or mortgaged within six months after his wife's death to raise a capital sum to compensate for loss of a marriage annuity for his brother Philip. After this the rents and profits passed for life to Gertrude Warburton. All rents and profits from leaseholds belonging to the estate were to go to Gertrude during her lifetime. Heirs after Gertrude's death were first her own children (her only child pre-deceased her), and, after a number of eventualities of which none applied because of deaths, would finally pass to Mary Allen if she was alive. Only at the end of Mary's life could the capital assets of the estate be sold. In fact Gertrude did not have that long to wait, as Elizabeth died in 1766 and Mary died in 1775 (www.thepeerage.com).

The Trustees had severe problems in supporting the legacies and annuity commitments of the estate, and it was only after the death of Allen's widow that opportunity to partly remedy the situation arose by allowing sale of personal assets. In the meantime Gertrude and the Trustees were clearly able to determine that the estate income, at least halved by the loss of the postal business, was unable to support Gertrude in the manner she hoped to be accustomed after liabilities were paid.

It was probably the problems caused by the will that prompted Gertrude and the Trustees to rationalise the quarrying business. They quickly ended anything of the building and blockstone export businesses, though St Bartholomew's Hospital was not finally completed until 1766. Chattels were not part of the real estate so a sale was being arranged as early as September 1764 and items from the railway and quarries which presumably belonged to Gertrude, were either sold on her behalf by Jones or had been sold by December at auction. Dealing with a now over-large household could be done quickly and after performing his last duty, the erection of the mausoleum, Richard Jones was dismissed, along with many other staff which Jones said numbered about a hundred.

In 1769 Gertrude went much further and took the drastic action (Jones thought wicked as contrary to her promise to Allen) of selling all the internal furnishings of Prior Park House which was then let, a year later, to Lord Kerry. In 1779 her husband Bishop Warburton died. She married again, about 1780, to Warburton's former curate (Boyce 1967, 275-58). This and the death of Mary probably encouraged her to move back to Prior Park since she could, through Chapman, then make modest disposals of property. Items began to be sold to Sir Cornwallis Maude from 1781. The major sale affecting Combe Down was in May 1788, when the Allen Estate Trust transferred much of the Combe Down area, including the quarries, land across as far as Entry Hill to the west and Kingham Field to the east, to the Trustees of Lord de Montalt, formerly Cornwallis Maude whom Mary Allen (by then deceased) had married. This portion was immediately mortgaged, to Henry Cecil for £9831. The mortgage was renewed for a third time in 1799 for a somewhat lesser sum. De Montalt, created Lord Hawarden in 1793, died in 1803 when he was succeeded by his son, who himself died in 1807. In 1808 his trustees offered the whole of the residual estate to John Thomas, a Bristol Quaker, for £28,000 which he accepted. By then however many of the properties, notably those including the old quarrying area and quarries, had already been broken up.

Developments in the wider area

The Turnpike Trust completed its roads around Entry Hill (entering Bath via the Holloway and Bath Bridge) and by about 1769 the Bradford Turnpike (along present day North Road) linked Combe Down with Entry Hill, which was winding and steep but passable for loaded carts. The Prior Park Carriageway was closed to stone traffic. The Blind Lane and Foxhill tracks, descending from Combe Down were precipitous and not suited to heavy traffic even had they been improved. East of Combe Down the new stretch of turnpike turned down Brassknocker Hill, but from the top of the road there, horse and carts could branch-off and approach Bath via Claverton or Bathampton. Improved roads and the ending of the wooden railway meant much greater competition was possible and quarrying gradually expanded in the wider area of the Downs, conspicuously so by the beginning of the 19th century.

Building in Bath appears to have gone through a major expansion in the late 1760s and early 1770s and again between 1785 and 1792 (Neale 1981, 168) before declining during the war years at the end of the century. Probably the Combe Down quarries, already in business in a developed form, were initially the principle beneficiaries of the boom, but growth of competition was the natural result once Allen's stranglehold was broken.

Leasing the quarries

At the sale in December 1764, Gertrude Warburton bought back a number of quarrying items. It is not fully certain why this procedure was followed, but almost certainly the purchase was intended to enhance her income by ensuring equipment was available to the new users of the quarries. The railway system had stopped being used by the end of 1764 (new style calendar by this date) and it is very likely that the new practice of leasing the quarries to independent quarry-masters or freemasons began at Christmas 1764.

It is possible, indeed the repurchase of items suggest it, that she took a certain rent quarterly or annually for the quarries including any equipment, which went to her directly, while a royalty payment for the actual stone (part of the estate) went to the Estate, from which there would be deductions before she received the benefit. The items sold at auction included:

A crane at Jones Quarry	£3.3s.0d
A crane at the Masons Crane House Quarry	£3.11s.0d
A crane at the Sheeps House	£5.15s.0d
Two crane chains and two stone chains	£5.13s.11d
A roundabout crane at Dolemead	£8.18s.6d
A stone carriage	£1.2s.0d
A smiths anvil, bellows, grindstone and board vise	£0.10s.6d

We have a little knowledge of what happened to two of the cranes, while that at Dolemeads seems to have gone to Bristol and that at the Masons Crane House possibly to Dolemeads, but at least one other was still at site and there were various parts obviously still available, including ironmongery from Combe Down sold in a second sale for £102.7s.0d (BRO. 667/5, 49) presumably from the cranes and carriages.

In June 1765 a payment of 10 shillings was made by Richard Jones and his son for half a year's rent of the crane: Richard and William Jones traded as masons and supplied block and wrought stone and 'measur'd and plan'd out' buildings (Pollard 1994, 9). The name Jones Quarry has survived for the Rockhall Lane site (Ralph Allen's Yard today) and though Addison (1998, 42) suggests a Herbert Jones may have owned the quarry, it is more likely a survival of the Jones' name from this time, or even earlier. If so, based on crane sites on the post-1764 Allen Estate map then, since the Masons' Crane House site is already known, the Rock Lane site was Sheeps House Quarry (this is the first time the quarry name could be located to the site, but has been used previously in this account to identify the quarrying based there).

Pollard identifies a further tenant in 1765, Richard Singer, a master builder (BRO D339/4/31); he – or a namesake – was still quarrying there in 1794 (see below). This can be inferred from a newspaper account of theft of stone from Singer by two of his men: Popjoy a freemason who had care of the quarry and Neat his wagoner. The latter worked Singer's horses night and day as part of his felonious pursuit supplying Prime Buildings in Bath at night (Pollard 1994, 9-10). In 1794 Popjoy was quarrying on his own account, so possibly he started his own quarry soon after 1765. Of other quarrymasters at Combe Down we have no knowledge until 1794 when data becomes available from an account for the by-then Lord Hawarden's Estate. Other separate quarries were, however working elsewhere in Combe Down and Odd Down, and there were also unlocated quarries in Combe Down known as Morris's and Sumsion's. The latter name was to become the best known in the district (BRO. D334/4/31).

This limited information and the data from a 1794 account and from leases after 1800, in light of information from the underground archaeological survey, allows some tentative conclusions to be drawn as to what was happening throughout the Allen Estate period of operation.

There was obviously no delay in implementing the new system at the old quarries on Combe Down. Jones was permitted to begin quarrying on some form of lease with his son almost immediately and while probably still working for the estate. At that time he was still presumably living at Prior Park. One can presume that other freemasons who worked for Allen were similarly treated, as it was advantageous to the Estate to act quickly, so they would have also leased quarries, the numbers being possibly determined by the number of entries and shafts to avoid conflict in working – possibly half a dozen or so operations. Singer may have been a new entrant, but conceivably was previously a building foreman for Allen who secured his stone supply by employing Popjoy to take care of the quarry. Popjoy's ability later to take a lease after his disgrace does not suggest much difficulty was put in the way of new tenants. There may have been an additional annual payment above Jones 'certain rent' of a pound per year for the horse crane, but later evidence suggests this would not have been onerous. The charge of £1.10s.0d a perch (equivalent to about £11.2s.0d for 1000 tons of 20 cubic feet of blockstone) was also very modest and, by its acceptability to the estate, does not suggest that direct management as under Allen was viewed as likely to be a more profitable alternative. Presumably the rent was set by comparison with other quarries in the area.

The use of levels or cartways certainly continued long past the Allen Estate period (Phase III) of operation, with new ones started and others extended after the estate breakup (see the early 19th-century section below). It is also clear, from the 3.5 to 4.5 m wide shafts (sometimes dubiously known as 'light shafts') found at, near or just beyond the Allen (Phase II) period working across the whole quarrying area, that winding of stone from shafts partly took over from hauling down the cartways. This may have been since the cartways were difficult to maintain, as can be seen by cleared collapses in them. But, more important, once well under the hill and without whatever advantages the railway had offered, it made sense to raise the stone directly from near or at the working place from the modest 10-12 m below surface. From the surface at the top of the Down it could be carted direct to its destination. It was a least-cost alternative and could be done with a simple hand winch or capstan or by horse crane, the latter much better for loading carts and capable of much higher output.

The availability of the sold-off parts of cranes suggests these would have been re-utilised on the shafts. At Jones' Quarry then Richard Jones and his son presumably had to re-erect one there since there was not one shown on the *c* 1764 Estate Map. The leases for after 1800 (below) strongly suggest the Allen workings at the Central Byfield area, with the diverging entries which the *c* 1764 map suggests were then abandoned, were not restarted until after 1800. It is just possible that the Jones' father and son worked on the west side of the Combe Road boundary pillar (Quarry 2398) in Lyncombe and Widcombe land acquired by Allen long before as part of his estate where the archaeology shows that workings of a suitable type for the time exist which are not otherwise easily explicable. They would have used an old entry into apophygate workings probably predating Allen, from just west of the Allen entries, possibly sinking the wide shaft for removing the stone at the roadside edge of what is now the William IV Public House car park.

Given that Allen's monopolistic methods no longer dominated the market, and demand was probably good at the start of the new regime, most of the leased ventures had a good chance of success, even with the existing or growing competition from others working or starting up at Combe Grove or the Lawns near Shaft Road in Monkton Combe lands, or elsewhere in Lyncombe and Widcombe, notably at Entry Hill and Crossways.

A set of accounts has survived for payments for

the quarries under Lord Hawarden from 1794 to the first half of 1799 (SRO. DD/X/LA/12). Unfortunately, as well as the pre-war building boom having fallen away by 1792, these were particularly bad years, as can be seen even in the small amount of data available. Even so, in the first year 7365 tons were produced, suggesting the 20,000 tons previously cited for Allen's likely peak production was not unrealistic (note that the ton is of 20 cubic feet, the contemporary measure. In modern tons, the figure would be 15-20% higher). Given the increased competition by that date and a transport system that imposed average tonnage costs rather than the near marginal cost of Allen's time, perhaps twice this level was appropriate to good years, shared between the freemasons. Some quarries produced nothing in some half-years, suggesting a direct connection with a particular building and implying that when it was completed, there was no general market for stone. The others with fairly high outputs like those of the Lancasheres (and Singers previously), Cromwell, Nowell, and Johnson and Pearce, were possibly master-mason builders.

The 1794 accounts show summer and winter production each year at Hawarden's Estate quarries, with total per year expressed both in the original perches, and derived, in tons of 20 cubic feet. Summer production was usually more than winter. In addition to the rental for the quarries, there was also a rental for Jon (sic) Godwin who had a stable adjacent to a quarry, who was charged at £2.2s.2d. This may indicate Godwin was working the old Masons Crane House Quarry, which a plan in a deed of 1805 shows as having a stable at the south-west corner of the quarried area. (LL. A/91/18/5/17).

Underground quarrying areas are defined for the Allen Estate (Phase III) based on technology changes considered to be post-Allen Phase II, and by inference based on the leased areas agreed after the estate break-up (see table below). The 1799 Charlton map of Lyncombe and Widcombe (BRO copy) suggests there was then no quarrying taking place in the Greendown late Collibees area off and to the west of Combe Road, though it very clearly shows the surface quarries a short distance away at Crossways and Entry Hill. Apart from the possibility of the underground quarry operated by Richard Jones and son in the 1760s, (Quarry 2398), and believed to have been accessed from the Jones Surface quarry (now Ralph Allen Yard), the granting of leases after 1803 suggests no other previous activity in Lyncombe and Widcombe. Similarly leases set for east of Combe Road, for the adjacent area in the Monkton Combe area of Byfield, and for the area between The Avenue and the Long Drung (East Firs), strongly suggest activity stopped with Allen's removal of the cranes and railway before his death.

The Layton and Hadley and other leases of the

Table 6.1 Underground stone quarries leased by the Lord Warburton, 1795-98

Areas quarried-out in (square) perches by individual quarrymasters and aggregate annual totals with conversion to tons of 20 cubic feet. It is assumed the workings were 5? yards (5 m) high.

Ladyday to Ladyday	*1795–1796 Summer + Winter*	*1796–1797*	*1797–1798*	*1798/1799 (half year)*	*Total*
Richard Singers	2.75+3.00	-	-	-	5.75
Richard Lankashere	-	4.75+3.00	2.50+2.25	1.75	14.25
Jas Lankesheer	3.00+2.50	3.00+1.50	2.00+3.00	1.75	7.00
Robert Simpkin	2.75+2.00	1.50+1.00	1.25+1.25	0.75	10.50
Peter Cromwell	4.50+ 2.75	1.50+3.00	0.25+2.50	0.75	15.25
Samuel Noel	2.50+1.50	2.00+0.50	1.25+1.00	1.50	10.23
William Hulance	2.25+1.50	2.00+0.00	0.25+0.00	0.75	6.75
William Popjoy	2.25+0.50	0.50+0.50	0.50+0.50	0.50	5.25
Jonathan Rudman	1.50+0.00	0.25+0.00	0.50+0.00	0.25	2.50
Joseph Smith	1.00+0.00	1.00+0.00	0.00+0.00	0.00	2.00
John Legg	3.75+1.25	0.50+0.00	0.00+0.00	0.00	5.50
Chas Godfrey	1.75+0.50	1.00+0.50	0.50+0.00	3.25	4.25
James Broom	1.25+1.00	1.50+0.50	0.25+0.25	0.00+0.00	4.75
Jon Godwin	2.50+1.00	0.75+0.00	0.50+1.50	1.00	7.25
William Johnston & Chris Pearce	3.25+2.00	3.00+2.00	0.50+0.00	0.00	10.75
Totals	54.50	5.75	22.50	9.00	121.75
Total tonnage at 20 cubic feet/ton.	7365	4831	3041	1216	16452

7.4 cubic perches = 1000 tons   Source: SRO.DD/X/LA 12. Also Pollard (1994 p11) with modifications

first decade of the 19th century (see Fig. 4.2 and next section for detail) affected the Firs quarrying areas. The Allen Estate period (Phase III) workings here seem to have commenced work in a strip which ran across the whole Firs site to the Long Drung, and at that time must have been marked by shafts at surface close to their southern limit. This was from near the Firs Shaft (as called in the Stabilisation Scheme) northwards in the Central Firs area, and by a new cartway from the west side of Sheeps House Quarry in West Firs. It left behind two or three unworked areas south or east of Sheeps House which were worked later, but generally somewhat more than the southern half of the whole site had been worked by the end of Allen Estate (Phase III) times. The worked area extended to between 40 and 100 m from the North Road of today, from which the new leases were issued after 1800. It is not known what work was done east of the Long Drung in these years, but the level from near the Brow extends north into a similar area, for which the Greenway family were the most likely quarrymasters. Much the same line of new development was also leased out after 1800 in the western side of this area, within Byfield, though this seems to have been almost a half century after work previously stopped there in Allen's time.

Underground quarrying and Combe Down 1801 to *c* 1838

Introduction

The first years of the 19th century saw the break-up of the Hawarden Estate into several parts. By 1800 quarrying on the southern part of the Combe Down area had virtually ceased while underground quarrying in the whole central underground area south of North Road was largely completed by 1838. The area between Combe Road and The Avenue became the core of the modern settlement owned by the Hadley family in what was virtually a planned settlement. This period also saw the beginnings of a great transition from underground to surface quarrying and large areas became occupied for this purpose on the periphery of the older areas (see Chapter 4).

West of Combe Road

West of Combe Road, only a small portion of the ground leased by Harry Salmon (see Chapter 4 for the other parts) was worked underground. The northern of the four southern plots was quarried underground by William Hulonce, and contains the larger part of the workings west of Combe Road. Even so the western end of his plot to the boundary was worked from surface, as can be seen today beyond the remaining entrance to the stabilised West Byfield workings now used as a bat sanctuary. In 1818 Hulonce sold part of his plot (copy documents supplied by Richard Irving), an acre at the western end including part of the extant underground to Daniel Byrne, and more to Sumsion later, whose portion on his death may have passed to Henry Street. Byrne's portion went through two changes of ownership before the remains of the surface quarry passed to James Byfield (1825), who left his name to the complex. His ownership too passed eventually to Henry Street in 1841. This multiple ownership explains the diversity of working methods seen underground, with at least three styles of operation in addition to any earlier working.

There is no doubt also that some working had taken place here long prior to 1805, since apophygate pillars have been found extending well into the area. It seems likely a cartway driven through these from the Jones Quarry entry to the north on the west side of the Combe Road Pillar was extended to facilitate the extraction from the area owned by Hulonce and, probably, two wide shafts, one in the William IV yard, belong either to the current phase (IV) or to the previous Allen Estate Phase III.

To facilitate underground working of the northernmost area an arched declined entry was driven in, to the rear of Tor View. It appears that it left just enough room to allow cart access to Rockhall Lane. (Professor Irving of Tor View has evidence the adjacent brewery (Hines Brewery at the William IV pub – Bone 2000, 42) used this entry via a track coming up against the back of his house: this was probably Hulonce's original route to the incline). The incline encountered bad ground at the bottom end, causing walls to be constructed and narrowing the access, so cannot have ever been as effective as was clearly intended by the size of its construction. From the base of the incline the workings extend to the still open (but gated entry) area into what is now the preserved bat area. A shaft was sunk at the back of the workings, in addition to that in the William IV yard, so in all this small area had the choice of some five entries. Whether all were used to draw out stone in this phase is not clear, but it remains a possibility, the surface being close with each owner using the nearest exit and making another if considered necessary (Case Study 7 in Chapter 12).

John Scrace worked part of his northern lease underground – in the area extending from a few metres south of the turnpike (thus excluding the roadside buildings from undermining) approximately to the modern service road off Combe Road to the western entrance of the Hulonce workings. It is possible the entry was from his or his neighbours adjacent surface quarry section since links with the southern workings of Central and East Byfield would not then have been established. The reason for underground working was probably the ground conditions, as in the northern section Scrace met a cockle bed (bearing calcitic vugs) of 1.0 to 1.5 m thickness, as revealed by underground boreholes, which caused him to work on two levels, leaving the useless cockle bed intact. This was a costly way of working for a limited height of stone, and in the Stabilisation Scheme it was found to be very unstable as the pillars appear to have been slightly misaligned. It is also likely that work by another freemason took place at the south end of this section (under the service road) since the style of quarrying in a small area there was very distinctively neat (Case Study 8 in Chapter 12).

Underground quarrying under the Hadley Estate east of Combe Road

As has been described earlier, the Hadley family gained lands in Combe Down (and elsewhere in Bath) about 1808 from the estate of Edward Layton. Layton had purchased them from Benjamin Wingrove, who had acquired the central area of Combe Down from the Thomas Viscount Hawarden's Estate.

The Hadley papers, in the Archives at Lewisham Library (LL.A91/18/—), have agreements relating especially to the quarrying of the 'understratum' of the land under the area between the Avenue and Combe Road (then Combe Lane) and on the south to what is now Combe and Church Roads. They enable a glimpse of the quarrymasters and their terms of working between about 1804 and 1815, with two substantially later agreements.

The agreements, or memorandum of agreements were probably drawn up by Jonathon Cruse, a Bath Land Surveyor. From about 1802 until 1804 a Jeremiah Cruse (possibly his father) was in partnership with William Smith, the geologist, as Smith and Cruse, at Trimbridge Street, Bath (Winchester 2001, 200). Jonathon Cruse, on his own or on behalf of the partnership, seems to have acted as agent, first for Layton, then Hadley. If so it seems likely that Cruse was not initially well acquainted with mining and quarrying leases as the early ones were particularly naïve. Successive leases gradually acquire additional conditions for which the most obvious source would have been Smith. Though accustomed to acting as surveyor on large estates, Smith was often away on his travels, which may have left the business without his continuous advice.

Mining and quarrying leases were essentially designed to protect the interests of the landlord. Broadly the landlord would wish to include some aspect of at least the following stipulations:

- The limits of the area to be worked and the type and quantity of stone (or mineral) to be worked within it and the period for which the lease was to run
- To ensure that the material extracted can be measured to allow payment and when, and to specify the price (royalty) to be placed on the material extracted, and what, if any, allowances might be made against the payment
- To ensure the income stream is maintained an entry fine might be levied and periodic, certain payments demanded whether material is extracted or not, and minimum levels of output required with penalties if this is not reached
- Specified minimum levels of employment and/or work to be carried out
- Payment of taxes and other expenditure incurred as a result of the lease are specified
- Penalties, notably recovery of debt from assets on the premises and personal effects of the tenant, voiding of the lease and return of the premises to the landlord if these above or other conditions are not met
- Provision for arbitration in case of dispute or to establish any allowable compensation to the tenant or landlord
- To protect the landlords' interest in continued working after the termination of a tenancy, and to protect the surface above from damage (other than expressly permitted in the lease)
- To prevent the lessee passing on his lease to another, without the consent of the landlord
- To specify the working methods to be used so as to protect future working, access and roof support and that of the surface
- To ensure access for the landlord or his agent without notice so as to ensure conditions of the lease are being kept

Other terms, especially for surface quarries would include the condition the surface would be left in – where it would bear on the infilling or conditions of the shaft top (*Bainbridge's Law of Mines and Minerals* (Brown 1900, Chapter 10; Appendix C).

Before the effective external control provided by late 19th-century Mines and Quarries Acts, the character of the lease and the inspections carried out by the landlord or his agent were also important in helping determine safe working conditions for underground workers.

The perch used in the leases for both linear and superficial perches of 5.5 yards and 5.5 yards square respectively (5.5 yards is almost exactly 5 m) with 40 perches to the rood and four roods to the acre. There are thus 160 perches to the acre.

Edward Layton died about 1807-1808 and his estate was inherited by his nephew Nathaniel Hadley of Lewisham in Kent, so after that date the leases are drawn up for the latter by the local surveyor or land agent. See Figure 4.2 for a plan of the quarry leases and Table below.

The first agreement was made on the 10th of May 1804 (LL. A91/18/5/25), when Edward Layton, now described as of Bath, let Samuel and John Pearce the understratum of the land adjacent to Combe Road. This was the land transferred by Salmon. It was described as one perch wide and forty perches long, extending south from the Turnpike and adjoining, eastwards, the adjacent quarry already occupied by Joseph Lankesheer. No lease seems to be available for this supposed

Table 6.2: Early 19th-century quarries and leases on Greendown, Combe Down (LL.A91/18/5/25 to 35) (see Fig. 4.2 for plan of locations)

Quarrymaster	*No.*	*Date*	*Size*	
Quarry leases sold by Harry Salmon in Lyncombe and Widcombe's Quarr Down west of Combe Road (Irving 2005).				
Samuel Nowel	1	1805	NK	Worked as a surface quarry
Jonathon Rudman	2	1805	NK	Worked mainly as a surface quarry. He died 1807 and his widow Sophia took over as quarrymaster, 1807-33
William Hulonce	3	26 Aug 1805	NK	Quarry worked mainly underground except at western end. One acre at the west end of the plot (including the west underground section) was taken over in 1818 by Patrick Byrne. Part also was sold to Abraham Sumsion in 1807 and presumably to Street after 1809. Byrnes portion sold (indirectly) to James Byfield in 1825, Richard Lankasheer, 1841and Street just after.
Henry Street	4	1805	NK	Worked as a surface quarry. Son Henry took over in 1814, retired 1851
Henry Street	5	1805	NK	Worked as a surface quarry. Son Henry took over in 1814, retired 1851. Passed to Phillip Nowell.
Abraham Sumsion	6	1805	NK	Worked as a surface quarry. He died 1809 and quarry sold to Street
John Scrace	7	1805	NK	Worked underground at the northern end on two levels. Shared crane with Sumsion in surface quarry
Quarry leases let by Edward Layton and Nathaniel Hadley (LL. A/90/18/5/25-28) between Combe Road and The Avenue, Monkton Combe.				
Samuel Pearce and John Pearce	8	10 May 1804	1 rood	Strip along the east side of the new Combe Road Wall
Samuel Pearce and John Pearce	9	15 May 1804	about 1? acres	Lease not available. This and the above quarry were taken over by Isaac Sumsion on 24 Sept 1808
Richard Lankasheer	10	15 May 1804	1? acres	The cartway and line of shafts run from the Rockhall entries almost to the Turnpike serving the adjacent quarries too.
Samuel Pearce Jun	11	15 May 1804	NK	Largely unreadable. later occupied by John Davidge and Abraham Sumsion with a moiety each
Ann Godwin	12	15 May 1804	1? acres	Widow of Jonathon Godwin.Originally drawn up for Richard Lankasheer, to whom it reverted if Godwin failed. Taken over by William Harold (by 1811) and on 1 Nov 1816 by Job Salter
Samuel Pearce and John Pearce	13	NK but by 1804	NK	Involved the ground on the north and north-east sides of Sheeps House Quarry. Largely worked-out.
John Greenway	14	22 Nov 1804	2 acres	Area to east of line perpendicular fromTurnpike at Farrs Lane ranging south. Already much worked.
John Greenway	15	26 Sep 1808	4 acres	This took the leased area in (14) across to opposite the Prior Park gates, and included the two acre quarry (14) above. It later became part of John Burgess' Three Acre Quarry
John Davidge	16	7 March 1827	1 acre 3 roods and 23 perches	In the semicircle just north of the southern curve of Combe Road
Combe Park Leases (Pollard 1994)				
Richard Lankasheer	17	17 Dec 1856.	Combe Park	Payment by the superficial yard
Andrew Greenaway	18	17 Nov 1809	? acre	Wingrove lease to Greenaway, passed to John Burgess in 1813.
Andrew Greenaway	19	1809	? acre	Mrs Drinkwater to Greenaway. Passed to John Burgess in 1814
John Burgess	20	1821		Bought and sold almost immediately, probably retaining underground rights
Main Firs Field Quarry Three Acre Quarry.				
John Burgess	21	October 1809	3 acres	Quarry ground advertised and sold to Burgess. Remnant unworked area of Greenway's 4 acres. No. 14
John Burgess	22	ND but c 1809	Approx 1 acre	John Houlton, c 1809 to Greenway and subsequently to Burgess.

Notes

John Greenway also leased part of the Sheeps House (surface) quarry area, in front of the two Allen entries for 99 years (LL. A91/18/5/30) presumably to secure level access to the underground quarries he bought or leased.

Abraham Sumsion was also quarrying at Lodge Style, Isaac Sumsion at Mount Pleasant and Mr Bennet was involved at Entry Hill.

Lankesheer quarry and, in fact the Lankasheer Quarry is later shown to be further away and the Pearces had altogether about two acres of reserves, with the ownership clear from other leases. The lessees could quarry and take away the blockstone and could use the same entry as Lankasheer, which was the eastern of the old Rockhall Lane entries south of Combe Road, which eventually ran north almost to the Turnpike. The Pearces' perch-wide holding had its own entry from Rockhall Lane, the west-side cartway running up the east side of Combe Road and under the old Telephone Exchange, now a church, but its usefulness had ended before the end of the Allen period because of instability. The quarrying was to be up to six inches inside the newly made wall, but not to go under it (which partly explains why the Combe Road pillar has been left at least for part of its length). Work was to commence the following Ladyday (March 23) and they were to pay three guineas a superficial perch (of ($5^1/_2$ by $5^1/_2$ yards or 5 x 5 m) – and no height limits were imposed. The workings would be measured by a competent surveyor, the Pearces would be barred entry if the rent was not paid and they had to sign a performance bond.

Commentary on the Hadley leases

This earliest of the Hadley leases (Samuel and John Pearce 1804) described the area involved, price of material from the 'understratum' extracted by superficial measurement (in the roof presumably, as the floors would have been obscured by spoil by time of survey) in addition to the dates on which it became due, and the termination as when the furthest boundary (the Turnpike, now North Road) was reached. There was provision for a surveyor to measure the stone extracted. The working conditions were limited to specifying the entry and underground level (ie cartway) and a requirement to support the surface and for the precise position of the Combe Road boundary. A performance bond, and reversion of the lease and barring of entry to the workings to the lessees if payment was not made were the only compliance conditions. Probably the landlord relied on custom and practice for his protection, but unfortunately we do not have earlier leases extant to compare. Undoubtedly, it was a minimal kind of lease, placing much on trust.

The lease for the larger, adjacent part of the two Pearce's quarry is not extant but it seems likely it would have been similar to those granted to Richard Lankesheer, Samuel Pearce the Younger and to Ann Godwin on the same date, 10 May 1804 (A91/18/5/26-28), less than a week later, on which the two Pearces' holding is shown on the plans. If so, the price was higher, £3.13s.6d against three guineas per superficial perch levied earlier. In each of the extant leases a light hole, or shaft was to be permitted of a specified size (14 feet diameter) and place at the tenants' expense, and spoil had to be removed. There was a somewhat grudging concession that more could be sunk 'if absolutely necessary', and there was provision for compensation if a gully, that is bad ground, had to be unprofitably worked through, and provision for arbitration. Work was to be done without delay or cessation and with all possible diligence and present roads to, and levels in the quarries, were not to be stopped. In the case of Ann Godwin, who was presumably the widow of Jonathon Godwin who had been a quarrymaster in 1795-98 under Viscount Hawarden, there was a condition that should she fail, then the quarry would be let to Richard Lankesheer, though for others the leases could not be passed on and there were provisions to cover non payment of the rent. In Lankesheers's lease Godwin had been substituted for him, but the succession of plans show ownership was as set out in the table and on the plan. Possibly Godwin's husband had had an earlier lease she intended to continue, or she was receiving help from her husband's previous quarrying colleague. Lankasheer was certainly one of the 1795-98 lessees, and perhaps pressure had been brought to bear on Layton the owner. There was also an arrangement for Godwin and the younger Pearce to share in the costs and benefits of driving an entrance and level into their two quarries.

This set of leases, plus that of the residual semicircular area to the south, bordered by the curve of Combe Road and let to John Davidge in 1827 (No. 16 on Figure 4.2 and in Table 6.2), took in all the ground bought from Salmon across to the boundary line running from the Turnpike south of Stonehouse Lane. Most of it was previously only quarried in the southern part, the Davidge quarry perhaps exploiting an unworked block most distant from the Rock Lane entries (probably Quarry 520 and/or 912).

From the above line, the area across to The Avenue and to opposite the Prior Park Lodge Gates (to where the Hadley Arms now stand), had been very much worked in its southern part, from the entries in Sheep's House Quarry, both from the two original Allen entries, and from a third on the west side, shown as extending in 120 feet northward by 1804. This western side entry appears to have been leased to Samuel and John Pearce (who are described as being west of Greenway and shown on his plan as so), and probably there was still considerable stone left unworked in the northern area up to the Turnpike (No. 11 on Figure 4.2 and Table 6.2), and likely to be attributable to Quarry 2342. The western Allen entry cartway had reached not far short of the Turnpike and had turned to the east in parallel with the Turnpike, at the southern edge of Quarry 2213, and the east Allen entry, from the archaeology below ground, would have been some way past the main Firs Shaft used on the Stabilisation Scheme and there were probably workings extending between them.

Greenway took a 99-year lease on the entrance area of Sheep's House Quarry in front of the two older entries, presumably to secure his access point and in 1804 took a two acre lease south of Farrs' Lane served by the two older Allen entries, which was stated as having much previously worked ground. Four years later, he negotiated a further lease with Hadley, which incorporated the earlier lease, as was described as being about four acres, to opposite the Prior Park Lodge gates. With this, all the area between Combe Road and The Avenue was covered by these leases.

The terms of the first Greenway lease were essentially the same as for the others except that firstly, the landlord reserved the right to enter and examine the workings at any time, 'without hindrance, molestation or intimidation'. Possibly this was the result of a bad experience in the previous half year, if so probably by the surveyor and agent, Jonathon Cruse of Bath. Secondly, that as a great proportion had been worked previously, Greenway was forbidden to pick or reduce in size any pillars remaining in such areas, which were to be left standing without any alteration whatever. There is, in the recent archaeological recording, plenty of evidence that such 'reworking', often by use of saws, was common in previous workings hereabouts and again this provision was probably the result of recent experience. The use of saws underground at this time, from the positions in which cuts have been found, was probably innovatory, although the landlord's concern was probably that he did not receive royalty on stone recovered from workings previously measured. Another possible change of working method (perhaps only as used in the other leased quarries), may just be indicated by a further provision, that removal of the blockstone by 'horse, cart, wagon or carriage' was permitted. This may indicate a proposal to use either horse transport or a rope-hauled system along the cartway. Evidence for rope haulage has been found in workings of this phase, though the practice was also used previously in workings probably of Allen's time.

Other evidence shows that a possible relative of John Greenway, Andrew Greenaway, who was also a mason in Walcot, Bath, was also accumulating quarrying rights in the adjacent area of the land between the Avenue and the Long Drung, the two of them possibly intending to sell them on. For the first time, perhaps significantly in view of the consolidations later, Hadley in the second lease imposed the condition that at least six perches of land were to be worked annually.

In 1808 there is also the first documentation of transfer of a lease (though another transfer prior to this is known, of Ann Godwin's to William Harrold). In this case Samuel and John Pearce's lease at the edge of Combe Road was passed to Abraham Sumsion (A91/18/5/31) whose family, from archaeological evidence (graffiti) and documentary evidence (Irving 2005; Addison 1998, Pollard pers. comm.), date back at least to 1750 (GRO. D2700/QP3/4/4) and forwards to after 1900. From the mid 19th century they became one of the more prominent quarrying families in the Combe Down area. In 1811 the quarry leased by Samuel Pearce Junior (No. 11 on Figure 4.2 and in Table 6.2) was also re-leased. It became held in two moieties (normally two undivided halves, but the plan shows Davidge holding the north-eastern half – Quarry 514), by Abraham Sumsion and John Davidge.The latter was also a long-time Combe Down quarrymaster for whom Davidges Bottom at Rock Lane would be named (for the area where he had a brewhouse and Public House. The 'lightholes' – as the wide shafts were sometimes known – in the lease were obviously by then considered normal, but a new provision was that they should be fenced.

In the 1827 John Davidge lease for the semicircular area he took effectively was the final major lease available in the Combe Road, Church Road and The Avenue – bounded area. It was an isolated block of unworked ground which was reached from the entries on the west side of Sheeps House Quarry by going south. Even after he finished work there, probably in Quarry 520, a small block within it was left unexploited until the end of the century (the later James Riddle's Quarry – no. 518). The price of any stone extracted had risen from the £3.13s.6d of the previous leases (bar the first) to £8.16s.0d per superficial perch – a very substantial change indeed.

The final lease, of 1856 to Richard Lankasheer, was for an area beneath Combe Park which is between The Avenue and the Long Drung and which would by then probably largely be worked out. This was perhaps the reason for pricing the stone produced, for the first time, by the superficial yard rather than the perch, which would be more suitable for reworking small areas such as large pillars. The rate of three shillings and sixpence was equivalent to around £5 5s $10^{1}/_{2}$d per superficial perch, a substantially lower price than in 1827, perhaps due to competition from the large Wiltshire quarries but, possible also, because of the dispersed nature of reworking or scavenging from old pillars (if that was what in fact taking place). The lease also specified a certain rent of ten pounds a year (a minimum rent regardless of production though otherwise the rent was based on production), with penalties if sufficient stone had not been cut, at three shillings and sixpence, to cover the half-year rent.

What effect might more carefully drawn-up leases have had, especially in the early cases? The obvious omissions are indicated by the additional clauses inserted in the leases as time went on. Some of these were very likely to indicate the learning curves of both landlords and their agent/surveyor in letting underground leases. The simpler leases were, perhaps a legacy of estate working under the Allen Estate and Hawarden resulting from letting, after Allen's death in 1764 to men who were likely to have been his trusted quarrymen previously, working to similar custom and practice. As new men came in, this would have proved less satisfactory, though the informal system had the possible advantage of permitting easy low-cost entry to quarrymasters, which might not have been possible (and thus disadvantageous to the landlord too in difficult times) with a more developed system with such impediments as entry fines and certain rents.

Not specifying the price or royalty of the stone quarried was an obvious defect, though perhaps less obvious before the inflation of the Napoleonic War years. Pollard has noted how the price of blockstone, seven shillings a ton, was the same in the early years of the 19th century as it had been in the 1730s. However he also notes that by the early 19th century some quarries, at least, were quoting 16 rather than 20 cubic feet to the ton (Pollard, pers comm.). There may thus be a hidden price increase. By pricing the stone got by the superficial perch, in combination of a time limit on working expressed only by the reaching the limit of the leased area (as opposed to a fixed and shorter lease period), the chance was lost to elevate the price when conditions permitted. Leases for an acre or so might well last for 25 years or more. The much higher royalty realised by the landlord in the 1827 lease illustrates how important this might have been to the landlord.

A further major potential defect was the lack of control (and thus pricing), over what the 'understratum' actually meant. It is not possible to be precise, but the working at that time had previously only exploited some five metres (and at times possibly less) of freestone, of which up to a metre might be the roof bed which was discarded either to give headroom or because of lower quality. Presumably this was the customary working height. However, by the 1820s, in Burgess' 'Three Acre' quarry particularly, up to some 7-8 m came to be worked, vastly increasing the amount of stone with far less than proportionate extra extraction costs, on long-held leases for the same royalty.

The requirements for supporting the roof and the surface above seem to have been satisfactory in practice. Despite modern health and safety concern at roof conditions and stability, after nearly two centuries of post-working stresses very little surface actually appears to have been let down. However had lessees decided, as they could under the terms, to work down to or even in the underlying Fullers Earth, at a depth from surface of only 20-25 m (working of the latter was later carried out at Odd Down and in the even closer Horsecombe), then short term stability during the actual leased period might have been followed, at the landlord's later expense, by substantial long term surface subsidence.

East of the Avenue and John Burgess' triangular 'Three Acre' Quarry

This section is a development of work done previously by Pollard (1994, 15-18). The Hawarden Trustees in 1805 conveyed the area east of the Avenue to the Long Drung to Benjamin Wingrove. While the above accumulation of leases was being developed by John Greenway, on the other side of The Avenue, Andrew Greenaway was also taking control of leases. On 17th November 1809 Wingrove leased to him for 29 years the right to quarry a triangular area of a quarter acre at the apex between the Long Drung and The Avenue. Greenaway agreed to drive a 'subterranean road or way' in a straight line from the mouth of his own quarry which was located on the Poor Tyning, just east of the Long Drung. It was to go across Combe Park to the 'rock head of stone lying under Firs Field'. The purpose was to convey away the stone belonging to the devisees of Edward Layton (Hadley), to be completed within ten years. Examination of the area near to the Long Drung and Greenaway's entry (perhaps the Brow Cartway which may have been owned by the family since Milo Smith's time) in the archaeological survey could find little trace of any such cartway in Quarry 2330, and it was probably agreed to simply to secure the lease; ten years was a long time, and the lease may have been part of a consolidation to be sold on (as in fact happened) and, in the event wide shafts were sunk in the target area anyway, making the idea of using a very long cartway entirely redundant.

In 1809 also Greenaway bought the adjacent $^1/_2$ acre adjacent to the earlier small purchase of $^1/_2$ acre at the top of the Long Drung, in between the Drung and Tyning Road, from Mrs Drinkwater: it seems likely from these two small remnants that the Combe Park and southern part of the Poor Tyning were largely worked out (Nos. 17 and 18 on Figure 4.2 and Table 6.2). In 1813 and 1814 the leases were assigned to John Burgess and in 1821 by default of an alternative buyer, Burgess was able to purchase the whole of Combe Park. He sold it very soon after (it passed, somehow, to the Hadley Estate), but it is extremely likely that he retained the right to quarry, rework and – possibly most important to him – use the previously worked ground to dump his spoil. The archaeological survey found considerable evidence of both reworking by sawing ashlar off pillars, and dumping of waste southwards using barrow-ways from the Burgess workings.

In October 1809 too, three acres of quarry ground upon Combe Down were advertised for sale 'at very reasonable terms'. This seems to have been the larger part (and, presumably, the unworked part) of the triangle of land between the Turnpike and The Avenue (No.14 on Figure 4.2 and in Table 6.2 – Quarry 2211), owned by Hadley and occupied by John Greenway. Whether Greenway and Greenaway were in cahoots, or if Hadley forced Greenway to terminate his lease (perhaps because of low production), this quarry too passed to Burgess. A combination of area calculation and archaeology makes it highly probable this three acres was unworked ground north east of a long boundary pillar about 50m east of the (Stabilisation Scheme) Firs Shaft, enclosing the ground from there to the

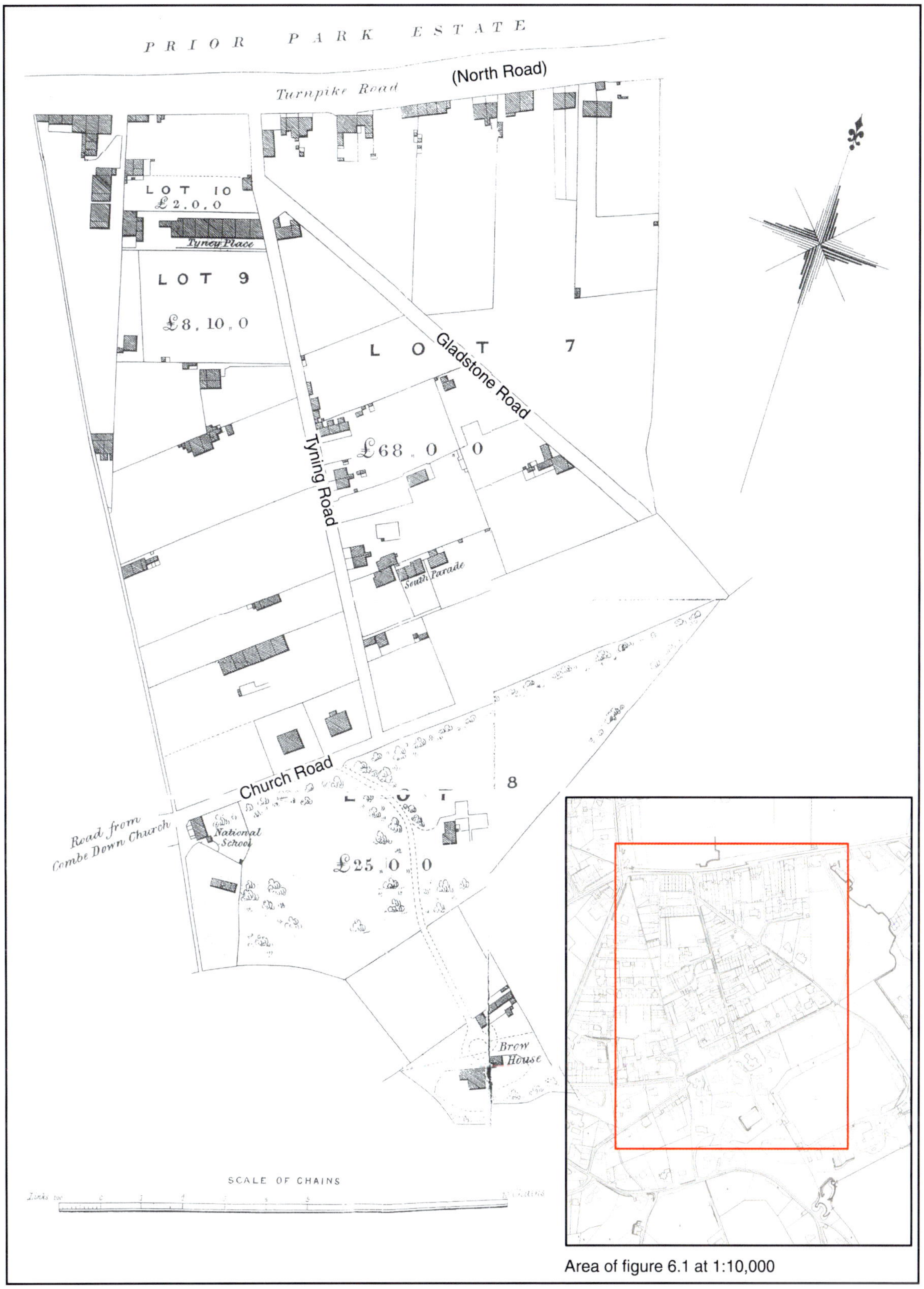

Fig. 6.1 Cotterill's plan of the area of Combe Down east of the Long Drung, with Tyning Road and Gladstone Road and the Brow in 1845. The plots for lease included the use of the stone beneath them (BRO)

apex opposite the Prior Park Lodge. The boundary shown (see Figure 4.2) avoids a later quarry operated by William Stennard at the north-west corner of Greenway's triangle, while part of his south-west corner had probably already been worked. Burgess and his son Harry appear to have worked this quarry through to the area beyond the Long Drung, apart from a substantial pillar left near the Turnpike road.

Burgess worked there, presumably to near his death in 1835, followed by his son Harry until the lease ran out in 1838. Previously John Burgess had been surface-quarrying behind the east end of the de Montalt Row and behind what later became Hopecote. It is clear that this quarry was either too small and/or was insufficient to supply what appears to have been a substantial market, including the Devizes Prison. It is shown as disused on the 1851 Tithe Survey, and was accessed from the 'Road to the quarries' alongside Claremont on Church Road. The move by him to underground quarrying, at a time when many others were beginning to work at surface was probably clinched by the leasing to him of three acres (leading to it being called by the archaeologists as the Three Acre Quarry) from Hadley or John Greenway.

The area involved in the Three Acre Quarry (2211) forms the largest single quarry underground at Combe Down, and has the appearance of being the best and most thoroughly worked with the greatest heights of stone removed comparable with anywhere else in the underground workings. Possibly Burgess had the highest output also, of any single quarry, surface or underground there. In production within the central area of Combe Down he was rivalled by Henry Street's surface quarry west of Combe Road, and underground output overall must have been greatly exceeded by surface working in the proto-parish as a whole.

Workings east of the Long Drung

In comparison to further west, very little is known about this area in documentary sources. A long cartway from the area later to become the house and grounds known as the Brow has a long cartway of rather different characteristics to those driven for Allen. The apophygated pillars along the entire cartway length through Quarry 2339 and 2350, suggest the cartway was begun at an early date. The most likely quarrymen for this were the Greenway family, one of whom was somewhere there in 1730 and another, who took the lease including the agreement to drive from the Poor Tyning to Firs Field, about 1810. This suggests the area known as the Poor Tyning, just east of the Long Drung was largely worked out, except for the piece owned by Mrs Drinkwater (née Houlton) at the top. Further east however there were number of quarries, most of them small and most of them apparently 19th century in date, working to about the end of the century.

The Poor Tyning had belonged before 1809 to the Houlton family and apart from that portion above, it passed to John Houlton of Farleigh Castle, and in 1825 he conveyed it to John Ovens Thomas of Prior Park (who had bought Prior Park after the death of the second Lord Hawarden in 1807). Thomas engaged Henry Fowler Cotterill to survey the site, then a farm, and to lay out private carriage roads, Tyning Road and Gladstone Road, obviously preparatory to development. Using 1000-year leases, the southern end, in 1828, was leased to Brow House (a large house with a fine outlook which would gain a premium price). The central part, of over 17 acres, went to Cotterill in 1830, including the freestone underground with rights to dig etc, and in 1831 and 1832 Tyney Place and land adjoining went to Jacob Wherret and Thomas Greenham respectively. Cotterill sub-divided his purchase into building lots and in 1835, for example, sold two houses built on an acre plot under 'which is an excellent freestone for building, with a well with an excellent supply of water'.

Thomas' leases paid a ground rent (£8 per annum in the case above), and in 1845 the ground rents of the properties were sold, for the Thomas devisees, by Cotterill. The plan of the area he produced (Fig. 6.1) showed some 50 houses had been built in the area between the Long Drung and The Lawns (BRO. 313). The properties gave opportunity for a number of quarrymen to secure their own quarries within their own boundaries, or to lease the stone from others which is very much reflected in the diversity of small quarries found under the area. Pollard (1994) on which the above is largely based, identified three such sales, including to Frederick Cross in 1831, mason: one acre in the fork between Tyning and Gladstone Roads 'and all stone quarries mines and minerals for 990 years'(possibly part of Quarry 2397); Harry Burgess of Combe Down, Quarryman also in 1831; and James Ford, Quarryman in 1840 – a plot 328 feet by 68 feet and the stone etc. He built two cottages there, now known as Richardson Place, and possibly part of Quarry 2351. John Davidge was also a purchaser but quarries were not mentioned.

Output of stone in the early 19th-century quarries

An acre of stone worked underground of 160 perches and 5 m high will weigh about 50,000 tons gross, and pro rata for other thicknesses. Taking the Three Acre Quarry as an example, perhaps 6 m high or more, it would thus gross some 180,000 tons less a 10% reduction for pillars left in place. Of the remainder only about 60% would be blockstone or other saleable stone, so in this case then the net output would be some 97,200 tons over some thirty years or about 3250 tons per year. A little higher yield is possible due to the greater height of at least of some of the stone worked. At this rate, the quarries, both west and east of Combe Road and as far as the Three Acres, might thus together produce some 10,000 tons as an annual average. Given that some years in the first third of the 19th century were particularly difficult, a maximum output of some 20,000 tons annually could have been possible. This is much the same as was possible for Allen's time, but there were at this time also substantial quarries in Combe Down north of North Road, and at Entry Hill, in Prior Park and Claverton nearby, at Kingham and Vinegar Down and near Lodge Style and Mount Pleasant. These were mainly surface quarries with less loss of stone (a little over 10% perhaps is likely) than for underground quarrying. Overall output capacity must thus have been at least double that enumerated above, possibly treble, in the early decades of the 19th century.

For the landlord this meant, in cash terms per acre, a gross revenue of 90% of 160 perches at £3.13s.6d a superficial perch, about £530. For the quarrymaster, per acre gross revenue, from 5 m-high workings at 14 shillings a ton (the price delivered at the Navigation in 1808), the yield is £18,900 and in workings 7 m high, up to £26,450. Thus on an annual basis, at the six perches minimum eventually demanded per year in the later leases, the landlord's income would be about £22.1s.0d per quarrymaster, and a one acre quarry would last some 24 years. The quarrymaster's gross revenue would be about £709, or in 7 m workings, about £992. Surface workings would gain at least 10% on this by not having to leave pillars, and there would be greater ease in extracting more blocks without breakage. The landlord would, of course, suffer surface damage, but, two centuries later, the surface worked ground would not need expensive stabilisation.

These rather modest returns demonstrate the problem to the landlord of having a large number of small lessees, creating substantial administration costs. A smaller number of larger producers with minimum output levels set to reflect the amount of land leased would probably be a more favourable situation. In the case of Burgess and his Three Acre working, it works out that he extracted about 16 perches annually over the thirty years. By his time the rent had risen. This would yield a rent of £58.8s.0d annually to the landlord and £2275 to Burgess for stone delivered to the Navigation or Kennet and Avon Canal. We have no idea of the level of profitability for the quarrymaster.

In the leases granted discussed above, only the later ones had a minimum stipulated output (six perches), and the turnover of lessees suggest that outputs were probably generally no higher than the six on average so landlord income was, at the lower level of expectation with eight leases, possibly no higher than around £170 gross. Developments after 1810, and the consolidation under Burgess and his probable wider market via the Kennet and Avon Canal suggests he had the potential to make a substantial contribution to the Hadley Estate. It was probably achieved, but at the levels of return shown above, the interest in property speculation, for other than or as well as quarrying, is easy to appreciate.

Working practices and areas involved

Layton and Hadley's leases generally directed that existing entries should be used where possible for the workings in the leases north of Combe and Church Roads and from the Sheeps House Quarry, allowing only one or two new entries at the latter which were within the same area as earlier. In two instances adjacent quarries were to use the same entries, an obvious economy imposed by the landlord. They were somewhat grudging about allowing shafts to be sunk and overall this probably suggests that an important concern to them was to protect the surface as far as possible from being affected by underground working. But, by being stipulated, it forced quarrymasters to rationalise their infrastructure and keep costs down, all of which would assist in their survival as tenants or lessees. Nevertheless, the line of shafts between Combe and North Roads in the Central Byfield area and one or two others on the eastern side, and the line of four shafts again from Firs shaft to the Hadley Arms and at least one beyond, suggests that regardless of the levels, the great bulk of stone was drawn out vertically. The exception may have been the largest blocks brought out from the western side, which might have been too heavy for the winding mechanism.

The proposed Greenaway long level was an unlikely venture to have been successful, as it would have had to be driven a long distance through old workings without the possibility of working stone en route. When he inherited the Greenway titles Burgess certainly used shafts for hauling stone out of the workings, and it is likely that other operators with 'light shafts' did similarly, in some cases (as the archaeology showed) using 'down-the-shaft' cable haulage to get the stone to the shaft bottom. It did not prove possible in the archaeological survey to locate any long section of cartway which may have been Greenaway's in the relevant area.

The plans with the leases (summarised on Fig. 4.2 and in Table 6.2) show that the Pearces' and Lankesheer's quarries used the old entries at Rockhall Lane. Already in 1804 the Lankesheer entry is shown as extended to the quarry south boundary, but only the starting point of the Pearces' entry is indicated; it had reached the quarry by 1808. In the case of the Godwin's and John Pearce the Younger's quarries, they were to drive a new joint entry and level, sharing costs and value of any stone sold en route, from just south and west of what today are the Rock Cottages, the surface entry-roads of which joined and curved to come to the general road level up Rock Lane. The route of the Godwin level underground is not clear, though presumably it communicated with two apparent light shafts shown on the plans, one near the southern boundary, the other at what is now Westerleigh Road. In Pearce's quarry the underground level is shown running west then sharply north, to communicate with a (probable) light shaft some 90 m from the southern boundary at the west side of the north branch of the Westerleigh Road.

The two leases granted to Greenway in 1804 and 1808 (and thus probably Burgess later), had an entry, probably used the western of Allen's original two parallel entries from Sheeps House Quarry, which by then had to pass through extensive areas of earlier working, to not far short of the Turnpike boundary. It was developed in a straight line just west of north for about 125 m, then turned north-east for a distance of about 30 m, at the southern lomit of Quarry 2213. Firs Shaft is some 20 m south of this. It is known from the archaeology to have met bad ground at the turning point, which probably caused the Allen Estate lessee to give up working further, but generally indicates the extent reached by 1800. The eastern of the Allen entries is also indicated on the 1808 Greenway Quarry plan, but without inside details. Since the last apparent use of this (based on the archaeological survey) was to connect by an eastwardly branch, it is possible this entry was in use for another area, most likely Quarry 2202. This may have started in the current period, but continued (on evidence of graffiti referring to the Hadley Arms, for a decade or so after 1838 at least. It is not known who operated this quarry, or the earlier quarry linked to it under Gay Staith to its south, 2340.

In 1804 Greenway was banned from picking or cutting into pillars in worked-out areas entirely, suggesting that the practice was common. If so it may have added a substantial extra amount of revenue to those quarrymasters close to older workings and (literally) prepared to cut corners. Although on each measuring day the surveyor would have little difficulty in monitoring the current workings, it would have been difficult indeed to check the maze of the older workings, so there was little risk of discovery. As observed in the underground survey the practice of working old pillars was widespread.

There is little information in the plans on underground boundaries, though archaeological investigation suggests these were often kept in the form of more-or-less continuous pillars. The only specific example in the leases is the Combe Road Pillar, whose position was carefully specified underground in the two Pearces' quarry, (no. 11 on Figure 4.2 and in Table 6.2). Underground the pillar was an obvious feature in the southern older workings (Quarry 505), and also appeared, less regularly, in the north (Quarry 507). It was hard however to distinguish in the central area at the south end of their quarry where work no doubt commenced in 1804, so the provision in the lease was presumably largely ignored.

No provisions was made in the leases about how cartways were to be laid out, or their size, except that they were not to be closed or obstructed, and the only issue in relation to them was that horses, carts, wagons, and carriages could be used. Only slight traces of the use of horses were found, although wheel ruts were fairly common. The term 'carriages' may however refer to the flanged wheel vehicles running on wooden rails, or may alternatively, as a somewhat strange technical term, refer to the use of cable-hauled vehicles, since their use was apparent in the west Central area of Byfield (Quarry 514), drawing over 40 m without evidence of rails, from a small dock formed of low walls, back to a shaft.

The sales and leases of quarries and other documents also tell us a little about the surface quarrying methods. West of Combe Road, Irving (2006) suggests that Nowell, on the southern outcrop of the Oolite, was the most fortunate in having the minimum overburden to remove. However as already noted, the overburden was fairly thin north of the underground workings owned by Hulonce. It diminished to around a metre in the area on Combe Road near the old Telephone Exchange (now a church nursery), making the cartway alongside the east side of Combe Road unsafe for use and causing the Pearces to share the eastern one of the old Allen cartways with Lankasheer. Removing this thin layer was obviously likely to be the most advantageous way of working. In other cases substantial amounts of overburden needed removal to allow the freestone to be worked economically, even if some return could be made for selling some of the Twinhoe Beds which could be used for specific uses, as Allen had found in the mid 18th century.

One fairly major purchaser of stone in 1808 was Westminster Abbey, for the Henry VII Chapel (Cottingham 1822, 10-12). Their mason Thomas Gayfere was authorised to visit quarries and report on useful stone – 'The finest were in the quarries of Samuel Pearce or for long and large blocks, the quarries of Mrs Radman and Samuel Newell.'

Despite misspellings of names, these are clearly the adjacent quarries east and west of Combe Road. The Abbey ordered 150 tons of stone from Samuel Pearce, to be delivered 'well and properly squared', on the quay at Bath at 14 shillings a ton by the 17th of March 1808 (note the doubling of price in the last decade or so). It reached the Bristol quay by May or June, but no boat was found to take it to London until December and it was then lost off Portland. Other stone was eventually used. This recalls Ralph Allen's earlier wartime difficulties (described in Chapter 5) and, possibly, the men were again pressed for the Royal Navy. It again emphasises the difficulties of marketing Combe Down stone before the Kennet and Avon Canal opened two years later.

Pierce Egan in 1819 was able to project a rosy view of the industry at that time (cited in Addison 1998, 32-33):

> *The vast depth of freestone which has been excavated from the earth, the lofty arches, or pillars*

remaining in a craggy state left by the excavators to let light into the subterraneous passages and caverns which extend for a considerable way under the earth, most interestingly claim the attention of the explorer.

James Tunstall in 1847 was equally admiring, though by this time the local industry was certainly being damaged by the Wiltshire quarries following the access opened up there by the railway:

. . . and now let us look into the Bath freestone quarries, now in full operation, immense blocks of stone taken out of the rock without the aid of blasting, and at once worked into convenient ashlar for the mason (Tunstall 1847).

From a local viewpoint, other than of the quarryman, there could be a different perspective; in 1829 in Bathford, just east of Bath, Henry Skrine, landowner, was approached by leading members of the community who were intent on limiting the damage done to the highways by the carrying away of extracted stone from Samuel and George Watson's recently quitted quarry. They persuaded Skrine to agree a lease for the quarry for fourteen years at ten pounds a year, which covenanted the new lessees, on behalf of the parish, not to quarry against penalties of an additional £100 a year, the latter of which was twice the earlier rent paid by Watson (SRO. DD/SK 8/2).

Some idea of the loads actually carried and how, which caused such damage, are contained in two inquests of 1825 and 1827. In the first William White had been crushed, dying later in the Casualty Hospital in Walcott, by a two-ton block of stone which fell off a truck, when the way-side collapsed. White was leading the shaft horse of the truck while the foreman (who was the witness), was leading the forehorse of the team of horses. In the second accident John Brewer was killed driving a stone carriage loaded with a block of stone about five tons in weight on the turnpike road. He fell and the fore-wheel passed over his body (BRO. Coroners Inquests 1814-27).

This first part of the century was probably the 'golden age' of quarrying at Combe Down. It was not the monopolistic and integrated concern which Allen had operated, but a period of widespread ownership often by master building masons, at many substantial and some smaller quarries cheaply and commonly enough available to be open to almost any mason to directly operate. Collectively these had the capacity of double or even triple that of Allen and may have actually produced at such levels, in the best years outside of the depressions around 1825 and again around 1830.

Combe Down itself also grew, its housing development closely associated with the quarry development, quite often involving the same masons either on their own quarries or sites purchased or leased from speculators. By 1838, with the exception of some small blocks of stone left unexploited underground, the central area of what we today know as Combe Down, once associated with Allen's quarrying adventure, was virtually worked out.

Combe Down underground quarrying after 1838

Quarries in the core area of Combe Down

For the core area of Combe Down by 1838, the ventures established under the Hadley leases were over, and the underground section under the Harry Salmon leaseholds had also closed, leaving only the 1856 lease from Hadley to Richard Lankasheer to work under Combe Park (at the east side of the Avenue to the Long Drung). This was probably for reworking pillars rather than any surviving block of ground and most likely was done using one of the old shafts there.

This left only four (or five if reworking of a substantial pillar is included) small blocks of unexploited stone, measured only in a few tens of yards, not acres, within the ground originally marked out by Allen for his quarrying adventure. There was further unworked ground east of the Long Drung and the Tyning as far as the eastern limit of older housing beyond Gladstone Road. As a finale, the activity from these remnant operations was however, fairly attenuated. Activity in the surrounding area however remained substantial, with a revival late in the century of underground working, though on a very modest scale compared to that at surface.

Thus the bulk of the underground quarries were probably fortunate in their timing, since the virtual exhaustion of substantial deposits in the core area coincided with a huge growth in competition from the Wiltshire quarries, competition which the surface quarries and the few later underground ventures had to face after the great Western Railway extended to Bath and Bristol, actually passing through the Wiltshire freestones at Box tunnel.

The five quarry sites

One of the larger blocks remaining was just east of the entries at the bottom of Sheeps House Quarry, and a second, just south of that almost under Gay Staithe (Quarries 2202 and 2340 respectively). There is no obvious reason why these, right in the core of the village, were left unexploited as the stone there appeared to have been of good quality. The former has a graffito mentioning the Hadley Arms and Coxes' and the relative qualities of the beer in those houses. The Hadley Arms was built in 1846 (Hadley Arms Deeds), so the graffito must be later. Other graffiti, possibly part of a later Bath Stone episode, and attributable to James Morris, was dated to 1863. The technology suggests working after the middle of the century. Both were worked using a branch out of the eastern of the two old Allen entries down

Rock Lane, thus ensuring the lane retained its industrial use at this late phase

The last (third) quarrying area carried out in the core area of the village seems to have been a substantial pillar or block of un-worked stone, located adjacent to the western edge of Allen's western cartway from the bottom of Rock Lane. This was worked via a shaft to the south of the block, presumably one sunk during or soon after Ralph Allen's time. The Quarry (2344), worked a large block of stone that was clearly demarcated with survey crosses and the stone was transported to the shaft via a barrow-way. There was probably a small winding apparatus at the shaft top. The shaft was subsequently capped in the mid 1850s, as later surface-derived pottery recovered from a shaft cone below the shaft dated to around the 1840s and 1850s. There was also later Bath Stone activity (2370), close to the base of the shaft where a large pillar was robbed and the former Allen cartway was cut through, this activity is thought to have taken place even later than other later Bath Stone activity in the 1860s, again attributable to James Morris, (Quarries 2367 and 2368).

In the northern part of the Firs Field, a fourth block had been left, probably because a fault just north of the Firs Shaft discouraged work beyond. This was close to but south of the Turnpike opposite Stennards surface Quarry on the north side of the Turnpike. The underground quarry (2215) was accessed from a small area of surface quarry (close to Selway Court today) left open after the rest of the quarry had worked north and the southern area was otherwise backfilled, and partly built upon. The entrance would have appeared as a deep hole surrounded by spoil mounds and was probably equipped with some form of crane. This was open on the 1882-84 OS 25 inch to a mile map, but was gone on later maps. This quarry thus seems to have been operational in the 1870s or a little later.

A further underground quarry, possibly worked by the Sumsions as owners of the surface quarry, is found largely north of the Turnpike under the original winding course of the surface quarry road there (now straightened as Stonehouse Lane). The quarry (2219) seems to have had a series of entries from the surface quarry along its length, which, from wrought stone found underground, may have acted as shelters for banker masons. It communicated with the workings under and south of the road, and exploited unquarried ground under the surface quarry road and the 18th century house (s) there. It seems likely to have worked in the 1870-80s and exploited (and perhaps exceeded) its rights to work to the centre of the Turnpike, since the trusts by a quirk of law only seem to have acquired the surface. Other surface and underground quarries along the road had probably done similarly. There was an obligation for them to support the road, but this seems only to have been seriously done, with a very substantial stone pack, in the 100 m or so near the Hadley Arms (Quarry 2211).

The fifth and final block was to be found just north of the curve in Combe Road near the junction with Summer Lane, apparently left unworked by John Davidge in his Hadley Lease. This was worked by James Riddle early in the 20th century, using a petrol winch on the 'Allottments Shaft' (Quarry 518). According to Addison it (1998, 43) it probably worked from about 1905 or a little earlier to about 1908, though it was still registered as his until 1914. The last quarrying, the fifth site carried out in the core area of the village seems to have been a substantial pillar adjacent to Allens western cartway from the bottom of Rock Lane. This was worked via a shaft, presumably one sunk at or soon after Ralph Allen's time. It took both the pillar and worked through the bottom of the cartway. Graffiti suggest this was done about 1912. There most likely was a small winding apparatus at the shaft top.

Underground quarries away from the core area of Combe Down

Entry Hill

A small underground working there (Quarries 2374 and 2375), infilled during the Stabilisation Scheme, may be of late 18th- or early 19th-century date. It initially seemed to have been superseded by the large, surface, Springfield Quarry adjacent which the Charlton map of 1799 shows was working on a considerable scale (Chapman 1996). However, the easternmost workings, (2374), that had mainly corbelled pillars, were likely to have been worked from the south-western edge of the surface quarry, using wedges and picks to extract the stone. Crane anchorage slots and wedge holes were noted in relation to other anchorage fixings in the surface quarry cliff face. The westernmost workings, Quarry 2375, were later in their working methods, and used saws to extract the stone, as well as picks and wedges, with evidence of several sawn faces and different crane anchorage holes. Lewis slots were larger than those in the former quarry workings to the east and were possibly driven from the Entry Hill road eastwards.

Foxhill

This fairly small series of four linked but separate phases of working was north of the Turnpike between and under the Foxhill MOD Establishment and the end of Combe Road (Quarries 2380, 2381,2382 and 2384). An underground quarry there – possibly incorporating Coxes Shaft Mine and certainly Tank Field Quarry – was infilled as part of the Stabilisation Scheme. At the corner (shown on the 1882-84 OS map) an underground quarry (2381) was, almost certainly, operated by Randall and Saunders, a Wiltshire based company, who worked a very small underground section at the end of a surface quarry noted by Addison (1998) as Gilks Quarry (now the Rugby ground). A crane was used to raise the stone from the north end of the underground workings which were there left opened out to surface (Quarry 2382). The Tank Field Quarry was adjacent to this, operated by the Combe Down (Bath) Freestone Co. Ltd, and had as owners J.A. and Lucy Tucker (the former a solicitor) with Frederick William Armstrong a builder and Frederick John Box a timber merchant as shareholders. The company aimed to take over an existing quarry (possibly 2380 or 2381) and operated from about 1909 to 1924 with the company dissolved in 1931. They employed up to 16 people in 1912 with 11 above ground, suggesting a masons' yard was in operation. The quarry stopped working in 1917 but briefly revived in 1923. The quarry extended underground almost to the west end of the MOD establishment. By about 1930 (OS 25 inch Map) the area had been designed Ralph Allen Park and houses had been built lining the road. The land was apparently requisitioned in the Second World War. See Case Study 13 (Chapter 12).

Lodge Hill (Style) Quarry – Shaft Mine

Quarrying was certainly being carried out there in the 1830s when the landowner was William Vaughn Jenkins, probably of what is Combe Manor (Hotel) today, and the operator was Isaac Sumsion. In the 1851 Tithe survey the quarry was owned by Isaac and Giles Sumsion. This is the longest-known family operating at Combe Down, with the first reference (above) in 1750 and which provided the chairmen of the Bath Stone Firms well into the 20th century. The quarry beds were described by the 1839 Stone Commission reporting on the stone for the new Houses of Parliament: there was ridding of 7 feet (Twinhoe Beds), then 'weathering beds of $3^1/_2$, 4, $4^1/_2$ and 2 feet with blocks 12 to 96 cubic feet obtainable' (Hemmings 1983). It is not known when underground quarrying began there. The northern part of the underground quarry (2372), was stabilised as part of the Scheme, and the depth of ridding found suggests that southwards the cover was very thin, reduced to only some $1^1/_2$ feet, including the road's asphalt covering. The overall height of 'weathering beds' (freestone), given in the 1839 account, corresponds reasonably with what was seen there recently. The Commission gave the cost of stone as 6d per cubic foot and a cost of 2d to transport this to the Dundas Wharf (about 2s.8d per ton). It closed around 1934-35 (Wooster 1978, 10). It was Philip Wooster (pers. comm.) who originally discovered the thin cover over the underground working under Shaft Road, while surveying the mine which by then covered some $6^1/_2$ acres, extending east towards Combe Manor. According to him it used horse-drawn tramways underground and, in latter years had three gangs of men at work, each using a face-crane to assist transport. In this it was more like the much larger Wiltshire quarries than most underground workings at Combe Down. At surface it still had the crane for loading, in disused state, in 1939 and seems previously to have used a steam wagon to convey stone to the railway station via Summer Lane (Midford Station, perhaps, or to Bath via the Prior Park Carriageway, paying a 2d toll. (Hemmings 1983).

Freylings Quarry

This was the name given during the Stabilisation Scheme to a small underground quarry (2373), bordering the Mount Pleasant Quarry. It used a decline for access (though the Scheme sank a new shaft to avoid dumped material) in the decline. It had similar features to the Shaft Mine section of the Lodge Style Quarry, above. Evidence was found for numerous underground cranes, represented by the chog hole loactions in the roof, and considerable amounts of contemporary quarrying graffiti from several phases of quarrying in the 19th and 20th centuries.

St Winifreds Quarry

Between Shaft Road and the road to Combe Manor Hotel is an area known as St Winifreds, bordering the area known as Combe Down Quarries adjacent to Shaft Road. Most of it has uneven ground produced by dumping waste as the quarrying faces moved from north to south, with a low face partly apparent at the front. An underground quarry of the same name is found in the south-west corner, worked in the 1920s and 1930s (Hemming 1983) and was approached by an external decline, and is of unknown extent but still open, behind locked gates, as a bat reserve. This ground belonged to Vaughn Jenkins in the mid 19th century, but the quarry company seems unknown, but was possibly Sumsions again, and later the Bath Stone Firms.

Even, however, if the surface quarrying activity was included, the activity after 1838 was meagre compared with the boom that took place in the 'Bath Stone' quarries of Wiltshire around Box and Corsham especially. The Oolitic limestones exploited in these areas are of later geological date than Combe Down, but are sufficiently similar to be termed 'Bath Stone' and be used as a substitute for Combe Down Stone. Depending on the source and its origin, they may even be better than some Combe Down, being free of clay and calcitic inclusions which were sometimes ignored by Combe Down in selling their stone. The Wiltshire beds are also far more massive than Combe Down, which was perhaps a disadvantage in the 18th century when the easy extraction of Combe Down's well-jointed beds was very useful, but an advantage in the mid 19th century when rail transport was easy for them (the GWR passed through the Box Tunnel) and working methods had evolved.

Nevertheless, demand up to the mid-century for Combe Down stone remained high, partly due to the success of Philip Nowell (Irving 2005, 76-79) and a stone business (also keeping several quarrymasters at work) driven by his architectural success, and partly to the local demand it was assured of from Bath. Combe Down also benefited, like the Wiltshire quarries, from the huge growth in building which took place in London and other urban areas, for which readily worked stone was a key element and where the Nowells had a (canal) quay devoted to stone.

Underground working today

There are no underground workings operating today within Combe Down, but in nearby Limpley Stoke an underground quarry is extracting beds from under the stone previously extracted, probably in the 19th century. These beds are harder than were previously worked, though this is not a problem for modern machinery. A modern mobile dimension stone-cutting machine can cut stone from the face to precise sizes.

Chapter 7

An outline of the technology used and the underground landscape at Combe Down

This and the following chapters use the archaeological survey in conjunction with other documentary or published sources to determine or deduce the techniques used at Combe Down, and either developed there or imported. This particular chapter will define the various types of quarries, attempt to describe the underground landscape as a whole and establish the basic terminology enabling an archaeological description of the development of the Combe Down Quarries to be made. Where possible terms used are those known to have been used by those who carried out the actual working of these and similar quarries, notably provided by David Pollard (pers. comm.). Detailed explanations of these terms are provided at relevant points in the text, expanding upon the brief glossary in Table 1.1, Chapter 1. In some cases when an appropriate term was not available to us, there has been 'borrowing' from underground terminologies established in various reports under English Heritage's Monuments Protection Programme (Cranstone 1992; Willies 1993; Chitty 1996; Ashbee 1996) and sometimes terminology from both sources has been given to improve understanding. The map on the inside rear cover shows the location of numbered quarries, and the chronological phase in which they were developed.

Why go underground?

It is perhaps not immediately obvious why Ralph Allen and (probably) his smaller-scale predecessors, decided on 'undermining' rather than surface quarrying, since surface quarrying was common and was used later elsewhere at Combe Down on a substantial scale. In stone quarrying, 'undermining', or underground quarrying, was commonly seen as a natural extension of surface quarrying, either to continue working beyond the surface boundary into areas where the surface could not be disturbed, or because of the amount of overburden that had to be dealt with in surface quarrying. This was probably greatest near (but away from) the southern margin of the freestone outcrop at Combe Down, with a consequent high cost of removal and a requirement for a large area for its disposal. Later phases of surface quarrying, at least in some cases, exploited areas with less overburden and had a wider market for the inferior quality limestone of the overlying Twinhoe Beds.

There were advantages also in maintaining the surface soil for grazing and building, and, in Allen's case, proximity to his great house, or in increasing the overall rental value to the landlord. Restoration was a feature of many later surface-quarrying leases, but this could not always be satisfactorily done. Working underground had advantages in providing a constant environment secure from inclement weather and was, perhaps, considered to be free of distractions for the quarrymen. Often the stone was less weathered, and without variations in colour induced by near-surface leaching. There was also a far greater willingness in the past to go underground where specific horizons were to be worked. Systematic stripping of overburden on a substantial scale was a development of the mid to late 19th century and, with some notable exceptions, was generally fairly rare before that period.

The main disadvantage was the support needed for the roof and loss of stone in pillars, and the need for lighting such as candles, and oil or acetylene lamps such as used at Combe Down in the later phases. Handling stone was often more difficult underground than outside and there was probably a need for greater skills, requiring more specialism by the managing quarrymasters or 'freemasons' and experienced men would be required to work with inexperienced labourers. But, in many respects the skills, techniques and effort needed to work underground were similar to those needed at surface and other skills were available due to long traditions of such work. Underground working may have been more dangerous, though accidents in surface quarries appear also to have been very frequent, and it is unlikely any comparison could be quantified before the collection of such records began in the mid to late 19th century.

Types of quarry

Quarries are classified mainly by the material extracted, here the freestone, by the mode of extraction (surface or underground) and by the method of entry. The types of quarry at Combe Down and other quarries that exploited Bath Stone below have been adapted from criteria established by David Pollard (1992, 26; Stanier 2000, 70). Six types of quarries, including those at surface and included here were recorded at Combe Down.

Surface (open) quarries:

Outcrop or hillside quarries

Outcrop, or sometimes hillside quarries were and are typical of the earliest quarries to be established in many areas up to and including the modern period. It was the earliest recognised quarry type within Combe Down from at least the late 17th or early 18th century. It involved the exploitation of beds of stone exposed on valley sides, especially where there were escarpment edges, which were the easiest of all types to reach the stone. Once the location of an outcrop quarry was selected, the topsoil and loose or 'brashy' subsoil known as ragbed or ridding was removed from the exposed edge or top of the hillside and deposited in the vicinity as close by as possible. This was often on the valley side behind the working face. It was sometimes used to create a level area or terrace to serve as a level working and loading area for the quarry. These levels often survive with the quarry face forming a cliff behind.

An outcrop quarry was developed along the escarpment at the southern side of Combe Down village, and others have been noted along both north and south sides of Horsecombe Vale. Ralph Allen's probable first surface quarry, Masons Crane House Quarry is now occupied by the houses and school in Quarrybottom and an unexplored mass of re-deposited topsoil and waste stone heaped into low banks within the Holy Trinity Church yard on Church Road is thought to have derived from it. The quarry was depicted by Robins in 1759 (Fig. 3.11) and, illustrating the appeal such places then had for the visitor, also on an 18th-century lady's fan (Fig. 7.1).

The primary joint structure at Combe Down is orientated NW-SE, with a secondary joint structure running at approximately 40° to the first. Quarrymen exploited the primary joint structure along the face of the hillside quarry edge by separating the blocks along these joints using a bar to ease them out, or by hammering a wedge if necessary into the bedding planes or joints to pry them out. Where the joint structure could not be readily exploited a 'wrist stone' – the least useful of the bed-blocks – was taken out to create a slot using hammers, picks, wedges or wedge and chips to break it out, freeing the blocks alongside it.

Hole quarries

Hole quarries were worked downwards from the ground surface where overburden was relatively shallow. Following removal of topsoil, the poor quality rubble stone beds, the ragbed or ridding, were removed to expose the underlying freestone beds, the different qualities usually being stored separately. After initial removal elsewhere, the soil and waste rubble was commonly deposited in the worked-out parts of the quarry and, as the works extended below the original surface, an access ramp would be used to remove the won stone.

The quarry located to the rear of the Old Rank or de Montalt Cottages was developed as a hole

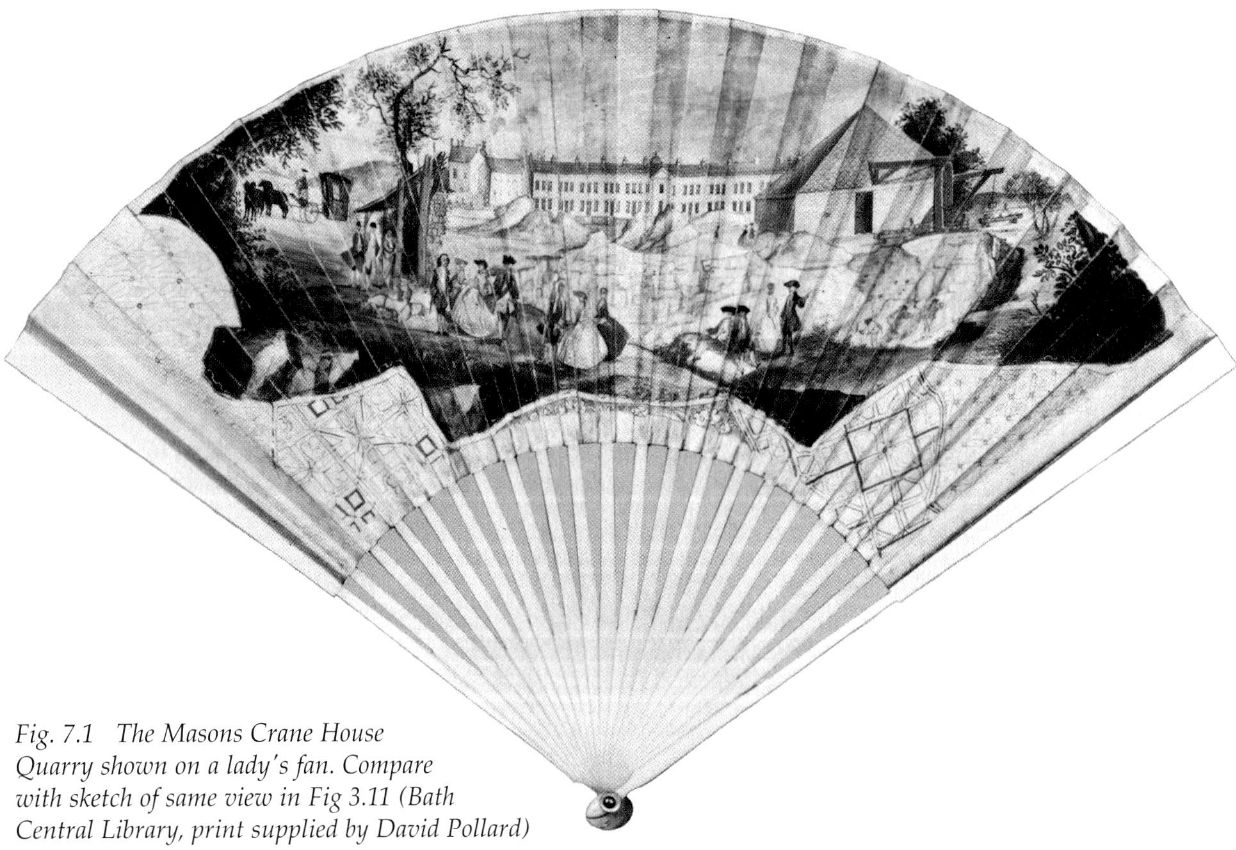

Fig. 7.1 The Masons Crane House Quarry shown on a lady's fan. Compare with sketch of same view in Fig 3.11 (Bath Central Library, print supplied by David Pollard)

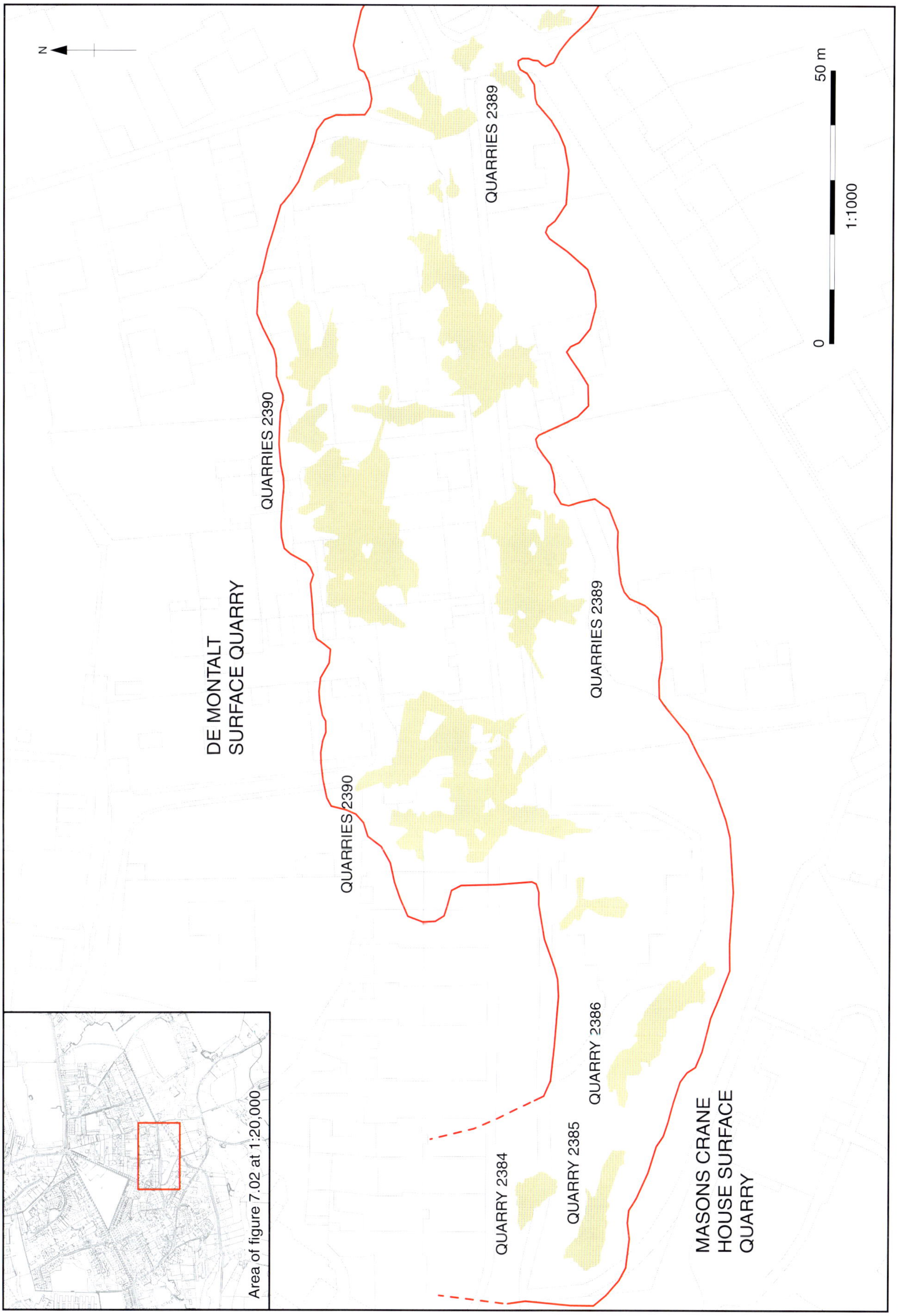

Fig. 7.2 Undermining quarries in the south of the core area, located by borehole and surveyed by laser scanning

quarry in the mid 18th century. It is referred to herein as 'Burgess Quarry' after the owner who worked the north-eastern end up to around 1815. It is some 150 m from the escarpment, and seems to have been developed this way to avoid the Old Rank of Allen's workmen's houses which had been built about 1730. The quarry located in Quarryman's Court, down Rock Lane, formerly probably known as Sheeps House Quarry, was also the hole type, possibly begun on a very small scale before Ralph Allen's developments, but certainly extended later. Both also acted as accesses for 'undermining' and, in the latter case for full-scale underground quarrying.

Where possible material from within the ridding was also sold, for instance for lime-burning or used for rubble walling if the quality was sufficiently good, and, as walling stone, for constructing the internal partitions of some houses in Bath. This clearly made the overall cost of removing overburden less, and, as noted earlier, Allen's move to surface quarrying probably came about due to concerns over safety underground, and the development of markets for the ridding.

Underground quarries

Legally, since the late 19th century these would be termed mines, hence Firs and Byfield Mines, though local freestone quarrymen referred and still usually refer to them as quarries as they produce squared or 'dimension stone'. The past usage is retained here.

Undermining quarries

This term was used locally for development of underground quarries accessed from a working

Fig. 7.3 Horsecombe Quarry, showing a large open joint or 'gull' which is parallel to the side of the escarpment resulting from cambering or land-slipping of the strata

surface quarry. Several outcrop quarries were expanded into small undermining quarries during the 18th century, as it became increasingly uneconomic to remove the overburden. In this report it is always used to mean small underground extensions to existing quarries, but it is and was sometimes used more generally.

Undermining quarries were identified by laser scan surveys carried out through vertical boreholes along the southern front of Combe Down village in the areas of Church Road, Belmont Road and the upper parts of the Summer Lane during the Stabilisation Scheme. The original accesses were found to be partially or fully backfilled with waste from the surface quarry, and their extents were plotted (Fig. 7.2). This provided or suggested the direction from which the infilled quarry spoil or later waste materials had been dumped and thus indicated from which of the surface quarries the underground entry had originally been made.

After initial surface exploitation of the quarry, an access point or entry was created to follow the desired stone beds below ground level. This undermining technique was convenient in locations where the surface had to be preserved because of existing buildings (for instance the Old Rank), but was also desirable where large areas and depths of overburden made surface removal less economic.

One of the earliest and easily visible examples of an undermining quarry was located on the southern escarpment at Horsecombe Vale (Quarry 2376) to the south of Bradford Road and the Brow, close to and just below a footpath on the cliff top known as 'Shepherds Walk'. The surviving outcrop consists mainly of vertical quarry faces extending for about 70-80 m (Fig. 7.3). Small openings were made into

Fig. 7.4 Cartway within an adit or level entry (Allen's east cartway, central Firs) before stabilisation (Photo, Paul Deakin)

the quarry face, but many were subsequently buried by successive rubbish tipping from houses above and to the north, so that only two entrances remained accessible prior to stabilisation. The outcrop quarry face preserves some small areas of bench working and rows of horizontal wedge-holes may indicate a phase of later extraction. The undermining was probably to preserve the footpath. The quarry had exploited the natural joint structure and 'gulls' that run parallel to the side of the Vale. It was driven northwards into the side of the hillside, then developed laterally to follow the natural joints. The joint structure was here so closely followed underground that many of the pillars did not actually support the roof on either side of the larger open faults, so that large areas of the roof collapsed and wide gulls were exposed with insubstantial pillar support on either side.

Some of the later Combe Down quarries, for instance at Stonehouse and Stennards Quarry (see Chapter 12, Case Study 10 and 12), north of North Road, were also undermining, and were developed from the much larger adjacent surface quarry on the north side of North Road..

Level heading or adit entry underground quarries

The term underground quarry is used where the main part or all of the quarry was exploited underground. In the level heading quarry, the passage, or adit beyond the entry was driven forwards into the workings (at the heading) either level or almost level with the slight slope draining outwards. These were often, at Combe Down, the cartway routes used for bringing extracted stone to the surface. In other cases a railway could be used or, as in the Stabilisation Scheme, trackless haulage was used. Level headings or cartways were predominantly used during the Allen Phase II period, when they were driven from four or perhaps five earlier small surface quarries or from external declined ramps made to expose the freestone beds. These include one located in Byfield, known until recently as Jones Quarry, at what is currently Ralph Allen Yard on Rockhall Lane, and Sheeps House Quarry at Rock Lane and Quarrymans Court (Fig. 7.4). Burgess Quarry may have been developed from the ramp at Claremont off Church Road leading to the earlier underground entries to cartways in East Firs. A fourth surface quarry with a level-heading entry, contemporary with the Allen period workings, was identified east of the Long Drung, Quarry (2350). This type of quarrying was also extensively used during the 19th century at Combe Down, where access points were made in the faces of pre-existing open or surface quarries, but usually in conjunction with winding at shafts.

Inclined or slope entries to underground quarries

Inclined entries, locally called slopes, usually with cartways or railways, were used to haul stone to the surface and to provide access to quarrymen. They were inclined down ('declined') below overburden or supported roof beds to gain initial access to workable freestone beds, after which conventional underground quarrying methods were employed as described later. Inclined entries were identified at three, possibly four locations at Combe Down. Three of these, associated with Combe Quarry and St Winifreds on the eastern side of Shaft Road, were operating during the 1920s (Addison 1998, 50). The other, the only one within Firs and Byfield, was at the Byfield Mine Quarries (503 and 504). This entry was referred to during the Stabilisation Scheme as the 'Irvings Incline', after Professor Richard Irving of Tor House which is located at the top of the incline (Chapter 12, Case Study 7). The structure had a stone-lined barrel-vaulted roof and was almost 25 m long. It was constructed in the first decade of the 19th century and included pieces of sawn stone, which was then still unusual at the quarry sites (Fig. 7.5).

Vertical shaft entry underground quarries

Quarries worked by vertical shafts alone were not common at Combe Down, though Coxes Vertical Shaft Mine and Tankfield Quarry, dating from around 1900 and situated under what is now the MOD Establishment at Foxhill were fully worked in this way (Chapter 12, Case Study 13). A number of smaller quarries in Far East Firs, near Gladstone Road were also worked from shafts, though most of these also had underground levels linking them to adjacent quarries and ultimately the surface. However, vertical shafts, designed to wind extracted stone directly to the surface from close to the underground working faces, were used in the 18th century onwards at Combe Down during and after the latter part of the Allen period. Though also linked by cartways, it is probable that the principle mode of working became vertical shaft in the workings as they approached North Road. This use of shafts can be recognized by cable grooves in the side of the shaft caused by dragging blocks to the shaft before hauling out (Fig 7.6).

The underground quarrying landscape

Introduction

Whereas the succeeding chapters have deconstructed the elements of the landscape so as to understand how the quarries functioned and changed, this short section attempts to model the underground landscape at Combe Down as it appeared while working and prior to stabilisation. In the same way as a surface quarrying landscape, a large underground area such as Combe Down incorporated a complex network of interrelated stone and spoil movement routes, working faces, dressing areas, spoil heaps and some of the usual detritus of human occupation on a disused industrial site. Unlike a surface landscape, however, the underground equivalent was far more restricted, lacking the direct interrelationships with human

Fig. 7.5 Reconstruction drawing of Irving's incline in use, early 19th century, West Byfield

Fig. 7.6 Vertical cable grooves in the side of a shaft, East Firs

Fig. 7.7 Inclined barrow-way over dumped rubble, eastern end of the Grand Canyon, Central Firs

Fig. 7.8 Stepped working face, south-west Firs. This was the only full example found, as usually faces were left vertical or the area backfilled with spoil

settlement, agriculture, vegetation and so on. As used here the term underground landscape simply reflects, like a painting, what was seen as a whole.

For someone entirely unfamiliar with the underground, the quarrying landscape can be likened to a rain forest kingdom in a rocky mountainous area. There is a wide expanse in the light above the canopy occupied by a large number and range of specialised creatures. Below the canopy the view of the landscape is more restricted, the range of activity less obvious. The trunks of trees, supporting the canopy generally prevent views of more than a few tens of metres and often much less, confusing and disconnecting the observer from the wider environment. The true floor may not be visible because of a layer of detritus. Movement and slightly longer views will be restricted to ways cut through the forest and, by following them the observer is likely to come across clearings with obvious signs of anthropoid activity and shafts of light penetrating from above. Failing to follow them will risk becoming lost. Given time and labour, removal of accumulated detritus from the floor may reveal traces of intense former human activity over smaller areas, and traces of activity over the whole accessible region.

The underground mining or quarrying area was usually more restricted and less visible than the surface equivalent and there was a further major difference. Generally it can be viewed as an area which expands over time, but which at any instant is totally limited by the boundary imposed by extraction. In some cases there may be after-use, for a different mineral perhaps or, for instance, as a mushroom farm. However, especially for most underground quarries, except for slow decay and closure over a very long, possibly geological time, the formation of the landscape within an area was usually, a single, if extended, event. Stratigraphically it will have a sequence of layers within it similar to those on surface sites but created in a relatively short space of time, and often the stratigraphy is as much horizontal as vertical, with spoil from one area being deposited within older, worked-out adjacent areas. In some cases the normal stratigraphic relationship is inverted, with the later workings at the bottom, though this is only occasionally so in the present study and in a very minor way.

Description of a typical working quarry

During the earliest working phase the quarries were very small and most would have been entered by means of either a slope, or by means of one of the level entries. This would probably be sited at the foot of a cliff-like face of a small surface quarry. In Ralph Allen's time it would have had a crane above loading on to his railway (see Figure 3.4), but at other times loading off the floor or a platform formed of waste stone onto carts would have been more usual. Other than at Allen's quarries the use of iron-railed transport and cranes at the interface with the surface was not common until the late 19th and early 20th centuries.

The entry itself led directly into the rock face, with sometimes a stone arch to reinforce it. The earliest quarries penetrated only a few tens of metres at most with rooms between pillars where the stone was worked and drawn back to the entry. In and after Allen's time larger quarries used levels or cartways, which were initially advanced in pairs a little distance apart to aid ventilation. These were typically left open two or three metres wide and high (up to five metres later), leading between solid rock pillars and rubble stone packs for 50 or 100 m. The floor was kept clear though the carts or wagons left shallow ruts. Any fallen stone or sludgy material was thrown on to the banks alongside the rutted track. Worked out areas – rooms between the pillars – were used mainly for disposing of the large amounts of spoil resulting from breaking the rock, but narrow views could sometimes be had over the spoil to the nearby cartway and the lights of its users (Fig. 7.7). The cartways led fairly directly to the working areas where there was open space and the pillars left to support the roof were up to eight metres high. Elsewhere only the tops of pillars were visible because of the spoil backfilled around them.

The working face itself was formed in large steps (Fig. 7.8), and it would there be apparent in the pools of light shed by the candles that the cartway was formed on spoil some two or more metres deep, leaving a trench at the front of the face to work the beds at the bottom of the face. The face might end just in front of the cartway, or the cartway could extend either side into a trench or gullet to allow a number of rooms to be worked in a line, with pillars left to support the roof between them. The face and gullet advanced as the stone was extracted. At the top of the face, pickmen worked at opening a gap below the roof to free the beds below. This was in the picking bed, and before the middle of the 18th century an opening was made a yard or so (0.9 m) high and a few yards deep and wide, the men forming the rather rounded and flaring-out pillar tops with their picks and hacks as they proceeded, between rooms of up to 10 m width. Later they cut out only a narrow jad slot at the top a metre or so deep and 10-15 cm high so as to free the beds below, leaving the pillars roughly squared. By using a weakness such as a wide joint, or by breaking into the face at each step or bed, the remainder of the bed-blocks at the lower levels between the pillars were fairly easily barred-out and slid down or raised up to the cartway floor level. The blocks slid fairly easily over rubble or on thick timber planks. Very large blocks, up to several tons in weight, were few but prized and a great deal more care and effort was used to load these straight out of the bed on to a carefully placed two-wheeled cart or four wheeled wagon. Blocks and rubble were roughly squared to send out of the quarry for building, the bulk of it smaller in size and fairly easily handled.

As the cartways penetrated further from the entries, then at around 50 m inside, shafts about a metre wide were sunk 6-7 m deep to the roof of the working to encourage more effective vertical circulation of the air (Fig. 7.9). Previously the air-flow had been horizontal and increasingly sluggish along one cartway level, through the workings and out another. From 70 to 100 m in, wide shafts were used, placed close to the cartways. These provided even more ventilation and it is very likely that most stone was thereafter hauled vertically up the shafts to surface, perhaps leaving only the larger blocks to be hauled out of the cartway on wagons. These may have partly been assisted by using cable haulage and a crane or winch from the surface.

Large amounts of waste, termed spoil, was created from breaking out the roof and the bed-blocks, the waste material from poor parts of the face, and from shaping or scappling the slightly diamond-shaped blocks into squared blockstone with a heavy scappling axe. Some of this spoil was used to build up the floor behind the face to cartway level and the remainder where possible was thrown either into adjacent worked-out rooms, or onto heaps held back by rubble packs behind the working face. Some spoil was thrown from the picking beds under the roof through cut-out openings or 'windows' (Fig. 7.10), or barrowed out, into spare space in adjacent areas. In any event, very little space was left under the roof away from the working face and the cartways. In mid and late 19th-century workings the amount of space in the working area did increase considerably, partly since better methods of breaking the rock using wedges

Fig. 7.9 Small ventilation shaft in the roof of the workings, with cap formed of stone slabs over timber supports, Central Firs

Fig. 7.10 Window used to throw out high-level spoil from Long Room working beyond, just north of Firs Shaft

and saws reduced wastage, but also because it could be packed some distance away into the wide areas left from older workings, for which long barrow-runs were sometimes used. In a few late instances at Combe Down, the extra space permitted the use of cranes and winches to handle the stone blocks more expeditiously, loading on to flat trucks running on iron rails. Like the earlier carts and wagons, these were largely manhandled unless a cable from a winch or crane was available. Some of these areas retained high spaces, with pillars and the final vertical face to which the beds were worked: these are the last to be worked in that immediate area and could be spectacular.

Fig. 7.11 Corbelled stone arch over the Hadley Arms steps. The stone pack was part of that supporting the North Road and the patina suggests it was built later than the steps (Photo by Paul Deakin)

Complexities at Combe Down

The simple model described above, replicated by the dozen or more entries and cartways, except for the large amounts of spoil, could describe many working underground stone quarries almost anywhere. In practice Combe Down had features which gave it an almost unique character and complexity in underground stone quarrying. Probably the most important was the character of the stone, which was closely jointed with up to three joint directions. It made it easier to extract, which was particularly suitable for the needs of 18th-century Bath and the massive quantities it required

Fig. 7.12 Partially sawn pillar with timber sprag (or scorter) to tighten blocks in the roof (Photo by Paul Deakin)

from a nearby source. But it also led to huge amounts of waste stone or spoil, perhaps 40% of the whole rock available, which, in its broken form, occupied half the space excavated, requiring a wide variety of strategies to get rid of it from working areas. It did have a beneficial effect however, providing much material for construction of rubble pack walls for roof support, and providing lateral support for the rather slender pillars, which would certainly have frequently collapsed, without this stiffening.

The respect the quarryman had to pay to the joints also led to very variable pillar spacing, forms and alignments. There is very little of the regularity of pillars, cartways and even working faces usually associated with stone mining, and as seen, for instance, in the Wiltshire quarries. The pillars at Combe Down were themselves very varied. A few were long and wide and seem to have formed boundaries or were left to prevent 'domino collapse', but most were fairly slender. There were several main forms, of which the apophygate, corbelled and direct are the three basic types, but the variations are many, partly since the quarryman worked back to the natural joints in the great majority of pillar faces rather than cutting through the bed-blocks. Some pillars, mostly due to faulting, were even inclined. Use of saws and wedges, used to cut vertically through beds, led to more regular variants, mainly of the direct pillar form.

The joints have also led to much weakness in the roof. This has led to some instances of 'heroic' underground structures, very heavy stone support packs over 10 m high at one shaft, and the occasional use of arches (Fig. 7.11), some again massive in their construction. Falls of stone have

Fig. 7.13 Section through a barrow-way with rubble thrown up onto dumped rubble and fines, West Byfield (Quarry 514) (Photo by Paul Deakin)

blocked cartways and either required strong support or caused diversions around them. Surprisingly perhaps, the amount of timber support used was limited, especially in earlier workings, but in later working areas, where as much space as possible between pillars was desired (Fig. 7.12), large numbers of sloping timbers between notches were placed in pillar sides supporting beds at joints in the roof.

Perhaps the greatest complexity came from the spoil barrow-ways which are found almost everywhere traversing the spoil tips in all directions and, from the occasional section seen through the tips, within the spoil tips also. Some barrow-ways just wander over the tips apparently tipping at random. Others follow purpose-made routes with roof-support packs alongside and have prepared surfaces for easy running, forming distributary patterns as they divide to dumping areas in abandoned rooms between the pillars. Though most are fairly level, others are inclined, and some, obviously due to a shortage of room, fill the barrow-way spaces to sometimes less than a metre high, then throw the spoil forward to stack right to the roof (Fig 7.13). Spoil dumping and barrow-ways did not respect cartways once their use was over, nor old working faces, nearly all of which were filled in with spoil. In some cases, neither floor nor roof was visible because of dumped, stacked or packed material totally infilling an area. Substantial areas at Combe Down came into this category, for example a large area between the two principal mines, Firs and Byfield.

Given the poor lighting in which the work was carried out, most of this apparent chaos was invisible during working. Approaching from an entry up the cartway, perhaps by a shaft with its light pouring down, and directly to the working face where the method of working could easily be seen with multiple lights, the scene would have appeared fairly well organised, if sometimes perilous, as heavy blocks were manhandled and carted away.

Evidence of the working methods was left on the worked rock and in the working areas. The use of chain or rope cables, fairly often passed down shafts and / or along the levels, left grooves on shaft sides or on pillars, there was evidence in Lewis slots of anchors for cranes and winches, and occasional groove marks on the floor and wall sides resulted from wheels. In one instance a crane anchor point was indicated by the impression of chain links in the soft rock (Fig. 7.14). Sometimes tools and more unusual items were left behind on a high ledge on a pillar. In one example, a pair of soft leather shoes was found (Quarry 2202), which could not possibly have been used by quarrymen – perhaps it was a joke. Other such placed finds included flat iron chips (for wedge and chip breaking), pipes and candles. Other tools were left behind packs or within spoil. Bottles were a common find, a few used to carry water to ease the saws, others to drink from, or simply dumped near entries. In the later working areas fragments of saws were found, and in the 20th-century workings on Shaft Road and North of North Road, full saws remained. In some places there old rails had been used to support loose material in the roof.

After-use of the underground workings added relatively little to the original working areas. The area of Byfield used as a WWII shelter from the Baedecker raids had a small area cleared of debris, with hooks to take an electric cord to a series of lamps. At the westernmost entry to what is now the bat conservation area of Byfield, the space under the high pillars had been used to store mushroom boxes, though it was not the actual farm. The same workings had a well, probably constructed around

Fig. 7.14 Mould of chain formed by links pressing into the stone. The chain, wrapped around a pillar, was part of a crane anchor, late 19th century, Central Firs (Quarry 2201)

1815 and later used for the Hines Brewery (it is now under the William IV yard) and the workings may have aided sinking of several other wells, or cleaning them out. The workings were also used as drains, for 'grey-water' in one case, and as a sewer in another.

In less environmentally-aware times, almost all entries left open were used either as storage or as waste management facilities. The Irvings Incline produced bottles and tins, a motorcycle wheel and similar debris. The Carriage Inn contained dumped 18th-century pottery in a drain, which was useful for dating. At the Firs Field 'Chestnut Shaft', finds at the bottom included a water tank, car body parts and a revolver (Fig. 7.15) among other objects.

Sawn pillar faces, because they were flat, attracted attention from graffiti artists, generally (except for recent examples) using a smutty coal as the drawing agent. These graffiti sometimes depict working information – names and types of stone to be produced or having been produced. Names and initials were common; those of one family frequently occurring underground are also found on a column of the Palladian Bridge in Prior Park. Some other graffiti were slightly risqué.

The underground landscape also had many other smaller significant features which illustrate the life and work of the men (never, apparently, women) involved. When working, an obvious requirement must have been lighting. Light from entries and from the wide, so-called 'light shafts' could only penetrate short distances. Most lighting until the late 19th century was by candles. Occasional crude lumps of clay, and spatula-like stones for holding candles were found attached to pillars, though usually a ledge or placed stone sufficed as support. In the late 19th-century oil lamps, like a small watering or oil can were used, with a wick for the flame. In some of the 20th-century workings around Shaft Road it is likely small acetylene lamps were used. None of these gave comparable light to that used today, and in the original working phases such lights, except immediately next to them, gave only a slight glow, if anything accentuating the shadows. Because of rats using tallow candles as food, in various places the quarrymen left behind candle boxes made of stone with a flat stone lid, and in one case a chest-sized box had been made of thin slabs.

The underground landscape just before stabilisation

The initial survey by Hawkins (1994) revealed there were at least some 13 hectares of abandoned stone workings in the main Firs and Byfield Mines. More workings have been found since, there and at other local sites now stabilised. Many pillars had suffered from post-working stress so had actual or incipient fractures or spalling on the sides, which would not have been seen during actual working. The beds between pillars were separating and the arch effect was thus lost, with consequently very many roof falls or potential roof collapses, sometimes over fairly wide areas. Hawkins, as surveyor, was the best person to appreciate what had been done there, and he said he could not understand the pillar system – for a long time the present project had the same problem). Most of the original entries and all shafts were closed.

In the Stabilisation Scheme, penetration of the workings had often to be away from the cartways because of old infilling or actual collapses at what were termed high hazard areas. The Scheme itself, due to safety requirements for the setting of temporary steel or timber supports, generally avoided the higher ground over cartways. Whereas the original quarrymen had followed cartways, as in the

Fig. 7.15 Revolver found within early 20th-century domestic waste dump in the Chestnut Tree Shaft, near the Memorial in Firs Field

workings described above, and largely moved south to north, in the Stabilisation Scheme the movement was normally west to east, across the 'grain' of the workings

The landscape encountered in those circumstances appeared very different from the original relative order of the cartways and working areas. As the original quarrymen had retreated, they had often destroyed the original working landscape by removing facilities and systematically or unsystematically dumping spoil behind themselves. In the past the workings were usually viewed from below, certainly by visitors. Viewing from just under the roof gave a very different view of the workings. During the Hawkins survey the routes often necessarily taken across rather than between dumped spoil made much of the exploration and survey difficult.

The workings were a confusing, almost threatening environment – in a desolation of huge areas of dumped heaps of spoil, many shapes and varied distributions of pillars, collapses, with view-lines often limited to a few metres and with the wider open spaces of straight cartway routes and open working areas rarely visible. Often the height below roof was under a metre and in some cases space was probably only present due to consolidation of spoil over time. The connection between the Byfield and Firs Mines was only a crawl-passage around 0.5 m high and had been left either as an emergency route or for ventilation. Some areas could not be penetrated at all without mining through spoil.

Discovering the actual landscape

During the Stabilisation Scheme the mine surveyor and archaeological survey gradually began to build a more logical picture of the actual working landscape. One overarching feature of this was that it was possible to identify something like a hundred different Quarry Areas separated by boundary pillars or otherwise identified by changes in technical methods or dumping from one area into another. Not all would have been be entirely separate quarries, as methods could change during the same phase of operation, but it is likely many were so, and it has been possible to positively identify some quarry areas and their owners and from this their dating.

Drainage

The only conventional aspect not covered in the technology chapters is drainage. The rock structure at Combe Down is both open and porous, with the water table some 10 m below the lowest workings. Apart from small puddles in clayey or finely ground-in material, water simply drained away immediately. The only known example was located on the northern edge of cartway 168 in Byfield Quarry 505. A small drain leading to an area of rubble used as a soakaway had been constructed besides a wheel rut to better dispose of water, and was contemporaneous with stone packs built to support the edge of the cartway in the 19th century.

Chapter 8
Techniques of underground quarrying: 1. The method of working

The following chapter describes how and with what tools the useful freestone was extracted from the natural sequence of beds at the quarry face, and how the face was laid out for efficient working with respect to the needs of supporting the roof, getting rid of waste or spoil, and removing the useful stone to the surface.

The backfilling of worked-out areas with spoil at Combe Down meant that there were very few examples of the working faces left in progress and the associated areas visible to us, partly because the usual benched or stepped face was finished by being worked back to vertical. It has thus been necessary to relate archaeological observations to known methods used in similar situations elsewhere. Only two or three stepped or benched faces were observed at Combe Down, none of which had obvious relationships to the bulk of the workings. In some cases low faces may have been worked vertically, but on faces four or more metres

Fig. 8.1 Remains of a picking bed entry at top centre, about 0.9 m high in East Firs (Quarry 2205)

high, benches were necessary for both access by the workmen and the handling of the stone blocks. Steps were made either bed-by-bed, or in steps rising over several thinner beds. Use of cranes, in the 'experimental stage' of the mid-late 18th century and during their more common use during the late 19th century probably used large steps or benches, allowing the crane to be closer to the face. It is not clear if the steps were kept covered with rubble as a wearing surface on which to drag blocks (as a possible example in the Grand Canyon may suggest), or if they were kept clear of debris, as the two or three smaller examples suggest and as displayed in a fine and large stepped face seen remaining in Browns Folly, one of the Wiltshire quarries at Monkton Farleigh.

The organization and layout of the working faces – room and pillar methods

Except in the sense that every pillar is part of a working face, very few fully visible working faces survived intact at Combe Down, although the survival of the top two to four metres was not uncommon. In many cases this was because the worked-out face was infilled by spoil from another working area brought in by barrows, and in others the face had been partly reworked by pillar robbers who usually only took the upper part above or just below the spoil levels. Isolated examples of nearly-full faces included a stepped face in south Byfield of undetermined but probably late 18th-century date (see Fig. 7.8) and small benched faces in East Firs. Generally, however, they were only seen at the stage where they had been worked back to a natural vertical joint at the quarry boundary. However, sufficient was seen, which combined with information about quarrying methods elsewhere, to allow deductions to be made as to the methods of working.

Fig. 8.2 Jad slot near the top of a pillar to allow the bed above or below to be wedged out

The working area of all phases had several common features. At the top there was either a picking-bed cavity (in soft, poorly weathering beds of low value) made just big enough to enter (Fig. 8.1) or, alternatively there was some form of jad slot or notch (Fig. 8.2), either made so as to 'free' and allow extraction of the beds below. Below that was the face itself, during working usually formed in a series of steps to allow access, developed in 'rooms' between rock pillars intentionally left behind to support the roof. A typical reconstruction of the

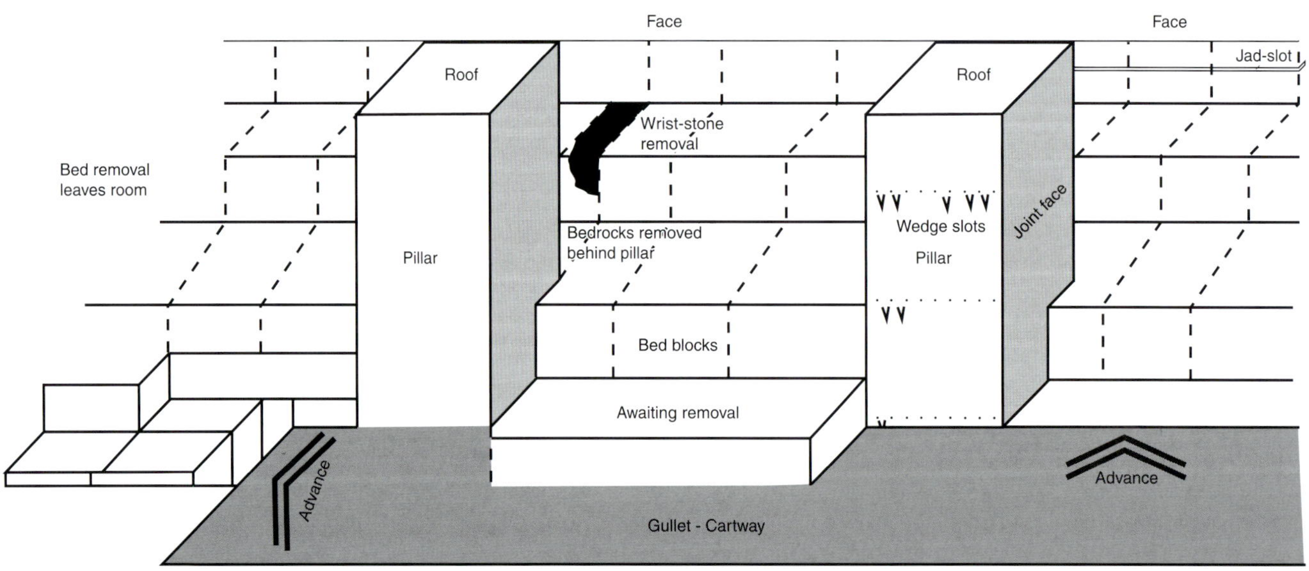

Fig. 8.3 Schematic illustration showing the development of the working face in benches to facilitate stone extraction

Fig. 8.4 Working on the steps of benches just below the picking bed of an abandoned face, Central Firs

layout is shown on the schematic diagram, Fig. 8.3. The beds below were progressively extracted to form the steps or benches that descended into a trench or gullet below the cartway or barrow-way level. The near vertical sides of the pillars were mainly natural joints, but otherwise the stone was picked or wedged (later sawn) off to form them (Fig. 8.4).

The working floor leading at the same level to the cartway was not at the bottom of the steps, but was formed with some of the waste or spoil produced, to a depth of approximately two metres. Blocks extracted above were slid down, while those below were hauled-up making handling easier. The working floor was used as space to drag stone, dress it and load it on to barrows or carts which were dragged back to the entry. This could all be very confined within a narrow gullet, or link to a worked-out area to provide more working space. In the case of Open Rooms, there was a substantial floor space. Any spoil other than that needed for the working floor was removed and stacked in worked-out areas, often at the back of the trench or gullet, held back as necessary by rubble-stone packs. Any space not used for working at the face or for transport was likely to have been used for spoil disposal.

Pre-Allen working (Phase I)

In the earliest undermining workings from an existing outcrop quarry, such as that seen at Horsecombe Vale (Quarry 2376), the opening passage was commonly made at right angles to the outside quarry face for a few metres, then turned left and right to form a passage behind the cliff face, using a gull or widened natural joint fissure resulting from cambering or landslipping of the strata, (Fig. 7.3) for ease of working.

From this the freestone was excavated as above described, to form rooms between pillars left for roof support again using natural weaknesses where possible to locate and form the pillars. The workings were usually developed inwards more or less equidistant from the entry (Pollard 1994). Spoil disposal, at least in the early crucial stages, was to the outside, forming a level working platform in front of the outside quarry, facilitating handling and transport. This led to a simple pillar-and-room system probably usually less than 3 m high including the thickness of any spoil left on the floors to form a barrow-way or cartway. Building-up of spoil behind the working face eased handling problems on higher faces.

This type of working was observed in at least two locations along the edge above Horsecombe Vale, and is also characteristic of some of the smaller apophygate pillar-worked entries of Phase I at Combe Down, except that it may have been necessary to drag the stone up an entry incline or slope to get it outside. Some rooms were left open to accommodate the most convenient routes for cart or barrow-ways and others were filled with spoil. Transport for both blockstone and spoil was probably all at one level.

Early Allen period working (Phase II)

The system used by Ralph Allen was characterized by a more systematic development. To open up more working places quickly and to assist in stone removal and ventilation, he developed pairs of fairly straight cartways, either in parallel or

Fig. 8.5 Apophygate pillar showing the area picked and broken-out at the top, with a small lip below where the freestone beds were removed (Quarry 505)

Fig. 8.6 Block of stone dressed in situ, showing the parallel jad slot used to free it

diverging out, with pillars and rooms developed between them and at either side up to two rooms deep (see Chapter 12, Case Studies 3 and 4). Cartway routes at this time had pillars 2.5 – 3 m apart, but their advance was done exactly as for other rooms. Typically rooms otherwise were around 4-5 m across but in the large underground quarry in Central and East Byfield (Quarry 505), the pillars became comparatively large and the longer spaces between up to double the usual length, with around 10 m and occasionally more between them. However, in the other direction, crossing the main joints, the spans were more normal with pillars 4-5 m apart.

In the majority of both Phase I and II workings, the picking bed was an opening including pillars of just under a metre high, and probably about five metres wide and deep (which is about a superficial perch, a very common measure used in the quarries) but had to be adjusted to the joint positions and any other rock weaknesses. The removal was done by first picking out a block of stone using a joint to start with, or by breaking with a hammer, after which picking or barring out of blocks was possible. The pillars left in solid rock were formed at the top using picks, so that the top was left spread out or apophygated, under the roof on all sides. In this type of pillar, the base of the picking was found marked by a small ledge or lip, from which the freestone below was broken out (Fig. 8.5). In this respect Allen appears to have followed the same practice as most of the previous, Phase I quarries.

As found nearly all faces seen were worked back to the final boundary at the vertical. However, the freestone beds, at his time as well as later, appear to have been extracted bed-by-bed, probably in steps descending to the front of the face, from the picking beds under the roof to the gullet or trench in the floor. By opening out a joint the first block of a row of them between pillars could be broken out after which the whole width of the bed could be barred out, or detached by wedging if tight, starting at the side of the pillar if feasible. By keeping the floor level, by dumping spoil, about two metres or a little more higher than the bottom bed, it not only helped dispose of spoil, but also meant both highest and lowest beds could be easily reached and the blocks dragged back away from the face. The level of the spoil was kept the same as the cartway back to the entry, which in Allen's Phase II workings was about 2.4 m from the roof.

The gullet was progressively filled with spoil as the system moved forward with a slope or a rubble pack holding it back. The side rooms were similarly worked both to left and right as the cartway advanced past their positions, probably opening the uppermost beds round the sides of the pillars while access to them was still convenient, and similarly behind the first row of pillars dividing to the left and right to form the next row of rooms. At the end of whatever extent was intended for quarrying, the beds were removed successively from the top layer down until the final face was vertical.

There was no positive indication found of linkage of picking bed cavities by a small opening or 'window' between them and adjacent worked-out areas, but unnecessary vertical movement of spoil was always to be avoided and some method may have evolved. When working in a room was completed it was generally filled with spoil, the workings near the entries particularly so, practically up to the roof.

Handling and moving the blocks of freestone released was only a considerable problem with thicker beds, and these could be cut into smaller pieces or dressed to size *in situ* unless the larger blocks were specifically required. Figures 8.6 and 8.7 show two examples of blocks *in situ* in a dressed state using two different forms of jad cuts used to free them. Most beds were fairly thin, below 30 cm thick, so were manageable, and thinner beds could be made into coursed rubble, suitable for the huge quantities of walling stone used at the backs and in partitions of buildings, especially in the mid 18th century. Larger blocks might be roughed-out to nearly the form required, for instance for lintels, reducing weight to be taken out. Where large blocks were specifically required, then timber slides and levers would be used to assist. The thickest bed generally seems to have been about two or three metres below the roof, so blocks from it could be slid or rolled straight on to a cart: it is possible the floor level was adjusted in height for this.

The probable advantages of concentrating on driving the cartway forward, as shown by the limited development sideways, were two. Rather

Fig. 8.7 Dressed block in situ showing use of a chamfered jad to assist removal. The use of the chamfered jad may be just the choice of the workman, but was probably mainly used for shallower depths

than 'fanning-out' the workings and utilising convenient rooms as a cartway, the cartway was purposely driven straighter, and, additionally, much greater numbers of working places were opened, though not necessarily being immediately worked. This was useful to cope with both the high production Allen aimed at and the peaks inevitable with the uncertain economic fortunes of the building industry. In the pre-Allen method the rooms were irregular and the route was likely to be far from straight. Under Allen, his fairly narrow pairs of fronts penetrated deep into the strata with straight cartways facilitating movement of carts from the quarry. The arrangement of two interconnected cartways provided much better ventilation through the quarry.

At some stage in Allen's time the use of picking bed removal was abandoned in favour of the face being holed at the top by the jad slot. This was not new, even at Combe Down, since the pre-Allen quarry 913 known as E4 Stub (Quarry 2347) and dated by graffiti and a clay pipe, had used this system before and after 1725. It is likely it was then familiar to Richard Jones, Allen's Clerk, as an apprentice there or nearby. The use of holing, locally called a jad, was an idea similar to, and possibly inspired by, coal mining practice. It involved cutting a long and fairly deep slot back to the next joint in one of the picking beds, perhaps two beds below the intended roof. This could be V-shaped notch or a slot with sub-parallel faces about 100 mm apart and could be up to a metre deep at Combe Down, depending on the joints. By inserting a wedge into the bedding plane below the bed below, the bed could be broken off back to the joint, or as far as the jad penetrated. The top bed under the roof was usually left slightly projecting, forming a corbel at the top of the pillar. This replaced the picking-bed-removal method, and also led to the rather laboriously picked out apophygate pillars being replaced with corbelled types with the projecting or corbelled top (Fig. 8.8). The corbel, like the supposed benefit of the apophygation, was probably considered to spread the load of the roof over a wider area, as did the apophygate pillar head. The freestone beds below could be then worked just as before.

This was probably a quicker method, and by apparently leaving one or two picking beds untouched in the roof, reduced the amount of spoil. Corbelled pillars were found either ahead of the apophygate pillars, or parallel to the cartway, but behind the (typically) first two rooms with apophygate pillars. Thus the side position shortened the travel distances back to the entries. The layout sequence was seen replicated in Central Byfield, Central Firs and in East Firs.

Adoption of new methods of organising the face layout

It is not fully clear why the system used in Allen's time changed, nor why two radically different systems were adopted. One possible reason is either that both the apophygate and corbelled pillars were defective in their design or that their

Fig. 8.8 Corbelled pillar in Firs Quarry (2201) Quarry Area 2211

design limitations were grossly exceeded, which will be covered in the chapter below on roof support. Several areas of the quarries with these types had either contemporary or subsequent collapse. The area east of the Avenue, for instance, with these pillar types had the largest number of 'High Hazards' found in the Hawkins survey (1994). A further reason may have been that after Allen died in 1764, the new lessees of the Estate included some who brought in new methods in an effort to increase efficiency. Both the Long Room and Long Wall with gullet-and-pillar systems had some clearly comparable features with methods used in coal mining in the 18th century (Flynn 1984, 82 *et seq.*).

Long Room system: Allen Estate Phase III working

This system was developed in Allen Estate times and was almost exclusively used in Central Firs east and west of the two Allen cartways and either side of the eastward spur (Quarries 2200; 2201; 2202). It was characterised by a Long Room developed along its length, rather like the cartway. However, unlike the cartway and its adjacent rooms, Long Rooms were developed in parallel with each other and their sides were often not completely divided into separate pillars, but had windows or doorways at a high level. It was a very systematic way of working, minimising the spoil-dumping problems. Particularly good examples were seen just west and north of the Firs Shaft (Figs. 8.9 and 8.10). Usually the rooms opened out from a cartway or a branch from it. The long sidewalls may have increased stability and led to improved spoil handling, but a disadvantage was the unworked stone found left in the side walls In East Firs, later 'pillar robbers' seem to have particularly favoured Long Room walls as the source of their stone, in one instance apparently removing the whole length.

The first Long Room in a sequence was developed just as a tunnel 4-5 m wide (Fig. 8.11). Because of the height worked it almost certainly used a stepped face with a gullet behind and a floor some 2 m deep in spoil at (or even higher to assist loading) than the outside cartway level. The next and parallel room used or formed windows and sometimes doors (rather like pitching holes in a barn) just below the jad-cut picking bed level, at intervals, into the earlier room, through which high level waste was thrown onto the former. already spoil-raised floor. Finally, as the stepped extraction system permitted, the windows were cut down between the new and old rooms, forming pillars between them. Accumulation of spoil may account for the windows found that were not created or fully developed into doors in the side walls to form separate pillars thus lowering the proportion worked. The process was repeated with the development of the next and adjacent Long Room.

Fig. 8.9 Long Room developed just west of Firs Shaft (Quarry 2200). Note the window in the left wall, and the almost continuous side walls and the end wall which would greatly assist stability

Fig. 8.10 Window used to dump spoil from the Long Room just beyond, north of Firs Shaft (Quarry 2201)

These Long Room areas seem to be associated with the use of cranes, and it is possible that the crane cable was used to drag stone from the face to the adjacent cartway or access. In Firs the cartway in the Allen Estate Phase III appears to have been somtimes at a lower level (with a greater height – 3-4 m to the roof), presumably to assist loading. The evidence for crane use was represented by the distribution of Lewis slots used as anchors and groove marks around and across pillars, used for the cables (chains) used to hold the mast. They are considered in more detail in the following chapter. The use of cranes, like the Long Room system, seems to have been limited, and it is possible the system did not fulfill expectations, perhaps because the space for cranes was limited and they required moving too frequently. The Long Room system seems to have been abandoned in favour of that described immediately below. However, in the late 19th and early 20th century a somewhat similar system was used with Open Room methods, and wider and higher rooms and with post cranes, leaving either solid walls alongside North Road or by leaving almost continuous 'boundary walls' between successive workings, as at Foxhill (see Chapter 12, Case Study 13).

Long Wall with pillars and gullet: Phase III to IV and later

From its relationship with the Phase II workings, this system was apparently developed at almost the same time as the Long Room system, but at the western side of Central Firs. It has parallels in the hand-working of surface quarries where there was an advance along a long face, removing a thin strip of stone along the whole length, with only a narrow trench or 'gullet' between the face and spoil dumped at the back. A degraded surface example occurs in the 'rough' at the southwest corner of the Byfield area, close to Shepherds' Walk. Clearer examples survive near the northern end of the Isle of Portland. The difference underground was that pillars had to be left to support the roof so the gullet had to be wider to allow this. The 'type site' at Combe Down where this method was first identified was at a face in north Central Firs (Quarry 2224), though as usual at that point the face, at the apparent boundary pillar, had been worked to vertical (Fig. 8.12). The system minimised the loss of stone in pillars by their more regular spacing (and by leaving less stone between when compared with the Long Room system above) and both the distance of movement and manual elevation of spoil required was reduced.

The cartway or stone barrow-way into the area could serve both a four-metre-high gullet (the working trench area) and faces developed at right angles to either side, several gullets at suitable intervals as it progressed forward. In front was a stepped face with a jad slot continually being developed forward under the roof as the bed-blocks behind were removed. Pillars were formed between the rooms being developed, using natural joints or by breaking using wedges, as distance and the joints required. Once beyond the pillar, the steps and removal of bed-blocks were continued behind the pillar as well, thus beginning the formation of the next room. The floor level in the gullet, at the same level as the cartway, was maintained using spoil from the working of the face. Thus blocks of stone from high on the face were slid or passed down, and those below the cartway had to be hauled up. Behind the gullet, which was kept wide enough for the barrows or carts used, along a length of some 30 to 40 m, was a continuously advanced bank of spoil, interrupted only by the rows of pillars left behind previously, and supported by a series of stone packs at the back of the gullet.

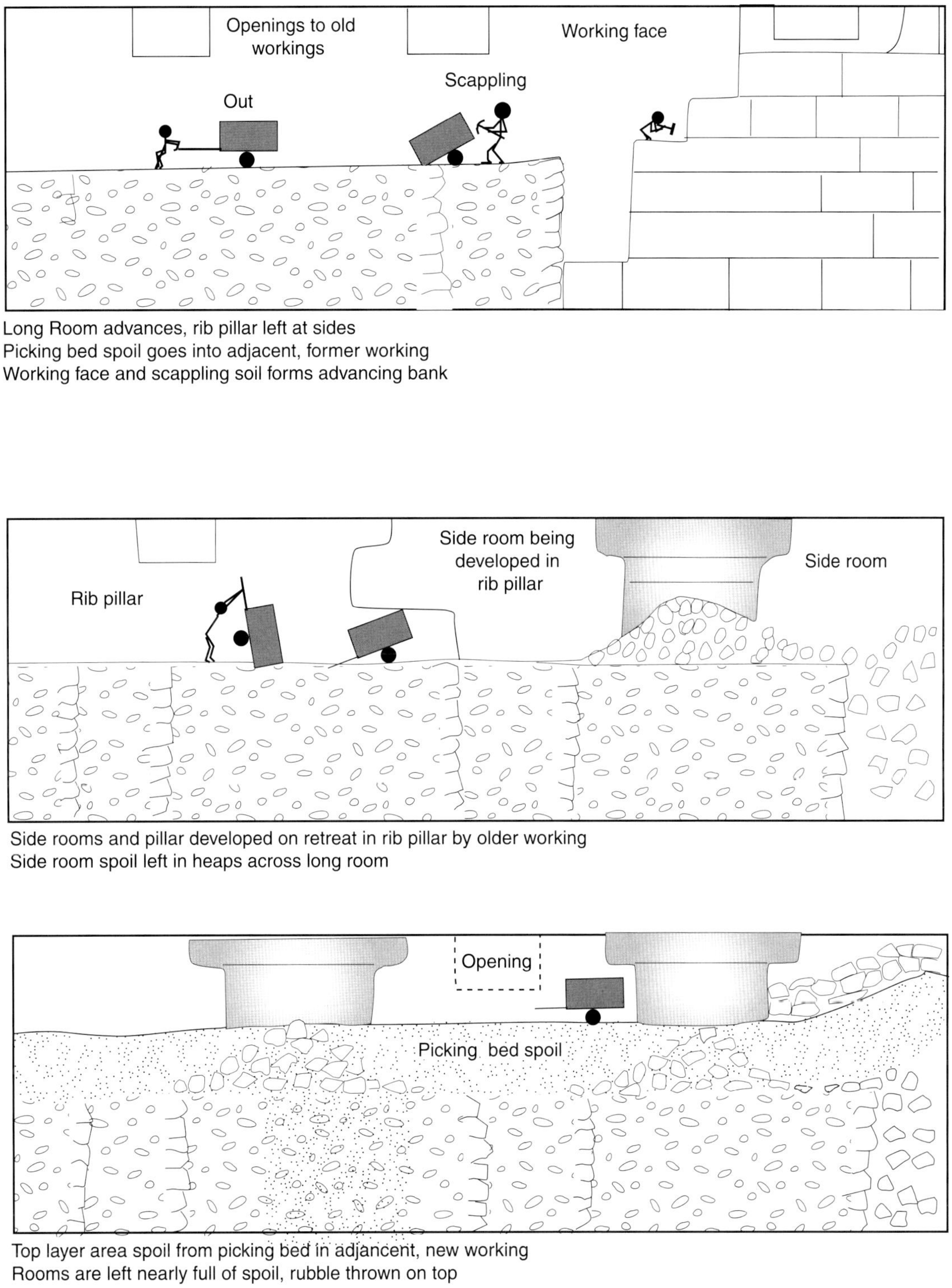

Fig. 8.11 Reconstruction of Long Room working method

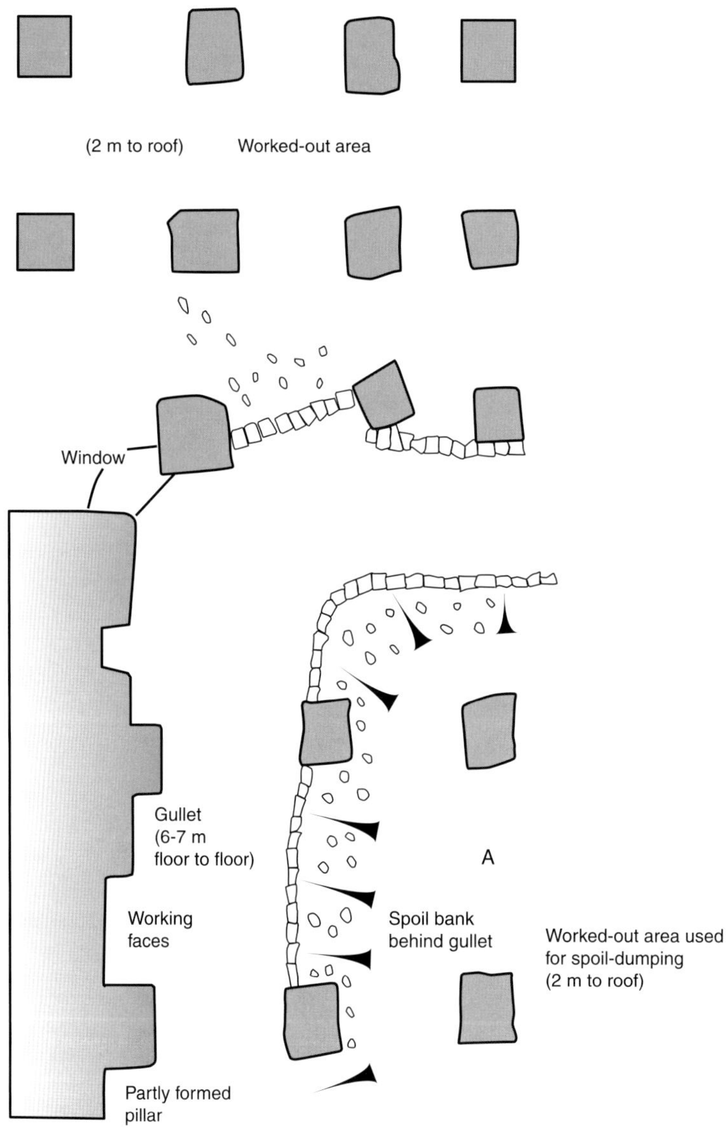

Fig. 8.12 Sketch plan of Long Wall, gullet and pillar working in Quarry 2224 in north central Firs. The area on the right (A) had been worked similarly, then was completely backfilled

Much of the initial spoil would have had to be removed as the long face started (into older workings elsewhere), until the first set of rooms had been formed beyond the gullet and pillars and the next set started. Then the set of rooms behind the pillars would form the new gullet and the old could be backfilled to a reasonable throwing-up height below the roof, typically 2 m or so. Thus a 6 m high face would have the lowest part some 2 m below the gullet, and behind the gullet spoil was stacked, contained by a series of advancing rubble walls to a height leaving 2 m or so open below the roof. High-level waste could be barrowed on planks over the gullet, or around the end on to the spoil bank behind the gullet, often being taken some distance away by barrows and dumped systematically to save space, but sometimes being just randomly dumped.

It was probably easier to work stone behind pillars by approaching from both sides and was substantially more economical in the labour cost of spoil removal by this systematic dumping, which involved some 40 % of the total material excavated. It also enabled easy supervision of several groups of men by a 'ganger', and made transport of the stone more systematic too as it was from the same area.

This multiple-face development was the dominant late 18th-century technique continuing into the 19th century and was, in slightly modified form, the technique used in the 19th-century Burgess workings in the Three Acre Quarry (2211), with continuous boundary pillars marking the end of each area of extraction and the starting position of the next, probably for stability with the high workings in such as Burgess' Quarry, which reached an extractive height of 8 m. This great height limited the capacity to throw material up the bank behind the gullet, and spoil management systems had to be developed to cope with lifting the extra two metres or so. The width of the gullet tended to be wider as the ceiling or roof height increased and the stepped area necessarily became greater to maintain a safe angle, so that planks across the gullet became less practicable. However, barrows could be moved around the ends of the gullet and, by use of inclines from the bottom or middle, to gain access to the dumps. The complexity of this is discussed for the Grand Canyon area (Chapter 12, Case Study 9). This same area had a face worked high up with a bank of rubble which may have covered the steps below. It was, unfortunately, not entirely clear whether the blockstone was dragged down on the bank of rubble to a barrow-way at the bottom or if some other mode of moving it was used.

Open Room working: Mid and late 19th century, Phase V and VI

This mid and late 19th-century development was designed to accommodate the use of cranes and winches underground to handle stone. It was partly made possible by either a reduction or better management of spoil. This had begun with the development of jad slots and, as the depth of freestone worked increased, the proportion of waste fell and the overall space available increased, in addition to any left from a previous phase of working. Use of more precise cutting of pillar faces using rows of wedge-and-chip breaking and sawing of faces also contributed (see below). It is also possible that all mining areas of the main Combe

Fig. 8.13 Open Room working area in Byfield Quarry (518). Cranes and winches were used to drag stone away from the working face across the floor. (Photo by Paul Deakin)

Down complex by the mid 19th century had plenty of adjacent older workings to accommodate spoil outside the current working areas. The system still had to use stepped faces to gain access but these could be steeper with use of crane or winch cables to haul blocks up or down to loading level. Given more room, loading could be done well back from the face by a simple crane lift.

In Open Room working, the floor was still built up of spoil behind a gullet as before, to be at the same level as the cartway, and was kept level by maintenance and by the dragging of blocks over it. Wider spaces resulted from keeping the spoil further behind the actual working area, and, where possible by greater spacing between pillars. Initially, since these workings were all surrounded by older worked-out areas, it is likely most waste was removed into the older workings until the need for handling room exceeded the space available. Typically the width between pillars was increased, requiring large numbers of timber sprags (locally termed 'scorters'), positioned from near the tops of pillars to the roof to tighten roof blocks against each other and possibly to reassure the quarrymen. Boundary pillars may have been carefully preserved and even buttressed by leaving rock at an angle to them, to add stability to the local area. Substantial open areas of some 10 m or so across seem to have been possible, sometimes with such two areas at right angles (Fig. 8.13). Some spoil dumping could still take place at high and intermediate levels into adjacent areas, but with lesser amounts it seems to have been possible to barrow it up on smaller and gentler inclines to flat-topped, bench-type spoil dumps within the worked-out areas as they became available. In many respects, the working underground was again following the methods used at surface, where hand methods had for a long time been supplemented by the use of stayed or Scotch derrick-type quarry cranes. Standard cranes of both stayed (using cables and Lewis bolt anchorages) and post-types were used, the latter becoming the norm. Winches, probably of the cast iron crab-type were also used for dragging, secured by chains and iron pins back to the wall. Cranes and winches are discussed in the following chapter.

Later Long Room working: late Phase VI and Phase VII

At Shaft Road and at the Foxhill workings, post cranes were used in conjunction with narrow-gauge railways to handle the stone blocks produced. This was done in large, very Long Rooms, 10 m and more in width. The usual stepped face was used, with a gullet behind, and a bank of spoil forming the railway foundation behind that (Fig. 8.14). Other spoil was stacked at the sides, so the railway effectively ran in the trench between. The crane was sited on the spoil bank, with the post inserted into a heavy stone block on the floor, and into a square chog hole in the roof. Blocks could be dragged some distance from the face, so the crane could stay in a single location for some time. Other cranes may have been used in conjunction as stacking cranes or to facilitate the transport of the block further towards the surface with the aid of further crane haulage. The railway was brought up behind the crane. Previous crane positions were obvious from

Fig. 8.14 Late period Open Room working method, showing use of a post crane to pull stone blocks off the face and load them on to a flat wagon (after Wooster 1978)

the chog hole in the roof. At Shaft Road two rooms ran parallel, a thick continuous wall between them except at a junction. At Foxhill, four separate Long Rooms (worked by three different ventures at different times) had boundary walls maintained between them. This method appears to be a modified Wiltshire practice.

Features on working faces

Picking bed removal

This was particularly used with forming apophygate pillars in Phases I and II. However the use of full picking bed removal continued until much later in some workings, notably in the quarry owned by Hulonce in West Byfield in the early 19th century, where it was combined with the jad slot, which more generally replaced it, in order to break into the picking beds. Evidence of its survival in the early 19th century was also seen on pillars in north Central Byfield, for example, some of which had been penetrated so deeply that they required stone packs or a timber props to support the roof (Fig. 8.15). In these later cases personal idiosyncracy of the quarrymaster involved was probably the key factor.

It seems possible that about a square perch of ground ($5^1/_2$ yards or 5 m square) was freed at a time as this was the common measurement used in calculating the value of stone recovered, but the actual size would depend on the weaknesses found in order to locate and leave the pillars. Though very few marks remained on the apophygate pillars, due to their tendency to spall, enough remained to show the intended pillar tops were carefully formed using picks with inch-wide blades, to either form an arch in some early examples, or alternatively to allow the top to bell-out, so as to provide a slightly wider area of support under the otherwise flat roof. It seems likely this particularly skilled work was done by a specialist part of the workforce almost certainly on some form of piecework since occasionally the area removed had been marked out by black lines indicating measurement.

Jad slots replacing picking bed removal

Jad slots were in use in pre-Allen workings *c* 1725, but the use then seems to have been isolated to one quarry under the south end of The Avenue. The advantage of the jad slot was that it produced much less spoil and generally only one or two of the picking beds was removed. The change was sometimes marked by an entablature between pillars at the transition between old and new methods, with the roof one or two beds lower on the newer side. The introduction of jad slots was associated with corbelled pillars (below), but they were later similarly used with direct pillars. In both cases the slot was driven in leaving a bed *in situ* below the roof, presumably to avoid damage to the workmen's hands, but in the latter case a wedge was driven in under the roof to bring down the upper bed so it was in line with the side of the pillar. The wedge often left a small mark where it was driven in (Elwyn Davis' observation). Occasionally wedge-and-chip splitting was used with roof beds stuck together with calcitic flowstone, the marks of which were seen in East Byfield and particularly in the underground Springfield Quarry workings.

The jad slot generally formed the first access close to the roof, what was, in later times at least in the Wiltshire Bath Stone quarries, referred to by the

Fig. 8.15 Stone used to pack up the top of a pillar where it had been intruded upon by the picking bed, east Firs

quarrymen as the 'breach' across the width of the room. In some cases, in what we termed chamfered jads, it was a wedge-shaped (flat-bottomed) or notched horizontal slot usually utilizing a bedding plane weakness and driven in until it met a vertical joint (Figure 8.16 shows working methods and terms used at the working face). Usually, the depth reached does not seem to have been large. The more conventional jad slot as used in 1725 (above) became routine after about 1750, usually by then with the more economical, parallel upper and lower surface. It was some 10 to 15 cm high and could be up to a metre or more deep although, at Combe Down, again because of the joints, it was generally substantially less. Vestiges of jad slots were mostly seen on pillars, but in a few areas, were found as continuous or semi-continuous slots across long lengths of a working face (Fig. 8.17). Variations in slot heights appear based either on local geological factors or personal preference, and perhaps the techniques required for extracting any given block efficiently, rather than development through time. Several different heights or types of slot could be used in a single quarrying area and across several periods. In later working, the jad slot was made wide enough to insert a narrow saw called a razzer,

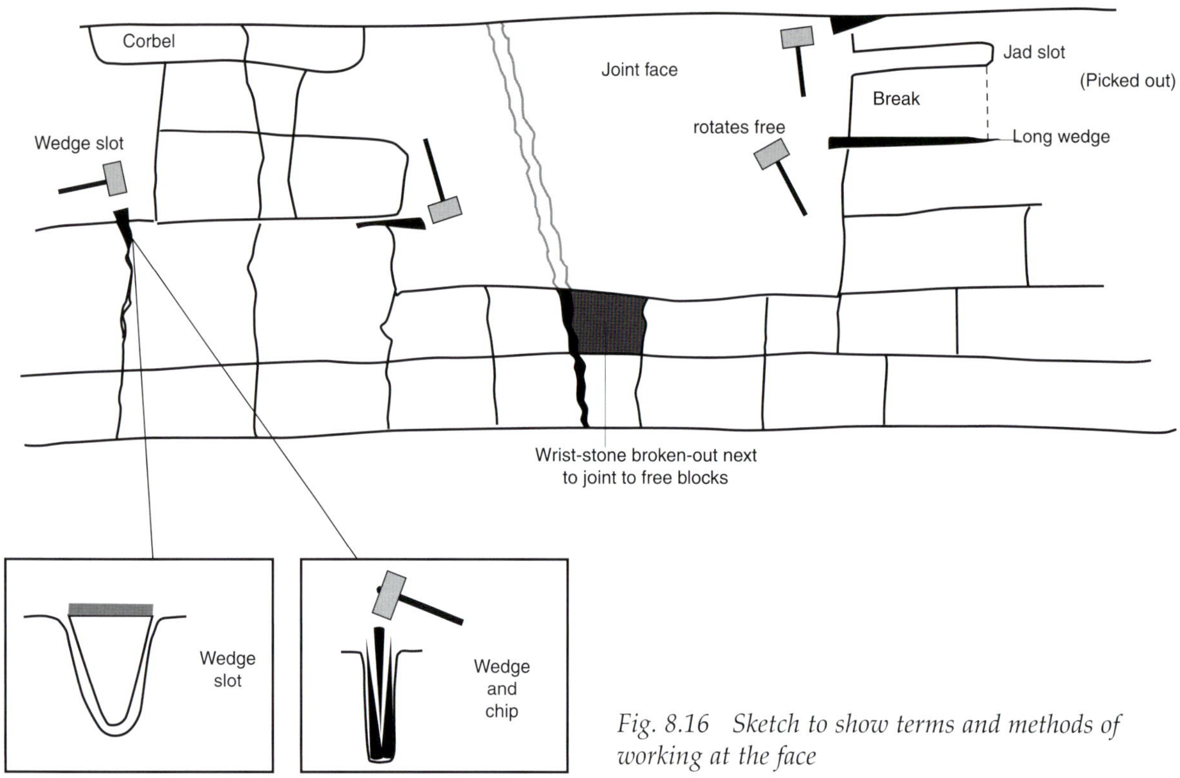

Fig. 8.16 Sketch to show terms and methods of working at the face

Fig. 8.17 Continuous jad slot across a face

until the saw cut allowed a regular frig bob saw to be inserted.

The jad slot was less labour-intensive than picking bed removal. There is some possible evidence, notably in southern Central Byfield that the two methods overlapped for a time with some pillars displaying one apophygate side and one corbelled.

The use of the term jad also refers to a tool, a 'jad' or 'jadding iron', a substantial bar with a wedge-like end. It was used to pare off upper or lower surfaces to keep them roughly parallel, or to square off inaccessible corners inside the back of a jad slot. The use of the term in relation to the horizontal 'jad' slot and the vertical or 'upright' jad or jad cut, within the context of the archaeological survey at Combe Down (so to easily distinguish the two types during recording) is somewhat anomalous in that most of the feature seems to have been cut by a horizontally swung blow of the pick in line with the length of the slot. The vertical rear ends of the slots, which survive on pillar faces, thus have rows of pick marks that can be attributed to the curved path of the pick blows. A description of this, as used at Box in Wiltshire, notes the long-handled pick, at the instant of impact, was dropped and then drawn out (Raymond 1870, 482), the dropping presumably to lessen the shock of impact on the quarryman. The tool-marks left by a jadding iron reflect blows delivered straight at the rock, usually on the underside of the cut, and have a different orientation, normally at right angles to the length of the slot (Fig. 8.18).

Price (2007), referring to quarrying near Cheltenham, cites a man as able to cut 13 square feet of jad slot a day (about 1.2 sq m) but in the context of the soft picking bed and shallower jad slots at Combe Down, substantially more than this would probably have been expected. The regularity of such work was probably an indicator of the skill and efficiency of the operator.

Cutting down the face

Cutting out of the stone either bed-by-bed using wedges or, especially next to pillars, by using long vertical cuts down through the beds was first done by picks or by wedges (Fig. 8.19 and 8.20) producing what we termed wedge pits to particularly distinguish this use. Later, in the early 19th century, within a very limited area of West Byfield, such vertical cuts (see Fig. 8.21) were extensively used and also termed jads (we used the term jad cuts to distinguish them), leaving curved pick marks on the full height of the pillar faces, and starting at about the same period, by use of saws (Fig. 8.22). Jad cut features, were sporadically evident elsewhere through the Combe Down Quarries. The earliest example was noted in the Stub E5, Quarry 2348, which is thought to have been worked contemporaneously to or just after the Stub E4 Quarry 2347, which was being worked in around 1725.

Fig. 8.18 Underside of a jad slot showing pick marks at the rear and jadding iron marks

Fig. 8.19 Wedge pits left in situ, East Firs

In the bed below the picking bed, between each pillar, either a natural weakness was widened by use of a heavy pick, or a wrist stone (Fig. 8.23), a convenient bed-block between joints, was removed by wedging and hammering to allow subsequent removal of adjacent blocks either side. Ideally a suitable joint could be located next to a suitable pillar position and the beds to be removed were worked from there. This continued until the whole exposed beds at a particular horizon were removed except at the pillars. At Combe Down the presence of wide gulls and joints sometimes made this particularly easy, especially near the outcrop.

Fig. 8.20 Wedge pit impressions left from cutting down a pillar

The substantial amounts of spoil produced by this breaking, and the subsequent squaring or scappling of the blocks, was removed and dumped as close to the working face as possible to minimise effort. Most of the stone sold (to judge by the contemporary buildings in Bath) was fairly easily-handled coursed rubble, but more substantial blocks had to be lifted, barred or dragged down onto or just above the gullet floor. Very large blocks, notably from the bed below about 2 m below the roof were slid or, sometimes, rotated out possibly directly onto a wagon positioned at the appropriate height next to it; scratch marks from re-working within Quarry 2202 show at least one example against the northern boundary pillar of an earlier Quarry 2201. After trimming of the smaller rubble or blocks, and squaring or scappling the larger, using a heavy scappling axe, the product was placed on barrows or carts for transport to intermediate stone storage or finishing areas or alternatively straight to the entry and onwards.

This somewhat crude system of block removal was assisted and somewhat refined by use of wedge pits using a wedge-and-chip technique and, in the

Fig. 8.21 The jad cut, used to cut down the side of a pillar, West Byfield

Fig. 8.22 A saw-cut pillar, mid-late 19th century, north Central Firs

late century onwards, their systematic use in groups was visible on pillars. By the early 19th century the efficiency of operation and the amount of spoil was also reduced by using either a jad cut or by using a saw, in either case usually forming one side of a pillar. Natural joint faces, however, remained common on other sides.

Working floors and storage areas

The existence of a flat and reasonably smooth floor, back from but still close to a working face and at the inner-end or next to a cartway, is an indicator of activity where either the crude blocks cut from the working face were trimmed or scappled with an axe

Fig. 8.23 The gap left by removal of the wrist stone, freeing adjacent blocks for removal. In this case the removal was probably part of the re-working of old pillars, Central Firs

Fig. 8.24 Stacked wrought stone, produced by a banker mason, East Firs

into squared blockstone or, in a few cases was used as dressing floor where stone was wrought or finished ready for building by banker masons. Stub E4 (Quarry 2347) and several places in east and Far East Firs, all quite small quarries had modest amounts of wrought stone stacked or on the floor and in these instances it is suspected the operator was primarily a banker mason rather than being a blockstone producing freemason (Fig. 8.24). Block stone was a rare find, with two sets of finds in Byfield and one in East Firs – the rarity suggesting perhaps it was usually trimmed near where found and it was then immediately carted out. It was perhaps because the Combe Down stone needed no drying before weathering, that the large over-wintering underground dumps of blockstone found in the Wiltshire underground quarries did not occur at Combe Down, though the space adjacent to the cartway near the Arched Shaft under Firs Field (Quarry 2211) may have been used for storage (Fig. 8.25). In later workings, using the Open Room method, much larger areas of floor were kept clear, but there it was probably for handling by winch and cranes.

Dressing areas for wrought stone usually had clear and well-trodden floors, sometimes with neat heaps of waste fines and chippings. All seem to have been associated with nearby working faces but again these were few. This waste, sometimes with broken carvings or mouldings, is termed 'banker waste'. Typical products were stone copings, but a wide range of mouldings were also recovered (see Shaffrey 2009). Effectively these were masons' workshops and at least two (adjacent to John Scraces Quarry (510) and Quarry (2219) under Stonehouse Lane, displayed an extreme neatness, quite unlike the rough and ready general working areas. Dressing floors as part of the working faces thus were unusual and all small-scale. For larger-scale operators the surface dressing floor appears to have had the advantage.

Toolmarks

Toolmarks provided important evidence for extraction techniques, and associations of toolmarks were often the best indicators, indeed, often the only way, of seeing how tools and other equipment were used. For the early-mid 18th century, however, the evidence (except in the E4 early workings) was slight. Hammers may leave curved shatter faces, but very few crush or prying marks were left from levering blocks clear, partly because, once the wrist stone was removed, the open jointing made other blocks easily removable. This lack of evidence was also due to the archaeologists never seeing the lower parts of early pillars and to the tendency of the contemporary apophygate pillars to spall, thus losing the original worked surface. The replacement of apophygate pillars with corbelled pillars and later, direct pillars left far more traces and from then onwards, tool marks were frequent and relatively easy to diagnose.

Fig. 8.25 Possible stacking area for blockstone in Central Firs (Quarry 2211)

Use of the scappling axe and hacks (and picks)

The scappling axe, of which an example was found, was a double axe, which took the form rather like the modern pick axe, but with the cutting-blade end in line with the swing of the tool. Scappling usually described the trimming of blocks to square them, forming blockstone. The scappling axe was, however, also widely used to roughly trim back the rock at corners of pillars to allow carts to pass, or to allow insertion of a saw or to widen or form a jad slot under the roof. They yielded characteristic wide 'chops' or blazes on the rock face (Fig. 8.26). Hacks were a similar tool except that the blade was at right angles to the swing, and left an indention cut the width of the tool on the surface. Although the term 'hack it back' remains even now in common use, the use of the hack, especially the narrow bladed (up to 2.5 cm) types, seems commonly to have been referred to as picking, though technically the pick is pointed. Many jad slots used the hack as well as the pick, though which can be easily identified. The most common form of pick type recorded at Combe Down was a flat-ended pick that had a width of 1-1$^{1}/_{2}$ inches, with the toolmarks often evident in the rear or the back of the jad slots. The narrow ended pointed pick was used in later quarrying after about 1870 and was usually still associated with the creation of a jad slot, but generally with a higher jad slot also for the subsequent insertion of a saw.

Fig. 8.26 Marks left by a scappling axe on a face, with wedge slots, south-east Byfield

Use of wedges

The wedge was either a fairly heavy, tapered iron tool (more sophisticated forms had a steel core) which could be driven using a sledgehammer into a joint or bed to lever them apart (above or below a jad slot), or the smaller more lightly constructed wedge-and-chip used in a pre-cut pit to split solid blocks. Various types of wedge holes (used as a generic term where a specific use was not determined) or pits (used with pit and wedge) or slots (used as anchors or to support a plank) were found with different uses. The terms for wedge pits etc. were necessarily used by us in absence of an established terminology to differentiate types, but as used here are not historically accurate and, for instance, Bath Stone quarrymen referred to the wedge pits as 'wedge holes' from the 19th century at least. A wedge hole is a hole made for a wedge and chips to fit into, in order to break out stone (Keith Palmer pers. comm.). As a young man Palmer worked with the last generation of quarrymen to use wedges and chips (Pollard pers. comm.).

The term 'wedge pit' was used during the archaeological survey to describe a triangular shaped slot, placed vertically or, more rarely, horizontally, into which two thin, trianguloid iron chips were inserted and a central wedge hammered in (see Figure 8.16) to break stone blocks from pillar faces, to lever a block from its bedding plane or to break ('rap') down the first bed of stone below the roof level and above a jad slot. The two slightly tapered chips were inserted in the pit with the thickest part at the base, then the thin wedge was driven down between them, exerting lateral pressure at the long sides. The pits were generally *c* 100 to 150 mm deep with parallel sides about 25 to 30 mm apart. These features were found in nearly all the quarries.

The use of chips, the commonest found iron artefact (see Scott, Chapter 11), was probably necessary because of the softness of the rock, spreading the taper effect of the wedge evenly over a much wider area. The slots were frequently recorded on pillar faces, mostly on the perpendicular faces set to the principal joint structure, and close to, but not aligned on, the secondary joint structure. They, along with many other tool-marks, are generally absent from the natural face of the principal joints, which usually required no additional working in order to extract the stone adjacent. Chip impressions left by this process were fairly common within the quarries but wedge impressions were less common because the wedges were placed between the chips and often had no direct contact with the face or the roof. However, in some places the wedges were driven beyond the limits of the wedge hole into the solid rock face, leaving a scar or impression at the base.

The evidence suggests that the wedge pit was not commonly used in the mid 18th-century Ralph Allen quarries (Phase II), an impression that is slightly ambiguous, as in fact, very few tool-marks of any kind survived on apophygate type pillars in Byfield Area (Quarry 505) and were apparently almost entirely absent in the Firs Ralph Allen Quarry Area (2200) where the pillars were well-preserved and any such marks would have survived. They were used more frequently during the Ralph Allen Estate (Phase III) periods based on evidence from Central Firs between Quarry Areas 2200 and 2201. Examples of wedge holes were found on pillar faces, in one case several being used together to split a block from its bed (Fig. 8.27). Wedges in combination with jad slots and barring seem to be the primary extraction method used during the late 18th century at Combe Down leading to greater regularity in pillars. This continued in during the 19th century, only diminishing in importance after the use of the frig bob saw became routine.

Some more complete examples of wedge impressions were found on the quarry roof and regularly survived better than examples on worked pillar faces, because they were unaffected by subsequent quarrying as extraction progressed downwards, were less impacted by subsequent later scavenging of pillars, and more importantly were more visible as they were not covered up by dumped spoil. Wedge pits and chip impressions within them were noted in only two early 18th-century quarries in the Firs and a number were recorded on the roof of the 1725 period E4 Stub Quarry (2347), with only a single example on a pillar face. Wedge-and-chip 'rapping down' was, however, used extensively at the late 18th-century Entry Hill underground quarry (Fig. 8.28) where the orientation of the wedge pits, in combination with the orientation of infilling with waste, allows inferences to be made about the directions of working.

Isolated groups of wedge pits were observed where blocks were removed from awkward places – around the tops of the pillars and in areas where the roof was bonded to the beds beneath, usually by calcite. They were sometimes driven in tandem to remove the same block, or singly in a progression of working across a quarry. In places, wedge pits faced each other and seem randomly placed, in others they illustrate how a temporary supporting pillar was circumvented and subsequently removed.

Working between pillars would have destroyed countless wedge impressions created on faces during the initial stages of work, and the impressions left on corresponding removed blocks would have been removed during finishing by sawing and/or scappling. However, a number of blocks which had been discarded as unfit for use bore single or multiple wedge holes. These were recovered from barrow-way surfaces and noted in banks of spoil. A large block (SF 753) had been abandoned on a barrow-way in an early 19th-century Firs Quarry (2203). The block bears the traces of two $5^1/_2$

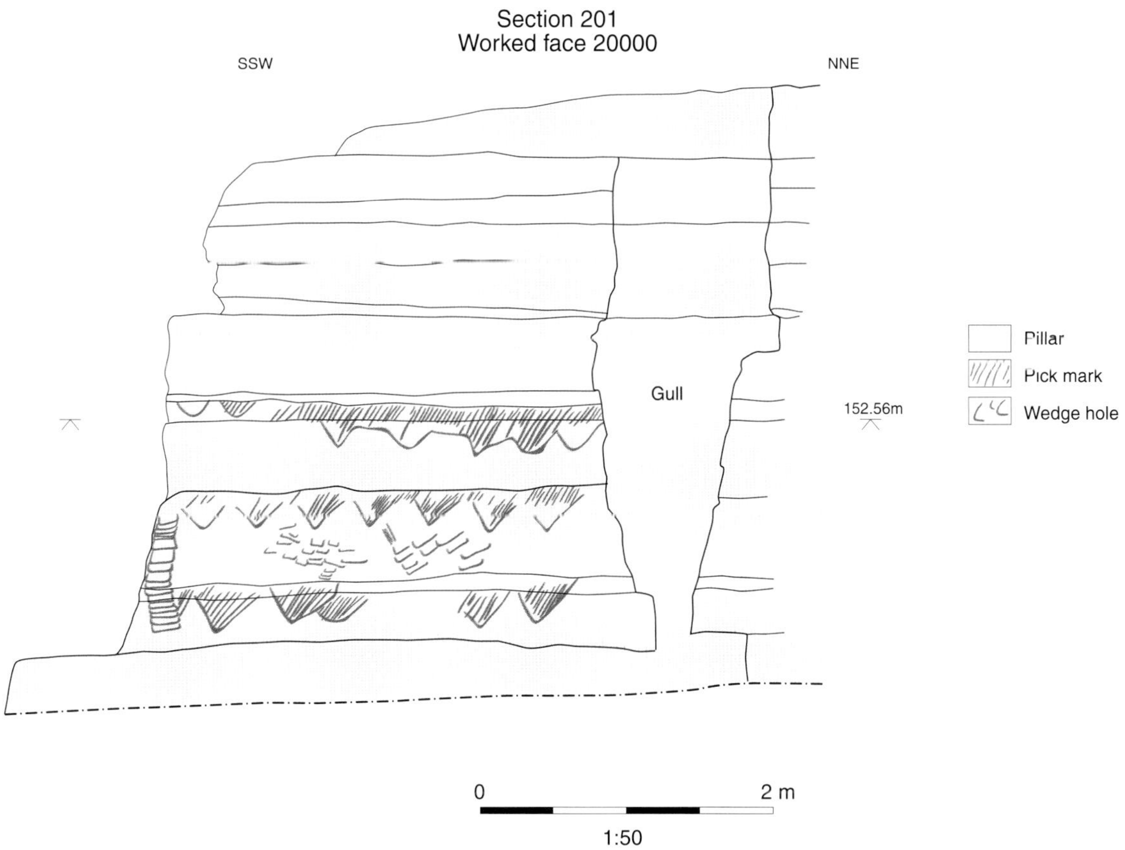

Fig. 8.27 Repeated rows of wedge pits on a pillar used to break successive beds

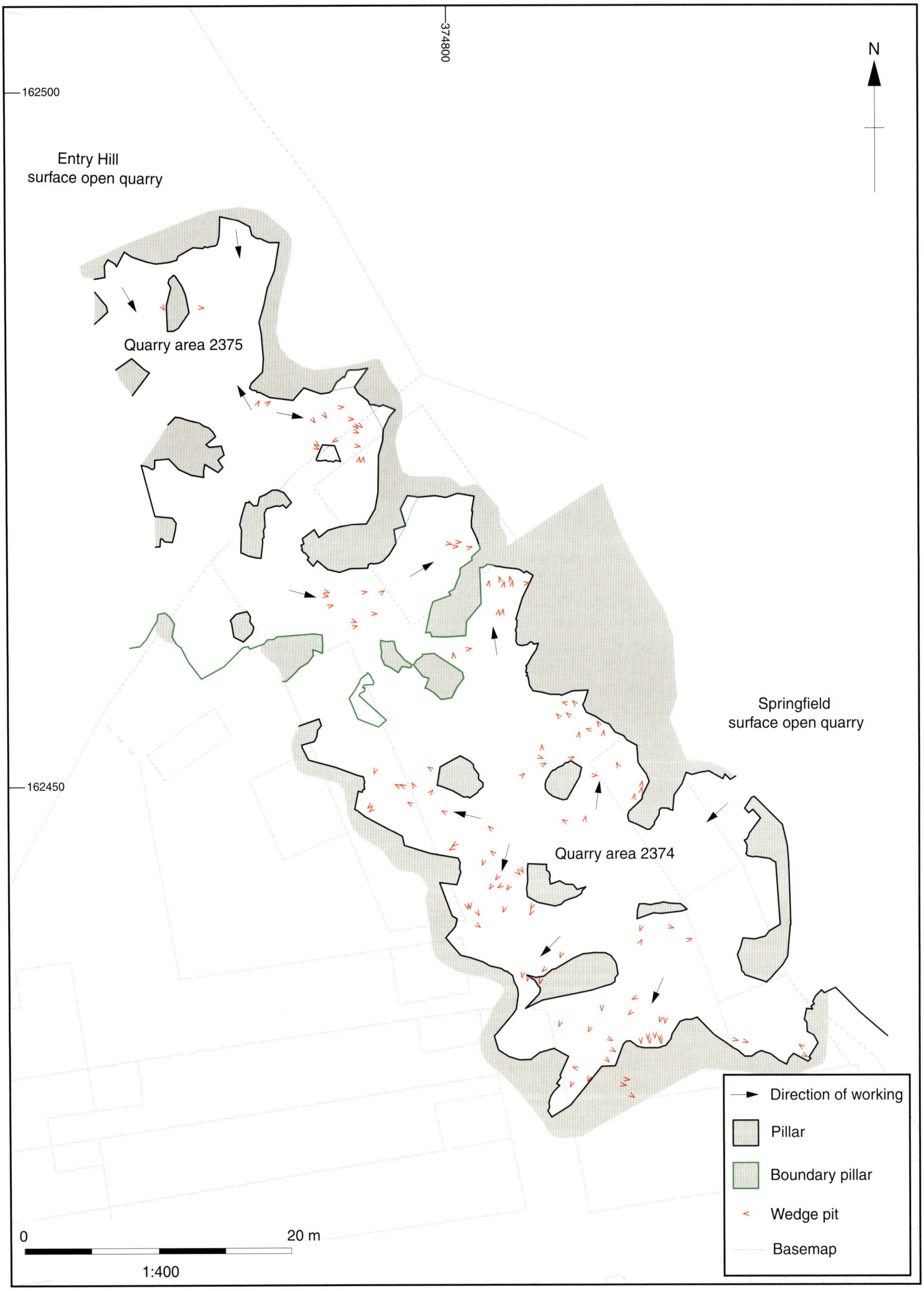

Fig. 8.28 Distribution of wedge pits in the roof of the late 18th-century Entry Hill quarry, allowing inferences to be drawn about the direction of working

inch (138 mm) deep wedge holes, spaced at 8-inch (200 mm) intervals, an example of what would have been a common temporary artefact of active quarrying.

The larger wedges found were probably only a small sample of the very many which must have been used which left little trace. These were not used with chips, but were driven directly either into joints or beds, or into the solid rock. These seem to have particularly been used to rap down a bed above jads, leaving the small impression noted earlier, and to lift blocks from their bed. Compared with the many chips found, large wedges (and hammers) were fairly rare finds, perhaps because they were less easily lost, or, possibly because they were part of personal tool kits.

Wedges were recovered from all quarry areas. Samples from Combe Down were all of a phosphoric wrought iron, easily available to a local blacksmith, and there had been no attempt to 'steel' them (Salter 2009). They can be classed as three basic forms distinguished by the shape of the base – rounded end; flat end; four tapering chamfered edges and reducing to a point. All types have relatively flat, broader, upper ends which received the hammer blows. Most of the wedges recovered had been bent or split by the impact of sledgehammers They are mainly 18th and 19th century in date, but one (SF1030) was recovered from the 20th-century Foxhill Quarry (2380). Their occurrence in the quarries can be attributed mainly to discard due to damage rather than accidental loss, and most of the 19th-century examples were recovered from spoil and barrow-way surfaces. Several wedges were found on pillar ledges, sometimes in association with chips. Some wedges and chips had been re-used to tighten the tops of timber sprags, or driven in at the base of the timber within the timber sprag recess on the pillar face.

The wedge slot is a superficially similar slot to the wedge pit, though was often larger. It is sometimes found at right angles to any feasible split, which does not always have an obvious purpose, but may have formed some type of anchor. Horizontal wedge holes survived less frequently within the quarries. These may also have been used to hold one end of a plank, either used as a cantilever on which to stand or for bridging the gullet. Lewis slots (see below) are easily identified by the 'dovetailing' or opening-out internally at the back of the hole and were used as anchors, notably for crane stays.

Jad cuts

Jad cuts was the term used in the archaeological survey to describe vertical cuts or long slots made by use of picks or sometimes jads, in order to cut out blocks or define pillars. Jad cuts were usually visible as repetitive, sweeping curved marks of picks on vertical faces extending from roof to floor. Their trace usually indicated removal of blocks bed-by-bed. They were not a common feature at Combe Down. Jad cuts were used within some of the earliest 18th-century Firs quarrying areas, including the 1725 period, Quarry (2347), in the Allen Estate Quarry (2201) and the Byfield Allen Estate Areas (913 and 506) dating to the mid-late 18th century. The largest numbers occurred in West Byfield in the quarry associated with Hulonce (Chapter 12, Case Study 7) and probably date to before 1830, though they may be by one of his sub-lessees. This limited use strongly contrasts with their use in underground Bath Stone quarries in Wiltshire, where literally miles of faces can be seen produced by this method, which continued in use there into the 20th century.

A 'doorway' next to the Irvings Incline had alternating use of jad cuts and sawing and splitting using triangular wedge-and-chip pits. Hacking and picking, or crude splitting with simple wedges also remained common in the middle areas of the same workings. This mix of new and old is hard to interpret as other than personal preferences of the workmen rather than strong management direction.

Subsequent to the archaeological survey a historically more accurate or apt term emerged for jad cut. A reference to them as 'not long since ... an improved plan of sawing the vertical – or, as they are termed upright jads became generally adopted' was found by Pollard (pers. comm. citing William Morgans 1871, 149). It may suggest an initial use of the technique, if only as an unsuccessful trial, at Combe Down at a much earlier date than used in the Wiltshire Quarries. The term 'jad cut feature' was also used during archaeological recording. Confusingly the term 'upright jad' may have been sometimes used contemporaneously to describe the vertical cut created by saws in the 19th century.

Use of saws

Sawing was a technique widely adopted during the later phases of quarrying and also often of reworking (see section below), involving cutting through individual beds or down the side of larger pillars. The evidence also indicates, however, that picks and jads or a scappling axe remained in use to clear away upper beds to allow saw access. The earliest use again occurs in the 1725 workings where sawn blocks and wasters (defective worked material) were found. It was of course a normal method of work used by banker masons at surface and the use in 1725 may simply be similar use underground. The use in a surface quarry was illustrated in an engraving by J. Hassall in 1791 (see Figure 3.12) with use of a large bow saw for cutting a block. It is clear that bow saws were also in use underground from occasional curved faces on pillars (Fig. 8.29). They were probably used for straight cuts also but the unwieldy frame must have made this difficult. The introduction of a cross-cut type saw, known as a frig bob would have made sawing a much more viable extractive

technique. It is likely that development took place in the first decades of the 19th century, from the availability of suitable steel (cementation steel) after demand for military edge tool materials had declined after the Napoleonic Wars. The first dated instance at Combe Down is an extensively sawn pillar in north Central Byfield with the date 1816 (Pollard 1994), and others followed in the 1820s and 1830s. Sawing as the primary means of extraction at the face was introduced, at Combe Down, about 1870 onwards and became common in the 1870s in several of the Firs and Byfield quarry areas with sawn faces extending the full height of pillars. At the end of the 19th and in the early 20th centuries sawn extraction was used throughout Long and Open Rooms, as in James Riddle's Quarry (518) in Byfield (Fig. 8.30) and Quarry (2213) in southern Central Firs. The pillars at the edges of these areas have multiple sawn faces and few natural exposed faces. The best examples were preserved in the Foxhill Quarry (2382) which had several sawn faces with hatching details. At the top of the same pillar was a good example of a picked jad slot created during the initial breach that allowed access for the razzer and then the frig bob saws.

Fig. 8.29 Curved face (at top) made using a bow saw, to allow access to vertical saw cut below. Probably part of a secondary working, pillar robbing activity, Central Firs Quarry (2203)

Fig. 8.30 Sawn pillars in James Riddle's Quarry (518)

The frig bob saw (Fig. 8.31) was available in lengths of between 5 and 8 feet (1.4 to 2.5 m) and, as used at Combe Down, was usually single-handed. The depth of the saw blade could be as much as 10 or 12 inches (25 to 30 cm) when new and the weight of the relatively thick blade (compared to the type used for cutting timber), was up to 56 lb (about 25 kg) providing additional pressure, which helped the sawyer to cut vertically downwards. A socketed wooden handle which was generally riveted to the blade could be maneuvered to project above or below the cutting blade to facilitate use in tight spaces. Sawing into a solid face was possible once the overlying bed had been opened far enough back. This could be, for example, below a jad slot made large enough for a smaller saw called a razzer (a worn down frig bob saw) to open it sufficiently for the high-backed frig bob to enter. The cut right to the back was maintained by a downward shift on the handle at the end of each stroke elevating the curved outer end of the saw. Cutting into a open joint facilitated this, and examples of utilizing such a joint were seen in James Riddle's quarry and at Shaft Road (Fig. 8.32). A narrow groove was left in the face beyond the joint. By making two saw cuts slightly angled inwards to a joint at the back, a wrist stone opening could then be created by wedging the block out (le Neve Foster 1901, 313). This meant a further reduction in waste compared with breaking out the wrist stone.

After the wrist stone had been removed, each vertical cut back to a joint would allow a block to be removed, but where the joint was too distant a saw cut had to be made parallel to the face from a somewhat awkward position at the back of the wrist cut. A rare example, because the joints were usually fairly close, with the last stone to be removed left in position, was noted at Foxhill Quarry (2283 see Fig. 8.33). An 0.9 m wide wrist stone at each bed was removed and the back face picked-back slightly to provide a hole for the quarryman to stand and to allow insertion of the saw behind the adjacent section of face.

Sometimes, leaving the trace preserved on pillars, each saw cut would be offset a few inches from the last, producing a characteristic zigzag-shaped or hatched face. Wedge-and-chip holes were also sometimes used in saw cuts to tilt the sawn block forward to lift it off its bed with least damage. The tilt forward of the bed was occasionally indicated by a curved groove where the corner grated against the remaining face. The cut produced was usually vertical or near vertical and it would not have been difficult to produce vertical cuts as was normal. However, in north Central Firs Quarry (2215) the pillars were sawn down consistently to

Fig. 8.31 A frig bob saw (late 19th century) leaning against a typical sawn face with graffiti

Fig. 8.32 Sawn face showing the slot left by a saw end penetrating into the rock at the far end (Shaft Road)

Fig. 8.33 Face in Foxhill Quarry showing picking-back behind the wrist stone to allow insertion of saw behind the sawn face on the right

have bases narrower than at the top, wedge-like. It is not clear why this form, sawn to a few inches wide in a few cases, was adopted, though no collapse seems to have taken place as a result.

When cutting stone from a pillar face, the sawyers would try to get as large a stone as possible and in order to insert saws the bed above was usually cut back by use of a broad-bladed scappling axe or hack (or occasionally sawn by a narrow-bladed bow-saw leaving a curved face). According to le Neve Foster (1901, 314) a workman could cut some 15 square feet (1.4 m2) of the softest beds in an hour. By our own experience, presumably on beds exposed so they were probably substantially harder, even allowing for our inexperience, the regular production would probably be substantially less.

Water was used to lubricate the saw and prevent it from binding with damp stone dust. In our own experience of sawing-out graffiti faces during the archaeological survey, water drips made the effort needed much less. Copious flows of water from a hose (unavailable to the original quarrymen) made it easier still. In historic times quarrymen used a variety of containers – tins/cans, glass bottles and even crucible-type vessels fashioned from Bath Stone to bring and hold water. Crude holes formed in benches on pillars formed basic reservoirs. Some tins were pierced with a small hole in which clay and a match stick were inserted to maintain a gradual flow. Water could be collected by placing the container or cutting a hole beneath a drip from the quarry roof. An example of water supply was

Fig. 8.34 A saw-sharpening stone in which the saw was held – teeth upright – for sharpening, Central Byfield (Quarry 505)

seen in parts of the Foxhill Quarry which became inaccessible after the 1940s. Here, four sealed glass bottles were recovered from the base of a vertical winding shaft in Tankfield Quarry (2383). These may have been bought down already filled from the surface, because no water was percolating through to the beds in this area. Several drip cans were also seen within the quarry, but consistent flowing drips from the roof were again not evident throughout the four quarrying areas.

Saw benches used for sharpening were a fairly common find, usually fashioned from a waste block of stone with a widened saw cut along its upper face. These were used to support inverted frig bob saws when the blades required sharpening (Fig. 8.34). Saw benches were sometimes also adapted from a larger block that was partially finished prior to the discovery of a fault in the stone. During the archaeological survey, the location of a saw bench often indicated the position of a banker mason's working area, in some cases where all other evidence of this had been buried or obscured by later waste deposition. The most obvious manifestation of saw sharpening, apparently done each Friday, were substantial numbers found of triangular files. These occasionally had been reused, their tang thrust into a crack or hole to form a coat hook, or, perhaps to sling their bait or lunch. Other tools required though not found would have included an opening hammer and saw teeth gauge to set the gauge and re-align the spacing of the teeth of frig bob saws. The correct gauge setting for each of the teeth, when accurately spaced, would apparently allow the passing of a match along the length of the saw blade down between the centres of the aligned teeth.

Blockstone was also sometimes cut by sawing into manageable sizes, which needed only single cuts to produce flat-faced building blocks or ashlar. The largest demand for blockstone must have been for ordinary stone-block sizes, which were more easily handled. Because of their fragile nature, these were sometimes supplied as 'double size' blockstone, before being reduced by banker masons on site.

Some insight into the tedious and heavy occupation of sawing is given by comments from Tanky Elms who worked at several quarries, including Springfield Quarry at Combe Down (Hall 1984, 99):

> *What we used to do was to get crooked handles instead of pulling the saw right tight which would rub the skin off your knuckles and arms. This gave you a couple of inches clearance from the stone. However, it made it more difficult to keep the saw vertical and so what he did was to get old bicycle tubes, cut off a length and wrap it around our arms tying it with a piece of string*

and with reference to taking a stone off pillars:

> *And you had to keep the saw moving all the time until the stone was removed. Was kept moving as the weight of the ceiling. What would happen was that the weight of the ceiling would come down on the pillar, the pillar would spread breaking the stone off to where ever you had reached with the saw resulting in half the stone being wasted . . . Two men would be put on the pillar stone, one would have to go and get the water and keep the bucket full which would drip continuously on the saws and the other would relieve him probably after half an hour.*

Reworking or scavenging/Bath Stone

Scavenging was an important part of the quarrying process. It typically involved removing thin slabs from pillars just before they were obscured by dumped spoil, or from those with easy access from cartways and entries. But there is also an abundance of evidence of later reworking, sometimes of a fairly substantial nature, usually involving saws to cut down larger pillars. There were also a very few small reworkings of blocks of stone for some reason not hitherto worked, perhaps left at the termination of a lease or depression in the industry,

Many of the widespread sawn faces that were seen resulted from reworking. The most commonly reworked elements of the quarries were the pillars located close to shafts and other entries or along major cartways. Pillars which bore the marks of older methods of working were found also to have sawn faces probably reflecting scavenging operations, sometimes by organised enterprises as the last stage in working an area, or in conjunction with working a nearby contemporary sawn pillar area, or individuals, local builders or other people making a small living. Evidence of subsequent dumping against the sawn pillars has been found, and in some areas the floor was dug out to gain access to the desired bed block and in others, filled back in again. Elsewhere dumping was clearly carried out as part of an adjacent major operation, and in a few cases in Firs, the sequence suggests very early use of saws apparently in conjunction with this.

Scavenging of the thick bed with its top some 2-3 m below the roof was an engineering problem. In the southern part of the East Firs area, this had often been done in a concealed manner, with spoil obscuring the true narrowness of the pillars below (Fig 8.35), though it was sometimes more blatant (Fig. 8.36). Colin Harris, site engineer for the Stabilisation Scheme (pers. comm.) considered it was possibly systematic Bath Stone at a late (retreat) stage of working. The apparent concealment may, however, have been an attempt to prevent lateral pressures causing pillars to topple, but if so, was not well done as substantial space was usually left under the upper, wider section.

The cartways often have pillars directly alongside which have been sawn on one face while most of the immediate area has an older system of working. Some of these surfaces are thought to reflect scavenging operations in conjunction with working a later nearby contemporary sawn pillar area, Many of the faces did not go below the spoil level which suggests that many must be later, sawn after the spoil was dumped. Some beds below spoil level were also worked but not consistently, so it is possible that sawing was also done where a finished set of pillars appeared strong enough to allow some limited further exploitation. The bed below the common spoil level was the thickest (we saw) in the sequence, and in various locations the spoil appears to have been dug away to reveal this desirable bed for reworking.

The process of reworking often involved use of picks and jads or a scappling axe to clear away upper beds to allow saw access and in some cases faces were split off using wedge-and-chip. In a

Fig. 8.35 Reworked example of removal of lower block of the best stone. Many examples of this appear to have been deliberately concealed in both East and Central Firs (2201)

Fig. 8.36 The most extreme example seen of the reworking of a pillar, in East Firs (2203)

very few cases, narrow saws (probably some form of bow-saw) have been used to take off a higher bed, leaving a curved face to allow a normal, cross-cut type of stone-saw to enter the uppermost of the best thick freestone beds. It was, however, likely that some examples of altered pillars in cartways may relate to the need to facilitate better accessibility to cartway traffic and/or haulage operations, whether contemporary or later, rather than being scavenging.

A small, undated example of where the floor had been removed for access to unexploited blocks of stone occurred in north Byfield Quarry (514). It involved the excavation of a trench measuring several metres in length, a couple of metres in width, and dug to a depth of about two metres. It was located against the western edge of a pillar and would have produced very limited amounts of stone blocks. Three good examples of more organised scavenging enterprises occurred in seemingly near-contemporary workings in Firs Quarry associated with the removal of a sequence of earlier Allen cartway surfaces. In Quarries (2367) and (2368) and the slightly later Quarry (2370), the lower limits of previously exploited pillar faces and larger blocks of un-worked stone were reworked following the removal of the earlier quarry floor and spoil over it, and also two separate cartway surfaces, with the remaining open routes of the earlier cartways used to facilitate the removal of the extracted stone (see also Chapter 12, Case Study 15).

An example of the removal of existing quarry floors during the subsequent working of a remaining block can be seen on the southern limits of Quarry (912) by James Riddle and Son (518) in the first years of the 20th century. Here, about 1.5 m of the former quarry floor level has been removed for the construction of the northernmost of two railways used to haul the won blocks of stone towards the vertical extraction shaft. The enterprise also exploited smaller individual pillar faces and reduced the size of other formerly larger long or rib pillars within the earlier Quarry Area (915). Evidence for secondary working is evident on some of the pillars in that the earlier corbelled pillars have been modified by the later sawing techniques into direct pillars.

Lighting

The main form of illumination used in all but the few late 19th- and 20th-century workings were tallow candles, of which several were recovered. Little detail has emerged about them and how they were managed in use, but associated finds included candle boxes, rudely cut out of a block of stone and covered with a flat stone lid (to prevent the candle being eaten by rats and mice) and candle holders, either of crudely formed, sometimes spatulate stone (see Figure 11.23) or or of small balls of clay stuck on pillar ledges or stood on the remaining stone benches. Several individual clay holders were recorded within Quarry 2224, and were placed on adjacent pillars besides a barrow-way route to illuminate the path taken by quarrymen either to transport produced stone blocks or to barrow spoil away. Others candle holders, clay and otherwise, were commonly located on pillars adjacent to vertical shafts in areas that were presumably more often used during stone transportation operations.

In Firs a flare-type torch formed by oil-soaked textile wrapped around the top of a stick was found, but its condition suggested it may have belonged to a later explorer of the complex. Oil lamps, probably using kerosene, became fairly common in the late 19th century. Home-made forms from tins are possible but a fairly standard type was probably similar in appearance to a small watering can with a wick in the spout, which gave a fairly good light compared to a candle.

The first carbide lamps would have been introduced in the early 20th century. They were made of brass with a body partially filled with calcium carbide onto which water dripped from a screw-on top, producing acetylene gas. The lamps were probably designed to be attached to a cap, but could also be set on a ledge. The lamps were small and cheap and the inexpensive carbide lasted for about four hours and the gas jet gave good illumination. At Combe Down, no lamps were found, and no carbide residues were noted. Larger hand-lamps with eight to ten hours life came somewhat later. These were likely to have been used at the Shaft Road (Combe Quarry) and Mount Pleasant Workings, and at Foxhill.

Chapter 9

Techniques of underground quarrying: 2. Transport and materials handling

Cartways and associated shafts

The well-organised and mechanised system of transport from the open quarries via the Allen railway was not extended below ground, perhaps because distances were much smaller and more dispersed with frequent route changes. Instead barrows, carts and four-wheeled wagons were used, and in most areas they were moved independently of rails. Where there was an absence of cart-ruts or barrow-ruts the method of transport for the block-stone from the working face to the surface was unclear, though dragging by winch and crane or on sledges or even wooden rollers were possible over short distances. It is possible that the absence of cart-ruts was more generalised than was apparent from the limited areas of floor that were open to investigation for this study.

Cartways were purpose-driven, fairly level routes (ideally slightly sloping toward the entries) within the underground quarries. They were used by a two-wheeled quarry cart or four-wheeled wagon with solid or spoked wheels. These left evidence of their use by ruts and hub-wear marks. Horses generally leave tread marks within the centre of the cartway but apart from one possible example (Fig. 9.1), these were absent at Combe Down, though their use was envisaged in a lease of 1804 (LL. A01/18/5/30). A small number of horse-shoes were found in the workings, and it is possible hoof traces were obliterated by later usage. Overall, however, the impression remains that horses were not commonly used and that the carts were moved by the underground workmen.

The cartways were the primary transport routes within the underground quarries, mainly used for

Fig. 9.1 The only example of a cartway where likely traces of horse-hoof impressions were noted, north-east Byfield Quarry (505)

the transportation of roughly squared block stone, rubble-stone for building and finished or wrought stone, either to the open surface quarries or a conveniently placed vertical shaft. There was also evidence of some cartway routes subsequently being used as barrow-ways, which often obscured original surfaces and filled the cart-ruts. The cartway system in the Combe Down Byfield and Firs quarries is shown in Fig. 9.2.

The earlier cartway levels had an unobstructed width of about 2.4 m between walls, pillars, packs or banks, rather more later, and a surviving height below the roof of about 2.4 m or a little more (typical of the earliest), to some 4.5 m later. Many had evidence of carts or wagon use in the form of ruts typically measuring internally 1.2 m wide and externally about 1.5 m wide. The ruts were left by the cartwheels which wore into the soft 'make-up stone' (fine chips or dust) placed on the floors to ease passage. The width is thought to reflect a cart width of the 19th-century type of about 4 feet as the earlier evidence for haulage of the 18th century (with one possible exception in the Byfield Quarry) was generally buried by later activity: This does not necessary preclude a similar width for early carts, but where earlier cartway sequences were recorded cart-ruts were generally absent.

The cartway routes followed headings developed from the surface entries. They were up to 4.50 m wide between walls or pillars, which was often the width of rooms worked, though often less for earlier examples. The method of driving such headings was very similar to that used on working faces (as described earlier) except that often one wall was not pierced, presumably for increased stability. Driving also yielded blockstone, helping the pay its cost, but near the entries some spoil would have had had to be removed outside for dumping. Cartway floors appear to have always been constructed on a thick layer of spoil. Some 2 m seems likely from occasional observations, but this could vary to allow construction of a slightly sloping cartway bed to help movement outwards of loaded carts. There was a trench or ramp at the inner or face-end up which below-floor bed-blocks had to be dragged, but the cartway level also allowed access to higher beds and, more importantly, reduced the height from which blocks from high beds had to be lowered. The height of the cartway above its spoil-floor probably reflected the total height of freestone and picking beds removed, so a 2.5 m average roof height above the cartway suggests a total of some 4.5 m of beds removed and greater heights may have resulted from deeper (and usually later) workings. Alternatively the height may have been adjusted to allow loading from the thickest bed, the top of which was two or three metres below the roof. It is possible that stone was dragged back from the face in working rooms, or barrowed back to the inside end of the cartway, perhaps up a ramp. In some cases, particularly in the late 19th century, cranes were used for both lifting and dragging of bed-stones along the floor and on to the cart or wagon.

The cartways seem to have been well-maintained. The ruts, though often visible, were normally only a few centimetres deep and often free of visible stones. The sides of the cartways were often sloping, covered in a fine dried sludge derived from clearing and maintenance of the cartway surface. The lack of deep ruts implies that the surface was dressed or made up with small stones and dust laid upon it to provide smooth surfaces. On curves within the cartway lengths the wheels of the carts had sometimes cut into the sloping bank, increasing the angle and leaving impressions of the wheel profiles; these were hard to measure, but suggest wheel diameters of about two feet (0.6 m) or a little more. Rub marks from the passage of carts and, perhaps, horses were found on the corners of the adjacent pillar faces. Haulage marks, including rope and chain grooves, were noted in places along the routes and at the vertical winding shaft locations which were positioned along them. This evidence assisted in the interpretation of types of haulage methods used and the systems employed.

The early to mid 18th-century cartways were driven from surface open quarries, and their routes were identified by the contemporary pillar types, generally apophygate with some corbelled elements, through which they had been extended. The Allen cartway routes were of this type and were all driven northwards from his surface quarries in the Byfield and the Firs quarries west of the Long Drung. Nearly all the cartways were re-used by later quarrying enterprises that succeeded Allen's direct management in the later 18th and during the 19th centuries. Consequently any clear early evidence for the use of the Allen cartway routes was buried or obscured by later transport activity. Because its use was abandoned before the end of Phase II and the area at the far end apparently not worked thereafter due to roof instability, the most likely un-modified Allen cartway was the westernmost of the two driven in from Jones Quarry in Central Byfield, which exhibits very prominent cart ruts (Fig. 9.3).

In the Central Byfield Allen workings (Quarry 505) the length of his eastern cartway extended over 120 m from his surface quarry operation at Ralph Allen Yard to the northern extent of the contemporary apophygate and corbelled pillars. Later cartway development, known to be mainly 19th century in date, extended this another 120 m northwards and, despite running along a series of wide shafts, may still in part have been extracting blockstone over the former length used in Allen operation. Some other earlier cartway routes not re-used in the 19th century were also of considerable length, with distances of up to 120 m and more. This practice of using cartways was generally absent in areas to the east of the 'Long Drung' (with the

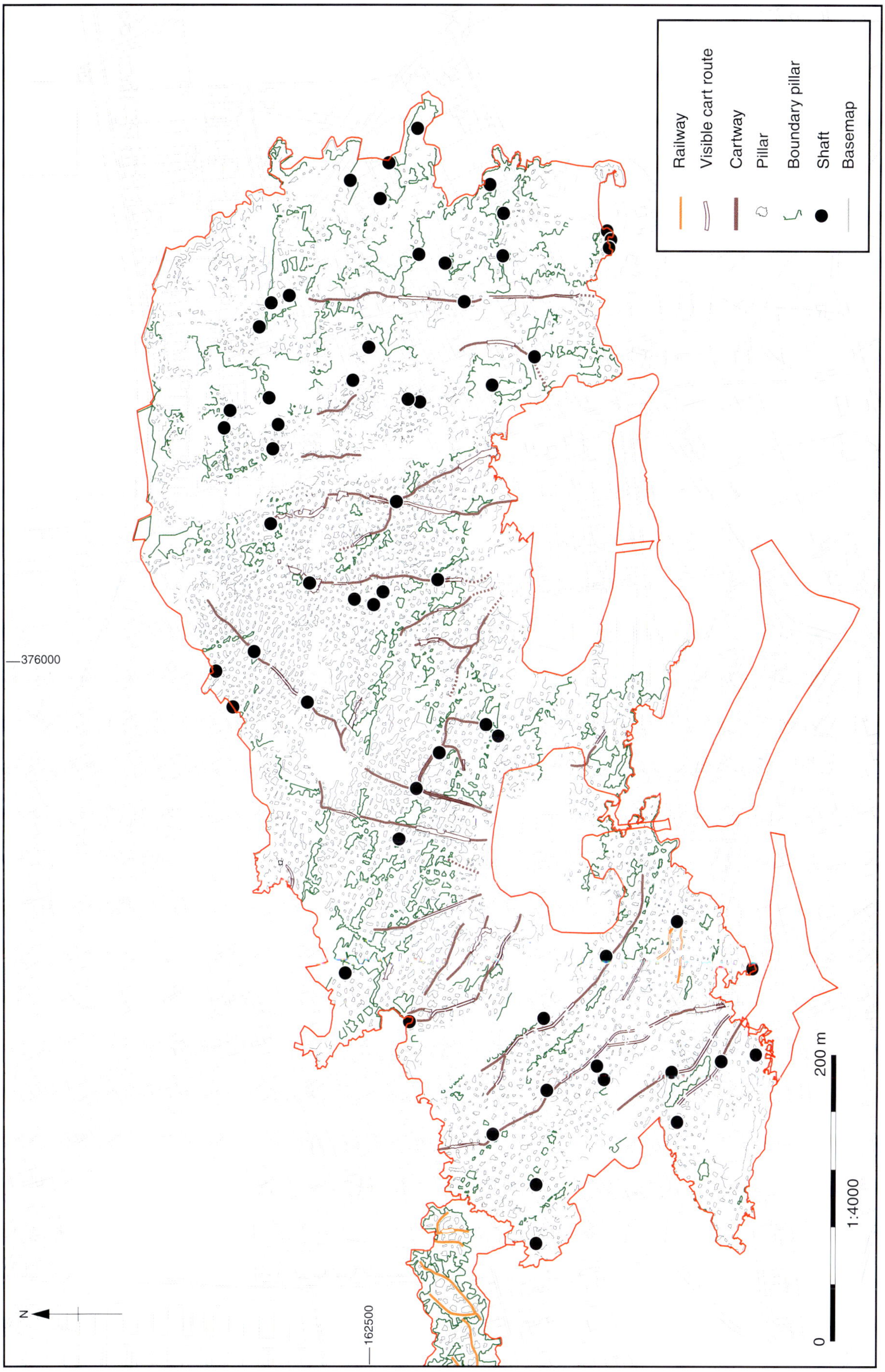

Fig. 9.2 Plan of known cartways in Firs and Byfield, also showing shafts

Fig. 9.3 Allen's western cartway in Central Byfield, near the entry. It appears not to have been used after abandonment in the mid 18th century because of bad roof conditions at its inner end, discovered after work restarted c 1803. The white reflection may be due to stone particles within the re-crystallising mud cleared to the side of the cartway

exception of The Brow cartways), where most quarrying areas were small scale and generally used winding shafts. Where cartways were used there, they tended to be in shorter lengths between the working face and the shaft, in some cases rendering the use of longer earlier-period cartways obsolete whereupon they became used as convenient dumping areas.

Allen period cartways

The Allen cartways (1730-1764, Phase II) were driven from at least two or probably three separate open or surface quarry locations, perhaps at a very early stage of the surface quarry development. The definite two were Jones Quarry at what is Ralph Allen Yard today and Sheeps House Quarry at Rock Lane. Other entries were probably located at the west end of the (later) quarry north of de Montalt Place, and the quarry behind Hopecote Lodge (later) worked by Burgess. A fourth quarry, located to the south of de Montalt Place 'Masons Crane House Quarry', was also operated by Allen and Richard Jones in his *'Life'* appears to suggest, using a simple sketch, that two parallel 'roads' were driven there (see Figure 5.6). As the roads cannot be any of the surface roads in the vicinity, Jones 'roads' can probably be presumed to be cartways. There was no corresponding archaeological evidence found to suggest that cartways were driven underground from there, though the possibility remains as much of the ground was later surface-quarried.

The initial Allen transport system seems to have used two cartways from each surface site, perhaps to provide better ventilation. In each case a surface crane was either definitely used or available (in sale details) for use above the entries to raise stone from the cartway level to the surface railway and down to the wharf at Dolemeads.

Allen cartways at Jones Quarry (Central Byfield)

At Jones Quarry the two original entries branched as a 'V' from the entries. They were each 3.6 m wide and both were driven for about 120 m on the eastern side of the Combe Road boundary pillar in Quarry 505, initially about 2.5 m below the roof. The western cartway (Fig. 9.4) was driven next to the almost continuous Combe Road boundary pillar and was that described above as abandoned in Phase II, and never re-opened due to instability. At the northern and the southern limits, high roof supporting packs had to be constructed. Instability also affected a wide shaft about 60 m along the cartway resulting in it being inaccessible to the survey team.

The eastern cartway had a much longer life after initially being driven in through an ashlar block, barrel-arched entry, then largely in rock for about 90m. Rope grooves were found on the wall near the entry, suggesting that, at some time cable haulage, probably from the crane, or if later, from a winch, had been used. Use of shafts further in may suggest cable haulage at the earlier time using the crane cable. It passed under a wide shaft at about 70 m inwards, but this was inaccessible due to partial roof collapse and a sub-circular pack built around it, so its possible functions could not be examined, though ventilation would certainly be one of them and the enclosing wall was so wide as to suggest a wide shaft suitable for raising block or stone. The cartway had been extended stage by stage, eventually almost to the North Road, the staging partly indicated by changes in pillar morphology, with shafts at roughly 50 m intervals. The shaft to the north of the first shaft (the walled shaft above), just beyond the last apophygate pillars, showed evidence for haulage as rope grooves were found on the underside of the shaft at roof level and near the floor, and in a pillar at the side, back towards the apophygate pillar area. They suggest that carts (or blocks of stone, on rollers if very large) were hauled to the shaft by a cable, but again the shaft's alternative use for light and ventilation may have been important. The shaft was filled with post-quarrying surface-derived material. The collapse at the first wide shaft and the adjacent small area to the east necessitated the diversion of the cartway to the western edge of the shaft; the former length to the north of the shaft was also covered by collapse. As the stones had been cleared to the edge of the cartway, it was clearly a contemporary fall but there had also been post-quarrying roof collapse. The date of the later collapse is unclear, but two scappled blockstones abandoned on the eastern edge of the shaft and surrounded by collapse material can be assumed to have been left prior to the collapse.

Allen cartways at Sheeps House Quarry (Central Firs)

At Sheeps House Quarry (2200) two roughly parallel cartways were *c* 25 m apart (Figs. 9.5 and 9.6), widening slightly further in. The slightly larger western cartway was 4.80 m wide between the pillars (though also partially lined with heavy roof support packs) and a late 19th-century excavation in the floor of the western cartway revealed it had an excavated floor-to-roof height of 5.50 m, the only example of an early cartway where the full height was visible. Rooms appear to have been made from either side and between the cartways as they advanced which would considerably improve ventilation and quickly provide a large number of working places. The original gauge is unknown as evidence of cart-ruts at the lower extents of the exposed sections was very slight and not measurable with any precision, and the transport system into the rooms alongside the routes remains unclear as the sides were usually packed with stone and spoil.

The western and eastern cartways were driven northwards for about 70 m, where shafts were sunk above each cartway, both narrow, at about a metre wide, so probably for ventilation. Some of the apophygate pillars had been partially modified, the cartway widened and additional support by packs was provided, probably in the 19th century, to prevent (further) roof falls. The apophygate pillars

Fig. 9.4 Upper section of Allen's western cartway in central Byfield showing a roof-fall which is below a wide joint or "gull" in the roof. The seated figure is mine manager, John Lister

Fig. 9.5 Allen's east cartway in Central Firs. Note the remains of an apophygate pillar, on the left, the original roof destroyed by a fall. The heap is a cone of debris from the narrow shaft above. The passage with light from the left links to the west cartway

Fig. 9.6 Allen's west side cartway in Central Firs showing apophygate pillars and support packs alongside

Fig. 9.7 Stepped section showing cartway and other development at the Allen western cartway in Central Firs

extended roughly to these first ventilation shafts, and probably indicate the limits of the first development phase. A series of east-west aligned large boundary pillars across the northern extent of Quarry Area 2200 where the shafts were located would also seem to indicate a northern worked limit. Beyond the shafts, alongside the cartways, the pillars tended to be corbelled, and probably belong to the later part of the Allen Phase II quarrying, extending probably just past the Firs Shaft used in the recent Stabilisation Scheme. The smaller depth of penetration of the apophygate pillars here suggest these workings were later than the Jones Quarry workings at Central Byfield.

The eastern cartway route was over 100 m long and 3.60 m wide between the apophygate pillars. The southern stretch of the route was re-used by a later cartway branching towards the east. A shaft thought to be contemporary with the early development of the cartway, located about 70 m from the entry (in an analogous position to that on the western cartway (below), was *c* 1.40 m in diameter. It was one of the few examples of a probable ventilation shaft and was also of a similar diameter to another located 30 m to the east above the branch cartway. The latter may have been contemporary with the later part of Allen Phase II Quarry 2200 or, more likely, was an early Allen Estate Phase III cartway, as it was associated with Long Room working and crane usage.

At a point 70 m from the entry shaft the cartway was extended northwards to close to the Firs Shaft and, judging by the corbelled pillars, probably just beyond during the late Allen Phase II working. This change was noted in the development of pillar type from apophygate to corbelled along the route. The shaft, though much modified by modern use, had Lewis holes positioned on one side suggesting a crane post support, and cable grooves were found on the pillars. Although no date could be confidently ascribed to these features, the areas adjacent to the shaft had Long Rooms similar to those further south-east associated with crane use, considered as probably Phase III.

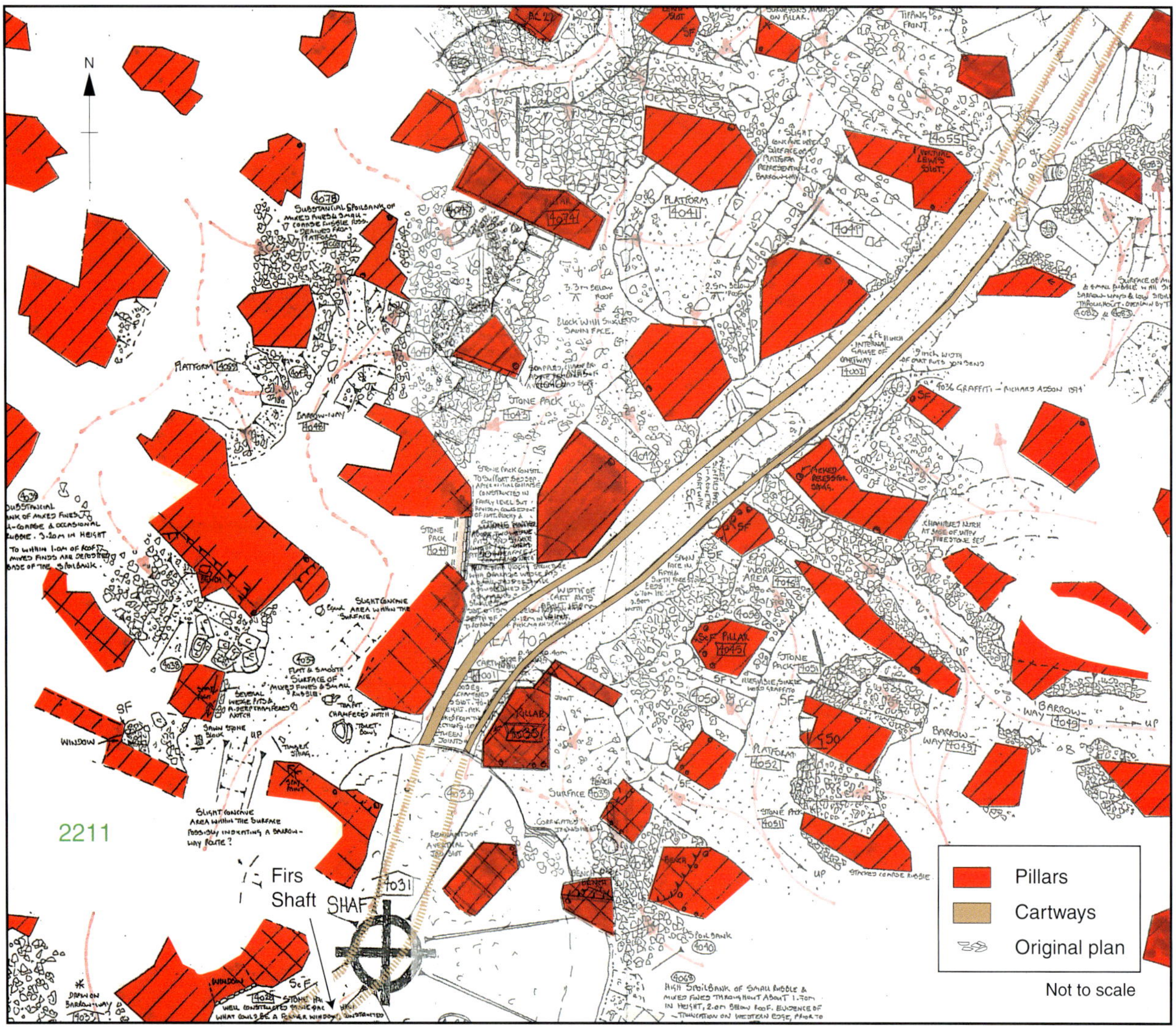

Fig. 9.8 Plan of cart ruts in the western Allen or post-Allen cartway at its north end near the boundary pillar

The western cartway was more fully investigated than the eastern, since the second shaft along it had been reused and the cartway excavated for stone in the late 19th century (Quarries 2367 and 2370). Uniquely this provided high stepped sections available for archaeological recording, extending halfway across the early cartway. Only the information directly affecting the Allen cartway is remarked on here, and the information revealed is more fully considered in Chapter 12, Case Study 15.

The southern extent of this cartway had passed, under a narrow shaft 70 m from the entry where there was no substantial evidence of cart-ruts. A 3 m-diameter shaft was located near the interface between the apophygate pillars and the slightly later corbelled phase of the Allen workings (Phase II, Quarry 2201), although several apophygate pillars were noted to the north of the shaft, abutted by later 19th-century stone packs. The shaft was probably used during the 18th century for winding stone out and was certainly used in the 19th century for localised Bath Stone activity associated with Quarries 2368, 2370 and 2344. From these northwards, later activity obscured the early cartway route.

The lower part of the stepped section (Fig. 9.7), records the lower sequence of cartway activity, with the uppermost cartway surface being the contemporary height of the cartway floor leading from this location southwards back to the surface entry. It is unclear how much further this cartway extended northwards beyond the section. However, it may have been extended during the Allen period to near the northern boundary pillar west of the Firs Shaft, and the line of the route was well-defined by the adjacent wider pillar alignment. A small area of cart-ruts were preserved close to the boundary pillar (Fig. 9.8) and a corresponding breach in the pillar suggested that the route may have extended to or even through this quarry boundary, into Quarry 2213 to the north. The workings hereabouts are Long Room and most likely Phase III Quarry 2201. The period of this breach could not be established archaeologically but the cartway alignment, covered with debris, had been noted – on one of the early visits to the site – as extending through the pillar face into two Long Room workings. By around 1800 the cartway was shown extending beyond the boundary pillar on a lease (LL. 91/18/5/30), which shows it then extending for about 30 m to the north-east, about 20 m north of the Firs Shaft. It was not possible to ascertain if there was post-1800 development beyond.

Apart from the ventilation shafts serving the cartway routes, the wide shafts that served the levels were of substantial diameter, in the order of 2.5 to 3 m or more. Precise diameters were difficult to measure because of walling and surface derived infilling, as well as access limitations. They were only *c* 6-7m deep to the top of the working. Any or all of the wide shafts may have been originally constructed to supply ventilation for which the metre width was adequate. If they were made wide, or later were widened, then undoubtedly this was to draw stone out vertically. The position of shafts probably indicates the distance dug before subsequent ventilation was required and they often seem to have been positioned at the interface between periods of quarrying activity.

Cartways east of the Avenue (East Firs)

Similar surface relationships as at Sheeps House Quarry at Rock Lane occur near Claremont on Church Road, with 'Quarry Lane' declining into a surface quarry (1851 Monkton Combe Tithe map), worked about 1810 by John Burgess. North of this in East Firs are apophygate workings with relationships that are very similar to those at both Sheeps House Quarry and Jones Quarries described above. However, very small workings, Quarries 2390, and 2346 displayed direct pillars and evidence of wedge-and-chip working under 89 Church Road occupying the ground between the Phase II Quarry Lane workings and the (later) outcrop-type Masons Crane House Quarry. Thus the entries to the Quarry Lane workings were not from Masons Crane House, so Richard Jones' sketch must remain a mystery, the probable ground involved having been later quarried away. Instead the entries must have been at the bottom of Quarry Lane. There are several possibilities, though these could not be explored because of safety restrictions.

Apophygate pillar workings in Quarry 2331, were located underground just beyond the north and north-eastern limits of the surface Burgess Quarry. The two parallel cartways, separated by *c* 25 m of quarried area in between, seemed to be typically Allen and are just within the area of surface tree planting that probably defined his quarrying area. The route of the easternmost cartway was located a few metres west of the Long Drung. It and the parallel westernmost route were accessible northwards for 25 m from close to the entry, though the southern stretches were buried by later barrow-way activity and spoil. The easternmost cartway had no discernible cart-ruts, only a smooth surface of fines, probably developed through later barrow-way activity. It was *c* 3.5 m wide and lay *c* 3.3 m below the quarry roof. The length of both the routes was estimated by measuring the plan distance from the surface quarry edge to the northernmost extent of the contemporary apophygate pillars, approximately 40 m. Unlike other Allen cartways recorded, there was no ventilation or larger vertical shaft at the northernmost ends.

Workings of a similar type also occurred at the western end of the same quarry. The area has been much affected by roof-falls, but the Hawkins (1994) plan suggests at least two other entries were physically feasible and probably likely. The apophygate and corbelled pillar association was observed in a similar association to two wide shafts seen in areas

Fig. 9.9 Shaft cap above cartway (2194) viewed from the underside, associated with Allen Phase II workings in East Firs (Quarry 2206)

west of The Avenue, but full details were not recordable due to collapse within much of these workings. The two shafts involved were capped by a particularly fine shallow-arched domes formed of roughly cut blocks (Fig. 9.9).

Pre Allen cartway (south-east of Sheeps House Quarry)

There were, however, examples of early workings with possible or certain cartway routes considered not to be Allen's but close in time to his. Quarry 2338 had a single apparent cartway length located on the southern extent of a Quarry (2347) which appears to have been driven southwards from the surface Sheeps House Quarry. It was exposed for a length of only 11 m as it had been tipped into at the northern and southern limits by later barrow-way tipping fronts. It was 2.50 m wide between flanking low stone pack walls and 2.5 m below the roof. A clay pipe of 1720-1725 was recovered from stacked rubble retained by a stone pack on the northern side. It is representative of the period during which the quarry operated and correlates with an incised graffiti dated 1725. The workings and possible cartway is particularly important because it is pre-Allen, and, as it almost certainly extended back to the surface Sheeps House quarry to the north, it suggests a date for the surface quarrying activity on Sheeps House southern boundary of the early 18th century or before (see Chapter 12, Case Study 2).

Contemporary non-Allen cartways east of the Long Drung

Two further extant sections of cartways were also surveyed in Far East Firs east of the Long Drung, within apophygate pillar working Quarries 2339, 2350 and 2369. Quarry Areas 2339 and 2350 are thought to have been contemporaneous although they were separated by a fairly extensive boundary pillar with few openings. One of the few located gaps identified during the Stabilisation Scheme was the continued drive northerly of the easternmost cartway route. This pillar extent may have reflected geological weaknesses and was left *in situ* to support the roof, rather than an operational or ownership limit; the alignment, if projected westwards matches the northern limits of the Burgess surface quarry face.

Both the easternmost and the westernmost cartways there appeared to vary from the Allen cartways by their roof being one or two beds lower in the sequence (that is, the roof was below the picking beds). Therefore, access from rooms at the side descended via a slope into it. The roof appeared less broken than those in the Allen workings, which may explain the choice of horizon. The easternmost cartway was developed northwards from a surface quarry south of Church Road in the area of the new housing development (west of Ashlands Cottage), and extended roughly below Tyning Road for about 200 m. An area of apophygate pillars was found on the western edge, with the cartway representing the eastern limits of the quarrying along most of its length and was located besides a fairly continuous pillar boundary. The northern extent of the cartway length was recorded in Quarry 2339 and the southernmost limits were surveyed in within Quarry 2350. The areas between were infilled with waste stone and/or collapsed roof materials and the location of a shaft could only be suspected to be at the southern limits of the northern open cartway.

The southern extent (Fig. 9.10) did not appear to have been extensively re-used in the 19th century, possibly because of poor roof conditions. It was open for about 16 m, otherwise sealed by roof collapse and accumulated surface- and roof-derived clay. The northern section was open for approximately 120 m and had been built over by a stone pack to retain waste stone. A masons' working area located 11 m to the south-west of this cartway length on the western limits of Quarry 2369, which is also thought to be roughly contemporary with Quarry 2350, contained a large collection of scappled and sawn blocks contemporary with the cartway, dated by graffiti on a block reading 'June 1730' (Chapter 12, Case Study 1). No cart-ruts were evident, as it had been re-used by later barrow-way routes and access for general spoil tipping. Several large shafts further along the route may have been either sunk or re-used for later quarrying.

The eastern cartway was 3.20 m wide between flanking stone packs, with the uppermost cart ruts recorded at a height of 2.30 m below the roof. The ruts had an inner gauge of 1.38 m and an external width of 1.76 m. The average width of the ruts at the base was 0.12 m, with a depth of 0.20-0.30 m depending on how deep the wheels had cut into the fines surface. The cartway sequence (Fig. 9.11) was recorded to a depth of 1.70 m below the uppermost exposed cart-ruts. An earlier set of ruts noted at 2.60 m below the roof and directly below the upper ruts represented the same cartway activity where the westernmost wheels had bedded into the soft floor level. Several small stones had been placed in the upper set of ruts, presumably to provide a more stable base for the wheels. The ruts represent the later use of the cartway route, possibly for the northernmost area of extraction.

The section exposed below the ruts showed no evidence for the use of carts. There were three compacted surfaces, possibly representing barrow-way use prior to extension of the cartway. The surfaces consisted of compacted fines with small lenses of clay in between. Below this a layer of larger rubble and fines included large blocks up to 580 mm in size. This overlay a very level and compacted floor of mixed fines and small rubble 3.40 m below the roof level, which represented the earliest of the surfaces on the route with no cart ruts. It resembled the deposit located within the lowest sequence of the western Allen cartway in Central Firs. This floor could have been used for the dragging or sledging of blockstone as it was very level.

This cartway had a possible length within associated apophygate pillar areas of approximately 200 m and was thereafter further extended to the north in

Fig. 9.10 Southern end of the cartway east of the Long Drung

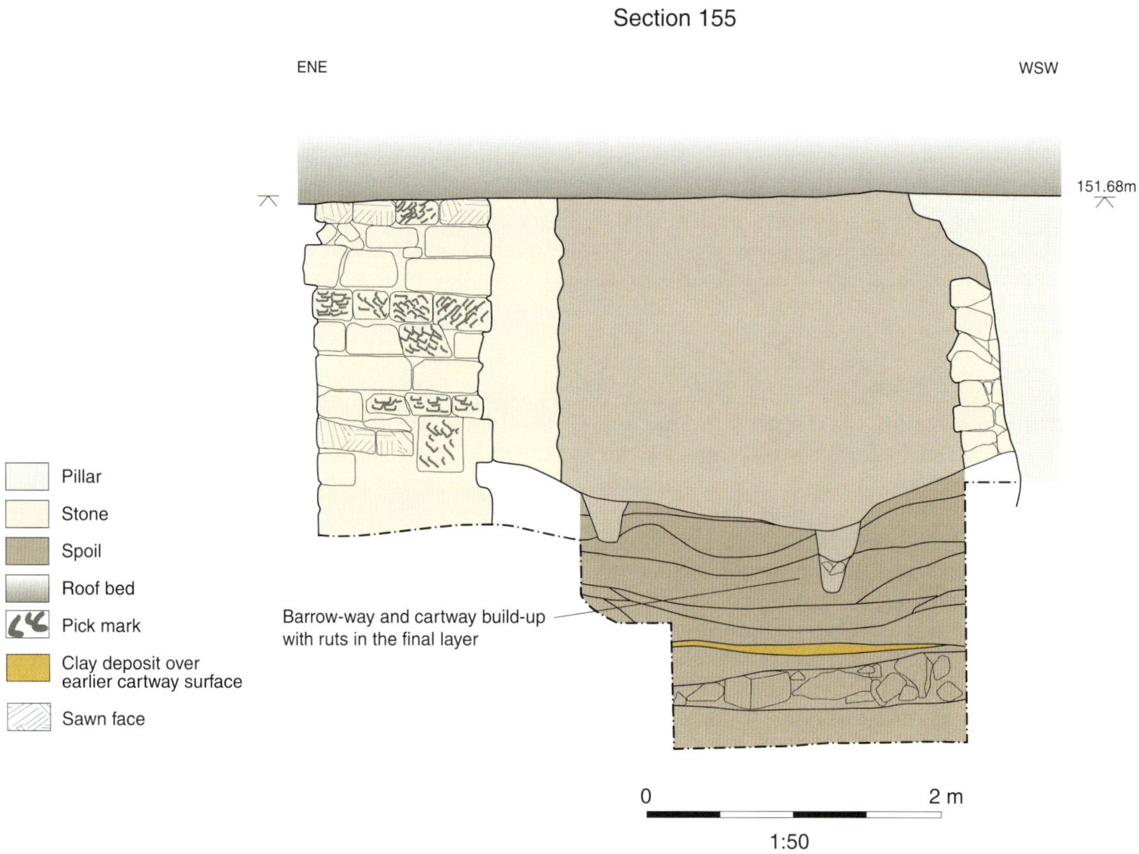

Fig. 9.11 The sequence of deposits in the easternmost cartway, east of the Long Drung (Section 155)

Fig. 9.12 Curving cart ruts cutting deep into the surface of the westernmost cartway east of the Long Drung

the 19th century within Quarry 2363. However this extended length was largely obscured by collapsed roof material. It was by far the greatest single length of cartway recorded at Combe Down. Its length may suggest that this form of pillar was retained much longer than in Allen's quarries, though eventually it was succeeded by just the same forms as used elsewhere at Combe Down.

It may be partly attributable to Milo Smith (who owned parts of Combe Down-Greendown just before Allen started there and worked there for a few years) and/or the Greenway family who were quarrying in Combe Down in the early 18th century through to the early 19th century, somewhere probably east of the Long Drung (see Chapter 5). The cartway system seemed, by its length, to have been the main transport for the Quarry 2339, and also the eastern part of 2350, which were probably part of one quarry reflecting its full period of operation. The cartway was well laid out, and although the former shafts along its length were inaccessible, their locations suggest that shafts were positioned at regular intervals of about 60 m, the first being *c* 60 m from the surface quarry at the interface between Quarry 2339 and 2350.

The westernmost cartway was developed northwards possibly from the same surface quarry as the easternmost cartway, to the south of Church Road. However, it had a definite curve to its course, extending at its southern limits towards the west, possibly towards the eastern edge of the open surface quarry behind Hopecote Lodge (Burgess Quarry). The cartway was associated with an area of apophygate pillars on the southern extent, with corbelled pillars on its northernmost observed stretch. The northern limits of the route, close to Sydenham Place on the surface, had largely been overlain on both edges with stone rubble, with about 1.5 m remaining open in the centre, or had been flanked with shorter sections of stone packs. Other high lateral deposits of mixed fines were representative of finer materials being cleared during maintenance of the cartway surface (Fig. 9.12). Where a short section of the cartway was open, curving cart-ruts had cut into the surface to a depth of 0.40 m. They indicated a gauge of 1.30 m apart and were at a level of 2.30 m below the roof.

Post-Allen late 18th- and 19th-century cartways

Cartway development undertaken by perhaps a dozen independently operated quarries continued after the death of Allen and the break-up of his monopolistic control. The existing cartways seem in most cases to have simply continued development northwards, the change apparent largely in the adoption of direct pillars replacing corbelled as the dominant form, and the height above the cartway floor becoming greater at around 4 m to 5 m, possibly reflecting a greater depth of stone being worked. This increased the recovery of stone, as becomes apparent in later leases (Chapter 6), within the same rental and royalty regardless of depth per superficial area. This was not an instant transition across the area, but a definite trend by the beginning of the 19th century.

Further cartways appear to have been started at the Sheeps House Quarry, with one on the western side of the earlier workings (near the steps on the footpath through to The Firs), and one at the Jones Quarry (Ralph Allen Yard today) passing through a stone arched entry and through the Combe Road Pillar and a short distance (some 20 m) of much earlier apophygate pillars under the William IV yard.

The new Jones Quarry cartway was carried through and along the western edge of the Combe Road Pillar to a shaft offset to the west side about 50 m from the entry. Once through the pillar, the cartway was driven within a 5 m-high, very open Long Room and was notable for distinct jad slots and use of wedges. It had very limited workings off. It appears to have been the only working in the Byfield area before the 19th century. The name Jones Quarry may tentatively suggest that it was worked by Richard Jones and his son, after 1764.

A third cartway was developed in the late 18th century on the western side of the two originals at Sheeps House, again through a small section of (again, much earlier) apophygates, modifying an existing entry (Quarry 2342). It was substantially higher than earlier cartways and had only limited development, penetrating about 20 m in 1800 according to quarry leases. There were faint traces of cart ruts on the surface. Leases also show substantial development took place in the north west corner of Sheeps House Quarry, which must have included the cartway there. This must have taken place in the late 18th century, as soon after 1800 a lease (Chapter 6) noted almost half the distance from the entry to the Turnpike Road had already been worked. Much of the area on this western side of the Firs and between there and East Byfield was inaccessible due to instability and backfilling with spoil, so archaeological examination was very limited.

The first decade of the 19th century saw very considerable development of new cartways in both Byfield and Firs, this time under new ownership but with only limited change of quarrymasters (see Chapter 6). In Byfield cartway development and quarrying had probably stopped before the end of Allen's life. In Firs it had continued, but in the older areas development was coming close to the Turnpike, nowadays North Road. In these cases, and probably also east of The Avenue, shafts and cartway development seem to have been integrated. The landlord, after initial reluctance, agreed to shafts specifically 14 feet (4.25 m) in diameter being put down as was reasonably necessary. Two lines of shafts shown on the first and second edition OS large scale maps ranged northwards in Central Byfield (to what is now Westerleigh Road and north east in Central Firs (over what is now Firs Field from Firs Shaft to the Hadley Arms.

An example of the type of cartway system that operated in the 19th century, and fairly representative of many other examples in form, was located on the eastern edge of Byfield Quarry (514). The cartway was driven northwards towards North Road from the south-western edge of the Sheeps House Quarry, from about 1808 onwards, as the quarry developed. Graffiti dates of 1818 recorded at the northern extent are indicative of the final stage of workings. The cartway extended for 120 m, with an average maximum open width of 5 m, at about 5.71 m below roof height. It was only possible to measure the cartway track on the northern limits of the workings. Here it had an internal gauge of 1.27 m (4 ft 2 in) with the ruts measuring 203 mm (8 in wide).

Fig. 9.13 Nineteenth-century cartway in East Byfield, showing rub marks caused by a cable on the right side pillar, and wheel ruts on the floor

Fig. 9.14 Cartway near Firs Shaft leading towards the Hadley Arms. Note rub-marks from a cable on the left side-pillar and a corbel with a post-working fracture on the right-side pillar

However, this cartway was possibly untypical in that it appears to have used cable haulage of wagons. There were no crane positions along its length, but there was evidence of horizontal haulage from cable grooves and other rubbed and eroded pillar faces (Fig. 9.13) where cable (rope or chain?) haulage is believed to have been used, and several other pillars were trimmed to allow better access for carts. The wear marks led back to a probable shaft location, suggesting a down-the-shaft cable from a surface windlass of some type. A rare example of a loading dock was also located at the northern limits. It consisted of a low stone pack wall constructed of roughly coursed rubble, measuring 3 m in width at the base of the structure. The cartway level had a surface of mixed fines and was loaded from a height of 0.94 m above its floor. A length of iron chain (SF 188) recovered from the edge of the loading dock may have been used for haulage or lifting, although there was no evidence of a crane location. The structure was noted by David Pollard when full access to the dock was possible. He records that the structure had 'accompanying tyre marks on the floor suggest the wagon was a low flat trolley or dray, with a fore carriage that enabled the front wheels to turn underneath the wagon bed. Certainly the fore carriage was able to negotiate a very tight turning circle around and almost touching a single pillar' (Pollard 1994, 42).

Besides the loading dock the main cartway had a secondary spur cartway on the western edge probably representing a turning area for carts. The carts may have been hauled by horse to this point as it was, unusually, large enough for one to turn so as to be facing the correct way, with the wagon in the loading bay structure for loading of the stone prior to the outward journey. This suggestion was by the late Mike Desmond, lead miner on the Combe Down Project, whose own coal mine in South Wales was the last to use pit ponies underground. David Pollard (pers.comm. 2002) had also independently suggested the same operation of the horses.

In Firs the major 19th-century development, other than the completion of older routes, was that driven from near Firs Shaft to near the Hadley Arms, under the north-east area of Firs Field, part of what has been (later) called the Three Acre Quarry. It was operated by John Burgess and son between about 1810 and 1838. It was some 5 m high and 4 m wide and passed next to or under four wide shafts in what was clearly a planned and integrated development. The section near the Firs main shaft cut for some 20 m across older Long Room workings, linking to the northern end of the quarrying area known to us as the Grand Canyon – Quarry area 2209, part of Quarry 2211. Smoothed sides of the pillars close to the Firs Shaft (Fig. 9.14) showed rope wear and it was evident, from cart ruts in the floor turning into it, that the first section last served the Grand Canyon (2209). Cable grooves on the shaft pillars may reflect another down-the-shaft haulage system, though these were destroyed before full recording was possible.

Past that junction the cartway ran in a deep cleft, formed on the south side by roof-high rubble stone packs and on the north, by 2-3 m high packs extending some 7 to 8 m over a platform approximately 2 m below the roof, and to a further flanking line of roof-high stone packs. This may have been an area associated with the cartway for storing or working on stone. The cartway then passed directly under two wide shafts. The first, known today as

Fig. 9.15 The arch supporting the shaft on the Firs-Hadley Arms cartway

the Chestnut Tree Shaft close to the War Memorial on Firs Field, was entirely filled with rubbish, and apart from determining this, its width could not be examined. The next shaft was clearly the main haulage shaft, more or less central to the Three Acre area (and was used by the Stabilisation Scheme as the Materials Shaft). The shaft (Chapter 12, Case Study 9). had been heavily reinforced by stone arches and rubble and block pack walls, apparently from the true floor of the excavated working to the roof and, on one side, to the surface (Fig. 9.15). The cartway on the north-east side had been blocked off and the cartway backfilled, suggesting that the north-east area was worked first, and in comparison to the clear floor from the Grand Canyon to Firs Shaft, the cartway floor on the west of the (arched) shaft had been allowed to deteriorate, suggesting the Grand Canyon section was among the last sections to be worked within the quarry. It was not possible to examine the cartway north-east beyond the arched shaft, but there is documentary evidence of a further shaft near the Hadley Arms and another was found in the area between the Long Drung and The Avenue, also worked by Burgess, so the cartway may have continued that far.

Barrow-ways

Throughout the workings there was an intricate network of barrow-ways used by single-wheeled barrows. These were used for transporting both rubble and blockstone en route to the quarry entry for sale, and for transporting waste material (spoil) to suitable locations that were more than handling- or throwing-distance from the working face. The type of wheelbarrow used, from the limited evidence of one survival-shadow (Fig, 9.16) and a few wrought iron wheel rims was similar to the usual builder's barrow, with the single front wheel running at one end between two long timbers with the handles at the other end. In between was either a flat bed for carrying blockstone or a bed with sides, called raves, for carrying spoil. In some cases the raves could be removed to convert to a flat bed, though specialised use of each type would have been simpler. As barrows often had to be operated under low roofs, causing the barrowman to stoop, shortened rear legs, in comparison to the builders' barrow, would have been advantageous. It is possible a 'mining barrow' (see Figure 9.17), with the body slung below the long timbers may also have been used (Derek Hawkins showed such a mine-barrow surviving at one of the Wiltshire Quarries at the 2010 NAMHO conference). Wheelbarrows could probably carry loads of up to 250 kg in favourable circumstances. Beyond that, two-wheeled quarry carts or four-wheeled wagons were probably necessary for all but short distances where dragging or rollers might be used.

The virtue of the wheel-barrow lay in its economy in first cost and in use, manoeuvring ability, and ability to move substantial loads with a single labourer. It required only small and narrow openings and could be used with low roofs (even below a metre were recorded, though these must have made for great difficulty) and can turn right-angled corners or worse with ease. It can be tipped over its front end, projecting material some height off the floor, with further height gained by using a small planked or spoil incline. Alternatively it can be tipped with great ease over the front corners or side, with care maintaining a fairly level floor with

Fig. 9.16 Wheelbarrow 'shadow' in East Firs, close to the Long Drung

tipping slope beyond. It was almost certainly the most used vehicle in the quarries. Its needs are few, but a level flat floor without even pebbles projecting make it much easier to use, and a reasonable roof-height above the floor (just less than human height) assists greatly, though the barrow with a willing operator could overcome almost any disadvantages if required (Fig. 9.17).

The extents and orientation of the barrow-way routes were recorded during the archaeological survey and plotted in relation to associated components, including cartway routes, shafts, spoil tipping fronts and tipping platforms and other deposits of spoil along their routes. Fig. 9.18 shows an example of the annotated drawn plans prepared during the archaeological survey with the barrow-ways and the pillars highlighted against the detail of surrounding features in Quarry Area 2332 in East Firs. Where possible the barrow-way routes were phased and related to the quarrying areas in which they appeared, and also recorded in relation to adjacent earlier or later sequences of routes. The direction of the tipping fronts and the use of the barrow-ways was recorded on the synthesised drawings of the quarries, and these were also used to relatively or absolutely phase some of the quarrying enterprises. The widths and the heights of the surfaces below the roof were recorded in order to interpret how easily

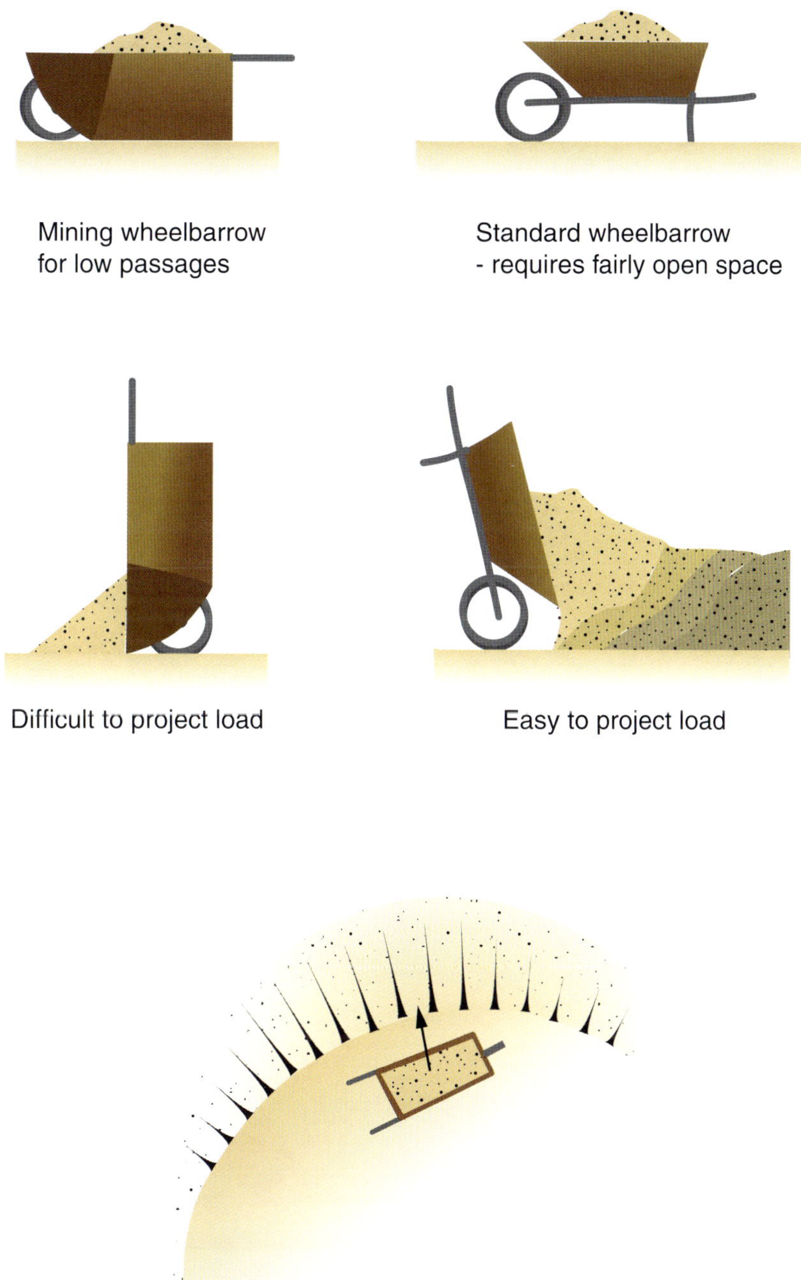

Fig. 9.17 Types of wheelbarrow and their use

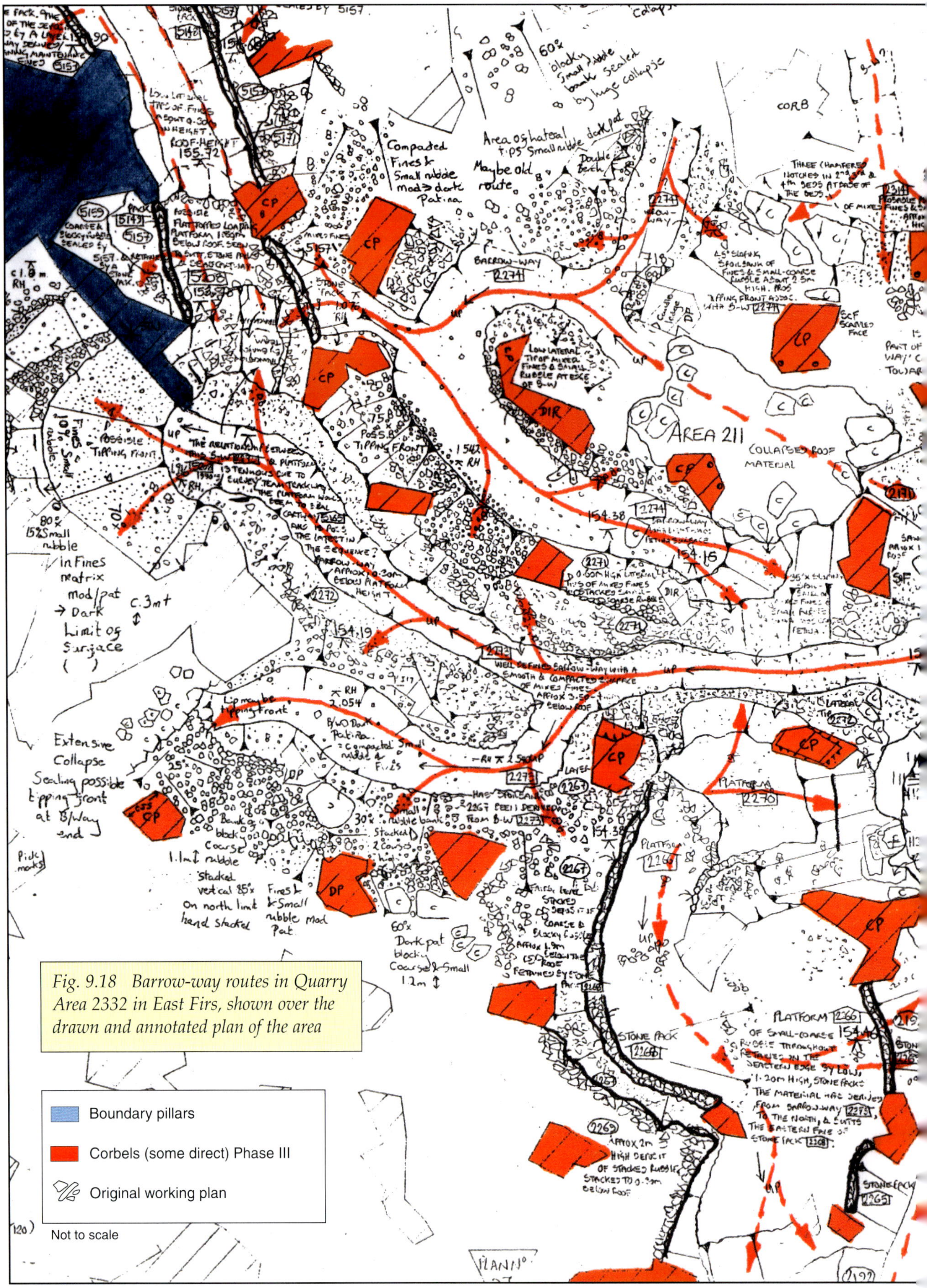

Fig. 9.18 Barrow-way routes in Quarry Area 2332 in East Firs, shown over the drawn and annotated plan of the area

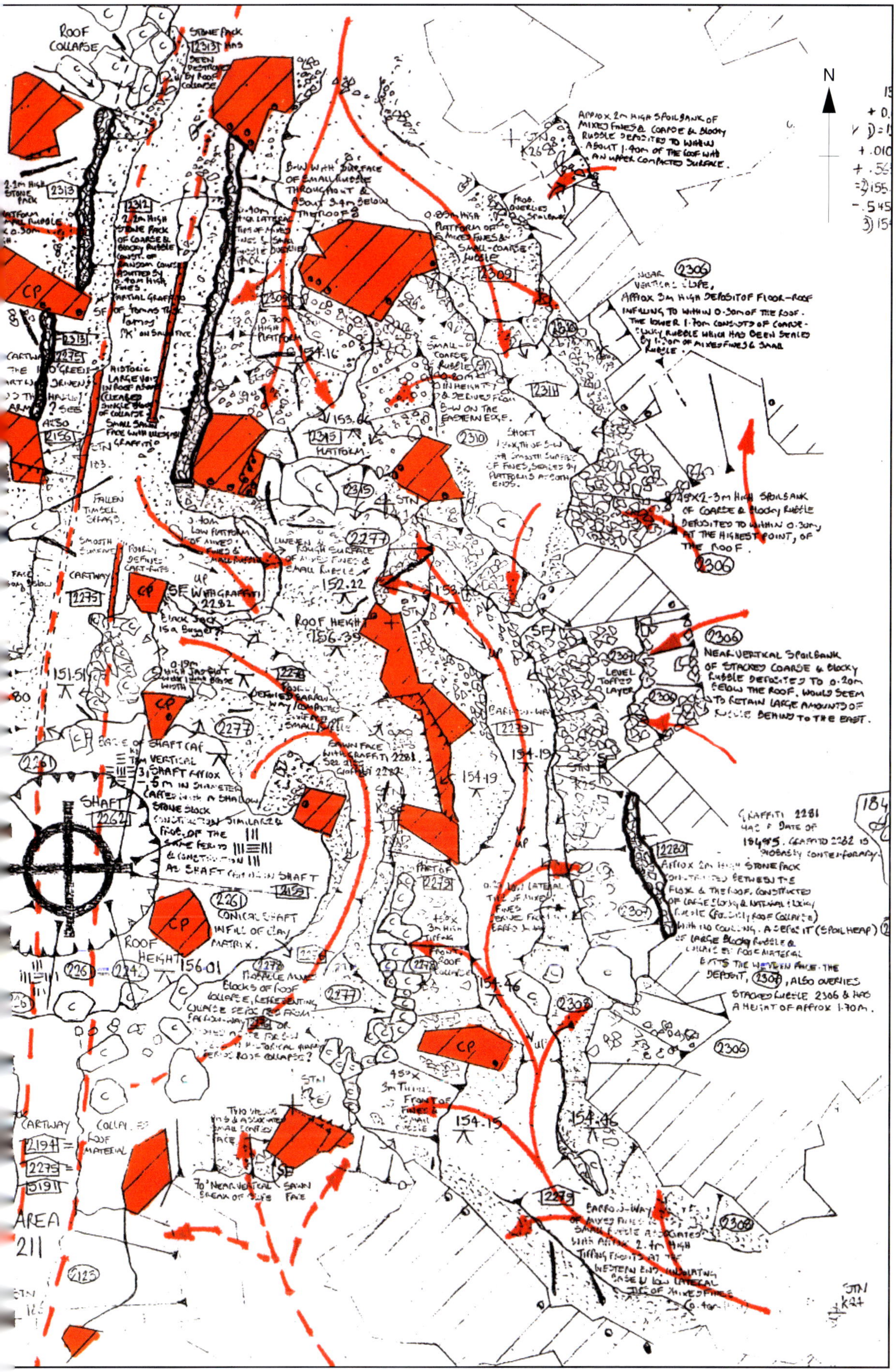

they could have been traversed by the quarrymen pushing loaded barrows.

Barrow-ways varied immensely in both form and use. Some cartway routes were probably used for both carts and barrows at various times and were frequently adapted to barrow-ways in later use. In open-working areas, the whole flattened floor area could have acted as a barrow-way route. Some barrow-ways functioned both to remove stone and spoil, though usually not both over their whole lengths – it was not always easy to differentiate the purpose, except where the barrow-way leads directly and only to a spoil dumping area or heap. Spoil-dumping barrow-ways had a particular value in that they may reflect opening-out of areas, so their position may indicate where work started: dumping from one area to another yields stratigraphic relationships which can indicate sequences and even allow derivation of dates or phases of working.

Barrow-ways used for access and stone extraction

In quarrying areas not directly served by the cartway network stone was transported either to the surface level entry or to a vertical shaft equipped for winding. Many of the longer (measured in a few tens of metres) barrow-way routes for stone extraction tended to be located in the quarrying areas to the east of the Long Drung, where smaller quarrying enterprises were established after about 1830. Access to the quarries was restricted to vertical winding shafts, with a reduced, if any, dependence on the larger cartway transport routes. Such barrow-ways became the main routes for the transport of the finished project, and usually were also used for the disposal of spoil. Within these quarrying areas where stone was sometimes quarried only for the construction of houses above, less of the total depth of commercially viable freestone was extracted and as little as 2 to 3 m taken. The routes often connected a working face at one end and a vertical winding shaft at the other. Such barrow-way transport routes often had flanking stone packs on either edge to protect the barrow-way from dumped rubble (of which a higher proportion was likely with lower depths of extraction) and for additional support for the roof.

A good example of probably the longest of this type was a barrow-way noted in Quarry 2356 (Fig. 9.19). It extended for at least 75 m, with floor to roof flanking packs on either edge. It was between 0.86 m and 1.20 m wide and had well-maintained and compacted floor surfaces. Several phases or sequences of barrow-way were noted, with the heights of the compacted barrow-way surfaces recorded at 1.50 m, 1.90 m and at 2.40 m below the roof level. Often several stone packs had been constructed on either side of the access routes. Sometimes they directly abutted the face of the previously erected pack and sometimes a thin deposit of mixed fines and rubble was stacked between successive builds. This barrow-way route and other examples of longer barrow-ways acted as access routes for the quarrymen and used, as conditions were adequate, to carry the finished stone, but they were also maintained for the lateral disposal of spoil in the flanking packs. This made them semi-permanent routes.

Fig. 9.19 Barrow-way in Far East Firs (Quarry 2356)

Fig. 9.20 Barrow-way in Far East Firs (Quarry 2365)

A shorter barrow-way of this type was recorded in Far East Firs in Quarry 2365 (Fig. 9.20). The barrow-way was initially 25 m in length and 0.80 m wide and was 1.80 m below the roof level, later developed to a length of over 50 m. Access was good throughout with level and compacted surfaces of fines. It had flanking packs along its length which retained discards on either edge, with adjacent rooms filled with spoil. The route linked the working face on the boundary pillar on the eastern edge of the Firs complex to a shaft in the north of the quarry.

Barrow-ways and spoil dumping

These showed the greatest variety, the workmen using any convenient strategy or tactic to minimise the effort of dumping worthless material. They utilised existing ways where available, sometimes on long disposal routes which took on a semi-permanent character with well-defined flanks, but in other cases, particularly in substantial areas available for dumping, they adopted a distributary (diverging branches) form (see Chapter 12, Case Study 7 for detailed plan of these). Often, however, they seemed less organized, dumping at any place in which spoil could be placed. Very often the distributary areas were formed on top of the spoil dumped or stacked directly, by hand, behind the face and gullet, and it was common for them to use adjacent quarries. Once open spaces had been utilized (possibly by several layers of spoil over successive barrow-ways amounting to several metres depth), the barrow-way routes were themselves filled in as a final act on retreat (Fig. 9.21). The barrow-ways often followed the former routes and were formed and built up over successive barrow-ways totalling several metres depth. This process of dumping over old cartways, spoil dumps and old barrow-ways led to very complex systems of barrow-ways one above the other, sometimes requiring inclines. What was actually visible was the last phase of barrowing over the immediately previous stage of dumping. Nearly every excavation (a near continuous process over 15 km of supported roadways) carried out by the Stabilisation Scheme that cut through spoil revealed sections of former barrow-way routes running below the visible surface.

One of the more simple sequential series of barrow-way activity was noted within the Central Firs Quarry 517 where three separate barrow-ways were recorded in section (Fig. 9.22). These are thought to date to the early 19th century, as the latest in the sequence at least had abutted adjacent sawn pillar faces, which would have needed to be clear of the floor in order for them to have been worked. The barrow-ways were probably all employed for deposition of spoil from the quarry workings of the earlier adjacent Allen period Quarry (505). The uppermost barrow-way had a compacted surface with low lateral tips of small rubble and fines. It was about 1.30 m below the roof, and sealed two earlier phases, located at 1.90 m and 2.10 m below the roof respectively. The lowest barrow-way exposed was seen to abut a rubble pack on its western face that had been constructed from an earlier unexposed barrow-way or quarry floor. The height of the uppermost barrow-way to the quarry roof was rarely below about 1.10 m, which provided a low but relatively normal (though undoubtedly uncomfortable) height to adopt a stooping posture while pushing a loaded barrow. This changed near the final deposition end of the barrow-way, which was a spur where the height was sometimes reduced to as little as 0.80 m, probably implying hand-stowing at this point.

A slightly more complex series of barrow-way routes, and representative of many of those investigated, was noted within a section (Fig. 9.23) exposed in the Firs Quarry (2347). The barrow-way was used to transport spoil from the main quarrying area to the west to deposit it in a Long Room. The barrow-way sequences were all attributable to the same phase of workings. The earliest deposit in the sequence was a surface of mixed fines located at 2.40 m below the roof. It also formed the point at which a lateral stone pack was constructed on the northern limits of the Long Room. A more extensive barrow-way or floor surface was also

Fig. 9.21 Barrow-way filled in on retreat in Central Firs. The rubble pack rests on the former barrow-way (Quarry Area 2347)

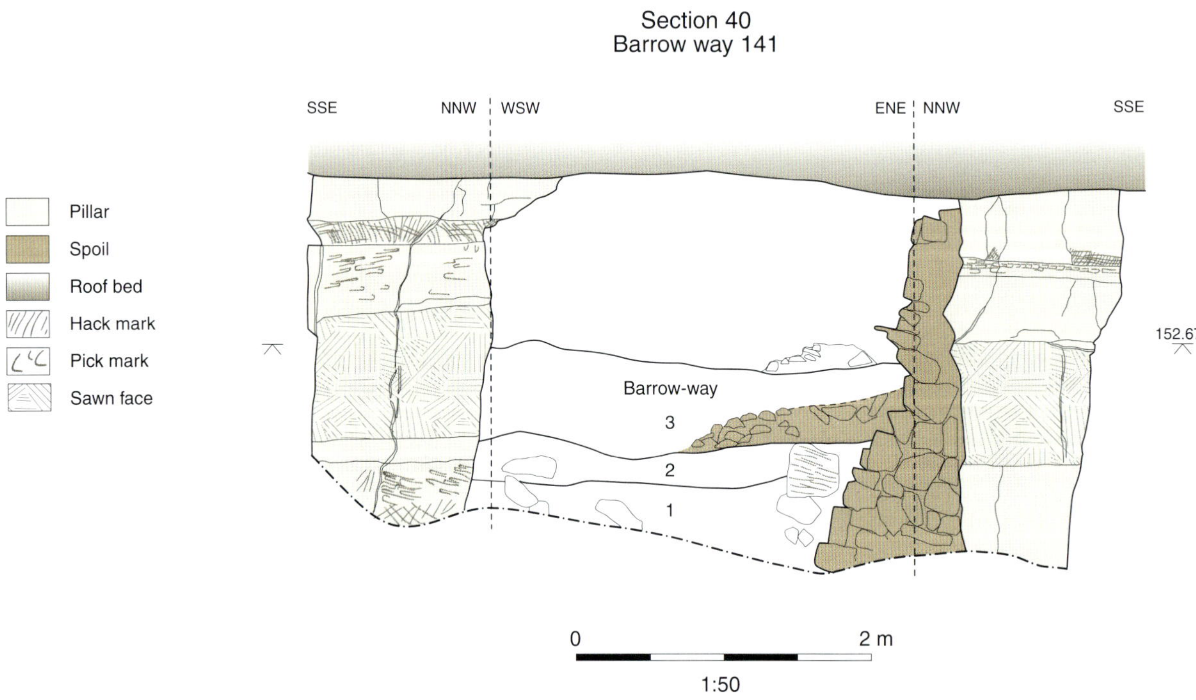

Fig. 9.22 Section showing the sequence of three filled-in barrow-ways in Quarry 517, East Byfield

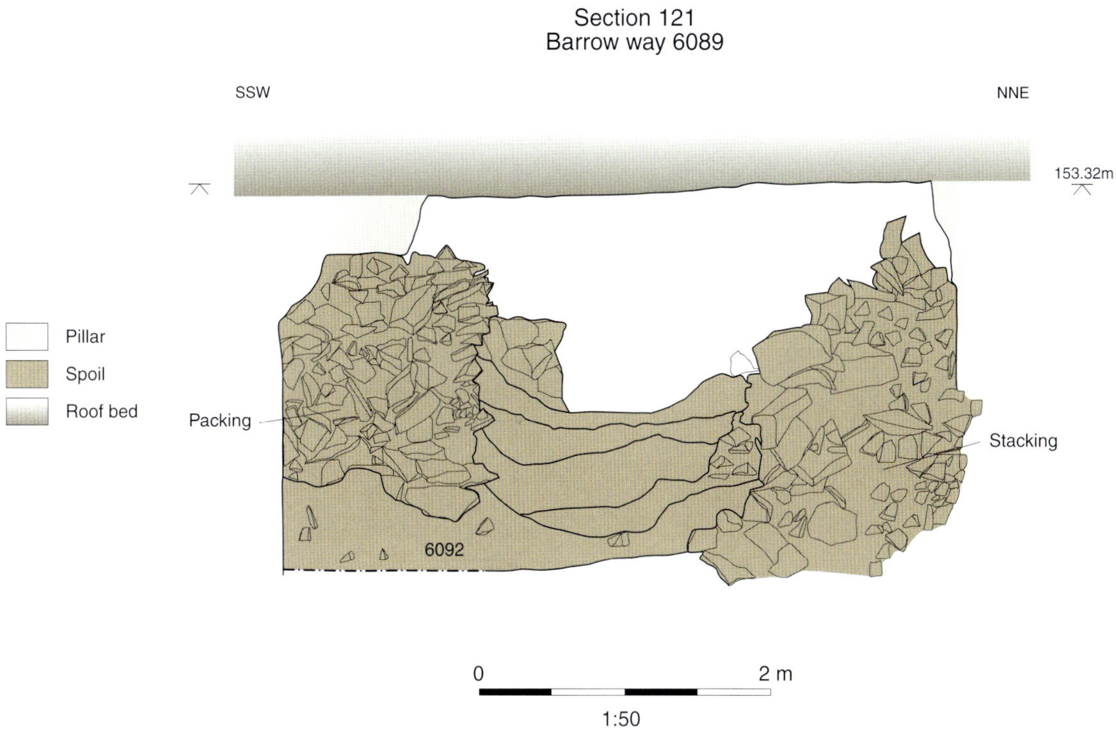

Fig. 9.23 Section of barrow-way in Central Firs (Quarry 2347) with lateral rubble-stone packs and stacking

noted at 2.40 m below the roof, from which a stacked rubble deposit on the north side was built. The floor was sealed by an undulating deposit of fines and small rubble 2.24 m below the roof, upon which lateral rubble was stacked. The uppermost barrow-way, which abutted both the stacked rubble deposits, extended another 15 m to the end of the room, and had four distinct compacted barrow surfaces located below the uppermost surface, at 1.42 m, 1.50 m, 1.72 m, and 2.04 m below the roof level. The deposits within the barrow-way sequence gave approximate depths for the deposited fines and rubble of between 0.20 m and 0.30 m, apparently fairly standard throughout.

A further example of barrow-way complexes is included in the case study of the Grand Canyon (Chapter 12, Case Study 9).

Features of barrow-ways

Barrow-ways averaged about 1 m in width, and had well-maintained surfaces of compacted fines or small rubble on the uppermost, often slightly depressed surface kept free of obstructions so as not to hamper the access by barrows. The quarryman had to navigate between the roof-supporting pillars, which thus always affected the routes. Especially in wider spaces, the orientation of barrow-way changed over time, which was sometimes the result of side tipping and subsequent wheeling over the later dumped material and sometimes from deposition of overlapping successive surfaces, or from raising the floor to deposit another spoil layer, perhaps at a late stage in the life of the barrow-way. These resulted in successive sequences of the barrow-way surfaces and their constituents, and also the lateral tips of fines and rubble deposited at either edge. The latter could include remnants of flanking stone packs.

Some barrow-way surfaces contained specific evidence for the use by wheelbarrows, though this was fairly rare with only eight examples of wheelbarrow grooves, all 19th century in date, recorded within Byfield and Firs. Such features were rare because they were prone to damage and easily filled with fines and clay washed in by water egress, or were trampled by quarrymen and later visitors or buried below later barrow-way sequences. Finds of iron tyres, and the impracticability of small wooden wheels subjected to such battering and wear make use of iron tyres almost certain on barrow-wheels.

In Byfield Quarry 514 a series of broken lengths of curving and straight barrow-ruts were preserved in the top of tipping platforms. They were 50 mm (2 inches) wide and represent the width of the wheels' tyres. The platform also had a centrally located groove where barrows had repeatedly traversed the same part of the barrow-way. This groove feature was also noted on barrow-way surfaces in the Byfield Quarry 510, and Quarry 2344. Barrow ruts 25 mm (1 inch) wide were also noted on the surface on the north-eastern edge of the Grand Canyon (2209). These linear ruts had were preserved in the centre of the barrow-way in wet, slightly clayey conditions and were attributable to one of the main phases of spoil dumping from the eastern end of the Canyon. Another feature seen near to the Firs Shaft showed the ruts curving up the lateral spoil dump

Fig. 9.24 Barrow rut preserved by calcitic flowstone Eastern Firs (Quarry 2337)

alongside the barrow-way, preparatory to tipping the barrow over its front corner. The survival of barrow ruts was assisted in some cases by calcitic deposition on the softer surface, noted, for example, in the centre of a barrow-way in the early 19th-century Firs Quarry 2337 (Fig. 9.24).

The make-up of materials within barrow-ways could include small and coarse fines (up to 5 mm) and rubble and coarse rubble 50 to 500 mm in the longest direction, with a thin layer of fines and small rubble (5 to 50 mm) forming the uppermost surfaces, which were compacted by regular barrow activity. The make-up materials could also be composed of larger pieces of rubble and even blocks (over 500 mm) or with clay deposits disturbed (usually in gulls) during the extraction of the stone. Clay was not a good material to barrow over and so tended to be deposited on the end of the routes on the tipping fronts and platforms. An example of this was contained in the tipping platforms in Byfield Quarry 514, where platforms contained a high proportion of clay-derived deposits buried below, while the barrow-way surface itself was relatively free. Where clay was deposited along the barrow-way routes as a sub-floor, it was overlain by a layer of fines to provide a more suitable surface. This method of using fines and/or small rubble for the uppermost surface to level it and provide a better surface for barrowing was also noted where larger rubble had been deposited in the make-up surfaces. (The floors of the Stabilisation Scheme roadways were prepared in a similar manner even though the tracked machinery was capable of dealing with far worse surfaces than a wheelbarrow).

Barrow-way routes were generally fairly level or had only very slight inclines or declines; understandably, severe inclines were avoided where possible. Where steeper inclines were recorded they tended to be in the areas with greater restriction of space. This could occur during the initial opening up of an area, for example, before the gullet moved forward to allow for spoil disposal behind. Inclined barrow-ways tended to be longer, again reflecting local shortage of space, while the longer travel distances would be less economic in time and effort to dispose of a barrow-load. There were thus few examples, perhaps less than half a dozen, of the steeper inclined barrow-ways of around 1:5 or 1:4 inclination.

An inclined barrow-way, recorded on the interface between Quarry 2202 and 2340, probably developed in the early phases of the quarry. It was about 38 m long and 1m to 1.30 m wide. It branched into two distinctive tipping platforms and other smaller barrow-way spurs and tips. One platform was formed by tipping into the deeper earlier quarrying area of (2340) and was *c* 5 m high. The barrow-way and tipping platforms may have been created when the initial northern extent of a cartway was developed at the position of the base of the inclined barrow-way for the disposal of spoil. The room in which the cartway was driven at this point may not have been much wider than the cartway, and of a similar width, about 4.50 m.

Another good example of an inclined barrow-way, which was carefully examined during the archaeological survey, is discussed in Chapter 12 (Case Study 9) at the Grand Canyon. It illustrates how spoil derived from different heights at a quarry

Fig. 9.25 (facing page) Barrow-way system linked to a cartway in Central Byfield (Quarry 505)

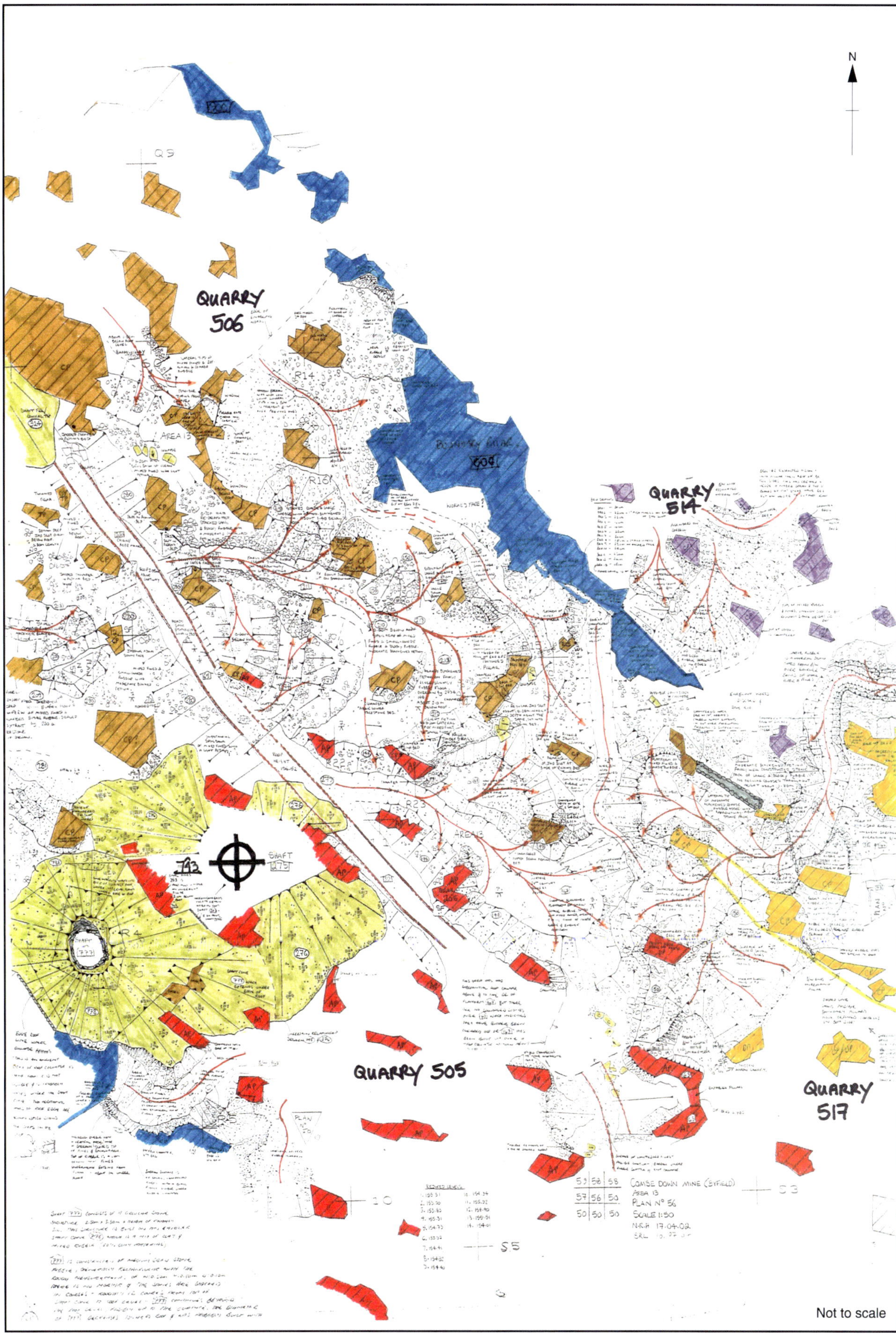
N
QUARRY 506
QUARRY 514
QUARRY 505
QUARRY 517
BOUNDARY MINE
SHAFT
COMBE DOWN MINE (EYFIELD)
AREA 13
PLAN Nº 56
SCALE 1:50
Not to scale

face was removed on the level to different spoil dumping areas, thus avoiding unnecessary effort in dropping spoil down only to have to raise it again via an incline. It is clear that the steep and long incline there was only used for spoil produced at the lowest levels, perhaps at a final stage of further deepening of the quarry, when nearby space was either exhausted or inaccessible. At that point the stone packs had risen to over 4 m high at the sides of the barrow-way.

Evidence for the contemporary use of barrowing from active cartways was recorded in the Byfield Quarry 505 where the mid 19th-century cartway had barrow-ways developed from the northern edges of the quarry, linking to tipping platforms (Fig. 9.25). The cartway continued in use as the primary transport route for stone leaving the quarry, for later quarrying at the north of the area, and transported stone past the barrow-ways. Some of the cart-ruts were partially filled with fines where the barrow-ways were active across its path.

A vertical sequence of barrow-ways in relation to earlier cartway sequences at what was originally Allen's western cartway was investigated in Central Firs in Quarry 2200 (Fig. 9.26). The sequence was in a vertical face of spoil where later quarrying activity in Quarry 2370 had truncated the barrow-ways and cartways and presented a high, free, stepped depositional profile. The barrow-way activity was associated with later deposition derived from a small scale 1860s scavenging operation at Quarry 2368, which included activity in the northern and central end of an adjacent cartway. The barrow-way activity within the former Allen cartway consisted of nine distinct barrow-ways of mixed fines and small rubble with occasional larger rubble and blocks. The uppermost

Fig. 9.26 Vertical section in Central Byfield (Quarry 505), showing successive barrow ways

barrow-way was 2.80 m below roof level and the lowest in the sequence was 3.90 m, directly overlaying the uppermost cartway surface. All the barrow-ways in the sequence abutted an earlier stacked rubble or stone pack on the eastern edge of the cartway surface. The surfaces varied in depth from between 0.08 m to 0.28 m and were fairly typical of the barrow-way depths noted elsewhere (see Chapter 12, Case Study 15).

Unusual forms of barrow-way

The main passage within the lower of John Scrace's unusual two-level workings in northern Byfield Quarries 508 and 512 contained a carefully maintained barrow-way. It was *c* 1.80 m wide and 3.30 m below roof level, and had an extraordinarily clean floor, devoid of debris, with an appearance of being literally swept clean. The route was used to convey stone from the working area at the northern end of the quarry to a possible entry to the surface open quarry west of Combe Road and south of the Bradford Road, which was active in the early 19th century. Low lateral packs had been constructed to retain varying sizes of waste rubble either side of the principal barrow-way. The rubble was well-sorted and neatly stacked. The quarry is considered below (Chapter 12, Case Study 8).

Haulage and winding

Much of the movement of materials was by hand, by use of iron bars (levering), lifting, carrying and throwing or by sliding, sledging or use of rollers, and, as described above by using carts, wagons or barrows. In order to move stone to surface, or in the case of spoil, to a distant dump, some form of mechanical equipment was used. The only firm evidence of manual equipment useful for these were finds of bars and two very long and wide and substantial timber planks and the shadow of a wheelbarrow noted earlier and what was most probably the handle of a two-wheeled cart. There was some evidence in wheel marks for wheelbarrows and in ruts for the four-wheeled cart or wagon, In the few later quarries there was use of narrow-gauge railways, in the form of sleeper-troughs, and for cranes, most commonly in anchorage points, cable grooves and impressions and in crane bases and chog holes in the roof. Winches were mentioned in early 19th century documents, and evidence of their use has been found in spigots, chains on pillars, a pulley, and floors over which stone appears to have been dragged.

As noted above, the wheelbarrow was capable of shifting loads of up to some 250 kg, though less was usual. It was cheap, effective and very manoeuvrable. For heavier loads the quarry cart or some type of wagon was required. The two-wheeled cart (Fig. 9.27) was available at least by the 19th century. It had a single long handle and effectively used its axle and wheels as a fulcrum to raise the stone from the floor prior to moving off. It consisted of a level platform mounted above the wheels and axle, fitted with a long and very strong handle with a cross-piece to facilitate pulling. To load the stone, the platform was placed vertically next to the stone, which was secured to it, probably by chains, with the handle raised vertically. By chocking the wheels and then pulling on the long handle the platform and stone was rotated to a position above the wheels and the cart and load could be pulled across a level floor. Typically only some 0.7m wide, based on Wiltshire examples (Pollard, pers comm.), the two-wheeled cart was capable of operation in narrow spaces such as the gullet, as well as on wider cartways. The device was capable of handling a ton with two or more men to pull it. The handle of such a cart was found underground near the Arched Shaft in Central Firs, worked by Burgess 1810-38 (Fig. 9.28). It was about 2 m long and some 150 mm square, heavily protected at the top, or handling end, by iron strip reinforcing, presumably to protect against hitting the roof.

The quarry truck or wagon, as used in the Wiltshire quarries was a four-wheeled, flat-bodied vehicle, with spoked wheels and a swivelling front axle. The width of cart ruts at Combe Down suggests a body width above the wheels of about 4 feet (similar to those used in the Wiltshire quarries) or slightly smaller, and it was probably about twice as long. No remains were found, but facilities were

Fig. 9.27 Example of a two-wheeled cart at Beer Stone Quarries, Devon (photograph by David Pollard)

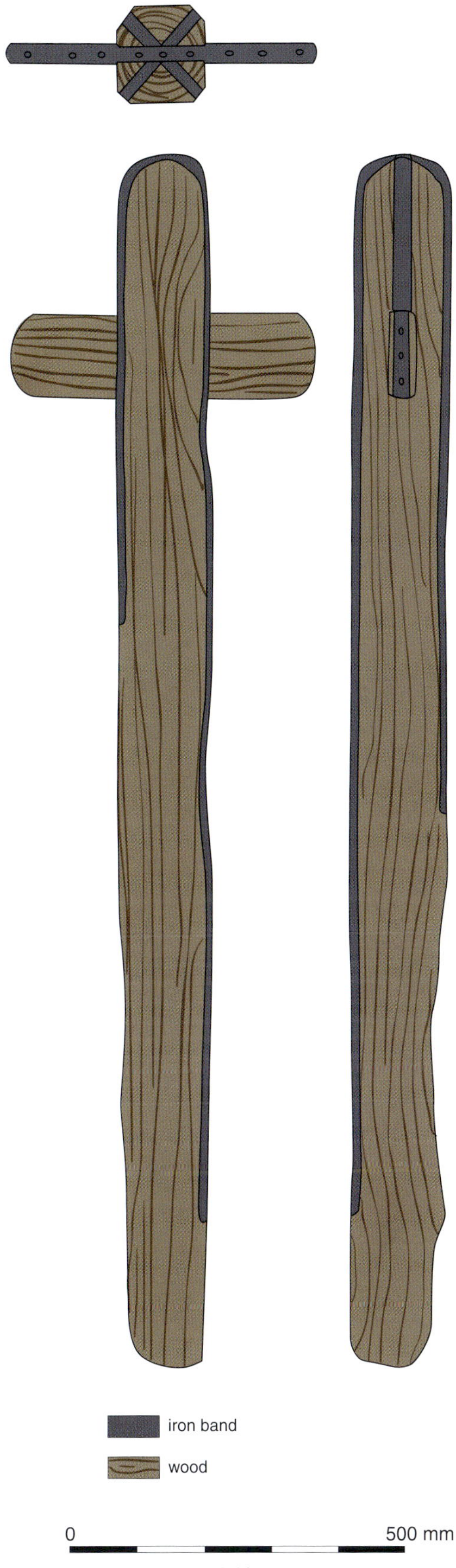

Fig. 9.28 Shaft from a two-wheeled cart, found near Burgess' Arched Shaft in Central Firs

provided at a loading bay in north-east Byfield (see above). It could be fitted for either horse-haulage or be pulled by men. The evidence for horse haulage underground is scanty. A few horseshoes were found, and one of the cartways within the Byfield Quarry 505 had what appeared to be horse hoof prints on the surface (Fig. 9.1). There was no evidence of facilities provided underground for horses, though in such small quarries this would anyway be unlikely. It may be that horses were occasionally used, but human haulage was the normal mode, as in many mines with small distances to cover.

A 1751 illustration of a large stone being hauled shows a carriage of heavy timbers with axles on which were solid wheels of up to 2 feet (0.6 m) diameter, without any form of steering, being hauled by use of a cratchet winch (see Fig 5.11). It would have been suitable for very heavy blocks of stone such as those seen at the Priory Park Gates, but although associated with other Combe Down equipment in the article, it is not certain that it was used there.

Railways

There was no archaeological evidence for the use of timber rails underground in a similar way to those used at surface. Evidence for use of narrow-gauge railways was very limited in the Firs-Byfield area of Combe Down, with use only apparent at James Riddle and Son's Allotments Quarry (518), linking the working area with a shaft in the early 20th century. It was possible to observe the track bed with timber-sleeper voids of about 0.85 m (2 feet 6 inches) gauge here and several dog or brog nails were found, of the type used to secure flat-bottomed rails to timber sleepers. It probably used a pre-existing cartway route, as it cut into existing spoil deposits. The working area was served by post cranes, which would have been used for loading the flat-bed wagons. Several chog holes from the cranes remained in the roof reflecting their shifting positions and probably that of the railway (see also Chapter 12, Case Study 14).

Railways were used in the areas at the Foxhill and Shaft Road sites that were archaeologically investigated, again in conjunction with post cranes (Chapter 12, Case Study 13). Illustrations of stone quarrying show a simple bogie with four wheels and a flat bed, and probably those at Combe Down were of a similar pattern (Fig. 9.29).

Capstans and winches

Various forms of capstan (vertical axle and drum) and winches (horizontal axle and drum) were used at surface to lift blockstone by means of cranes, both

Fig. 9.29 Flat-bed four wheel bogie stone wagon (photograph by P. Wooster)

Fig. 9.30 Grooves in the side of a pillar at Allen's east side entry at Jones Quarry (Ralph Allen Yard), the nearly horizontal grooves suggesting the crane cable was diverted up the cartway to draw-out carts or wagons

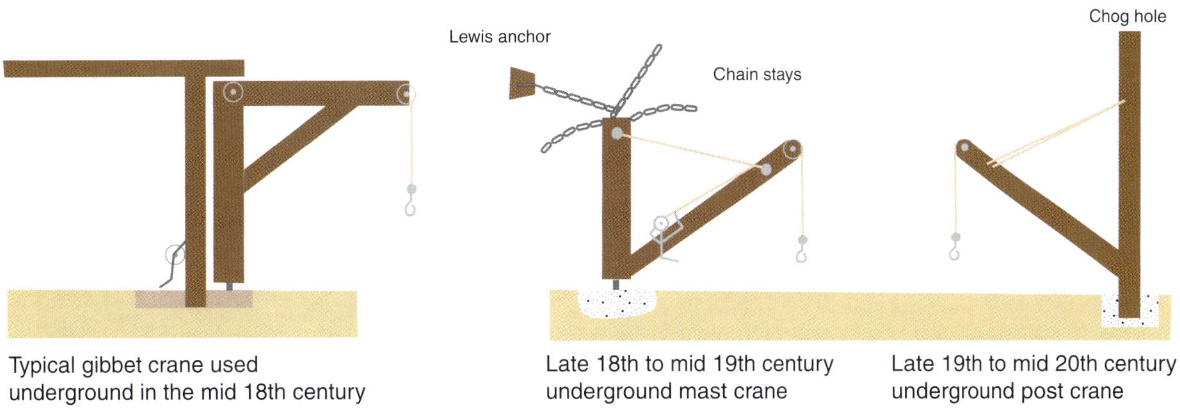

Fig. 9.31 Types of crane used underground

Fig. 9.32 Crane-cable grooves in the branch cartway out of the east Allen Cartway in Central Firs

Fig. 9.33 A Lewis in situ in north Central Firs, used to hold a crane stay

man-operated and, from description, horse-operated. In all the likely Allen Phase II quarries and later workings, there is evidence in the form of cable grooves, of the cable being extended down shafts to either winch stone or to haul a wagon along a passage (Fig. 9.30). In one instance west of the Long Drung (Quarry 2361), the cable was clearly drawn up the shaft (see Figure 7.6), to pull blocks of stone from some metres away from the shaft bottom, where they were presumably stored.

The capstan and possible early forms of wood-framed winches or windlasses probably gave way in the mid 19th century to the winch (crab-type) with a geared cast iron barrel with ratchet, wound by one or two men using chain or, later, iron or steel wire rope. The sparse evidence of this type of mechanism is in a sale of John Scrace's quarry (508 and 512), in the corner of the junction of Combe and Bradford Road in about 1809, which included a share in two windlasses (Pollard 1994).

Lifting and cranes

Cranes were introduced by Allen, to the design of the engineer Padmore (Chapter 5) from *c* 1730 at surface. These used a capstan and a rotating gibbet and raised stone from the entries to the cliff top and railway. It seems likely that these, or parts from these were mounted over shafts when they were first constructed, perhaps late in Phase II (late Allen period). The only surviving detail is cable grooves at the bottom of shafts worked at that time. Cranes of perhaps a similar but smaller type appear to have been used underground in the mid or late 18th century, probably early in Phase III. They were clearly, from anchorage evidence, of a stayed mast type, but it is unclear whether these were jib or gibbet types. In either case there must have been some form of bearing stone for a pivot, though none was noted during the survey. The variety of types is shown in Fig. 9.31.

At least three fastenings and chains for each crane would have been required, but, by analogy with later surface cranes, there may have been more in case one stay broke. Traces were found of Lewis holes from anchors on pillars, grooves from cables wrapped around pillars, and, in one instance, of the chain impressing a negative mould of itself in a pillar face in the Allen Estate Quarry 2201(Fig. 7.14). The use seems to have been an experiment, used only in one area on the east side of the Allen cartways in Central Firs, where they seem to have lined the cartway. Blocks were probably dragged from nearby Long Rooms and loaded on to carts. Cable grooves at an angle indicating an origin near the roof are characteristic evidence of their use (Fig. 9.32).

Crane anchorages

No complete examples of cranes survived at Combe Down, though examples survive in some Wiltshire underground quarries. The usual evidence for their

Fig. 9.34 A post crane base-stone, with a pillar behind and a chog hole in the roof with a Lewis slot

use was the crane stone or base stone, the chog hole or anchorage points (usually Lewis or wedge holes), cable grooves, or haulage grooves and rub-marks from the movement of the jibs.

The Lewis (Fig. 9.33) is a keystone shaped three prong iron or steel lifting and anchoring device that fit into a 'dovetail'-shaped Lewis hole in a stone block or pillar. It was often referred to as a pair of Lewises. The dovetail hole in the stone can be cut easily and quickly by the quarryman, using a holing pick, after which the two outer prongs are inserted and fixed in place by the third, central prong and fastened in place using a U-bolt to which a cable is attached. It can withstand a direct pull. The simple straight-sided wedge hole was also used as an anchor, often placed around the back of a pillar so that the pull on it was at right angles to the wood or iron device in the hole. Its virtue was probably to lessen the length of chain otherwise needed to wrap right around a pillar, a method that was also used.

Both mast and post cranes required a basestone, on which the post could pivot. The only basestones found related to the later post cranes were located in the early 19th-century Riddles Quarry (518). The example (Fig. 9.34) was formed from a substantial block of stone up to about 0.9 m square and 0.4 m or more deep. In this a slot of the width of the post was cut, leaving the rear of the stone intact. This allowed

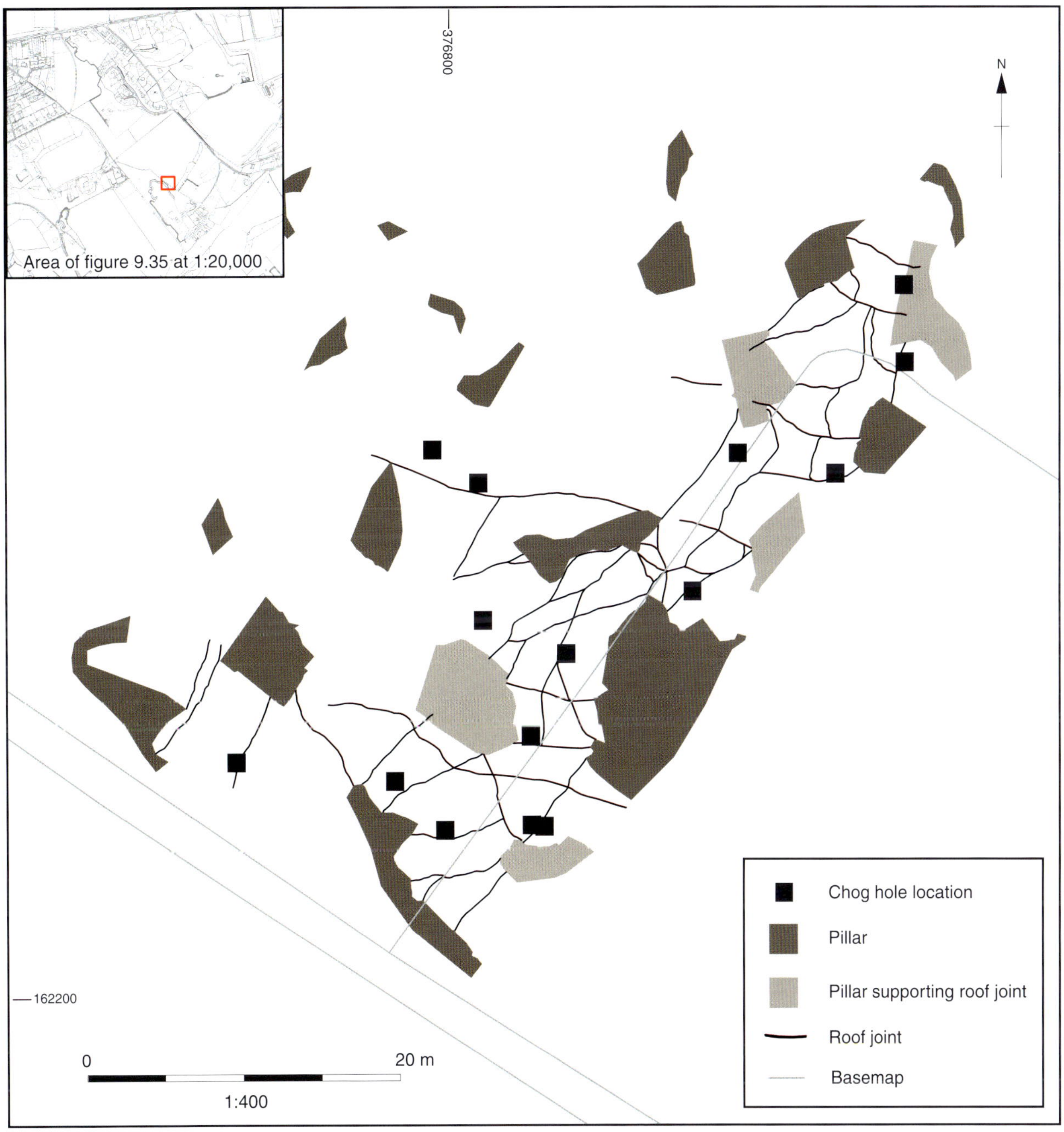

Fig. 9.35 Chog holes and their relationship to roof joints at the Mount Pleasant underground quarry

a wood bearing holding the iron pivot at the base of the post to be slid into position, while grooves in the sides of the stone slot allowed an iron plate or keeper to lock it into position. The post could be sunk into the floor or, as at Shaft Mine (2372), was supported on a coursed pile of rubble stone, presumably to accommodate a short post.

Mast cranes had cables or iron rod stays to suitable anchor points or around pillars, to hold the mast. There was a jib, rather like that sometimes used on ships, fixed near the bottom and held in position by an iron brace from near the top of the post. They were used before 1880 in Quarry 2215, where a Lewis anchor point was found *in situ*. It seems likely these cranes were similar to the standard quarry cranes used at surface, with a central mast standing on a pivot or basestone, anchored by four or five iron stays to anchor points (Pollard, 1994, 47). The mast had a jib on which was mounted a two-man winch, normally using chain cable at that time. Such cranes could both drag and lift blocks of stone and might even employ a 'muck box' to move spoil. No evidence of this is known from Combe Down, though Pollard (pers. comm.) reports the use in other Bath Stone quarries.

On the post cranes, the post was secured by being let into the roof using a chog and chog hole. The chog was a square block of elm, which would be particularly good at resisting damp conditions. It was bound with two or more iron strips fastened to the external face, and a central hole to accept the metal pivot pin projecting from the top of the crane post. The chog hole was a corresponding square-sided hole cut into the ceiling or roof to accept the chog and was placed immediately over the centre of the basestone. The post was hauled up into position using a Lewis let in adjacent to the chog hole. Occasionally two slots were found, which may indicate the failure of the first hole. Where no Lewis was found in the roof besides a chog, there was generally a corresponding Lewis in an adjacent pillar face. The crane had a post attached or pivoted at both a basestone on the floor (crane stone) and at the roof, where it was fixed into a square recessed chog hole. The jib was fixed near the base of the post with a barrel winch or windlass mounted on the jib with an iron tie-bar located to close the hypotenuse. A number of post crane sites were found in James Riddle's Allotment Quarry of *c* 1910 and also at the off-site underground quarries at Foxhill, the Tankfield Quarry/Coxes Vertical Shaft Mine, (Chapter 12, Case Studies 13 and 14) and the Shaft Road and Mount Pleasant Mines (Fig. 9.35). All these had associated cable-haulage grooves from wire rope or chain and marks from where a block of stone may have rubbed against a pillar face. Those recorded were all either very late 19th or early 20th century in date.

Fig. 9.36 Preserved underground crane at Combe Manor Hotel

The early types of mast crane were located along cartways and, in at least one example, appear to have been positioned a little way back from the cartway in 'crane holes' between pillars. Ideally a minimum space of some 4-5 m width was necessary for jib movement, but the example cites probably had less, which was perhaps one reason why the use was later abandoned. Later mast crane sites were within open-room working systems where there was substantially more space available. The general form of these crane types is known from other locations where some are still *in situ*. Examples of post cranes are on display at the Bath at Work Museum and outside the Combe Manor Hotel, neither apparently from Combe Down, but of a similar type (Fig. 9.36).

Archaeological evidence of the cranes

Mast cranes

The locations of mast cranes were determined from Lewis hole group distributions (Fig. 9.37). Several locations also contained wedge holes for anchorages, but these were less common and represent only a small fraction of the anchorage types. One of the earliest known locations of the mast-type crane was identified alongside cartways in Central Firs, probably associated with the eastern development of Allen Estate Phase III quarrying within Quarries 2201 and 2202. The latter was developed along a

Fig. 9.37 Probable mast crane position inside the Firs Shaft indicated by wooden slats projecting from an arrangement of Lewis slots

branch cartway out of the original eastern Allen cartway, the furthest extent of which has been partially blocked by a substantial fall. Both quarries are associated with Long Room developments and remains were found on both sides of the fall.

Evidence for at least two crane locations was identified at the eastern end of the branch cartway (Fig. 9.38). The northernmost of the cranes had several wedge hole anchorage positions and was associated with incised chain groove impressions on the adjacent pillar faces, suggesting that mast-cranes were used. A later stone pack was constructed, probably in the mid 19th century, to support a collapsed roof at the eastern crane location, isolating the former crane position from its contemporary cartway length. Similar wedge and Lewis holes were found the other side of the fall from here, and a chain under compression had left moulds of its links on a rock face (Fig. 7.14), illustrating the stresses imposed on it. The southernmost crane location was associated with the same length of cartway and must also have facilitated the lifting of blocks on to carts for transport either to the east (where there was a wide shaft) or to the surface quarry at Sheeps House. Several anchorage wedge holes, similar to those noted in the crane location above were noted, along with a 25 mm wide chain groove. There is evidence for several Lewis anchorage slots in the vicinity also, which probably relate to the crane locations. These, therefore, represent some of the earliest recorded Lewis holes in the Combe Down complex, although the Lewis itself is known from medieval if not Roman times.

Later mast crane sites are both believed to be attributable to the mid 19th century are also located in the Firs quarry; found within Quarries 2213 and the south side of 2202. In both these areas there is an absence of the crane base stones in the floor, possibly obscured by later waste stone fines on the floor and from barrow-way activity, and the primary evidence for the locations are the surviving anchorage slots in the pillar faces. Quarry 2213 contained about a dozen individual Lewis anchorage slots that are associated with mast cranes, in the sawn and natural pillar faces throughout the extracted area. In contrast to Quarry 2202, not all the pillar faces were accessible to survey and thus, probably, not all the anchorage slot locations were seen and these and other related haulage marks may not also have been measured. The marks are likely to be attributable to crane anchorage because of their close proximity to the quarry roof with an average distance of 1.0 m below the roof. Several wire rope haulage grooves, measuring 25 mm in diameter, were also noted on the pillar faces, with many angled grooves on the faces aligned between the quarry floor, from where a block had been hauled. They were projected towards the roof, which may illustrate the locations of the end of the crane jib, or perhaps a single pulley block used to divert the cable.

Quarry 2202 contained over 75 individual crane Lewis anchorage slots in the sawn and natural pillar faces throughout the area, though some caution is needed since for as well as anchorage points, Lewis's anchors were also used to change cable directions and also to fix winches in position. Most of the quarry was easily accessible to the archaeological survey and most of the pillar faces were investigated for anchorage slots and other haulage features. The majority of slots were located at varying heights between 0.28 m and 1.40 m below the roof, and were often found either in pairs in close proximity to each other on adjacent pillar faces. The pairing and close proximity would suggest that more than one slot was used to anchor the same crane stay. The cranes in this area were likely to have been moved and re-located as the

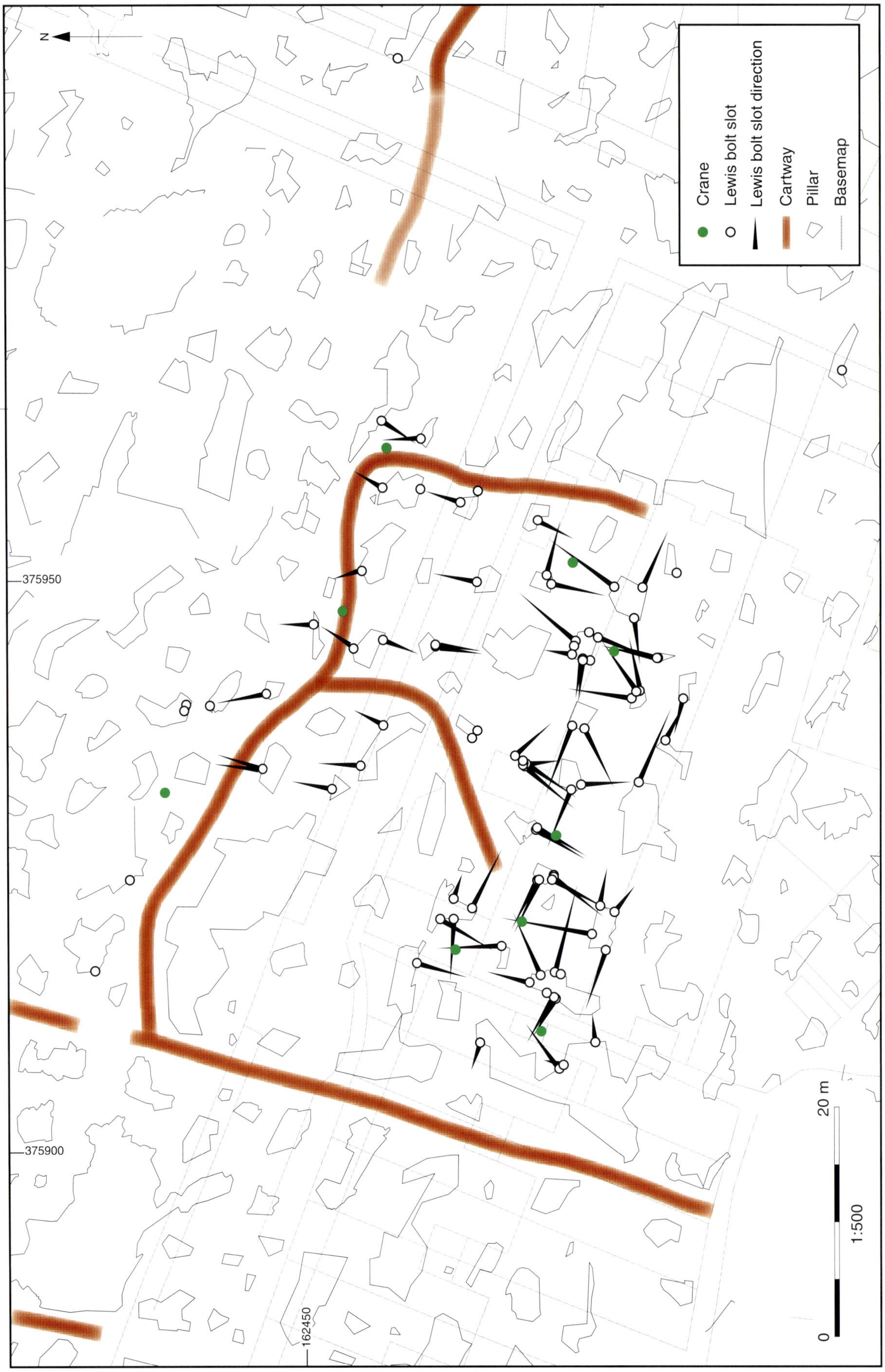

Fig. 9.38 Crane positions in Central Firs (Quarry 2202) indicated by positions of Lewis anchorages

quarrying operation was extended westwards and southwards and as the former locations were superseded. It is not known how many of the cranes were being operated in the area during the same period or how often they were moved during the quarrying operation.

Placing of the anchors was probably done in the early part of the development of working faces, when the benches were high enough. This suggests a reasonable degree of prior organization was necessary, and of course, geological conditions were also a constraint. A projected line from the mouth of each of the Lewis holes in some locations was also plotted in order to ascertain potential crane positions, though the exercise met with only limited success. This may have been due to contemporary problems and caution in their placing so as to avoid a direct pull on the anchor.

Quarry 520 in West Byfield had several crane locations along the length of a single cartway ascertainable from Lewis holes. They may have been used both for dragging and lifting extracted blocks of stone and possibly for haulage of carts along the length of the cartway.

Post cranes

Quarry 518 in Allotments Quarry had Lewis slots and crane anchorage associated with use of post-type cranes. Six crane locations were identified from chog holes and another two suspected but not accessible for survey. All were associated with two lengths of railway used to haul extracted stone to a vertical haulage shaft. Other features relating to the operation included rope marks or grooves on pillar faces, illustrating their use for dragging as well as lifting blocks

Crane locations had the top of the post anchored in a square-cut chog-hole with a Lewis slot beside it, and both cut into the roof. Beneath the post location, a single large stone crane block or basestone (SF 670), was found half buried and fixed in the floor of the quarry (518). This had an entry slot and slots for a gate. It was the only recorded crane basestone within the Combe Down Mines (though another was distantly seen in the Shaft Mine) and was recovered and videoed as part of the underground survey (Chapter 12, Case Study 12).

Chog hole dimensions varied throughout the quarries, reflecting the size of the chog and were (in contemporary measurements) half an inch to an inch (13-25 mm) larger than the chog dimensions. The hole depth varied from a minimum of 7 inches (170 mm) up to an inch or two (25-50 mm) deeper. The external dimensions of the chog hole had a greater variation and the recorded sizes ranged from 9 inches square (225 mm square) to $10^1/_2$ inches (265 mm), and the most common dimension was $12^1/_2$ inches (313 mm) square. All examples were recorded from within quarries started or operated within Phase VII (1867–1938). They were seen within the James Riddles Quarry (2202), the Foxhill Quarries (2382 and 2383), at Shaft Road Quarry (2372) and in the Mount Pleasant Quarry (2373). At the latter up to 15 chog holes were recorded in the roof, spaced at shorter intervals than noted elsewhere.

Chain and rope grooves

Grooves were created by iron chains or hemp (later, iron or steel ropes) either moving against or cutting into a face or corner of a pillar. They indicate use of pulley blocks, winches or cranes. The use of chain 'cable' in the quarries long pre-dated the use of wire rope for haulage, and chains continue to be used for lifting operations today. Chains of probable late 19th- or early 20th-century date were recovered from the Foxhill quarry complex and those quarries located off Shaft Road on the western side of the Combe Down Quarry (Quarries 2372 and 2377) and a short length of another was observed high on a pillar in Quarry 2202) held in place by spigots driven into the rock. Its purpose was unknown.

Chains left different impressions depending on whether they were fixed or moving on the pillar face. Actual impressions and outline of the shape of the chain itself may be pressed into the face or, where the chains moved across the pillar face, the impression would be an intermittent line of smoothed grooves, where every other link within the chain length had made contact with the face. The general width of the chains at the quarries was about 1 inch (25 mm) in diameter. Examples of chain impressions were particularly noted at two locations within Quarry (2202) where Open Room working had allowed extensive use of haulage equipment.

Other types of haulage marks were recorded underground. A common feature noted in the Foxhill Quarry complex (Chapter 12 Case Study 13) were grooves and marks left where a stone block rubbed against a pillar face. The marks were more intermittent than the chain and the wire rope marks and more diffuse in appearance. A good example of this was seen scored into a face with graffiti in Foxhill Quarry (2381) close to the roof level. This type of feature may be associated with the use of the crab winch, which dragged the blocks horizontally and about 500 mm above the quarry floor (Pollard 1994, 47). Alternatively, the Foxhill examples attributed to use of the post crane may, be related to the use of crab winch. Similarly, two spigots at low level, inserted into circular drilled holes in Quarry (2202) may have been for a crab winch, though securing a diverting pulley is also a possibility.

Vertical shafts

Vertical shafts – both narrow, and intended just for ventilation, light or access, and wide, to lift extracted stone directly to the surface from underground working faces – were used in the 18th century onwards at Combe Down in and after the latter part of the Allen period. They may have developed from earlier 'light-holes' or ventilation shafts and were used increasingly during the 19th century. They were circular or semi-circular in form, either sunk above or to the side of cartway routes for efficient transport of stone from cartways to the surface or. Those in east and Far East Firs were sunk independently of the cartways. Vertical winding shafts appear to have been employed throughout the 19th century in almost all areas, and it is possible even old cartways linking to them effectively replaced by them. East of the Long Drung from the 1830s, most smaller quarrying enterprises were accessed from the surface, perhaps because entry from the edge of the hillside was impossible due to restrictions due to landownership. Over 50 known shafts were constructed at Combe Down and were used until about 1923.

All open shafts were useful for ventilation and the first, fairly small diameter shafts – about 1-2 m diameter – were probably sunk for that purpose, for example to the cartways developed from Sheeps House Quarry by Allen, perhaps at points marking the extent of the initial development phase. The first

Fig. 9.39 Shaft (12062) capped with stone blocks

Fig. 9.40 Underside of barrel-arched shaft-cap in East Firs

shaft was located about 30 m from the southern end on the eastern cartway, at a probable spur. There was a similar shaft in the parallel, western cartway and another on the spur.

Some smaller shafts, notably in the eastern areas of the Firs quarries, could certainly have been used as man-access shafts, though no footholds or remains of ladders were noted. A few seem to have been used for haulage, again particularly in the eastern areas, which was evident from rope or chain grooves in the walls. This would have been mainly suitable for very small scale operations hauling small blocks or worked stone. These and all narrow shafts recorded, other than wells, were unlined and tended to belly-out where they passed through particularly weak horizons within the maximum depth of some 7 m of Twinhoe Beds. Apart from examples fill with rubbish or soil, all were capped using stone slabs, for instance shaft 12062 (Fig. 9.39), some supported with iron bars or timber (see Figure 7.9). Other shafts that remained inaccessible to the survey, were infilled with surface-derived rubbish, or were surrounded by collapsed roof materials.

Larger shafts ranged from around 2.5 m diameter to just over 4 m, perhaps widening a little as they passed through fragmented beds. With one major exception – the arched shaft in the Burgess Three Acre workings (Case Study 9, Chapter 12) – all examined shafts were unlined, though several had either solid faces or pillars left, or packs below for support.

Wide shafts were often referred to as 'light shafts', a term possibly first coined by Pierce Egan (1819) since from below they indeed were conspicuous for this. However, although effective at the shaft bottom the light falls away rapidly within a few metres and the contrast makes it difficult to see inside the workings. All of the earliest wide Allen shafts, typically placed some 70 m from his level entries, had rope marks below the shaft, indicating stone or carts being hauled to the shaft bottom, and from hence to surface. There is no evidence of the lifting mechanisms, but some adaptation of either the Padmore cranes (Chapter 4) or the usual mining horse gin seems likely. The shallow depth and weak rock made it simple to sink wide shafts, which allowed larger blocks or lengths of cut stone to be brought up horizontally, simplifying both lifting off carts below and on to carts at surface, sometimes of considerable loads.

Firs Shaft had Lewis slots about half way down, radiating to a common centre, suggesting some form of post had been supported. There were also cable grooves on a pillar. No clear use was determined, but some form of down-the-shaft haulage system is possible, to drag carts or stone to the bottom. The Firs Shaft, another shaft located beneath the Chestnut Tree, and a further shaft with a well were part a series of shafts located on a single cartway attributable to Quarry 2211. They followed a line across the north-east of Firs Field (the Arched Shaft and Chestnut Tree Shafts are others). Together they were associated below with a cartway, so were clearly part of a combined system of haulage and winding. A similar line of shafts and an underground cartway was seen in Central Byfield ranging north under the Westerleigh Road.

The only quarries apparently totally dependent on vertical shafts were those at the Foxhill Quarries, of the late 19th and early 20th centuries, which apparently used a steam engine for winding (Chapter 12, Case Study 13). In the early 19th century, leases began to specify the shafts should be fenced, apparently done with a circular wall some 3 feet (0.9 m) high. A recent small excavation of the Chestnut Tree shaft on Firs Field exposed a wall built about 3 feet back from the shaft edge. However, since that time most of the wide shafts had been filled, some carefully with quarry waste, others with domestic rubbish. Several had rubble or block stone caps; two in probable Allen workings in East Firs had splendid low arched domes, perhaps rising just over a foot in the centre, and others had barrel-arched variations (Fig. 9.40). Firs Shaft had concrete slabs, and another wide shaft under the garage on The Avenue had iron girders and concrete.

Chapter 10

Techniques of underground quarrying: 3. Support of the roof and spoil management

These two aspects are considered together since spoil management was an integral part of the roof support, either directly providing support by being banked, packed or chocked to the roof or, indirectly, by providing lateral support to pillars which supported the roof. It is unlikely with the relatively high workings at Combe Down and often slender pillars, that without substantial lateral support, they would have been as stable as they were.

Support of the roof

The quarry roof

The roof, often locally termed the ceiling, was ostensibly the simplest feature of the quarries. It was usually either one of the picking beds or, except where it appears to have been absent in the north-eastern areas, the bastard or capping stone. The latter was a hard bed of limestone formed along bedding planes that provided a clean parting. It provided a fairly strong roof under which the quarrymen could work in relative safety where gaps between pillars were not too extended. The first bed of freestone beneath the roof in most of the quarry was usually thin and unproductive for quarrying, and in parts of the quarries, for example above the Brow cartway in East Firs and sometimes where corbelled and direct pillars were used, it was left in as the roof.

Observation of the roof strata can provide clues to quarrying methods. In both Firs and Byfield a change in roof bed (leaving part of the former picking beds *in situ*) resulted from the introduction of jad slots and corbelled and direct pillars, while the bastard bed of hard but well-jointed stone in the roof of the Byfield and West Firs area seems to have been replaced by thin-bedded limestones in the East Firs, in which frequent collapses occurred. Variations in roof height were used during archaeological recording throughout the quarry complex to determine the limits of individual quarrying areas and changes in quarrying practice.

Natural features of the roof, such as the joint pattern, could create problems for quarrying and thus often dictated the working process. The joint pattern at Combe Down had two principal systems, the major set orientated roughly at 70° across the Down, and the other major set at 140°. Further joints appear to have run parallel to the escarpments, possibly due to cambering of the strata. These directions dominated the pillar orientations, sometimes aligned, more often crossing the alignment. The fractures often formed a close interlocking pattern on the roof underside, but observation of pillars suggests this pattern did not always extend downwards through the more massive freestone beds, nor, where visible at falls, upwards into the bastard bed. The thin beds above the worked stone seem to have been particularly fractured and thus the situation as seen during the Stabilisation Scheme may have been due partly to post-working stresses.

The combination of joints and weak bed partings led to delamination of beds and, if the roof spanned a large number of joints, this could cause roof-hanging and either slow or catastrophic collapses. This must always have been a major problem, and there were many instances of roof collapses which had occurred during and after quarrying, leaving roof voids and the gradual onset of the void working upwards through each of the sedimentary beds, or through to the surface forming a crown hole. Evidence of contemporary collapse, over, for example, a cartway, took the form of removal of a collapsed material or diversion around a fall. It was obviously the function of pillars to support the roof, but often substantial effort was expended to reduce the incidence of falls of roof, including the use of rubble stone packs, infilling with spoil, and timber sprags and chocks, and in the post-quarrying period features such as road, house or drainage pipe support packs or walls were installed.

General aspects of pillar design and spacing

There must always have been careful consideration as to what to leave behind as pillars, which typically took up 10-15% of the total stone available and whose function was to secure the roof. Except in the true arches of closely spaced apophygate arches, seen at some entries and occasionally in workings, the pillar spacing would impose tensions in the 'beam' of the ceiling beds exposed between them. Because of the small spacing of the jointing normally found, this would tend to delamination and downward arching and collapse if excessive. What was considered excessive was determined by experience. The spacing would also determine the load upon pillars which were prone to failure, because of close jointing and up to three sets of joint directions as well as faulty execution of design. Fortunately the process of failure of roofs and pillars was often slow, and many weaknesses were of marginal importance during actual working, though becoming obvious a century or two later.

The early apophygate pillars (see below), which seem to have developed from arches by stretching the gap between the arcs, aimed to reduce the effective beam span by spreading the pillar support at the top. This unfortunately placed a heavy load on the thin extremities of the apophygate pillar, which eventually led to spalling, though it may have been obvious during the time of working. Introduction of the various forms of the jad slot to replace picking bed removal left the topmost picking bed or beds in place, which was probably more stable than the overlying bastard bed. This was used in conjunction with the corbelled pillar, which clearly was intended again to reduce the effective span between pillars, but was equally prone to failure during working or under post-working stresses. Towards the end of the 18th century, in Allen Estate Phase III times, this was addressed by the use of direct pillars with vertical sides up to roof height which seem to have been proved the best type. Pillars which were in a state of structural failure with faces spalled, and pillar fractures developing were described during the Stabilisation Scheme as degraded pillars, from which failure (collapse) could be prevented only if the roof load was carried on adjacent pillars.

With all the pillar forms, an equal spacing along lines of equally sized pillars enhanced stability. Because joint problems sometimes prevented this, over-sizing was necessary for permanent stability. Ideally the pillars would be regular, probably rectangular, with the long direction placed at right angles to the predominant joints to reduce the buckling effect. In practice the predominant joints were used to assist extraction, so the long sides of pillars frequently were bounded by the predominant joints. This could be ameliorated by staggering successive pillars along joints, as was the case in some areas of pillars in East Byfield, whose plan (see Figure 1.9) shows a slightly curving delineation as a result. This staggering would have prevented collapses along lines of particular weakness.

This may also be the reason for limited length of views in many areas, with pillars deliberately left, possibly as the end of working faces, more or less across the room being developed. What we have termed long and boundary pillars – long lengths of pillar which often seem to denote different workings or phases of workings, as that under Combe Road – may have had a similar function. Conversely, in James Riddle's workings, Quarry 518, developed soon after 1900 in East Byfield, the pillars were carefully aligned along a major open joint, probably to facilitate the use of saws. The lack of the collapse this practice might have created there may partly have been due to the regularity of shape and spacing which sawn pillars promoted.

Pillars were the most variable conspicuous features of the quarries. Archaeologically they were a major source of information about quarrying methods, each pillar in effect representing a small portion of the actual working face. The setting and size of pillars appears to have been entirely empirical, depending on the experience of workman and supervisor and it is clear that this was often barely adequate and certainly not consistently applied. Changes in pillar forms and in their setting, where these could be isolated from natural features, indicated changes in the quarrying method and could, with caution, allow definition of separate working areas and phases of activity.

There was clearly a limited variety of pillar forms aimed at, and within those varieties some were more or less regular, others highly irregular. Pillar forms and the degree of regularity resulted, to a considerable degree, from the local jointing characteristics due to the two principle systems of joints, the third subsidiary system, and the angles at which they crossed, and because they were utilised by the quarrymen both to break into the beds and as natural faces to work to. The extreme cases were pillars which may also have been affected by minor movement, possibly from faults. These included the inclined long curved surfaces on sailed pillars, a characteristic form from faulting (Fig. 2.7), and others where successive beds have 'stepped aside', possibly the result of post-working movement. Most pillar surfaces or faces were either natural or were slightly modified joint faces which were relatively uninformative, though highly variable.

Although arched forms of support in pillars are found in underground quarries elsewhere, for instance at Browns Folly at nearby Monkton Farleigh in Wiltshire and at Beer in Dorset, the particular extended apophygate form at Combe Down led to a highly idiosyncratic landscape apparently not described elsewhere. Other types can be seen in many locations, in both underground mines and quarries.

Pillar classification during the archaeological survey

A wide range of pillar forms were initially classified in order to describe what was being observed. This was subsequently simplified to a few basic forms, with a range of what might be termed 'irregularities', largely resulting from joints and fractures or, occasionally, mistakes. The principal pillar types comprise; arched and apophygate pillars; corbelled pillars; direct pillars and long, rib or boundary pillars. Variations noted within the quarry complexes included entabled, sailed, tapered and irregular pillars (Fig 10.1).

Apophygate pillars

Aside from the unusual and precocious use of corbelled pillars in the *c* 1725 workings of the E4 Stub, Quarry 2347 (see Chapter 12, Case Study 2)

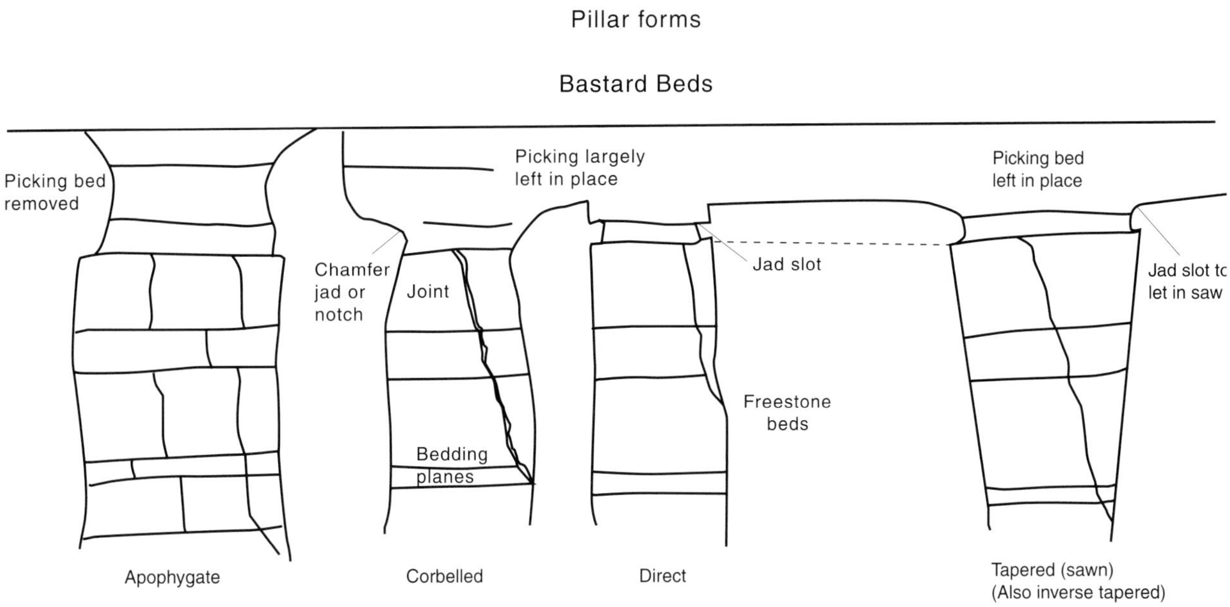

Fig. 10.1 Various pillar types

under the south end of The Avenue, the earliest pillar form generally used at Combe Down appears to have been the apophygate. This form seems to have resulted from the extension outwards of an arched form. Several areas of such pillars are found near to the southern outcrop which, because some cases appear to have been abandoned soon after Allen's takeover, pre-date his involvement, so it seems he may have continued rather than initiated underground quarrying at this location. Their use seems to correspond to the pre-Allen or Allen Phases (I and II), up to around 1750 (Figs. 10.2 and 10.3). A stone block abandoned close to one such pillar had the date 1734 written on the face and the date 1730 occurs on another in East Firs.

Though apophygate pillars were most common at the southern margin, they also occurred in underground quarries on the west and north side of Sheeps House Quarry. Likely early examples generally seem to be those in small working areas, where work was abandoned after Allen's takeover. Allen may well have extended other existing areas. Other examples were found in the Byfield Quarries 505, 516, 519, 910, 911, 914, 2349, and 2398 and Firs Quarries 2200, 2204, 2331, 2339, 2341, 2345, 2346, 2350, 2352, 2360, 2369, 2371, 2384, 2385, 2386, and 2399.

It is a highly distinctive pillar form with a slightly hourglass shape with a top resembling the *apophyge* or 'curving-out' seen on classical columns (Fig. 10.4). The shape was intended to increase the bearing area under the roof and was formed by hacking out the three or four picking beds from around the pillar just below the roof level. Between the top outer edge of the apophygate pillar the roof was flat, following a bedding plane. The beds below the picking bed appear to have had joint faces predominating, with a characteristic narrow lip or collar at the intersection below. The pillars could be apophygate on one or many sides and a few were roughly corbelled (see below) on other edges of the pillar as well as having apophygate features. In the more northern areas of their occurrence in Central Byfield, the curvature of apophygate pillars seems to have been reduced, so that some pillars appeared almost direct in style.

Apophygate pillars were sometimes worked in such a way as to form crude arches with one or more adjacent apophygate pillars. The true arched form was found close to entries, for instance the eastern entry at Jones' Quarry at Central Byfield. It would have been especially useful near the margin of the outcrop where weathering and opening of joints was most critical. In the outer areas of Byfield the room openings between pillars were only 2-4 m, and the apophygate pillars formed an actual, or close to actual, arched roof for the rooms. Further within the quarry the distance between pillars increased to at times over 10 m, so the continuous arched form was lost. The wide spacings and over-narrowing of pillars also seen in some parts were probably unwise, sometimes resulting in considerable instability. It was apparent in Byfield that collapses had taken place before underground haulage ways went out of use. In Firs Quarry the apophygate pillars were again close enough together sometimes to form arched roofs, but, especially in East Firs the gaps between pillars were widened and this area had the highest density of 'high hazard areas' noted in Hawkins (1994) report (Fig. 10.5). It is possible that collapses seen notably in apophygated areas of East Firs caused Allen sufficient concern over deaths around 1755 to begin surface quarrying, as John Wood claimed in 1765. It may also have stimulated the change to corbelled pillars likely at about the same time (see below and Chapter 3).

Fig. 10.2 Apophygate pillars within Central Byfield (Quarry 505). The pillar on the left is severely degraded

Fig. 10.3 Apophygate pillar in East Firs close to The Brow cartway in Quarry 2339. This is an example outside Allen's quarrying area

The distinctive hourglass pillar shape was sometimes emphasised by the deterioration of the pillar face, either because where they were located close to the quarry mouths or to vertical shafts, where flows of cold air produced frost damage, or from degradation due to roof load. Spalled, small tile-like limestone fragments fell from the pillars and were deposited around the base. Evidence for spalling of this sort and the damage to possible toolmarks by degradation of pillar faces was recorded within Allen period Byfield Quarry 505. Of the 70 or so apophygate pillars located in the central part of the Allen area there, only 10 had surviving evidence of tooling on the faces and well-preserved toolmarks were found on only one apophygate pillar. The toolmarks were represented by an area of curving pick marks on the underside face of the apophygate curve. The toolmarks emphasised that this form or shape of pillar face was chosen by design, rather than as a consequence

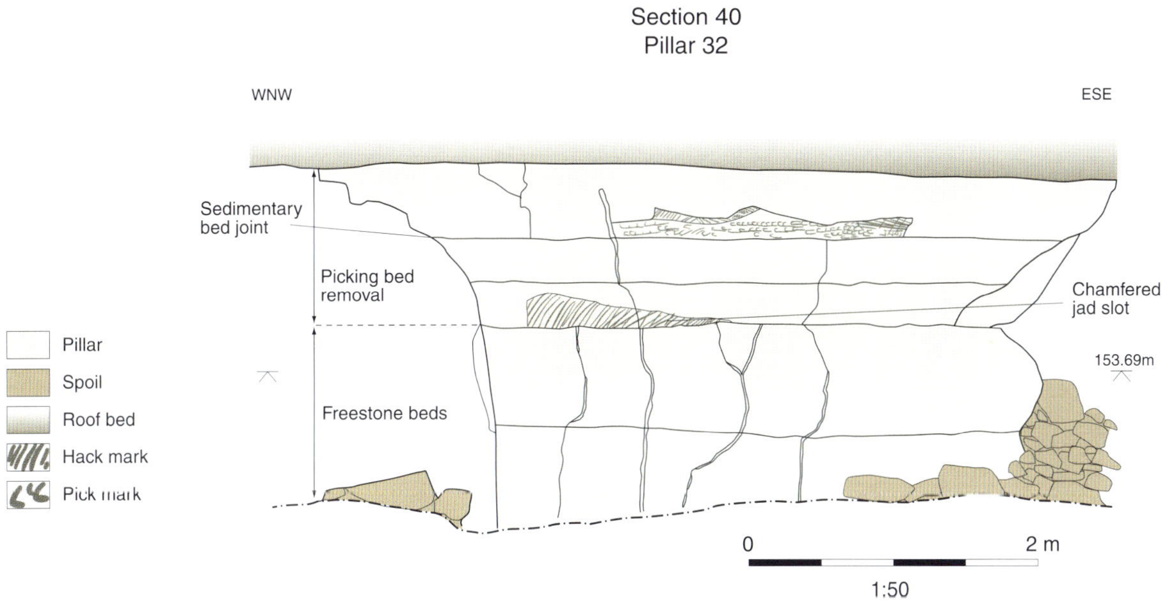

Fig. 10.4 Section of the top of an apophygate pillar including a small chamfered jad

of post-quarrying frost damage or spalling or delamination through pillar load. Toolmarks would also have been lost to the reduction of pillars during subsequent later quarrying and/or Bath Stone.

In the succeeding phases of working, joint faces were left on pillars as the norm, but this had declined before or by the 19th century, first with two faces normally cut by vertical jad cutting or otherwise breaking or sawing, increasing the regularity of the pillars, and, finally, by sawing on all, or nearly all faces, providing, in the final stages of working, almost symmetrical pillars. A regular vertical form of pillar will usually provide the most efficient form of support. Because of the joints and fractures and the methods used to produce them, the desirable regularity in pillar form was rarely achieved, and a very wide range of distortions existed which either required larger or more pillars, or led to weaknesses in the degree of support.

Corbelled pillars

Corbelled pillars usually occupied very limited areas as an outer ring to apophygate pillars, though their earliest introduction was in the E4 stub Quarry 2347, where the corbels projecting at the top of the pillar were more substantial than was usual later (Fig. 10.6). In Byfield Quarry 505 a transitional form of pillar with both apophygate and corbelled style developed (Fig. 10.7). The wider use of corbelled pillars may have resulted from the general intro-

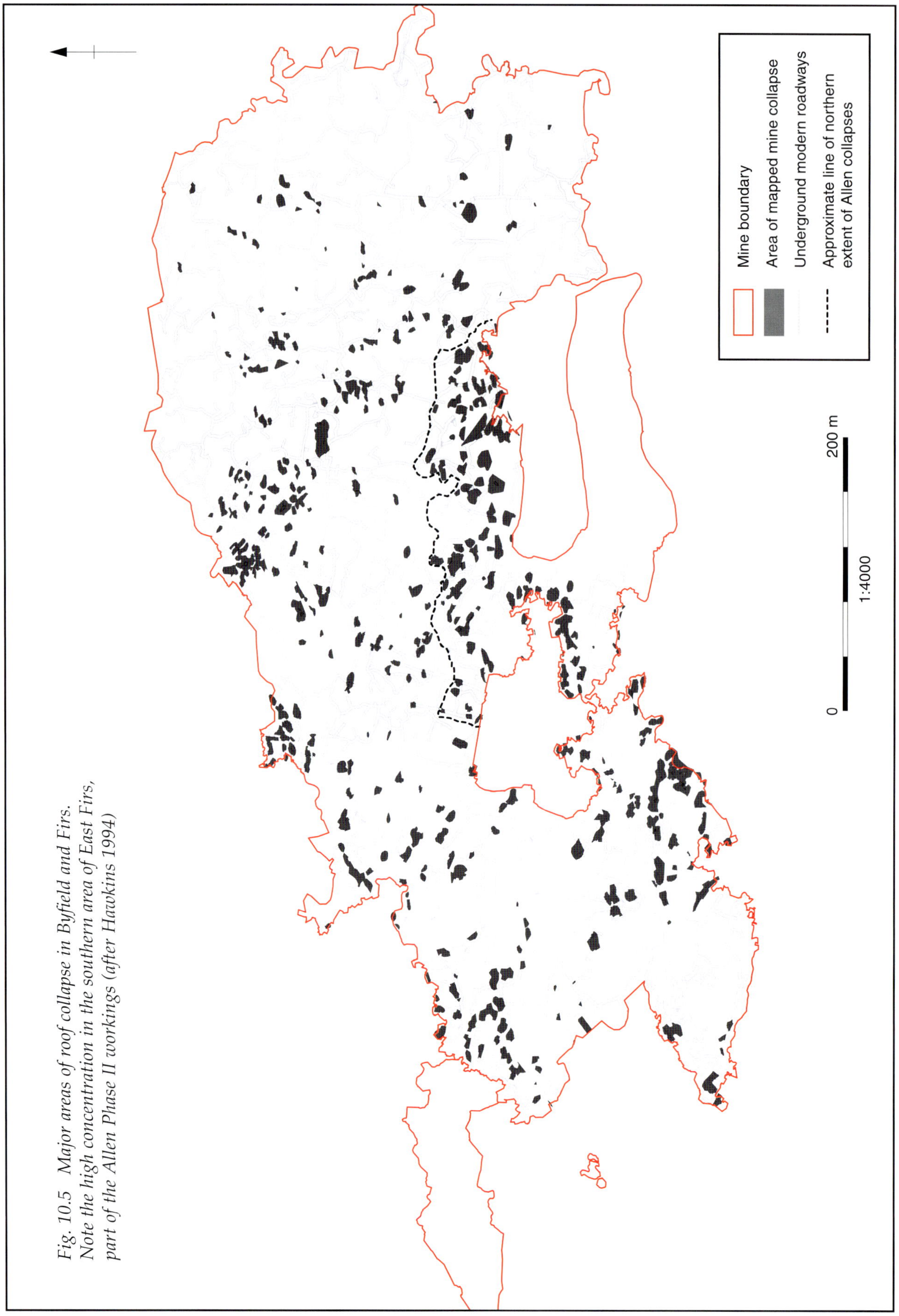

Fig. 10.5 Major areas of roof collapse in Byfield and Firs. Note the high concentration in the southern area of East Firs, part of the Allen Phase II workings (after Hawkins 1994)

Fig. 10.6 Corbelled pillar in Byfield

duction of parallel or chamfered jad slots and notches as part of picking bed removal and perhaps also from a desire to continue the supposed advantages of apophygate pillars. In the parallel jad slot the near horizontal top and lower surfaces went back usually to a joint with the top surface a single bed or two beds below the main roof level. In chamfered forms the horizontal lower part of the slot was cut into the middle or lower of the three beds typical of the picking beds, with the top-surface angling down to either a joint face or the lower bed surface, allowing easier access for a short-handled pick. The term 'notch' was used for when both surfaces of the cut diverged significantly from the horizontal, and where the jad slot was reduced in height to generally a single, or perhaps two rows of pick marks. By driving a wedge into the bed parting below, a section of bed was lifted and/or broken at the back of the cut and could be removed. The section of bed above was left protruding as far out as the next joint as an 'ear' or corbel. Its use seems to have led to the gradual abandonment of removing the full-height picking bed for access and the junction of apophygate with corbelled areas was sometimes seen as an under-slung projecting ledge across the intervening room. Such pillars were described as 'entabled', and good examples of this type were noted on the periphery

Fig. 10.7 Transitional form of apophygate and corbelled pillar in Central Byfield (Quarry 505)

Fig. 10.8 Corbelled pillar just west of Firs Shaft, with inadequate support in the core width below

between the Quarrying Areas 505 and 517 in the Byfield Quarry.

The use of the chamfered or parallel jad was probably speedier than than fully forming the pillar in the picking bed. Structurally the corbelled pillar acted in much the same way as the apophygate pillar. A wider area around the top of the pillar was protected and in similar fashion this may have led to a tendency to provide too small a pillar below (Fig. 10.8). Corbels frequently developed a split in line with the main pillar-side, along the natural joint face, rendering them structurally useless and often dangerous. It is possible the vertical split was not developed fully during mining and was a post-quarrying fracture as the superimposed loads adjusted. In some cases corbelling may have resulted from Bath Stone rather than original design and many other examples are possible but hidden by dumped spoil. The insertion of a vertical timber prop, or a small stone pack positioned under the corbel sometimes provided additional support and is a clear indication of concern during actual working.

The introduction of corbelled pillars generally was difficult to date, but considerations of the position of later leases and known periods of inaction at Byfield suggest their introduction in the middle of the Allen Phase II, probably continuing into Phase III when they were gradually superseded by direct pillar forms, with only the occasional occurrences, in isolation, of the formerly widespread corbelled pillar use.

Direct pillars

Direct pillars were the most common type found in later workings (Phases IV-VI), and proved the best type structurally (Fig. 10.9). Ideally they had vertical sides with either a joint face or a carefully vertically broken side with bed end faces in line from floor to roof. Many cases, however departed from this ideal and overly narrow or other irregular forms were common (Fig. 10.10). The introduction of the direct pillar, alongside corbelled pillars, appears to have coincided with the common use of horizontal jad slots with near-parallel faces and usually, as with corbelled pillars, the abandonment of the full-height opening-out of the picking bed. With the almost simultaneous systematic use of wedge-and-chip method for the breaking of sedimentary beds and for the removal of blocks from the pillar faces. Fairly regular shaped direct forms could usually be achieved and the later introduction of saws allowed for very regular forms, as was observed in the latest worked areas of the quarries in the later 19th and early 20th centuries.

The introduction of the direct pillars using the jad slot caused a substantial reduction in spoil from the picking beds above, and also largely coincided with the use of the Long Wall with gullet method of working. There appears to have been a transitional period in which both corbelled and direct types co-existed in Phase III, corbelled pillars disappearing almost completely by Phase IV. It is possible this was largely a local development as the structural

Fig. 10.9 Direct pillars in Central Firs (Quarry 2340)

advantage of direct pillars and their simpler method of formation became apparent.

Boundary, rib and long pillars

These terms described pillars which were larger than the normal pillars, particularly in length, but also width. Boundary pillars, as defined during the archaeological survey, were substantial lengths of either lines of long pillars or boundaries un-pierced by full room-sized openings (but sometimes by door-like and window openings) over stretches of linear un-worked ground. Typical rib pillars usually had a square or rectangular section, with the long dimension around three times the shorter, and the shorter usually 1-2 m wide. Long pillars had substantially greater ratios. They were typically found across the line of general working or at abandoned working faces, and may have been deliberately left to provide buttressing as well as roof-support to small quarrying areas.

Division of the wider area of the quarries seems to have been into separate working areas, perhaps on the basis of work teams or, later, leases or direct ownership by individual quarrymasters. Boundaries of areas can sometimes only be indicated by differentiated working methods (for instance types of pillars, or tool markings), but more commonly such divisions were also worked back to a 'long pillar' or 'boundary pillar' pierced only by small access or ventilation openings or 'windows' or 'doors', used either to dispose of spoil into adjacent areas and/or for access or ventilation. These, with a combination of other changes would determine a quarry boundary pillar. However, caution must be given to the fact that some of the perceived boundary pillars may also have been left *in situ* to support a localised geological variation or weakness such as a roof gull or to give additional support to a principle cartway route.

Boundary pillars also served as ownership markers and/or sometimes defined features that were noted on the surface, such as parish boundaries and roads and footpath orientations. An obvious boundary pillar reflecting (in this case) a surface boundary, ran under Combe Road from the south of the freestone outcrop to North Road and divided both Allen's quarry and later quarrying interests in the Byfield workings; during the Stabilisation Scheme it was called the 'Combe Road Pillar'. It was pierced by one or more 'doors' which may have been used for pedestrian access, and was robbed on both sides for much of its length but generally without cutting through. . At the southern end it was much wider and more well-defined, with the northern limits having more piercings and remaining no more than a series of slightly larger and more prominent pillars in the general pillar plan. The earliest Allen working in Byfield was apparently limited entirely to the eastern side, but probably during the later 18th century a new entry was made (from Ralph Allen Yard) and the pillar pierced to form a trackway to serve the western area, even though it was only a few feet away from the earlier entries. The

Fig. 10.10 Section of worked face 3007, an irregular form of direct pillar with narrow core, showing picked and wedge and chip broken surfaces

quarrymen were forbidden in their early 19th-century lease (Chapter 6) to cut through the Combe Road boundary under the northern end of Combe Road as defined by the new east side roadside wall. The terms in that case do not appear to have fully been adhered to. The quarryman, Tanky Elms, (Hall 1984) remembered -

> ...*that there was a quarry boundary running along* [and under] *the road at Westwells* [Odd Down], *Bath and Portland on the one side and, Sheppards on the other side of the road. Sheppards went over to the other side of the road and had to pay I believe seven hundred and fifty pounds compensation.*

In Firs the equivalent boundary was the Long Drung. The longest rib pillar there as surveyed by Hawkins (1994) seems to mark the 18th-century northward limit of working in the central area of the quarry west of the Avenue, with subsequent working on the east and west margins. The quality of stone in the pillar was lessened by minor fracturing or faulting and it may have been initially a convenient place to stop development northwards. Areas to the north of the rib pillar were subsequently worked southwards during the middle years of the 19th century by entries driven from open quarries north of the North Road. Quality, however, can hardly have been the reason for its survival to the south, where it also formed the northern boundary of 18th-century workings and the south west boundary of the early 19th-century Three Acre Quarry. It bordered a high pillar area with little lateral support from spoil so it is possible there the rib pillars were left to improve stability.

Several other boundary pillars identified during the archaeological recording were used for the hypothetical division of the underground workings into separate quarries (see inside end cover), though in some cases the function may have been to prevent a lateral movement or 'domino effect' and rather than being a legal boundary, they were simply a convenient boundary between separate areas of the same quarry workings. This seems especially true of several such 'boundaries' in the Three Acre Quarry near the Haley Arms. This implies that long and boundary pillars were probably fully understood at the time to enhance lateral stability. Long pillars often formed the sides of Long Rooms and lengths of cartways were also commonly located besides the pillars, undoubtedly for the stability offered.

Irregular pillars

The term irregular is used to describe pillars of unusual form that resulted from either particularly poor, or misconceived workmanship, or from difficult natural circumstances, notably jointing. Saws allowed pillars to be very regularly produced, but in Quarry 2215 in north Central Firs, the pillars were carefully produced with a downward, wedge-like taper so regular and frequent that they must have been deliberate, but were clearly misconceived, as they placed the load on a very small area (Fig. 10.11). Saws were also frequently used for scavenging, cutting into otherwise regular pillars to remove the best thick bed just above or below the spoil-dump level. Many pillars appear to have been severely cut away just below the spoil to a highly dangerous degree which, subsequently hidden, was hard to detect. Particularly bad examples were found in the southern part of East Firs where nearby apophygate pillars also led to instability. Joints produced some of the greatest irregularities, including pillars with major open joints running through them. A pillar in the Grand Canyon in Firs, for example, from one angle appeared twinned, but was regular from the other.

The side piercing at the edges of individual quarry rooms developed windows and openings close to the roof level. In the case of some pillars, additional openings were sawn or wedged below these creating H-shaped pillars. One example was recorded in the mid 19th-century Quarry 2211 and another, in the early 20th-century James Riddle's Quarry 518 was the clearest example seen at the Combe Down (Fig. 2.4).

In Firs Quarry (2215) strongly inclined joints intersecting at a small angle produced long, thin pillars termed by the archaeological team 'sailed'

Fig. 10.11 Tapered sawn pillars in north Central Firs (Quarry 2215)

pillars the ends coming to a sharp edge (Fig. 2.7). They could be part of local fracture systems. Such inclined pillars were liable to be very unstable, and it was perhaps for this reason that work was suspended in that area. Another example of a sailed, sawn, pillar had a deeply picked jad slot in the first bed below the roof to allow access for the frig bob stone saw. The full height of the exposed pillar face was sawn with over 12 beds worked to the contemporary floor level. Other pillars in the quarry had both natural and sawn faces and adjacent sawn pillars showed evidence of horizontal wedge holes and timber sprag supports. Not uncommonly pillars were also cut into at the top further than was probably intended, and rubble-stone packs and timber sprags were then used to support the roof above the damaged section.

Pillar spacing

The spacing between pillars varied greatly, but seems typically to have been about 4-5 m with the largest spans seen in East Firs of up to 13 m. Spacing of pillars and their distribution along joints usually showed regularity on the overall pillar distribution plan produced by Hawkins (1994). However, there was no overall regularity in the workings, due either to variable geological circumstances or to the particular working practices of the quarryman, quarrymaster or freestone mason involved. In some cases there was a surprising lack of common sense and appreciation of personal risk.

Roof and lateral support (other than pillars)

Pillars provided the principal support for the roof and long or boundary pillars often additionally provided lateral support, preventing progressive collapse due to failure of one or more pillars. Given the dubious stability of many pillars with heights of up to 8 m (unusual in mining generally), it is clear that the spoil infilling also provided lateral support, sometimes to the roof, even if only by preventing loose roof blocks descending and destroying the arch effect. Spoil infilling was frequently placed around or between pillars and kept in position by rubble-stone packs, which was clearly done to improve stability as well as the management of spoil.

Carefully built, coursed rubble-packs occurred throughout the workings, but less commonly than in some underground quarries, so it was probably regarded as a method to be used in specific circumstances rather than routinely. Some were contemporary with initial workings, but many seemed to result from re-opening of old cartways for later working. A good example was exposed at the side of the branch cartway along Allen's east cartway in Firs Quarry 2201 when it was re-opened to work Quarry 2202 in the mid-19th century. Here a major collapse was removed, and a massive 1 m-thick by 5 m-high stone pack was constructed by the cartway to lessen the roof span and hold up the remaining roof.

Other less carefully constructed packs of roughly stacked and coursed rubble seem to have been used

Fig. 10.12 Substantial rubble and blocks used to construct the wall near the Hadley Arms under the south side of North Road

Fig. 10.13 Rubble stone mortared packs under a municipal drain running across the surface overhead in West Byfield

to separate areas of quarrying, possibly for boundary purposes or to direct air-flows, and many small packs were built to provide local support where roof stones had slipped or slumped. In one instance in north Central Firs a pack had removed the arch effect and an adjacent keystone block slipped as a result, remaining hanging above the floor. One of the most major coursed-rubble and block packs or walls was built near the Hadley Arms Public House, running below the south side of the North Road. It was in an area noted by Hawkins (1994) as having a roof thickness of less than four metres and sometimes below three, and where a very substantial fall appears already to have taken place. This filled the space between pillars to a height of about 5 m and was chocked under the roof by small stones (Fig. 10.12).

In the area of substantial collapse and risk to the adjacent area, the North Road wall was temporarily discontinued, considered too dangerous to remedy, though it is possible the bad section was under a different quarrying operation. Spoil was used in conjunction with rubble-stone packs to underpin many parts of the North Road. In some cases the packs were built as square 'cribs' and the box-like structures filled with rubble. The road had been undermined largely by entries in the surface quarries to the north. Multiple packs were built perpendicular to and parallel to the road and infilled between with spoil. Some of this was done inefficiently: in one case loose bed-blocks in the roof were held by a pile of small stones resembling a particularly insubstantial way-marking cairn. Substantial packs were built between pillars at Stonehouse Lane Quarry (2219), probably to retain surface quarry spoil in position while a line of mortared-stone packs were built to support the length of a drain in the mid 1920s in West Byfield (Fig. 10.13).

Roof collapse contemporary with working

Roof collapses occurred throughout the Combe Down Quarries. They could involve the falling of an individual block (Fig. 10.14), such as that noted in Quarry 509. More serious collapse due to roof bed separation was noted (Fig. 10.15) above the cartway in Quarry 2211, where one or more of the sedimentary beds of stone just above the roof level had separated from the bed above, but remained *in situ* at roof level. Some collapses were contemporary with the quarrying periods but could also represent post-quarrying episodes. Partial collapses were sometimes held at roof level by the construction of stone packs, stone chocks or timber sprags, and could also be alleviated naturally by the proximity of roof supporting pillars or hinged on adjacent roof blocks.

Roof collapses noted included materials of any size that had fallen to the quarry floor, including localised falls of single roof blocks, other small areas collapse from a single bed, or hundreds of smaller blocks or several roof beds spread over large areas. The depth of collapsed roof bed could

Fig. 10.14 Fallen blocks in Byfield (Quarry 509)

be relatively shallow and spread over a large area or more localised but completely filling a room. Some falls were clearly catastrophic, with a large quantity collapsing at once, or progressive, the roof falling portion by portion over a long period. The former probably often involved thick bastard beds delaminated at roof level, the latter possibly developing later as the void, or 'crown hole' developed upwards into the thinner Twinhoe Beds (Fig. 10.16). Roof collapse was regularly noted at the limits of the underground and the surface quarrying entries whether level or shaft. A good example was recorded in Quarry 2371 where the collapse extended beyond the quarry roof, with roof materials slumping into the underground quarry.

The clearest evidence for contemporary roof collapse was a void in the roof and an absence of associated roof bed materials on the quarry floor, showing that the material had been removed during the period of quarrying. An example of such collapse was recorded above the cartway in Quarry 2200. A flanking stone pack on the western edge of the cartway had been constructed up into the crown hole created by the collapse, suggesting that the collapse had occurred during the mid 19th century, or was consolidated during this period (Fig. 10.17). In Quarry 2211, a stone pack was constructed on a pillar bench at roof level to support an area of roof bed separation. The block still rested upon the pack and the pack had obviously been built to support the roof but it was unclear whether the block had actually separated from the roof during quarrying or later.

Fig. 10.15 Delaminating roof bed in Central Firs (Quarry 2211)

Fig. 10.16 Collapse of Twinhoe Beds in East Firs

Fig. 10.17 Support pack in Central Firs

Fig. 10.18 Pack built on top of fallen blocks in Central Firs (Quarry 2200)

Roof collapse and/or instability surrounding a vertical wide shaft was recorded in the former Ralph Allen Quarry 505 in Central Byfield. Here, the shaft was consolidated by stone packs, and a cartway was diverted around the shaft and collapsed materials. The shaft and cartway were probably already in existence at the time of the roof fall, as the diversion cartway and continuation of apophygate pillars to the west and north of the shaft took place during this early phase of the Allen period (see Case Study 4 in Chapter 12).

In the case of some contemporary roof falls the collapse had been cleared and the material stacked to the sides of barrow-way or cartway routes. Other contemporary roof collapses were completely or partially buried below subsequent deposition of spoil from later quarrying or further falls. Quarrying tools and/or finished stone products had, in some cases, been left in stacking areas waiting for removal to the surface, but were buried by roof collapse before they could be moved.

Only two crowbars from the 19th century were recovered. One of these, discovered during excavating for the modern Stabilisation Scheme, was from below an area of roof collapse in Firs Quarry (2360), and is believed to have been lost during quarrying. The collapse had sealed the quarry floor and the bar with it.

On the floor of the cartway in Quarry 2217 there was evidence for both contemporary and post-quarrying roof collapse. This was a main transport route for the movement of stone products to the surface quarry north of North Road. Flanking stone packs show that the roof required consolidation during the period of quarrying and post-quarrying collapse was noted at the southern end of the cartway (Fig. 10.18). Some 10 m to the north between packs there was evidence that collapsed roof blocks had been broken into smaller pieces and moved on to the edges of the cartway.

In recent years former quarryman Tanky Elms worked as a safety man in Spring Quarry, which involved supporting and safely bringing down weakened roof beds in quarries that were requisitioned for use during WWII. He said that:

> *by tapping the ceiling one could tell how far one could proceed in safety, and avoid a large fall, especially if the ceiling had joints across… If any doubts were entertained about the depth of the ceiling, we'd pick a hole through the lower bed.. insert an iron and tap the bed above to test its strength and then you know whether to pull it down or keep it there and support the lot.* (Hall 1984, 105)

Timber supports

Timber was only rarely used during the early quarrying periods, though lack of survival may account for the general absence of 18th-century supports. During most of the 19th century timber was commonly used for support but it seems that this was largely on an ad hoc basis rather than as part of a systematic procedure.

Timber sprags

Timber sprags (known locally as scorters or scaulters, cleats and props) were most commonly used in 19th- and 20th-century quarrying at Combe Down. Very few timber supports were vertical or horizontal, but were generally positioned diagonally between vertical pillars and roof beds to give localised support. These sprags were placed in small picked notches, ledges or depressions on pillars (probably called 'holes' or 'scorter holes' originally, though termed 'sprag recesses' during the archaeological survey) and angled against potentially loose bed-blocks in the roof, usually at a joint (Fig. 10.19). The effect was to provide some vertical support, but, more effectively, to provide lateral pressure to prevent keystone blocks from dropping out of place. In later quarries large numbers of sprags were used where long spans of major joints were left otherwise unsupported for sawing purposes, or around pillars in Open Rooms where spans were considered large. Evidence of fallen or rotted sprags was noted as a darkish organic residue in the notch or depression that held them. Sprags were sometimes tightened by using redundant iron wedge chips and, probably, thin wood wedges.

Fig. 10.19 Use of timber sprags for roof support in West Firs (Quarry 2342)

The most frequently observed type of sprags often measured 2-3 m long with various diameters. The sprags varied in number from an isolated sprag, to those employed in pairs, and also to the use of multiple sprags where several were sprung from the same pillar. Quarries 2202, 912 and 2215 often used multiple sprags sprung from a single pillar and more elaborate support mechanisms were employed where main access cartway and railways were in operation. Quarry 912 had one example where the pillar had four timber sprag recesses cut into it to necessitate the support of localised bed separation, with two of the timbers remaining *in situ* on the pillar (Fig. 10.20).

In another example in Quarry 2202 up to nine timber sprag recesses were used for bed support and all nine had fallen to the quarry floor in the intervening 150 years without collapse taking place.

The sprags were held in vertically-backed, square cut recesses. The sprag recesses, cut into the pillar face, were large enough to accept the sprag and were generally well-executed features. The top of the sprag was also held where it engaged the roof bed with a smaller recess which was no more than a small hole and may have been more difficult to access by the quarryman though generally they

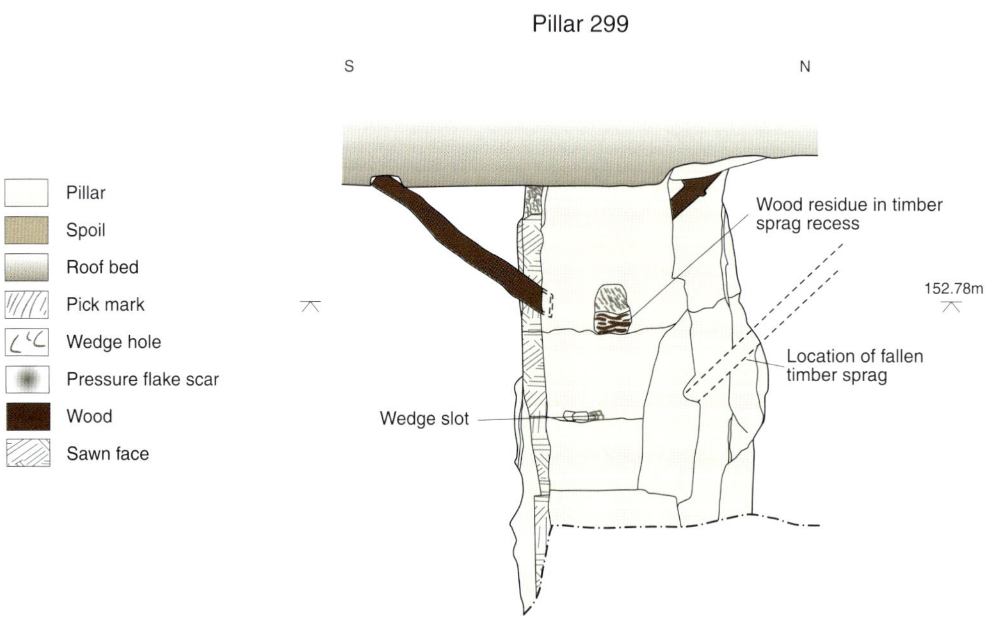

Fig. 10.20 Section of pillar 299, showing timber sprags from pillar supporting the roof

Fig. 10.21 Wooden wedges or cleats inserted to tighten roof blocks

must have been inserted when the face was there still high under the roof for access. There was also a possibility that by extending the recess that it may also have further weakened the block that it was intended to support. The recesses were executed with a 25 mm (1 inch) wide pick at the back of the slot, although a few examples had also been cut with a wider 75 mm or 100 mm bladed hack, and a pick was also largely employed to cut the less well-formed roof recesses.

In a number of the quarries, (the early 19th century Quarry 509 in Byfield being a good example) broken pieces of iron chips were used to tighten up the base of the timber sprags where they meet the sprag recesses so to prevent them from falling from the roof. Several of these were recovered from the timber sprag recesses and this seems to be a particular common quarrying practice. In another quarry in the Firs complex (2342), iron chips were recovered from the roof where several of the timber sprags had subsequently fallen, and where the chips and timber had been calcified together. This practice does not seem to have been used throughout the quarries as chips have not been located on the majority of the timber sprag recesses which were often accessible where the timber had fallen to the ground and only a residue remained. Whether small wooden wedges were alternatively used was not determined. They would generally have rotted so as not to be distinguished from the timber sprag remnants.

Wooden wedges or cleats, often made from softwood, were also used to support the roof-beds and were placed in the cracks between the separating roof blocks so that they would provide some support and help to tighten them where the roof had weakened. They may have provided an early warning of any movement in or of the roof beds, with wedges subsequently falling to the quarry floor. The occurrence was rare in the Firs and Byfield quarries – excepting James Riddle's 20th-century quarry (518) – but common at Shaft Road quarries 2373, 2377 and 2382 and at Foxhill, suggesting it was generally a later practice (Fig. 10.21). In many instances the individual blocks were generally keyed together, but where they had got slightly parted or sounded 'hollow' and had separated from the sedimentary bed above the insertion of the wedge or wedges prevented any of

Fig. 10.22 Horizontal timber used for roof support under Shaft Road.

the blocks from falling. The function was similar to that of sprags and may have been substituted once this was realised.

The placement of horizontal timbers was a slightly more common practice by the end of the 19th century and examples of their use were seen in all the outlying quarries. Timbers were used to support roof beds by wedging them between two pillars or placing them in horizontally cut recesses. Timbers were also used as supporting beams across the entire width of a quarried breach or Long Room or positioned where there was a roof gull and from where smaller rubble and materials that had migrated or that had been washed from the upper geological beds may have fallen. The timbers in these situations were often used in conjunction with other smaller cross timbers. A particularly long example was found at Shaft Road, where it spanned the whole of the room width just under the roof (Fig. 10.22).

Vertical timber props were fairly unusual, with only a few examples, less than a dozen, being noted, some in early 19th-century contexts, though again becoming more common late in the 19th century onwards. Smaller lengths of timber were occasionally noted placed within small jad slots or breaches in the working face to support a single block of stone or give additional support to a roof bed above the slot. Other examples have been used to prop individual weakened or displaced blocks of roof bed including an example sited on the top of a stone pack (Fig. 10.23). Sometimes at the top of the prop or vertical sprag a 'cap' or flat-sided block of timber was placed on top to wedge the timber to the roof or act as a platform. At Mount Pleasant Quarry (2373) in an exceptional example, the roof bed surrounding a collapsed roof had been held with eight or more large diameter props that were fixed into picked recesses in the roof and dug into the quarry floor.

Modern mining engineers have been sceptical of the use of timber supports, considering them ineffective. However, many timbers that still remained *in situ* were particularly solid. On removal for analysis (with modern supplementary support provided) they were found difficult to saw through. The original size was probably more substantial than seen in recent times because of rotting exteriors. The timbers were probably used for a combination of both support and a warning mechanism of ground movement by either squeaking or by splitting. Several sprags were seen to have split under the weight of the settled roof from above within the double height Quarry 512. The inclusion of underground quarries under the mining acts and effective inspection by Mines Inspectors, which was acttive by the late 1880s, may have led to the more systematic timbering seen in the later periods of working.

Steel and iron rail supports

Iron or steel rails were used to support the roof beds at Combe Down in the Foxhill Quarry (2383). The rails had been used in a similar way to horizontal timber sprags but gave a more reliable and permanent support to a poor roof. The three rails were placed across the width of a quarry breach in jad slots between two or more sets of pillars above narrow gauge railways. Three pieces of rail were recorded and each was securely held within two opposing horizontal cut recesses on either side of the room or heading. Two of the rails measured $3^1/_2$ by $3^1/_2$ inches in profile, with the remaining rail $4^1/_2$

Fig. 10.23 Vertical prop placed on a pack

by $4^{1}/_{2}$ inches. They were all flat bottomed and had slightly curved and convex upper surface and would have presumably have been re-used from an underground railway that served this area and the other three quarries at Foxhill.

Stone roof supports

Stone arches

Arches have traditionally been used for roof support at mine and quarry entries because of their inherent strength, quick erection and use of onsite, cheap materials, though sawing into voussoirs would add considerably to the expense where they were used. They were not a common occurrence at Combe Down, but two, at the Arched Shaft under Firs Field (Chapter 12, Case Study 9) were part of a spectacular feature, and there were others at both inclined (slope) and level entries in Byfield, and in Firs under the side of and near North Road of a corbelled type. Dome forms as well as simpler barrel vaults were also used to cap shafts, supporting a burden of weak material a few feet overhead. They were also used to support individual pillars. In a stone-using area, arch building would therefore have been a common skill among masons and other stone workers and it is perhaps somewhat surprising they were not used more often given obvious problems with support over cartways. It is possible other arches were built at entries not available for archaeological examination.

The entry incline known as the Irvings Incline in south-west Byfield, Quarry (503) was probably constructed in the early 19th century and was used for access as well as to bring stone to surface. It linked via a sloping area and ramp of mixed fines at its northern limits which appears to have served the high pillar Quarry 503. There is, however some doubt of the actual date, as in the original lease of *c* 1805 (Irving 2005, fig. 7). Access to the same quarry was also provided along an adjacent cartway and through one of three contemporary stone arched entries from Jones Quarry, via what is now Ralph Allen Yard, and it is just possible that the Irvings Incline was constructed slightly later, replacing the earlier entry, as it was not used by the adjacent (surface) quarries as originally anticipated.

The arched Irvings Incline linked the underground quarry workings and an inclined spoil ramp that was used to transport stone blocks to the surface. The incline measures 1.30 m from the roof of the arch to the current floor horizon, is about 3 m wide and 20 m long and inclines at about 1:4. It was built to support an area of poor roof in the vicinity of a large gull located above the central section of the arch and to hold loose stone-blocks and overburden at the mouth of the incline at surface level. It used coursed rubble blocks in the walls, some with sawn and scappled faces, and there was shaped and sawn block in the barrel-shaped arch, apparently placed in a lime mortar. The northernmost extent of the arch was constructed 0.30 m lower than the southern extent so as to go under a lower section of roof where it met an earlier sub-

Fig. 10.24 Stone arch at a West Byfield entry (Quarry 505) with sawn voussoirs

parallel ribbed pillar that remained from earlier 18th-century workings where the incline entered the former underground workings. Later deterioration of the quarry roof conditions, probably after the principle working of the quarry, then necessitated the building of a stone pack at the base of the incline, partially obscuring the original wide access to the incline from the spoil ramp beyond in the quarry workings.

Three other stone arches were also constructed in the southern part of the (Byfield) Jones Quarry at Ralph Allen Yard. where three converging cartways emerge. Because of the use of sawn blocks to create the interlocking voussoirs, they are thought to have been constructed at the same time in the 19th century when the much older cartways were still in use for extracting stone to the surface, but an earlier date is not impossible. All three largely remain buried by post-quarrying surface-derived debris. One arch has been preserved behind a layer of sand, prior to concrete consolidation (Fig 10.24).

This feature was constructed in a quarry located below the Firs Field. The arch was built between two pillars to support the edges of a wide shaft above a cartway that was driven between about 1810 and 1839, as part of the Burgess' Three Acre Quarry (2211). It was approximately 7 m high and about 4.5 m wide. The principal western and eastern faces were constructed of level courses of roughly squared and natural blocks located above the bottom course of voussoirs on the underside of the structure. There were over 150 individual principal voussoirs within the bottom course, arranged in 35 adjacent stones and between three and four blocks deep between the principle western and eastern faces. The majority of the voussoirs had their principle edges fashioned by sawing with only a few blocks having un-sawn faces. The voussoir course had been sprung from either side of the cartway on benches cut into the faces of each of the two pillars. Eight additional courses of sawn and squared blocks were also erected within a fissure or natural joint below the voussoir course on the northernmost pillar.

The shaft was adapted as the main supplies shaft for the Stabilisation Scheme, and a scaled section of the southern face and the underneath of the arch have been included derived from laser scanned data. Concerns there over the roof's stability also led to the space between pillars on the shaft's northern extents having a massive, coursed-rubble pack erected, with other packs constructed on the western edge also.

Six examples of smaller roughly coursed rubble arches have been noted in the northern part of the Firs Quarries (2211 and 2391), constructed to support areas of poor roof where the Twinhoe Beds have been exposed in this part of the quarry. These were located between separate areas of later Bath Stone episodes, and were used to allow quarrymen safe passage beneath areas of collapsing roof bed. The corbelled and rounded arches (Fig. 10.25) were contemporary with several other exceptionally tall stone packs, and all the features were constructed in the vicinity of the Hadley Arms.

The stone arches were integrated within and located at the bases of the much larger stone packs constructed of smaller rubble and material derived probably from the collapsed Twinhoe Beds at the roof level. The packs range in height between 4 m and 7.5 m and were roughly coursed throughout, with rubble blocks within each of the successive courses being slightly stepped out from the course below to form either the corbelled arched openings and/or the rough barrel-shaped vaulting. The arches contain no specifically cut or shaped voussoir key stones and had been predominantly constructed out of naturally derived rubble. One of these covered what were known generally as the Hadley Arms Steps (Fig. 7.11), though the steps appear to predate both the arch and the Public House. They may be associated with contemporary quarrying in the area as part of the Three Acre Quarry, or more likely, associated with a quarry located to the north of North Road. This operated prior to 1850, and again from the 1870s (under G. Mann) and into the early 1900s. Graffiti of 1887 and 1888 found within this area may be contemporary

Fig. 10.25 Rounded arch 4142 in Central Firs (Quarry 2391) under North Road, probably dating to the early-mid 19th century

with the corbelled arches and stone packs and Quarry 2391. The larger and taller stone packs are not thought to be part of this quarrying and may be part of the later consolidation of the Turnpike Road located above this part of the quarry, now the North Road.

Roof support packs

Packs in general were the most common means of roof support, other than timber sprags, and provided a more permanent support than timber. However, the stone packs were used to a greater extent to support, retain and to store spoil, and this is considered in greater detail in the section on spoil management below.

Roof collapse threatened serious adverse impact on quarrying, access and transport within and beyond the directly affected area. The principal working areas, transport routes (including barrow-ways and cartways) and access points located at the junction between underground and surface quarries were more robustly supported than other areas. In these locations additional consolidation and support was provided by flanking stone packs along cartways, stone arches and/or more numerous timber sprags. The principal transport routes were well-maintained and it was likely that smaller roof collapses would have been quickly cleared.

Roof supporting packs do not appear to have been routinely constructed as work progressed, but seemed to only occur in locations where the roof was specifically considered to need supporting or where localised collapses had already occurred and additional support was needed. Many were probably largely built to safeguard the quarryman during the period of extraction and though leases of the early 19th century placed a duty on the quarry-master to (permanently) safeguard the surface, in the main this seems largely to have been only observed where there was a highway or buildings or for a cartway or barrow-way route, in addition to at the quarry mouth where future safe access was required.

Where the packs were used to support the roof they were generally vertical in form, which increased the load supporting capabilities in comparison to an ordinary rubble pack that had a battered profile. They were generally also constructed of better selected materials that included larger blocks as well as rubble. Some of this material derived from beds of poor weathering quality, such as the picking beds or the overlying Twinhoe Beds where there had been a collapse, and some was sawn or roughly squared blocks that had been discarded after a flaw was discovered. But in some locations the quantity used was so great, and of so good an apparent quality, that it was clearly stone that otherwise would have been marketed and thus had an economic cost above that of the labour exerted in its movement.

Depending on the location, the packs were built single-faced against banks or solid rock, or double-faced where they had to stand independently, with coursed rubble or block facing around a filling, if especially massive, of carefully placed rubble, or tipped-in rubble if less so. The stone packs were also occasionally used in combination with timber sprags, as noted earlier.

The majority of the longer or larger supporting packs are thought to be contemporary with 19th-century workings, or in those areas and especially along those routes attributable to the 18th century that were still required for the transport to the surface. Since time appears to be a major factor in the progressive collapse which took place, it is likely that some newer sections of cartways, often virtually if not entirely abandoned by the mid century, never needed such substantial support, though, as seen above, in the use of arches and heavy stone packs, the large scale and generally higher workings in the Three Acre Quarry and adjacent quarries were very heavily supported.

Reinforcement of old cartways

The roof support packs in older areas were principally west of The Avenue in the Allen and Allen Estate (Phase II and III) areas, Quarry 2201 for example, where cartway development was fairly continuous through them, the packs largely being built between the pillars. East of the Long Drung, roof support packs, notably along the Brow cartway appear generally to have been contemporary with the 18th-century working, and were nearly continuous over substantial sections in front of the pillars. Otherwise only short lengths of stone packs were seen during the survey although they flanked much of the length of visible cartways. They were well-constructed with fairly regular courses built from blockstone, and extended from the quarry floor at cartway level to the roof (Fig 10.26).

The northern extent of the most western of the two Allen cartways in Quarry (505), roughly under the old telephone exchange (now a church) on Combe Road, seems to have had pronounced instability in the late 18th century, preventing its continuance around 1804 though this was originally envisaged in the lease. Here the roof at worst is only a metre or so below the surface, and to safely stabilize the end section a 3 m-high vertical pack has been built on the north eastern side in very short sections, stabilizing a section of roof bed and a localised slipped block of stone that protruded from the roof bed, before the next section was emplaced: this has left vertical gaps of unevenness in the pack.

The eastern of the two cartways in Quarry 505 did remain in use and short lengths of pack assisted the arched support near the entrance. It required substantial reinforcing of the roof around the first of the wide shafts on it, adjacent to which a substantial roof fall had caused the original cartway route to be diverted. A sub-circular rubble pack had been built around the base to consolidate the lip and the shaft

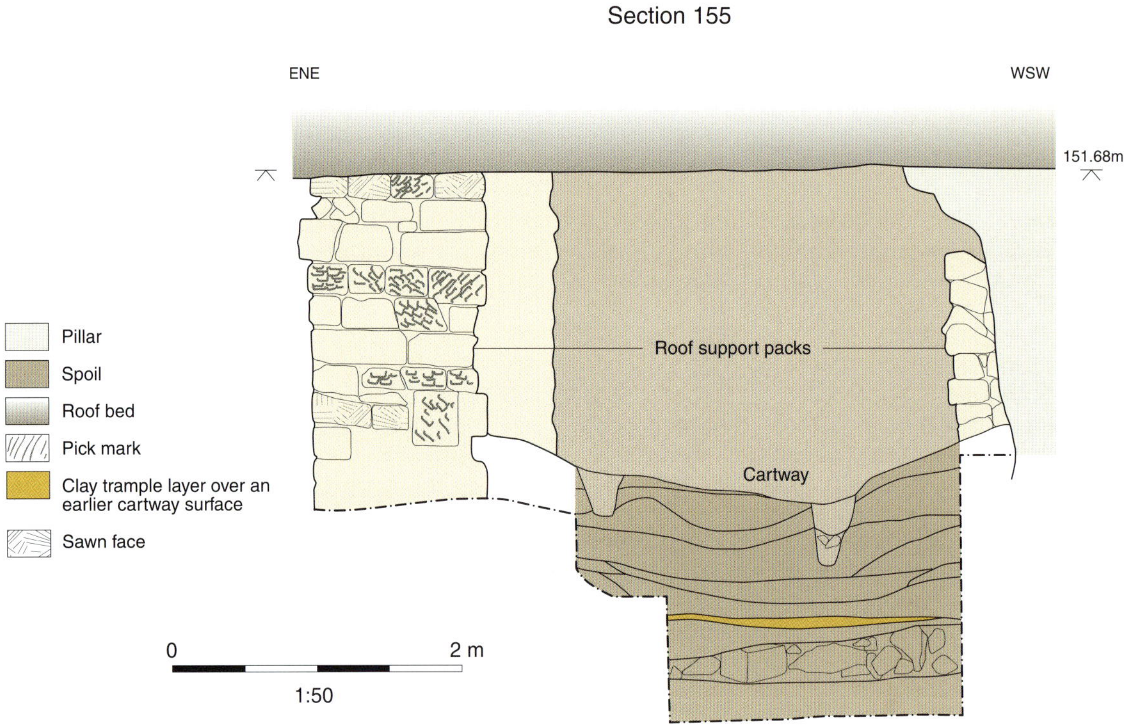

Fig. 10.26 Section showing roof-support packs alongside cartway (Section 155) in the easternmost cartway east of the Long Drung

had been filled. The pack was constructed over the remains of the roof collapse around the shaft and contained large blocks of collapsed roof materials within the first four or five roughly placed courses for approximately the first metre lift. The upper metre of the pack used rubble stone that contained some scappled blocks and re-used smaller roof collapse material.

A similar but much more substantial sequence of consolidation took place along the two principle older cartway routes and the eastern branch from them in Central Firs driven northwards from Sheeps House Quarry. The first 50 m-extents of both routes from close to the quarry entries were supported on either side by (early to mid) 19th-century flanking support packs (Fig. 10.27). In the eastern route, this had followed either roof collapse or deliberate stripping out of the roof between the apophygate pillars, leaving only residual features. In the western route the roof conditions were particularly bad despite the heavy support by provided by regular-coursed stone blocks. The branch cartway over which a large fall separated it by the 19th century from its more distant section, was adapted to link to the mid 19th-century Quarry 2202. Again the roof was stripped out, or had previously fallen, and a floor-to-roof pack of massive stone and rubble 1 m thick and some 4 to 4.5 m high had been built for some 15 m along the northern side. The two main cartways further in seem to have required much less support, possibly because they were not reliant there on apophygate pillars.

19th-century working roof support packs

In Firs Quarry 2217, north of North Road near Farrs Lane, a 19th-century cartway had been supported on either sides of the route by five individual segments of stone pack which were in places built into the voids of an already partially collapsed roof. They measured up to 15 m in length, with an average width of 2.20 m, and were built from the floor to a height of 3.70 m. The materials used with the construction were generally stone rubble and small blocks up to 0.95 m x 0.14 m, some of which had scappled faces. A stone pack, also incorporating several large fallen roof bed blocks in its construction, had one block measuring 1.60 m length and 0.90 m wide built into the pack 3 m above the cartway surface – a testament to the ingenuity and determination of the quarrymen. Smaller roof falls had been broken up and stacked alongside the cartway, but a larger collapse in between the packs finally made the cartway inaccessible.

The south edge of North Road near the Hadley Arms Quarry (2211) was supported by a 110 m length of roof support packs (Fig. 10.28). The pack was constructed generally 4-5 m high, utilizing block and rubble stone of good quality, including scappled and sawn blocks. Several arch forms were built into the pack, as noted earlier. The wall is believed to have continued to the west, following the same alignment, where a similarly constructed wall was seen in Quarries 2212 and 2216, and in places was used to retain spoil on the northern

Fig. 10.27 Flanking roof support packs in the first section of Allen's west cartway from Sheeps House Quarry

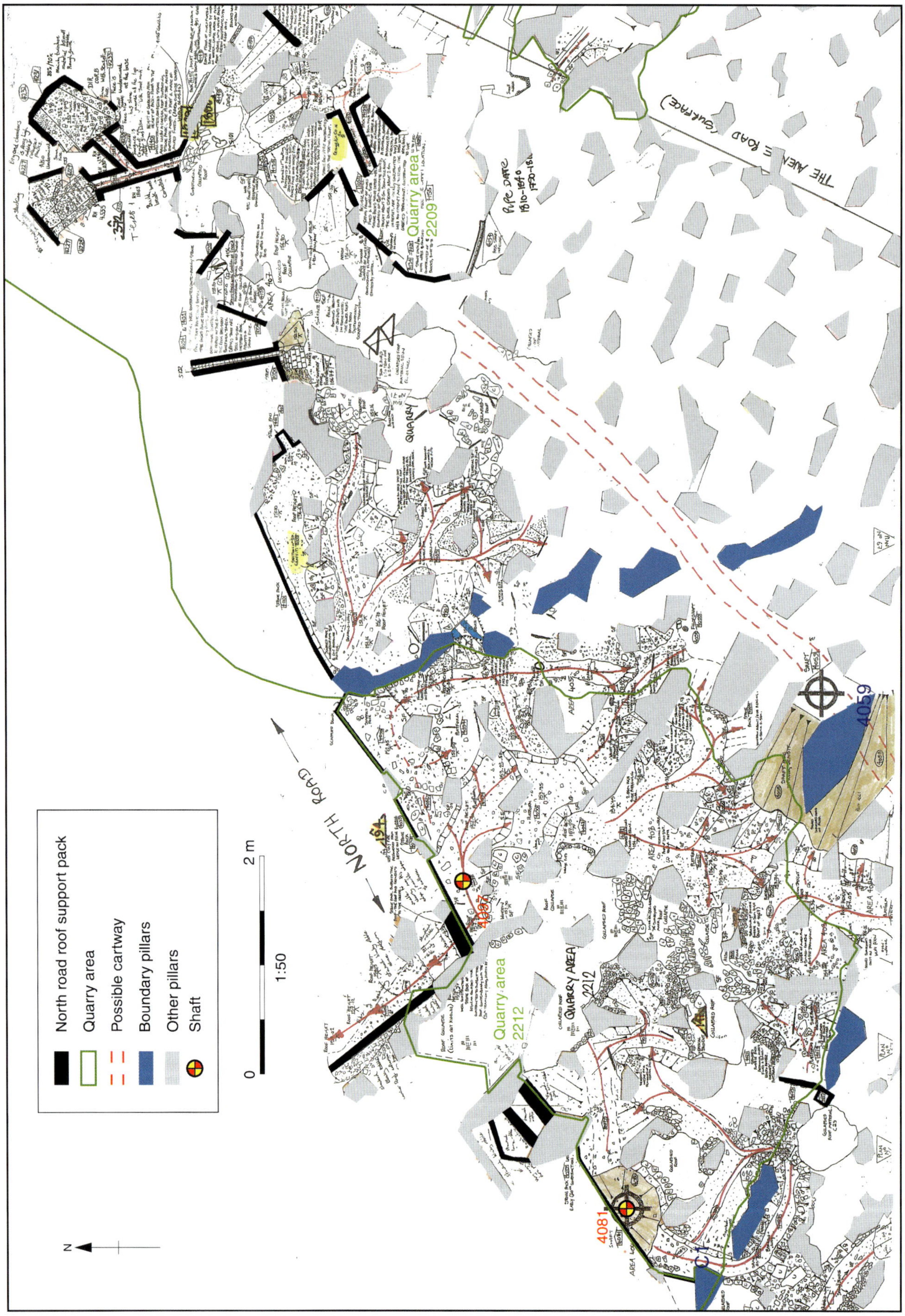

Fig. 10.28 Plan of the North Road roof support pack

Fig. 10.29 Bathite blocks used to support roof under the rear of the Hadley Arms

edge, below the North Road. All the segments of the walls are believed to be contemporary and were part of the consolidation of the former Turnpike Road when it was strengthened to accept tramway traffic in the early 20th century.

An unusual use of materials within the same Quarry (2211) was a stone pack located to the southwest of the Hadley Arms, constructed to consolidate and retain the backfilled surface-derived materials in one of the former extraction shafts. It also provided a room underneath the Public House for entertainment in the 1980s (Hadley Arms landlord pers. comm.). The lower courses utilised stone from the quarry whereas the upper courses used discarded 'Bathite' blocks, a faced concrete block made by Tarmac at Mount Pleasant Quarry in the 1970s (Fig. 10.29).

East of the Long Drung are examples of well-constructed stone packs, in regular courses using roughly squared blockstone and some rubble. These were built under a house to give support to the foundations. Several other instances of pack building occurred in this area which feasibly had the same main purpose.

Stone walls

The distinction during the underground survey between stone walls and stone packs lay in their construction, and the use of bonding materials such as lime mortar or cement. The two types of construction were generally built of similar materials, although the walls often had better constructed courses and more carefully chosen and more regular blocks. The structures were generally placed where the roof was poorly supported and which needed the additional integral strength that a wall bonded with lime mortars (or cement mortar in the 20th century), provides. They were generally located along the edges of the cartway or railway routes at the mouth of the quarry entries. The vast majority of the mortar bonds were consistent throughout the recorded quarries, and consisted of a weak mid gray, lime mortar with few visible inclusions.

Several such walls were used to seal up the mouths of the former quarry workings where they were no longer used for access. This type of wall can be seen within the Firs quarry mouths at the southern extent of the former Allen cartways at the open Sheeps House Quarry (Quarrymans Court). The walls were constructed with rubble stone and thinner material from fallen roof beds. More recent blockages have used concrete blocks.

Conical packs

Small stone packs within the open areas of the rooms between the pillars were occasionally built, with perhaps a dozen examples found. Most were square or rectangular in plan but two examples were conical in structure. One located in Firs Quarry (2203) was constructed next to two other stone pack roof supports (Fig. 10.30). It was built to a height of 2.30 m and supported an area of roof that (unusually) had been previously supported by vertical timber props, which had failed. The other, located in Byfield Quarry (514), was noted previously by David Pollard during his 1994 survey (Pollard, 1994, 86).

Chocks

The simplest form of chock was either a timber or stone block used to prop a roof bed or individual roof block. They were only rarely found, but were occasionally used on top of packs or on top of spoil heaps built close to the roof to close the gap under the roof. Another form of stone chocks, also referred as cogging or cribbs, were constructed of selected blocks of stone that were placed in position to form a more solid mass. The area near the Hadley Arms

Fig. 10.30 Conical pack in East Firs (Quarry 2364)

close to the North Road included chocks that were either stacked singularly or in combination with other stones placed within a few feet of each other. Stone chocks were placed against the roof above rooms filled with spoil and were sometimes tightened with small timber wedges. Some spoil-filled rooms had a series of rectangular cribs built with rubble stone walls, perhaps a metre across and placed one after another to fill the whole space back to solid rock, and were filled with rubble, from floor to roof. The multiple walls gave much more resistance to crushing than a standard roof support pack. Some of the features at North Road, may however, be post-quarrying, designed to support the road.

Fig. 10.31 Sawn stone used as a sprag in West Firs (Quarry 2342)

Another type of chock was a variant of the sprag, with the insertion of an individual block placed at an angle or near-vertically. Two examples of this type of support have been seen within the Firs quarry complex. That from Quarry 2342 was a sawn stone support measuring 850 mm x 200 mm x 113 mm in blockstone, which had been sprung from a pillar face into the roof. A sawn block in Quarry 2211 was placed on the southern side of the cartway to support a failing roof block (Fig. 10.31).

Support value of all packs and spoil

With relatively high and narrow pillars and often their wide spacing, had the Combe Down Quarries been largely emptied of spoil, then it is fairly certain the whole complex would have suffered severe 'domino-type' collapses affecting much of the surface. In the few full-height pillars seen, up to 8.5 m, the pillars can be seen to be riven with a series of joints little more than a foot apart. These have opened slightly and collapse was likely within a relatively few years or decades. That this did not happen was due to two important factors. First, that all spoil as well as artificial features assisted in providing lateral support, so that the bed blocks of stone forming the pillars could not normally be forced sufficiently out of position for the pillar to collapse. Second, because the top of the spoil was frequently close to the roof, any collapse of the roof, typically bed-by-bed due to delamination, could only be for a relatively small distance, as the void which developed above the fall 'bulked-up'. Any void would soon be filled by progressive collapse preventing upwards migration and crown holing.

Disposal of the waste

Spoil

This is the general term for the economically unwanted material, otherwise termed discard or waste, 'gob' or 'deads', which was produced during the production of saleable stone. It could be material which was found with the stone, such as clayey material in fissures, stone of low quality, for instance a poor weathering stone, or was too fractured or contained cockles (calcitic vugs), or had been broken in the process of quarrying or squaring of blocks, or coursed rubble. At Combe Down, something like 40% of the total amount of quarried material was discarded as spoil, occupying somewhat over half of the excavated space. Most of it was oolitic limestone from the quarried beds. The proportion was probably higher in the earlier periods and less in the later. Its disposal or dumping was a major problem, both for the expenditure of effort and for leaving sufficient space to effectively continue working. It is clear that a great deal of consideration was given to spoil disposal, and working methods evolved first to reduce it to a minimum and secondly to stow it in suitable places with maximum efficiency. However, a great deal of spoil was not carefully disposed of and this led in part to the apparently disorganised appearance of both spoil heaps and the general underground landscape.

Sources of spoil and developments towards its reduction

Once the quarry had developed, the first source of spoil was encountered just beneath the roof level, the picking bed. It was just under a metre thick and of soft and poor weathering quality, not generally saleable by reputable quarrymen. In the earlier quarries this was totally removed and most discarded – amounting to some 20% of the total excavated material – although some may have gone for interior walls of buildings where weathering was not a factor. This proportion was vastly reduced with the use of the different forms of jad slots and especially of the regular parallel-faced jad used to undercut the roof. The use of the parallel-faced jad below the roof negated the necessity to remove the picking beds, and in consequence reduced the spoil produced to perhaps about a half or third the former. This spoil was all produced at a high level in the workings and ideally was kept at and disposed of at that level. This may not have been easily possible with the simple method of working by pillar and room before the mid-late 18th century, but the development of Long Rooms, with their high level windows and Long Wall, pillar and room systems with planks to cross the gullet, meant this became feasible. The elaborate barrow-way systems seen around the Grand Canyon (2209) allowed spoil at different levels to be taken for considerable distances without losing significant height (see Chapter 12, Case Study 9).

Individual beds between pillars were freed in a similar way, but vertically by opening out a joint or by breaking one of the less valuable bed-blocks or wrist stones. Stone was inevitably broken as bed-blocks were extracted. Because the cross joints were not at right angles, there was further waste at pillar faces and joint intersections, and some beds, because of lateral variation in the deposit, were probably of little use either because of quality, thinness or presence of vugs or cockles. Finally the stone had to be squared from the usual diamond shaped bed-blocks, roughly in the case of production for coursed rubble, more accurately by scappling for blockstone. These losses probably produced the bulk of spoil. The remainder was from joint or gull infillings, notably clayey material or small stone from close jointing, or fracturing from rock movements, and from erosion derived by water draining down and from roof falls migrating into the beds over the freestone. There was some scope for reduction of some of this wasted material. Use of lines of wedges and chips for breaking and, above all, the use of saws resulted in the amounts of spoil again being moderately reduced. By taking more beds from below, maximizing the take of stone up to the 8.5 m depth seen in one or two areas, then the proportion of waste or spoil again fell. Further spoil came from the banker mason activity in some areas in which blockstone was worked into wrought stone or items such as ridges, mullions and other decorative or ornamental work.

This wide range of origin produced many types of spoil. There was plenty of stone of block size, some from overlying dropped beds such as the ragstone or bastard stone, with much not obviously different from material being sent out, except perhaps being somewhat more misshapen. Indeed it is difficult to see why much of it was not broken to produce coursed rubble for building. Perhaps this was somewhat reserved for thinner beds of good quality, despite occasional comments about the willingness to supply poor stone given the opportunity. Stone was, of course in plentiful supply and any losses were those of the master rather than the workmen, so the apparently good quality stone seen in the better quality packs may sometimes have been chosen for the ease of workmen rather than using the misshapen real waste material. Fragments of rubble size, below about 500 mm to some 50 mm, and small rubble for lumps below this were the most common, but there were very substantial quantities of fines, below about 5mm, typically produced by pickwork or banker mason activity. Often these different sizes could be separately dumped, with advantages, for instance of better barrow-ways and cartways.

It is clear, from the patterns of spoil, that organised systems of dumping evolved, both in the design of the working methods, and down to the actual individual dumping of the barrow-load. The disposal of the spoil was thus an important part of the quarrying process. It was visible in the large number of barrow-ways at different levels, in the 'windows' found at picking bed level through which spoil was dumped, and in the tactics adopted, presumably to minimise effort, to stow and stack the spoil in an organised fashion. This resulted in common features such as lateral dumps alongside barrow-ways, packs supporting large banks of waste and gently rising banks and inclined barrow-ways leading to arcuate tipping fronts. Because of the need to use wheelbarrows, spoil disposal almost certainly had an effect on the height of roof above the floor, either over about 1.8 m (full height walking) or somewhat below this at about 1.4 m (crouched walking) near the final disposal area. An indication in the reduction of amounts of spoil was provided by the higher roofs over dumps in some of the later quarrying areas, though this was difficult to assess as almost everywhere around them by this time had already been worked and there was no need to keep any but the most direct required access ways open.

The initial opening out of a new quarrying area generated large amounts of spoil which had to be removed from that area to provide working space. In an established working this was dumped into other adjacent worked-out quarrying areas. Many of the longer-than-average barrow-ways were seen to be related to this initial excavation, and were essential to the opening-out process. An example is Quarry 503 in Byfield, where the abundant fines produced by the jad cuts and slots in the high pillar area were dumped in the adjacent older workings on the north side. The fines were dumped to within a few tens of centimetres from the roof, partly by 'shooting' it over the end of the barrow, after which rubble was thrown-up by hand to the roof level (Fig. 10.32). This dumping continued so as not to intrude on the open, worked-out areas to maintain handling space before its eventual abandonment.

The most constant feature throughout the quarries for dumping involved the development of cartways and other access ways to just behind the front of the face on top of a bank of spoil, which was kept just behind the working face as it advanced. This formed a working floor at the end of the cartway where stone could be scappled and stored prior to carting away. Initially the floor of the cartway was some 2.5 m below the roof or ceiling, found in many areas of the unaltered original Allen cartways with up to some 2.5 m of spoil under the working floor. The higher space in later cartways perhaps reflected a greater depth of beds extracted, but the key reason was probably to place the carts or waggons next to the prime freestone bed for ease of loading. Greater depth, however would allow a similar quantity of waste to be stored in the cartway as previously. The spoil dumped as the face moved forward allowed the low front or gullet (below cartway level) to be cleaned up with minimum effort and any shortage of material for this was easily made up by material from higher-up. This probably included finer material, which was used to produce a good surface for the cartway. This specific use suggests such material was often stored or kept in specific dumps to be available for this purpose, rather than being dumped indiscriminately with other sizes.

Fig. 10.32 Miner seated in original position demonstrating waste-stowing to roof level

Due to lack of archaeological access to full working sections, it was generally only possible to hypothesise about dumping of other spoil in the early phases of working. At the upper levels, visible in workings alongside cartways, for example in the western Allen cartway from Jones Quarry 505, the spoil was dumped in low banks alongside the cartway and was covered with fine dusty material. This was clearly ground stone slurry scraped off the cartway, implying its replacement by good fines as a routine and a degree of cartway or roadway maintenance. On this, or built off the floor were low rubble packs behind which spoil had been stacked almost to roof level in places, in the form of irregular humps. The material below was not visible but modern roadways which cut through such banks revealed repeated sequences of barrow-ways overlain by layers of spoil. It is reasonable to think this probably meant a fairly orderly system of barrowing from different levels as the working face moved away, with final dumping either by 'shooting the barrow' over its front, or by hand shovelling or throwing-up. Perhaps because of the use of jad slots and less spoil in the adjacent parts, the top surface had long heaps of rubble and fines between barrow-ways, with a good height for barrowing, though the heaps between the barrow-ways were piled to within a metre or so of the roof.

It seems likely that a system evolved in the development of side rooms for dumping spoil. At the base of the sequence was the floor formed behind the gullet. On top of this was dumped material, which could not be accommodated at the subsequent current working area. Finally, spoil from the picking bed was dumped through 'windows' at the highest level, opened into the worked-out earlier room alongside. Direct evidence of these windows had largely gone with the removal of the lower beds, but they were sufficiently frequent in almost all areas to suggest this function, probably as well as improved ventilation and access. This dumping process was not completely exposed in any archaeological sections as rooms of deposited spoil were usually only partially exposed by the Stabilisation Scheme, but the hypothesis explains features seen in different areas, notably within the Gullet workings in Quarry Areas 102, 201, 302 and 305. This hypothetical sequence was initially understood from observations of the Long Room system, where in Quarry 2201 material dumped through windows was visible lying on roughly levelled spoil overlying the top of a former cartway floor (see Figure 8.10).

In what became the most extensive working method – the Longwall with gullet and pillars – most spoil was, after the initial phase, dumped behind the long gullet (see Figure 8.11) first at the working floor level at the gullet, then behind a low rubble pack to a height 1.8 -2 m below the roof, (thereby leaving a useful walking height). Finally, dumping from a high level using barrow-ways and possibly planks across the gullet took place. Examples of suitable planks were found, but not *in situ*. Subsequent dumping from this or other quarry and secondary dumping often reduced this height to around 1.4-1.6 m, the likely but uncomfortable height needed for barrowing with a bent back, and in some areas dumping was done up to roof level. Figure 10.32 shows a modern miner seated in an surviving original position demonstrating how material would have been handed up and stowed at the back to roof level.

The first typical roof height of about 1.6 to 1.8 m provided a useful walking height behind a barrow, sufficient to allow both tipping over the side of the barrow on the margin of heaps, or projecting the spoil forward over the front end. There was certain evidence only of the builders'-type barrow at Combe Down, which could be used for either purpose. Such a barrow would, though, be difficult to use in low-roof areas, and a modified form of barrow with low rear legs may have been used, or no barrow at all. In many mines a mine barrow with a body slung below the 'styles' was used, its base sitting on the floor, and something similar may have been used in difficult conditions at Combe Down, and the mine barrow was certainly used in the Wiltshire Quarries (Derek Hawkins, pers. comm.).

Always subject to individual idiosyncrasy, spoil heaps developed in many ways, but several

Fig. 10.33 Inclined barrow-way

common types evolved. Those recorded were all at upper levels in the dumping sequence, but it is likely that similar methods of easy dumping at the top resembled those lower down, which were subsequently buried.

Either the barrow-way was developed on top of an existing bank or, especially in high workings where space allowed, a single barrow-way, inclined as necessary, was developed out of the working area at grades typically between 1:4 and 1:6 (Fig. 10.33). To achieve this spoil was dumped over the front of the barrow to form a tongue ahead, with coarse lumps thrown to the side. Fines seemed to have been selected to form the surface for the wheel to run easily.

Once within the dumping areas, along the barrow-ways as they passed through worked-out rooms, two types of dumping applied, the first perhaps where large amounts remained to be disposed of. Lobate tipping developed a narrow gentle incline or ramp to the furthest point, providing a good barrow-way surface as the incline pushed forward – typically of fines dumped and graded over a rubble base. Dumping subsequently took place laterally over the side of the barrow (always the easier mode of tipping) along the sides of the gentle incline, either as it was formed, or subsequently, preserving a level or sloping surface with the barrow-way at the higher level. The supply of materials, notably of fines for the surface, possibly decided this. The second method may have applied only in the final stages, with lateral dumping alongside the barrow-way so the barrow-way was then below the level of the dumps. Sometimes this was carefully done back to the pillars to maximize use of space, sometimes only immediately adjacent to the barrow-way itself. At the end of the run, spoil was similarly dumped and, in many cases, rubble or fines thrown up to just under the roof. Shrinkage seems often to have left a small gap under the roof – if not, deliberate stacking or packing of material to roof might be suspected.

Lobate dumps were often found in either wide areas or in a long linear sequence of rooms. In other cases the tipping took place on a wide front, across rooms, with a distributary system of barrow-ways (see Chapter 12, Case Study 9). These were developed similarly to the above but along the 'dumping-margin' a series of ever-widening arcuate fronts were developed, with the barrow-loads clearly having been run around the near circular front tipping over the edge, until the margins of the area were reached. A highly developed sequence of this type was found south of the Grand Canyon (Fig. 10.34). It may be more suited to dumps where there was a high proportion of fine material, allowing a good surface over the whole of the dump.

Roof heights requiring a severely bent back suggest a local shortage of dumping space. Spoil in such passages and rooms seemed often to have been stacked at one side, leaving just space for the barrow before final filling of the remaining space to 0.80 to 1 m or so from the roof. Wheel marks were sometimes visible in fines and rubble at the side where the next load was dumped. In many cases rubble stone had finally been thrown up on to the top of the earlier mixed fines and rubble and stacked rubble.

The first stage in many sequences seen seems to have been dumping by roughly stacking rubble at one or both sides, placed so as to keep the barrow-way clear. Then the barrows were dumped over-

Fig. 10.34 Lobate tipping in Central Firs south of the Grand Canyon (Quarry Area 2209)

the-front at the end and often rubble thrown up over it to the roof, with a gradual withdrawing from the filled area. Sometimes where space was less critical the withdrawal phase took the form of a series of ramp-dumps along the barrow-way, like a series of slightly inclined long steps. In proximity to boundaries, and especially near cartways where space was retained in the adjacent worked-out side-rooms, substantial rubble-stone packs were built to hold spoil back.

As the amount of spoil diminished with improved working methods, new forms of dumping were used once space was established, presumably with the object of providing more space in the working area as well as reducing the distances of travel for dumping. In some cases this was by placing spoil behind large, carefully-built rubble-packs, in others by dumping to form level platforms in a series of broad steps or platforms with short ramps between. A notable example of the first was seem lining the cartway between Firs Shaft and the Hadley Arms, with a wide space created (for storage?) on one side of the cartway, and beyond and on the other side high rubble packs to the roof holding back spoil. In some cases whole areas were apparently devoted to spoil. South of Stonehouse Lane, the gap on Hawkins (1994) map between the modern Firs and Byfield Mines was almost totally infilled with floor-to-roof spoil. The quarrying area 2397 appears to have been worked in advance of the workings on either side, owned by different concerns and used by them both as their dump. There was no provision for this in the leases soon after 1800, but either the landlord had permitted it on the basis that it provided support for the surface or the two leaseholders worked systematically in conjunction with each other. A similar walled-off dumping area (2395) appeared to be under the Long Drung in East Firs, and, indeed, much of Far East Firs was packed with spoil in this way, though that may be a consequence of small-scale, low-height working.

Open Room quarrying in the latter part of the 19th and the early 20th century developed another form of spoil disposal. This may have been partly possible because all adjacent areas offered space to allow the working area to form a flat space at cartway level between three or more rows of pillars. When this space was no longer required for the working floor it could be more systematically filled. Examples were seen in James Riddle's quarry (518), and also in what was probably Stennards Quarry 2367 south east of Farrs Lane south of the Turnpike, both late-stage working . These took the form of a series of broad flat-topped platforms linked by ramps. Neither was very large, partly because the quarries themselves were small, partly because they were only involved with the last stage of the working.

The overall impression formed from the archaeological investigation is of generally carefully considered strategies for spoil disposal minimising the effort required to dispose of anticipated necessary amounts in older working areas, with disposal clearly preferred immediately adjacent to the working face. These strategies were modified by those who actually carried it out. Particularly where there was no pressure on dumping space, in perhaps the last stages of operations, which are what were left for recording, a less disciplined form of dumping may have prevailed, leaving the apparent chaos observed by modern investigations as the over-riding characteristic.

Spoil as evidence of associated activities

The study of barrow-ways and spoil dumps developed into one of the most fascinating aspects of the workings. It was one of the best indicators of the sequencing and can even allow deduction of dates or phases of working. The spoil itself carries evidence of the methods of working employed in given areas, and the type of spoil could sometimes reveal its origins and, sometimes, secondary activities. Examples include use of jad cuts in Quarry 503 and dressing or bankers' waste in Quarry 2347 (see Working floors in Chapter 8). In two cases greyish clay related to wells sunk to the Fullers Earth below. Openings into the quarries – shafts and side entries from surface quarries – had waste with a patina subtly different from that derived from underground. Dumped rubbish was even more distinctive.

There was an obvious stratigraphic relationship in dumping since it could only take place in an earlier worked-out area. Whereas the source area was often totally concealed, barrow-ways leading from it to dumps indicated the concealed area was newer. In many areas it was possible to plot these relationships, which were shown on the site archaeological plans with arrows. These confirmed earlier and later relationships with pillar forms, for example, thus allowing broad phasing of the workings and sometimes close dating in the relationship with a well-dated areas known from leases or other documentation.

The use of packs for spoil stowage

Stone packs were one of the more common quarrying features. They were located throughout the Byfield and Firs quarry complexes and within outlying quarries. They were used to support the roof and to retain waste stone discards. Their use for roof support has been considered above, and it is their use for stowage of spoil which is considered here.

In quarries, the ability to build packs – drystone walls if used at surface – was a skill most labourers, let alone skilled quarrymen would possess to some degree. Only substantial packs and arches would require skilled supervision. It was therefore useful to discern the original intent in order to differentiate between stacking and packing.

The stone pack was self-supporting and able to withstand pressure from spoil dumped behind it, or to support a roof sometimes to heights of several metres or even higher in specialized applications such as at the Arched Shaft under Firs Field. It was normally carefully built with placed, often coursed block or rubble stones, following a regular line and with a plane front. It was often battered slightly, to better withstand pressure. Stacked rubble was defined principally by function and capable often of only supporting its own mass. Usually, however, the stones were unevenly placed and the front irregular. Stacks were fairly low and typically rather carelessly placed, the only requirement being to ensure they did not fall back on to a barrow-way. Lateral dumps of rubble alongside barrow-ways were often slightly 'tidied' by stacking the front to keep the barrow-way open. Stacked dumps were found throughout, whereas packs were generally located with a distinct purpose, lining cartways, laterally supporting pillars or holding back substantial banks of spoil. When reinforced by mortar, as next to entrances or, in one case, for support under a main sewer, they have been distinguished as walls.

The stone packs and walls were generally recorded as solid black rectangular areas during the Hawkins survey in 1994. These were generally fairly accurate in both their orientation and extent and were used as general baselines in the archaeological survey. They were adapted to illustrate each of the structures individual facets and their relationships with other quarrying features in the landscape.

Stone packs were regularly used alongside cartways and the more permanent barrow-ways to provide support for the spoil discarded beyond. They were either built-up in advance, so that dumping could take place against them from the rear side or, more commonly raised as material was dumped over or on top of them, using selected pieces of the dumped rubble or block in their construction. Notable packs lined the Firs Shaft-Hadley Arms cartway (2211), and the Grand Canyon (2209) which branched from it, had packs over 4 m high placed between pillars (thus giving them support too), and built-up in a few places to within a metre or so of the 8.5 m high roof. Some packs had a dual or triple function – to support spoil banks and the roof, and even to laterally support pillars.

Small packs were used to hold back spoil behind the gullet, one pack succeeding the next as the face advanced. They formed loading bays or walls and, at Shaft Road for example, to build a platform on which a crane base-stone was placed. In Far East Firs a small pack formed a base for a saw-sharpening stone bench. They also protected wells, built either from inside the well, or from within the quarry workings. In one case a pack was built like a helter-skelter around the well, to allow climbing up to place the top stones.

Chapter 11

Other structural and artefactual evidence in the underground quarries

Introduction

This chapter is in two major parts, the first focussing on aspects such as the after-use of shafts and, and those parts of the workings that were non-quarrying in the technological sense, together with wells. These features were all prolific providers of artefacts, which, along with the many examples of graffiti also included here, inform us of social aspects of the life of quarrymen and Combe Down people as well as their work. These were recorded in the archaeological survey simply as contexts, though the term 'large find' would be equally appropriate.

The second part concerns what were defined for recording purposes as 'small finds' which have been the subject of specialised post-excavation reports. These covered metal, stone, wood, glass, pottery, and bone. One of the most unusual items found was a revolver and bullets found in the surface-derived material dumped within a former vertical shaft located on the Firs Field, which now has a Chestnut Tree growing from it. However, after being photographed (see Figure 7.15) the revolver and bullets were handed over to the Avon and Somerset Police and destroyed before further examination was possible.

All the finds have intrinsic interest, and in some cases have a wider application, such as the recovery of numerous 18th-century clay pipes, which now form a major part of the reference collection for Bath (Lewcun 2009). These, as well as the pottery and graffiti are particularly important in that they help determine the chronological dates of workings and quarrying and also can suggest the periods where quarrying operations ceased. Graffiti, for instance, can not only indicate when the initial quarrying began but also where secondary Bath Stone episodes were carried out underground. Other artefacts, such as the quarrymens' tools and equipment or examples of the carved stone itself can also determine periods of working and the different quarrying methods used to extract the stone. Graffiti and some artefacts and other features tell us something of the miners and their activities and interests, or of social conditions in the neighbourhood, such as in the Second World War.

Non-quarrying features and quarry after-use

Wells, disused shafts and drains

Several wells were identified and more are known to have existed. Only four were open the full depth to water and only three could be examined in any detail, one only by camera through a small hole made in the lining. Shafts observed to go below the quarry workings were considered as wells even though not open to water, to avoid confusion with the narrower ventilation shafts. The shafts were usually lined with rubble above and below the underground quarry workings. Several wells were collared in the workings by circular stone packs, in one (presumed) case by the helter-skelter pack referred to earlier (see Chapter 10), presumably to prevent adulteration or accident. At least three were associated with Public Houses reflecting that brewing was once usually done in-house: the Carriage Inn, Hadley Arms and the William IV. Others served private houses or groups of houses. The Carriage Inn well may have done this too, and possibly also watered horses as its top stood close to the wooden railway.

A well that seems to have been sunk for a house in about 1818 (Irving 2005) but which became Job Salter's Brewery and, later the Hines Brewery, had a prominent circular stone pack-wall which was located on the cartway under what is now the William IV Public House (Fig. 11.1). It is thought to have post-dated the quarry working (504) where it had been constructed. The internal parts of the pack were inaccessible to survey as it was filled with post-quarrying surface-derived materials. The pack was constructed of well-coursed rubble stone throughout and had larger blocks, some sawn, placed to form an opening or door within the eastern extent. The lower extents of the pack had a darker patina on the rubble stone, with the upper four courses above the door a much lighter shade. Several of the courses around the door and at the same level around the structure were partially bonded with mortar. The presence of the door within the quarry suggests that water was accessible to the quarrymen for application during sawing, and/or that it was convenient to the brewery for cleaning the well of debris as this part of the quarry was easily accessible from the surface.

The well was preserved *in situ* on the edge of the Bat Preservation Area and was laser-scanned as part of the archaeological survey.

The most important well for dating purposes was at surface, located between the backyards of the Bank and Langham House near the junction of Church Road and The Avenue. It would formerly have served the Carriage Inn and the Old Rank of Allen cottages. Underground it lay within Quarry Area 2347. The well was marked on the 1st Edition OS Map but is believed to be of early 18th-century origin as grey clay, probably the Fullers Earth clay, was raised from it and dumped in the adjacent working underground. This and an adjacent

Fig. 11.1 Job Salter's well, later the Hines Brewery well, in West Byfield, under the present William IV Public House yard

Fig. 11.2 The Carriage Inn well, in the pre-Allen E4 stub workings, (Central Firs, Quarry 2347). The two pump rods and pipes are visible, but the pumps themselves appear to be buried

deposit from a drain contained pottery of early to mid 18th century (see pottery report below). The well was lined with medium to large limestone rubble both below the quarry floor and above the quarry roof (Fig. 11.2). A circular stone pack of stacked rubble largely surrounded the well. The quarry roof was some 4 m below the slabbed cap, and 15.45 m from the roof to the bottom of the well, with about 0.3 m of water at the bottom. There were two pump pipes and two rods down the well, but neither of the pump barrels was visible, indicating the bottom was partially infilled with clay, either washed or squeezed in. This suggests the original total depth from surface was about 20 to 21 m. The well was at the northern extent of the probable pre-Allen (Phase I) working, but whether the working or the well came first is uncertain.

Another example of a previously known well in the Byfield Quarries was just east of Ralph Allen Yard, located within the 18th-century Quarry Area 911. The well is thought to have served a house above and was open to water at the base. Its protective stone wall was visible in the workings as an irregular, slightly conical feature made of thin, tile-like stones. This was the outside of a lining placed from the inside (Fig. 11.3). It was possible only to safely pierce this to allow a camera and the distance measurer to be inserted (Fig. 11.4). Photographs showed the inside to be fairly smooth-lined, about 1 m in diameter, but bellying-out somewhat.

A well was located at the top of Combe Road in the side road to the remaining Byfield entrance (now the bat entrance). The well was about 1 m in diameter and passed through the double-height workings of Quarry Areas 508 and 512. It was lined with rubble throughout its depth, was circular in profile grading to oval, and bellied-out after a metre or two. It also had a protecting rubble pack in the quarry workings to stop materials entering the water below.

It was sunk through the lower Combe Down Beds (Ragstone) and into the clays of the Fullers Earth formation below, with this clay dumped over a small barrow-way which previously served the lower Quarry Area 512. The deposit may have been dumped over the barrow-way in a well-clearing operation that post-dated quarrying as no other shaft sinking debris was present. It would be attributable to the house that the well served, probably dating to *c* 1829.

Another well-shaft in Quarry Area 2219 on the northern side of North Road east of Stonehouse Lane, belonged to early 18th-century houses so possibly pre-dated the underground quarrying. A debris cone largely obscured a protecting, low, rubble pack-wall around it as it passed into the lower beds, but a pump rod and pipe were visible, hanging in the upper part of the shaft.

One of the deeper wells was located within Quarry Area 2330 close to Richardson Place on the surface. It had an internal diameter of 1.40 m between the rubble lining. The depth was 16 m and

Fig. 11.3 The well in Byfield (Quarry 911) just east of Ralph Allen Yard, was sunk through pre-existing workings but, unusually, had been lined from within the well shaft, resulting in the 'inside-out' pack seen here

the depth from quarry floor to an infilling in the base was 9 m. There was an opening of 1.5 m from the quarry floor to the roof and a further 5.5 m from mine roof to upper surface where the shaft had been covered with a tin sheet. The shaft was lined throughout with fairly level courses of small to medium sized limestone rubble. The well is shown on the 1884 OS map and contained late 19th- and early 20th-century material derived from the surface.

Drains

An 18th-century drain was located a few metres to the north of the Carriage Inn well. It measured 1.10 m east-west and 0.70 m north-south and had been lined with sawn blocks throughout its height. The drain continued up towards the surface for 1.50 m before surface-derived grey ash and blocks restricted a view.

In the area close to the Hadley Arms (Quarry 2337) sewage had been allowed to drain into the

Fig. 11.4 The Byfield well inside the pack shown in Fig. 11.3 was not seen other than via this photograph. It was taken with the camera pushed through a small hole made in the pack

Fig. 11.5 The probable drain in Quarry 2392 in Far East Firs, projecting through the roof. Made of thin-sawn stone slabs, there was no sign of it ever having been used

workings, though the actual drain itself was not inspected. In Far East Firs Quarry 2392 a chute-like feature made of thin sawn slabs projected down into the workings, but there was no trace of water, effluent or deposited solids below it (Fig. 11.5).

Disused shafts

There seem to have been three possible fates for shafts after their quarrying usefulness disappeared. Some were capped over, others were filled with what appears to have been quarry waste, and those left open became the village refuse amenity. Most smaller shafts were capped over, though sometimes only with a thin cover of corrugated metal sheet, more often with stone slabs. Larger shafts, perhaps especially where they were fenced, were commonly used for refuse disposal.

Two shafts which seemed to relate to the end of Phase II in East Firs had particularly spectacular caps, shallow domes, seen from the underside. One adjacent to the branch cartway from Central Firs into East Firs in Quarry 2203, had a cap *c* 3.5 m across, which arched only some 70 mm measured from the centre to the outside. The stone blocks used in its construction were arranged in the form of repeated squares. There was no sign of how the arch was supported during its construction. The second shaft was a short distance further east. The two shaft caps were so similar and both had the same relationship between apophygate and corbelled pillars that they surely belonged to the same period of construction and possibly were the work of the same mason (see Figure 9.9). Only one other shaft cap had a similar domed form, but it did not display the same fine construction. The shafts are believed to date to the first half of the 19th century. Other caps were barrel-arched or, if more recent, of concrete, as under the Garage in The Avenue and, in the form of beams, at the Firs Shaft.

Shafts filled with rubble included several ranging north of the Ralph Allen Yard (Jones Quarry) entries, towards North Road. They seemed to have a clean clay and stone fill, possibly derived from nearby surface quarries. The southernmost had a circular pack (see Figure 10.19) surrounding the base so the fill could not be seen.

At least two shafts beyond the Firs Shaft ranging towards the Hadley Arms were filled with rubbish and probably a third and fourth also. This was derived from the surface and attributable post-quarrying debris, including much village rubbish, such as bottles, tins, car parts and the revolver and different gauge bullets, which had formed into cones of deposits which completely obscured the quarry floor below. They apparently filled the shafts, one of which still has the chestnut tree growing from it and is situated close to the War Memorial on Firs Field. Firs Shaft itself had considerable rubbish on the floor 2 m or so deep, but did not appear to have been filled. It too had a tree growing over it until blown over by a storm. This may suggest a cap of timber had been installed before rubbish could fill the shaft, or that the rubbish had got hung-up in the shaft, forming a platform.

After quarrying had ceased, Irvings Incline into Quarry 504 with its link to the workings associated with Quarry 503, may have provided an access for the benefit of the William IV Public House. This (or its brewery predecessors) also had an entry to the workings in the pub yard via a shaft, and also had a vehicle way (*c* 1870) located immediately behind

Fig. 11.6 The Irvings Incline, a slope entry into the West Byfield, Hulonce Quarry (504) made early in the 19th century and variously used for storage and an air-raid shelter after quarrying finished

Tor House to the incline. It is possible that beer brewed there or brought in was stored underground. There was plenty of space for this but no obvious remains (Irving pers. comm.). The incline appeared to have long been used for storing items such as bottles, tins and old tyres, but during the Second World War the incline (Fig. 11.6), which gave access to the immediately adjacent part of the Byfield Quarry, (503) served as an air-raid shelter, particularly during the Baedecker bombing raids of April 1942. These hit Bath severely and left a trail of bomb craters a short distance away across the back of Foxfield. A line of screwed metal eyelets placed in the roof of the incline and extending into the quarry, originally provided support for an electric cable from the surface for lighting during the raids. Heating was provided by 'Coal Devils', and both were apparently provided by Mr Vernon Bishop.

It is possible the other Byfield Mine entry, now the grilled bat access, was similarly used. The floor of this had also been cleared for 'industrial' use, reportedly for storing mushroom boxes. Today this whole area of West Byfield is a bat reserve, including a heated breeding conservatory, with the two original accesses open but grilled. The inside areas, however, have largely been concreted over. The Irvings Incline arch and the Hines Brewery well have both been consolidated.

The area of Firs Mine under the Hadley Arms was briefly used in the late 20th century for 'partying', with a somewhat inconvenient access via the Hadley Arms steps. The area was fitted with a number of floodlights.

Graffiti and inscriptions

Introduction

As well as features and artefacts directly related to the quarrying activities, there were also many examples of graffiti, some related to quarrying, some of miscellaneous chatracter. The graffiti took the form of lettering, pictures and marks that had been written, drawn or scrawled on surfaces. Inscriptions were far rarer, either initials and dates or surveying marks carved into the rock. Both provide, with due caution, a useful source of dating evidence as well as providing valuable insight into the social context in which the quarrymen worked and lived.

All the graffiti observed was recorded, whether contemporary or post-quarrying. Selected samples were either recovered by sawing-off the slab on which they were drawn, or stripped off using a silicone rubber 'peel' (the methodology is described in Chapter 1). As in limestone quarries elsewhere, the graffiti varies in its subject matter, style, form and ways reflecting the period when it was done. The survey of the surviving graffiti at Combe Down has captured material which was originally written either by the quarrymen contemporaneously with stone extraction, by visitors to the quarries during their various operations, or by visitors in the subsequent post-quarrying periods.

Graffiti drawing materials

The graffiti was mainly drawn with a soft coal marker then locally available, or with soot produced from the tallow candles and/or oil and acetylene lamps which were used for lighting. The ship graffito (see Figure 11.12) used a dark brown mud for the masts and later graffiti included chalks and sprays for colour. The position and materials used to draw the graffiti partly determined both the survival and the clarity of an image. For example, graffiti formed with candle or lamp soot, which tended to be written on the rough surface of the roof of the quarry, generally had wider lettering, and had more blurred edges than the lettering performed by hand-held markers that have a narrower writing surface and are easier to use. Typically such marking was done on smooth, vertical, sawn faces.

Graffiti included drawn or incised or scratched crosses on the pillars of limestone, probably survey marks, used to determine locations or calculate areas worked. In some cases this was very neatly done, perhaps not surprising in a quarrying district. Examples include survey crosses which seem to have been introduced in the early 19th century and have been noted within Firs Quarry Areas 2344, 2211 and 2342. As a modern comparison, the Hawkins survey identified pillars by sector and number, and survey points, using spray paint.

Incised graffiti

The earliest incised graffiti found were located below the end of the Avenue and Church Road junction in Quarry Area 2347. This included a pillar with the initials 'JM 1725', the 'J' in the form of a crossed 'I'. A few metres to the south of this a floor-to-roof stone pack contained incised graffiti on two sawn blocks, one with the initials 'RJ', the 'J' again a crossed 'I', carefully placed at eye level either side of the pack. A further sawn block, placed less conspicuously at the side of the pack, was lightly inscribed with 'XX IX', just possibly indicating a date (see Chapter 12, Case Study 2). There is a possibility that the RJ stood for Richard Jones. Some years before 1725 he had been apprenticed to John Pitcher, who had leased a quarry at Combe Down from about 1715. As described in Chapter 5, Jones in 1729 began to work for Ralph Allen, becoming his clerk in 1731, and effectively manager of Allen's stone quarrying and masonry business. A somewhat equivalent graffito, though drawn rather than inscribed, was found in Far East Firs near the Brow. This had the name 'Francis Oliver' and a 'June 1730' date (Fig. 11.7). There seemed to be no reason to doubt the dates inscribed and the two sets of graffiti were important in establishing the dates of working in

Fig. 11.7 Francis Oliver graffito of 1730 in Quarry 2339 (Far Eastern Firs), strongly suggesting a pre-Allen date for the workings there

Fig. 11.8 Tapered pillar in north-east Byfield Quarry (308) with a wide range of inscriptions including the date 1816 (faint and high to the right), giving a fairly reliable earliest date for the sawn face and other activity in the area

these areas in association with the methods used. Apart from surveyor's crosses, inscribed graffiti subsequently was very rare and it was not possible to confidently assign 18th-century dates to any other graffiti found.

Graffiti becomes common

The earliest saw-cut workings are located in areas dated to around the early 18th century but are rare and it was not until the introduction of stone saws in the early 19th century, leaving plane surfaces, that graffiti became more common. One of the earliest examples of written graffiti dated to 1816 (Fig. 11.8) and was noted on a sawn face in north Byfield, although generally graffiti dates ranged from around the 1830s and 1840s and later. Two other early examples of dated pillar faces were noted in Quarry Area 500, for 13th September 1823, and September 1823. The pillar faces have subsequently been used for later graffiti ranging from the later 19th century, and continuing into the later 20th century with visitors from post-quarrying periods leaving their initials and impressions (Fig. 11.9).

Fig. 11.9 Layers of later graffiti on a sawn face in West Byfield Quarry near the entrance

Fig. 11.10 Successful stripping of the Knowles and Prescott graffito using silicon rubber. Far East Firs (Quarry 2363)

In the later periods names and initials of individual quarrymen occur, some recognizable as local, for instance Arthur Knowles and Charlie Prescott, June 19th 1900, who lived and probably quarried near Gladstone Road. Their graffito was used for a (successful) trial of the silicon-stripping method of preservation (Fig. 11.10). Well-known national figures appeared; one dated to 1845 recorded the name and social comments relating to 'Black Jack', possibly the same man as was employed as Queen Victoria's 'rat catcher and mole executioner extraordinary' in the mid 19th century. He was featured in Henry Mayhew's (1851) *London Labour and the London Poor,* which was being serialised in the 1840s (www.wikipedia.com).

At the Foxhill quarry which commenced in the 1870s, an assemblage of graffiti includes one which was simply a list of numbers from 1 to 19. This seems most likely either to be practice in arithmetic, relating to a personal lesson. The modern equivalent used on the Stabilisation Scheme was the 'toolbox talk' with skills being imparted for personal working skills improvement.

Cartoons and sketches

There are a substantial number of cartoon-types or caricatures, presumably often images relating to men then working in the quarry. A recovered item of graffiti from Foxhill Quarry 2380 depicts a quarryman with pipe and hat (Fig. 11.11) and dates from the late 19th century. 'William Bishop 1898' appears above the graffiti and may have been attributable to the image. Several other images of quarrymen, appearing to be in the same artistic style, are noted within the vicinity. The Bishop family lived in the village. The name and image seems likely to be William Vernon Bishop, the great grandfather of Steve Bishop of Combe Down who worked on the Stabilisation Scheme for several years.

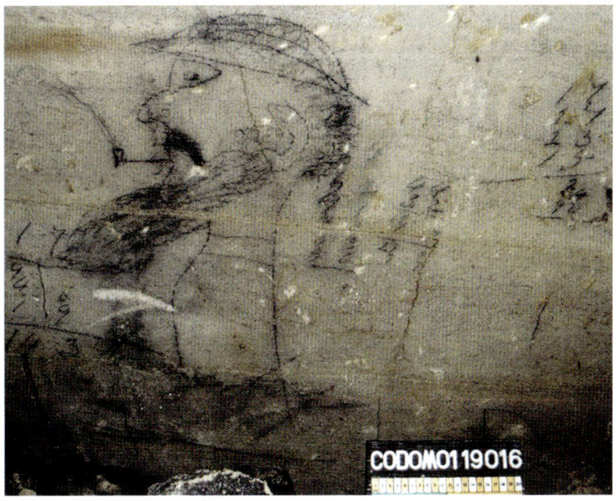

Fig. 11.11 Graffito of man with hat and pipe (Foxhill Quarry 2380)

The finest of all the images found was probably 'The Ship', a three-masted vessel, possibly a type that came up to Dolemead on the Avon Navigation (Fig. 11.12). . The slab it was drawn on was sawn out and preserved. It was clearly a labour of love, with the masts coloured with a brownish clay. The vessel is towing a small tender which, it was initially suggested, might be a Severn trow. However, as depicted, it has three masts which identifies it as a lugger, a more glamorous craft which carried less cargo but made more speed. Goodburn (2009) describes the image as primitive, stylised and of two phases of production combining engraving

Fig. 11.12 'Ship' graffito, depicting a lugger, a type active in south-western waters, including perhaps, the Avon Navigation. Other graffiti in the area is dated from the 1830s (Eastern Firs, Quarry 2203)

Fig. 11.13 The Hadley Arms graffito, dating to after the building of the pub in 1845-46 (Central Firs, Quarry 2202)

with smudged-on pigment. There are three masts, the largest set in the middle or aft (on the right side of the image) with a four-sided rig of sails on each mast, and a long bowsprit set for'ad for a triangular head sail. The apparent letters along the side may indicate cargo owners' space in the hold. Luggers were renowned as fast seagoing ships in the channel and western approaches. Goodburn suggests that, as an association of tunnels and shipping was commonly linked to smuggling in the late 18th to the early 19th century, it might be part of Georgian Bath's black economy. A smoked candle version of a ship of a much cruder application, was also found under the roof in Phase I workings under the south end of the Avenue.

Beer, sex and politics

Beer, sex and politics were as attractive to the common people as they are today. One graffito apparently involves an argument over the respective prices of beer at Cox's and Cluny and/or at the Hadley Arms pubs (Fig. 11.13). Coxes may have been on Bradford Road, possibly the establishment now called the Foresters. The Hadley Arms was built about 1845-6 (Hadley Arms Papers), so the graffito will post-date this. Probably the same individual who discussed the price of beer also commented on the price of the local cigars on an adjacent pillar, though the latter was not accessible to photograph due to a substantial roof fall. The John Sims section may have been added later.

Good Beer $3^3/_4$d per qt'
at Cox & Cluny (?)
Hadley Arms"

Good Cigars
1d/4 each"

John Sims
September 6 1876
Bath

Tanky Elms suggested that quite a few of the gangers or quarrymasters in the quarries were also pub landlords (for example Salter at the later William IV, Davidge at Rock Lane and the Lankasheers who had a Public House at the bottom of Rock Hall Lane, all in the early-mid 19th century)

Fig. 11.14 'Holly' graffito, Eastern Firs (Quarry 2206)

Fig. 11.15 Graffito possibly based on 'Jane' the mid 20th-century Daily Mirror cartoon character (Foxhill Quarry 2380)

and that when pay came the gangers always got their money from the quarrymen first. The quarrymen ran up slates in the pubs, and they sometimes used the pillars in the quarries for reckoning how much they owed (Hall 1984, 92), though this last has not been identified at Combe Down.

Several items are slightly risqué and, perhaps because of this, their discovery attracted attention from the *Bath Chronicle* and *The Times* and *Daily Mirror* national newspapers. 'Holly' appears to be the name of a buxom and well-dressed young lady seen in two graffiti (Fig. 11.14) in East Firs, and what may be a scantily-dressed 'Jane' from the mid 20th century *Daily Mirror* cartoons (Fig. 11.15) occurs in the Foxhill underground quarry. An undressed lady, nicknamed 'The Naked Lady', shows a full frontal figure and details with emphasized mammaries and sexual organs on a large sawn face in Central Firs (Fig. 11.16). The graffito may have formed part of young quarrymen's informal sex education, although its educational value may not have been the original intention. A present-day formal mining equivalent can be found at the Bulga Pit near Sydney, Australia, where, *The Times* informed us, classes for mine staff include advice on sexual foreplay which are available to help keep miners happy (Steele, 2007, 2). The graffito was preserved *intacto*.

The name and date 'Sumsion, 1836' had been scrawled over the Naked Lady, which suggests the face was sawn and the graffito added around the end of John Burgess and Son's working of the nearby Three Acre Quarry. The Grand Canyon's long incline for dumping spoil ended very close by. There is no data for Sumsion, a local quarrymaster working within this area, though Isaac and Giles Sumsion had surface quarries on the north side of North Road, and near Shaft Road.

Another graffito, located within Quarry 518 was a caricature of a top-hatted but otherwise naked gentleman smoking a cigar (Fig.11.17). The cartoon,

Fig. 11.16 The 'Naked Lady', dated 1836 or earlier by a scrawled date over it (Central Firs, Quarry 2201)

Fig. 11.17 Graffito of top-hatted but otherwise naked man, probably early 20th century (Central Byfield, Quarry 518)

probably from the first decade of the last century, may suggest class antagonism and could be directed at a member of the local (quarrying) business establishment. It was within James Riddle's quarry. He was a small independent quarrymaster, and it may, just possibly, have been poking fun at Isaac Sumsion, chairman of the then dominant Bath and Portland Stone Firms.

In 1838 Chartist supporters had met in Combe Down's Widcombe Field for a mass meeting to hear Henry Vincent (Swift 2006, 61) and graffiti suggesting the reading of Mayhew's articles in the mid-century and the above anti-capitalist cartoon may reflect a continuing radical tradition at Combe Down not unusual in mining and quarrying communities, especially in times of economic stress.

Tally graffiti

Another type of graffiti underground was the 'tally tables' or 'tally graffiti' which record the different sizes of the blockstone and the amounts of block produced by the quarrymen. A catalogue with the relevant illustrations of the recovered tally blocks is located on the web site. Quarrymen used ready reckoners, referred to as 'cubing books' to calculate the volume of the extracted blocks (Pollard, 1994, 39).This was especially important to the quarrymen as the amounts of block and the sizes of the blocks produced determined the wages of quarrymen as they were paid by 'piece work', their wages determined by the amounts of blockstone produced and sold.

Fig. 11.18 Tally graffito, early 20th century, near to the West Byfield entrance in Quarry 501

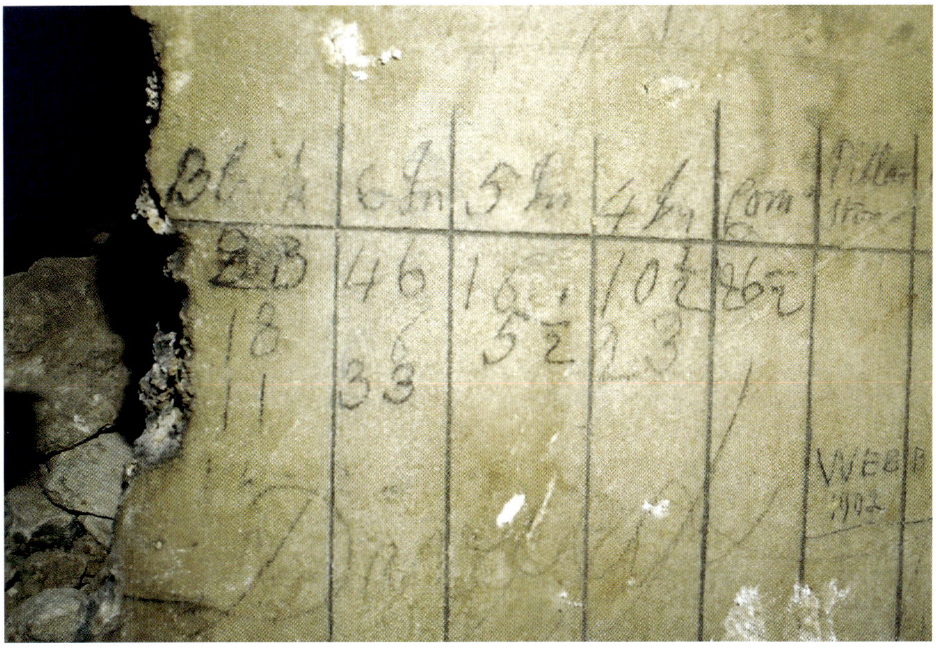

Fig. 11.19 Graffito of tabulated form of types of stone produced or to be produced, c 1820s (West Byfield Quarry 501)

Tally boards recovered were generally in a poor condition and were often smudged where they had lain within the spoil or were otherwise faint and often illegible. Where they can be read, often they include a simple arrangement of vertical columns with a single horizontal line at the top for separate information or title, or they can be arranged in a chequer board way.

Two good examples of the type of tally graffiti seen on the sawn pillar faces appeared in West Byfield near to the entry to Quarry Area 501. The first dates to the early 20th century (Fig. 11.18) thus long post-dating the main primary extraction. The names of several quarrymen including Ellis, Smith and Walters are depicted and there are accounts of the amounts of blockstone that were produced, with other initials of quarrymen. The same pillar face was used several times to tally the amounts of blocks produced and some of the same individuals were also recorded a number of times. The second example dates from about the 1820s (Fig. 11.19) and was set out in table fashion with columns indicating the quantities and the types of blockstone and wrought stone that were produced, including listed 'Block', 'Coping', 'Pillar' and 'Ridge' Stone, and the rows below listing what is thought to be the dimensions of the stone produced, measured in inches. This table would seem not to have been used for any great period of time as many of the columns remain empty. The rows within other similar tally tables often have the name of the individuals that have cut the stone. The tablature form of tally graffiti is less common than smaller lists of tally numbers on sawn pillar faces, although both forms probably indicate similar types of information. The lists have not been fully interpreted but are generally thought to be either dimensions of individual blocks produced or the size of the blocks of sawn material to be recovered prior to cutting of the block, and may sometimes be recorded in (cubic) feet of blockstone.

Tally graffiti were probably most commonly seen on small pieces of sawn block again depicting the names of the quarrymen and the amounts and the size of the blocks produced. An example of an early sawn stone block containing tally graffiti was recovered from a stone rubble pack near to a small contemporary masons' working area, in Far East Firs, Quarry 2352. The area contained numerous sawn blocks that had been stacked ready to extract to the surface via an entry located a few metres to the south. The graffiti block possibly of 1730 date, reads: 'June 7 30 44 cut offs 75? Frank Oliver'. The '44' and '75' may refer to the amount of blocks cut or produced.

Some also note the type of stone that was being produced and examples have included: 'block' stone, and 'coping' and 'roof ridger' stone used for the apex and pitches of the house roofs in Bath. Another includes '46 of Banker', possibly referring to banker masons. One tally board had a depicted hangman image but no additional tally information.

The Dowling graffiti

A few names and initials were repeated: one or several John and James Dowlings got to most places in the quarries, and inscribed their name(s), all with a fine copperplate hand. J. Dowling also left his inscribed signature on the Palladian Bridge in Prior Park (Tom Davies pers. comm.).

A Dowling graffiti dated 1886 was a version of the tally graffiti. It had seven rectilinear boxes drawn on a pillar face with each box containing numbers written internally and externally. Pollard (pers. comm.) has suggested it could be listing the dimensions of the blocks within the boxes, with the two figures in each indicating the dimension of each block in increments of inches, with the number of the blocks produced, written externally on the right hand side of the boxes.

It seems certain that a 'Je. Dowling' was a quarryman who worked in several parts of the Byfield and the Firs Quarries. His signature has also been recorded at the Foxhill quarry complex, and has close associations with the quarryman W. Hole, with his signature being seen in many instances on the same sawn pillar face. The majority of the sawn pillars containing the signatures of Je Dowling and W J Hole were located on pillars which had already been partially worked and were therefore easily reworked. They were generally located on the principle cartway routes. In other words, they were part of a pillar robbing enterprise worked in conjunction with later small-scale quarrying. Dowling and Hole graffiti were also located within the Foxhill Quarry 2381, one reading 'Je Dowling W. Hole June 18th 1882'. The graffiti may indicate contemporary primary quarrying activity as this quarry is thought to have been operating between 1875 and 1884. This is one of their earliest recorded graffiti dates and in the later 1880s Dowling's name is especially common in the Firs Quarry. The latest known piece of graffiti containing both their names is thought to represent pillar robbing activity or a small quarrying operation at the northern end of the Firs Quarry 2211 and includes the names 'Walter Jeo.rd', 'Je Dowling', 'A. Moody', 'W.J.Hole and 'Diamond Jubilee 1897'. The Diamond Jubilee was officially celebrated on 22 June 1897.

Post-quarrying graffiti

Graffiti dating to post-quarrying periods are also numerous throughout the quarries, especially at places underground which were easily reached. The graffiti ranged from the names of individuals and members of school parties to caving clubs and even an army regiment. A Foxhill Quarry (2382) example relates to the Bristol Exploration club and was dated 1944. The date has been useful in determining the period in which the Foxhill quarries were accessible from the surface and are thought to have been infilled soon after it was written. Members of the same club have been recorded within the main Byfield quarry entrance dating to the same period. One of the underground quarry enthusiasts who left his signature was John Richardson, who regrettably passed away in 2008. He and his associates Dave Bishop and Rob Jones passed their 1970s and 1980s photos and surveys to us. Other graffiti from post-quarrying periods record the details of visitors from the 1940s to the present day, including those who assisted in the survey of the quarries before the Stabilisation Scheme began. There were also the unsurprising collections of sexual images and crude profanities. A 'rather hermaphroditic' female form, as she was described by the recording archaeologist, Peter Burge, was found near the Byfield entrance and it is clear that in some respects graffito art has changed very little over the last century or two.

The artefactual evidence

Introduction

In collecting artefacts, which in some cases involved very large numbers of similar items, the aim was to form a representative collection suitable for dating, exhibition or which told a specific story relating to Combe Down miners, families and the wider society. Most were recovered through Stabilisation Scheme operations, and though the (modern) miners were increasingly aware of the need to preserve the contexts of finds, this was not always entirely practicable; it was not, for example, usually feasible to excavate dumps so as to preserve the stratigraphy, so a 'whole dump' context was often all that was possible.

The largest collections were of pottery, metals and glass, clay pipes and wrought stone, with wood in much smaller quantities. These have been examined, partially during archaeological recording, but mainly after its end, with the results presented in a series of specialist reports that are available in full on the website: (www.combedown-stoneminesproject.com). There follows summaries of these reports. The illustrated items in the figures (11.20-27) represent a selection of the full archive of illustrations available on the website. The identifying numbering in each category is that of the full archive.

Metalwork by Ian Scott
including ***Metallurgy*** by Chris Salter

Introduction

Some 581 metal finds were examined (Scott 2009) of which 398 were tools and 286 were related to mining, a further 21 were probably masons' tools. Some tools clearly had nothing to do with quarrying: a scythe, fishing reel, a revolver and darts, probably part of the domestic rubbish discussed earlier.

All the tools were found within the mines. Because the finds were not sealed stratigraphically or buried, but rather found lying discarded within the workings, it is possible that tools and other items had been moved after their original deposition. Some finds were recovered from debris dumped into old workings and in some but not all cases it was possible to date the dumping episodes. Any dating based on location must be viewed with caution and data available on tools is generally too limited to date the workings.

The mining tools, masons' tools and general tools are considered first, and then the finds associated with moving and manoeuvring the quarried stone, and finally equipment, such as lamps, tinder boxes and bottles, used by miners themselves.

There are examples of mining tools from each stage of the process. The most common finds in terms of numbers were chips, wedges and triangular files, but while this reflects the numbers of these tools used, it also reflects the fact that both chips and wedges regularly broke and were discarded, and that files wore out or could be easily misplaced.

Picks

Only five examples of pick heads have been found. The included three large examples which had curved tines of a type used for general (including non-quarrying) use. One example and a broken tine are slightly smaller and are likely to have been specifically designed for quarrying (or mining), for example, for cutting jad slots. The sample of surviving picks, however is too small to give a true picture of the potential variety of picks used or of any possible changes in the forms through time.

Saws and associated objects

A single probable razzer saw has been recovered and at least eleven frig bob saws. The latter range in length from 920 mm to 1850 mm. The width of the blades also varies from a minimum of 245 mm to a maximum of 303 mm. Some variation in width is undoubtedly due to wear in use. The probable razzer has a blade which was probably a worn-down frig bob blade.

The razzer and most of the frig bobs that were recovered lacked handles, but had mounting plates for the attachment of handles. In most cases these were formed from a strip of metal folded and riveted to the end of the saw blade. The outer end of the folded strip was formed into a tube into which the handle could slide. The one exception has its handle plate, which was not riveted, attached to a tongue extending from the back edge of the saw. The attachment then formed a tube into which the handle could slide. The frig bobs were all from Phase 6 locations; this is consistent with its known widespread use from the 1840s (Fig. 11.22).

In addition to the complete or near-complete saws, a number of saw blade fragments were recovered and also vertical saw handles from stone-cutting saws. The handles could have been fitted to frig bob or razzer saws but their form was quite different from the handle fittings on the extant razzer and frig bobs. These wooden handles were attached by tangs to plates which were slotted unto the ends of the saw blade or were slotted and riveted directly to the blade. All were vertical handles and the evidence suggests that they were attached to the top edges of the saw blades. The finding of a pair of handles may indicate that these were used together on one saw, but this is by no means certain. The fact that these handles came from earlier workings dating to Phases 2 to 5, suggests that they *could* be handles of earlier forms of stone cutting saw, predating the frig bobs.

Three saw-setting tools, all of quite different forms, were found, one from a Phase II location. Fifty one files of triangular cross-section used to sharpen large stone cutting saws were found in some numbers and were mainly datable to later phases of quarry working. The files are all very similar although they do vary a little in length.

Wedges and chips

These were the two most commonly found mining tools, used together to split rock.

Sixty-six wedges or possible wedges were recorded. 16 were fragments, mainly pieces from the battered heads of wedges or broken off blade ends. It is clear that they broke frequently in use, for not only were there the broken fragments, but one was split at its head or striking end. Four of the possible wedges are not identified with certainty; they may be too small or their form suggests that they may have served another purpose.

The remaining 46 examples can be divided into five types. Type 1 is parallel-sided but has a square end. The most common form is Type 2, which is parallel-sided and has a rounded end. Type 3 is similar to Type 2 with parallel sides and a rounded end, but differs in that the top is chamfered and slightly narrowed. Both Type 2 and 3 can be found in long and short forms. The shorter examples are about 200 mm and the longer are usually between 230 and 250 mm. Chronologically Types 2 and 3 may have been used at Combe Down at the same time. Price (2008, 96-7) writes that at Whittington underground quarries, near Cheltenham, three types of wedge were found. One form was like Type 1 with the square end. The two other Whittington examples differed from Types 2 and 3 at Combe Down in having rough points instead of rounded ends. Type 4 had a narrowed and chamfered top, with a square ended blade and was slightly narrowed. Type 5 was large with a round end, narrowed with a chamfer to the head. These two latter were each represented by a single example. It is possible these variations were local, perhaps even to specific suppliers, and that the last two types were less commonly used at Combe Down.

Chips were by far a more common find, with 168 examples being recovered. They are remarkably uniform in appearance, probably because they are very plain and simple in design. Many of the chips were broken, the majority of them being split down the middle, which means that the overall length of many of the chips cannot be measured. Less than half could be measured across the full width.

The longest example measured 170 mm, and the shortest 95 mm. The average length is 132.6 mm and the majority of the measured chips (n = 87) is between 120 mm and 140 mm long.. Their width ranged from 55 mm to 98 mm; the average width is

78.3 mm, and most (n = 7) are within the 66 mm to 86 mm range, with only two measuring between 55 and 57 mm and six between 90 and 98 mm. Chips were predominantly found in Phases 3 to 5. Nearly half the examples are unphased or are of uncertain phasing.

A small number of sledgehammers were found, four complete and one fragment, ranging from 154 to 170 mm long and between 80 and 88 mm wide. A further small hammer was found in a domestic dump context.

The tools used for cutting and handling freestone blocks

The following is a description of the methods used in the 19th and early 20th centuries, and a selection of illustrations of tools recovered from the quarries. The full photographic archive is available on the website: (www.combedownstoneminesproject.com)

The first stage was for a picker to prepare a breach at the working face, that is creating a slot at the top of the stone about 6 in (15 cm) thick to a depth of about 5 ft (1.5 m) using picks of different lengths. Then the sawyer would use a razzer saw with a shallow blade to start the downward cuts of the stone blocks. Once a good depth of cut had been achieved a larger frig bob saw with a blade up to 8 ft (2.4 m) long would be used. Water was dripped into the saw cut from a can for lubrication.

Once the saw cuts had been made, the first block (the wrist stone) was removed by using wedges and chips (also called feathers) to break between the beds. The block would then be dragged out using a chain and Lewis bolt fixed to a Lewis hole cut in the front face of the block. The back of the block would then be cleaned up with a scappling axe. The back of the cut where the wrist had been removed would be cleaned up using a pick, then the remaining blocks could be removed by sawing down at the back with a razzer saw, and then breaking the blocks from the bed with wedges and chips.

The cut blocks could be manoeuvred using pinch bars and crow bars, and trimmed and tidied using scappling axes. To lift and transport the blocks Lewis bolts were employed. A Lewis hole was cut into the block into which a Lewis bolt was inserted. The block could then be lifted by a crane onto a wagon or truck or dragged by chains and winches.

A selection of tools found in the stone mines is illustrated in Figures 11.20-22, including:

Wedges (SFs 35, 530) of varying sizes for splitting blocks from stone beds in combination with chips or feathers

Chips (or **feathers)** (SF 455) used in combination with wedges to split stone from beds.

Sledge hammers (SFs 93, 624, 807) used with wedges and chips for splitting blocks and with chisels for trimming block and cutting Lewis holes

Frig bob saws (SFs 445, 196), for cutting away large stone blocks, maintained using

Saw setters (SF 5998)

Scappling axe (SF 614) used for trimming and tidying stone blocks

Pick axes (SF 779) used in preparing the cutting operation and for trimming and cutting faces

Spokeshave (SF 870) for shaping a moulding on a stone block

Three-legged Lewis bolt (SF 474) used for lifting blocks of stone by crane using –

Chain and hook (SF 1029) dragging blocks by a winch

Horse bit (SF 519) for occasional use of horse and cart for transport underground

Small oil lamp (SF 281) Typical source of illumination underground

Comment

Other than chips and wedges, the number of iron tools found was remarkably low. Partly this may be because wedges and chips and files are fairly small when compared with sledgehammers, picks and crowbars. A further possible reason has been put forward during a conference discussion, in respect of the few hammers and large numbers of gads (pointed wedges) found in mines near St Marie-des-Mines, France: the gads, which required regular smithing, belonged to the masters and the hammers belonged to the men. Since a substantial number of chips were used to tighten roof-supporting sprags, this may suggest wedges and chips were supplied by the masters, and the hammers crowbars and picks were provided by the workmen themselves, and accordingly taken care of more assiduously.

Metallurgy of wedges and a saw by Chris Salter (Oxford Archaeological Research Laboratory)

Seven of the wedges and a fragment of a frig bob saw were examined. All the wedges were of phosphoric wrought iron, indicating the original iron ore was derived, most likely, from a Coal Measure ironstone source, rather, than, for instance, the nearby Forest of Dean or more distant Furness District. The consistent levels of sulphur in all samples points towards coal-fired smelting, a process initiated in 1709, at the cast iron stage rather than the charcoal blast furnace process. Such iron then required refining to produce wrought iron.

There had been no attempt to harden the faces or 'cutting' edge by carburizing. The use of phosphoric iron would have made an attempt to do so difficult, as phosphorus slows the diffusion of carbon into iron. In most cases, the hardness of the combination of phosphorus content and work-hardening would have improved the properties considerably above those that would be obtained from a plain wrought iron.

Two of the wedges which came from earlier (Phase I and II) contexts had a higher level of potassium which probably derived from charcoal ash used in the finery furnace, where the iron and charcoal were placed together. A third sample was indeterminate. The other four, with much lower potassium probably derived from the coal fired puddling process of *c* 1783-4, where the iron was kept in a separate hearth from the coal. The wedge composition and production process for the two 'early' samples are thus consistent with an origin of prior to soon after 1783-4 when Cort's puddling and rolling process begun to predominate because of much reduced costs. They may thus belong to Phase I, II or III though a later re-use of the metal is not entirely precluded. The other four samples will all date after the introduction of puddling.

A number of the samples showed one of the problems with the very directional nature of wrought iron, which is cracks and corrosion running along the planes of high slag inclusion density. Mostly this was in the form of cracks running back from the blade. However, one sample suffered transverse cracking and subsequent corrosion along the slag inclusions exposed. The transverse cracking is likely to have been a result of the heavy mechanical misuse that this sample has suffered. It had been bent in the plane of the wedge, no doubt due to the wedge having been used as a lever rather than a wedge.

Given the size of the sample it is difficult to determine if a phosphoritic iron had been deliberately selected for the manufacture of these wedges, or if it was used because it was the cheapest local supply to the manufacturer. The use of phosphorite iron would have certain mechanical advantages as long as the wedges were used in the manner intended (in compression).

The fragment of frig bob saw was the only artefact made of steel which was studied. Two possible interpretations are offered, the first that it may be a high carbon shear steel, which was commonly used in edge tools, and produced by piling and forging blister steel produced in the cementation furnace.

The second and possibly more likely interpretation offered is that it was the product of a liquid steel-making process (cast steel) from iron produced in a coke-fired blast furnace, which must be after 1709, but as the iron from the early coke-blast furnace was not considered suitable for steel making the saw is likely to have dated from much later. The fact that there was a high proportion of silicate and other non-sulphide inclusions present and that there was not enough manganese to entirely prevent the formation of iron sulphide would suggest a mid to late 19th-century date for this sample which would possibly have been made from a form of blister steel produced from puddled iron, remelted to give it a more homogenous composition. Use of the much higher quality crucible steel was probably too expensive for such a lowly artisan's tool.

The relative softness of the rock would have made harder grade materials unnecessary for successful use as wedges. Cast steel was introduced in quantity in the late 19th century, but until then it would have been substantially more expensive than the traditional wrought iron. As an explanation for the absence of steel wedges in the last phase of working, it is possible that any steel wedges introduced and used were more highly valued, and less damaged and thus more cared for.

The use of cast steel for the frig bob was probably not without an alternative, especially in the earlier part of the 19th century. Converting to steel by carburisation. would have involved the cementation process to convert wrought iron to a steel in which harder, higher-carbon layers provided a cutting edge. Cast steel at an earlier date might also have been derived from the Huntsman process, but this was comparatively very expensive, though it began to be used for hard rock-cutting tools about 1840. It is more likely that either Bessemer of Siemens-Martin steel, both bulk liquid steel processes, was used for the late 9th-century saw.

The use of saws in the quarries seems to become fairly frequent around about 1820 and common by the 1840s. Possibly this was due to searching for new markets by the Sheffield industry after the end of wartime demand after 1815. The use of different forms of iron and steel for quarrying tools is a suitable subject for further research, preferably on better stratigraphically secure examples than those used here.

Illustrated finds (with archive catalogue numbering)

Fig. 11.20 Splitting the stone from the bed

39 Wedge Type 3, large with slightly tapered blade, round end; Narrowed and chamfered at the top below battered head. Fe. L: 363 mm; W: 83 mm. Sf 530, Mine area 514 or 2221, Phase 3 or 4. [ID 251]

51 Chip, complete. Fe. L: 124 mm; W: 78 mm; Th: 12 mm. Sf 455, Mine area 2337. Phase 4. [ID 228]

52 Chip, complete. Fe. L: 123 mm; W: 76 mm; Th: 12 mm. Sf 455, Mine area 2337. Phase 4. [ID 229]

35 Wedge Type 2, complete, with round end. Fe. L: 217 mm; W: 83 mm. Sf 623, Mine area 2217, Stub C15, context 1200. Phase 4 or 5. [ID 478]

59 Small sledgehammer head, oval eye. Fe. L: 165 mm; W: 57 mm. Sf 93, post quarrying from Shaft 4031 (G2 supply shaft) context 4034. [ID 42]

57 Sledgehammer head with circular eye and body tapered to a narrow striking face. Fe. L: 168 mm; W 86 x 85 mm. Sf 624, Mine area 2359, Stub O10. Phase 3 or 4. [ID 462]

58 Sledgehammer head with circular eye and broad striking face. Fe. L: 168 mm; W: 88 x 76 mm. Sf 807, Mine area 917 tipping? Stub R8. Unphased. [ID 487]

Fig. 11.21 Stone cutting and shaping

11 Frig bob, long blade. Complete blade with curved worn cutting edge. Now lacks handle mount, but has at least two rivet holes. Fe. L: 1795 mm; W: 265 mm. Sf 445, Mine area 2379, localised quarry, context 1002. [ID 348]

18 Pair of vertical saw handles probably from a two-person saw rather than from frig bob or razzer saws. Wooden handles attached to tanged iron fittings and secured with collars. Fe and wood. L: 290 mm.

Sf 196, Mine area 2217, context 1089. Phase 4 or 5. [ID 20]

27 Saw setting tool, pointed handle or tang and round bade with 3 keyhole shaped slots. Fe. L: 182 mm; W: 78 mm. Sf 598, Mine area 2200, Stub A30. Phase 2. [ID 469]

64 Scappling axe head, complete, with part of handle in situ. Fe. L: 314 mm; W: 125 mm. Sf 614 Mine area 2365, Stub M14, Context 11022 Phase 4 or 5 [ID 476]

1 Jadding pick, with short stout tines. L: 500 mm. Sf 779, Unlocated. Unphased [ID 557]

89 'Spoke shave', with U-section central blade. Tapering pointed arms, one with right angle bend at the end. Fe. L: 320 mm;. Sf 870, Mine area 2369, Stub D29 Set (7) LH side. Phase 2. [ID 430]

39

51 and 52

35

59

57

58

0 250 mm

1:4

Fig. 11.20 Metalwork: Splitting the stone from the bed

11

0 500 mm

1:10

18

27

64

1

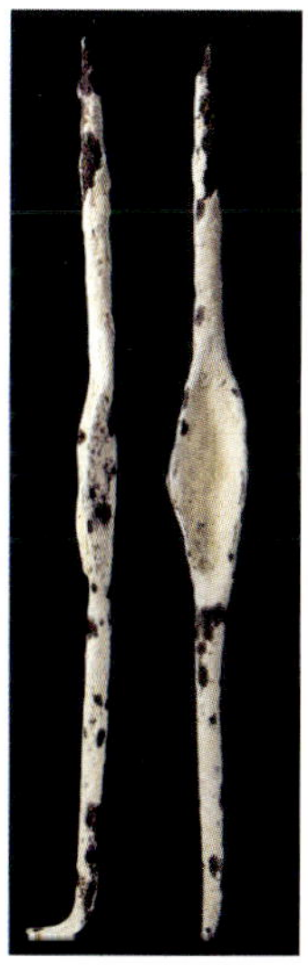

89

Fig. 11.21 Metalwork: Stone cutting and shaping

0 250 mm

1:5

Fig 11.22 Haulage and lighting

128 Bridle bit with hinged two-piece mouth bar with S-shaped side bars. Fe. L: 190 mm; 128 mm. Sf 519, surface debris from Quarryman's Court, Stub E4. Unphased. [ID 254]

160 Crane hook with attached length of chain with 6 extant links. Five smaller links (L: c 80 mm) and one longer link (L: 100 mm). Fe. Hook L: 210 mm; W: 135 mm. Sf 1029, Foxhill, Mine area 2380, Road. Phase 6. [ID 530]

169 Three-legged Lewis, comprises hooked yoke, one side leg, the central leg and the connecting pin. One leg missing. Fe. Overall L: 614, Hook L: 320 mm. L: 614 mm; 186 mm. Sf 474, Mine area 2215, Context 1127. Phase 4 or 5. [ID 354]

191 Small hand held tinware oil lamp, comprising sheet metal saucer with sloping sides and holding a low cylindrical oil reservoir with domed lid. There is a small hole in the lid for the wick. The lamp has a sheet metal loop handle. Fe. D of saucer: 104 mm; L with handle: 124 mm; Ht: 30 mm. Sf 448, unlocated. Unphased. [ID 465]

192 Small tinware benzolene lamp, with simple loop handle. Domed resevoir/body with slight moulding. The central hole lacks its threaded collar. Slightly recessed base, joined to body with a soldered and rolled seam. Fragment of cloth (wick?) inside reservoir. L: 110 mm; D: 98 mm; H: 42 mm. Sf 281, Shaft Road Quarry Mine area 2372, Context 5000. Phase 6. [ID 138]

128

160

191

192

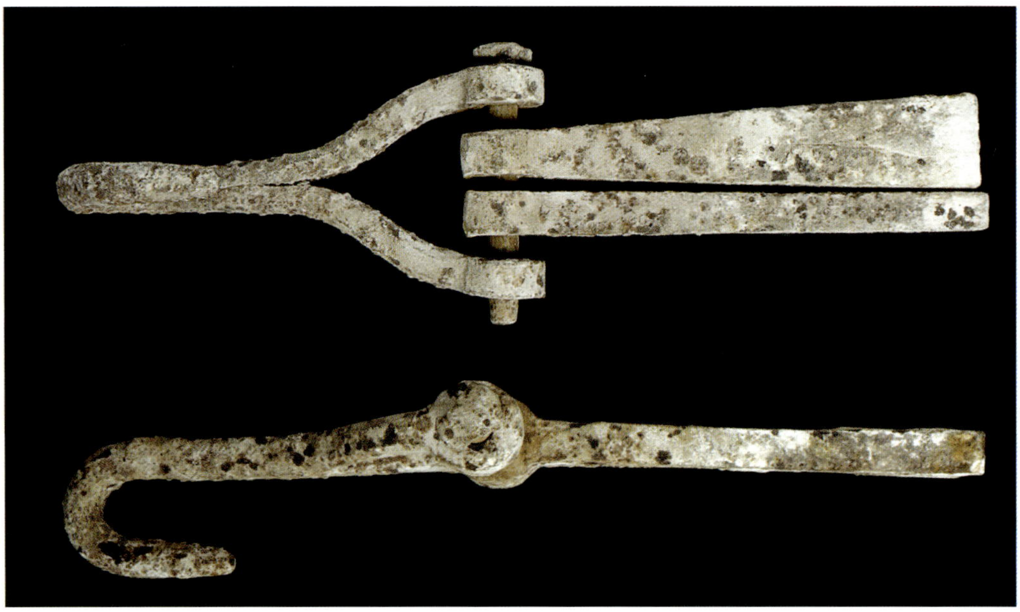

169

0 250 mm

1:5

Fig. 11.22 Metalwork: Haulage and lighting

Wood finds and identification of species
by D M Goodburn

Timber examined was relatively young, between 100 and 250 years, for which the humid, cool and dark conditions had aided preservation (Goodburn 2009). However many examples were either reduced to heartwood, or to a peaty mush. Proximity to metalwork aided preservation of the adjacent timber. A total of twenty pieces of worked wood or composite assemblies such as barrow wheels were examined of which eleven were sprags or planks. Dendrochronology proved impossible for the young coniferous material, and normally datable species such as ash had too small samples. Species identification (Challinor 2009) showed three sprags were of pine (Pinus sp.), two were of spruce (Picea sp.), and another was elm (Ulnus sp.).

Tools

A simply made wood gauge was made from a plank of softwood measuring 0.45 m (18 inches) long with a series of 1 inch-long notches set in its side at 8, 10, 12 and 14 inches. Its likely use was to measure stone sizes.

A partially decayed wheelbarrow, little more than a 'shadow' except where the wood was close to iron parts, was found in 19th-century workings (Fig. 9.16). Complete, and made of elm, it would have been about 1.76 m long and 0.7 m wide with a wheel diameter of 0.42 m. The woodwork was reinforced by iron tie rods, one at the front end and one at the rear of the box. Smaller strap sections were found at the wheel end of the nave.

Oak and iron cart shaft

This was a cruciform of two pieces of oak, *c* 1.78 m long and 0.4 m wide. The main beam is a box-quartered section of oak with bevelled corners, protected with thin iron bands fastened on with iron tacks. One end was broken. The cross-piece was plank-like with rounded ends,, set 0.25 m from the rounded and iron-reinforced end of the beam in a through-mortice. The most likely function of the shaft was for controlling a two-wheeled handcart (Fig. 9.28).

Barrowing boards

The largest was 4.23 m long and 0.28 m wide, 0.07 mm thick, tangentially sawn along lines of three inches. It was cut from a slow-grown tree, one of the spruces (Picea sp.), and had about 150 growth rings but no clear sapwood or bark. This precludes a local origin, in favour of alpine or northern. The second plank was a little wider at *c* 0.3 m, 3.68 m long and slightly thinner.

Shoring timbers – sprags

A substantial proportion of the shoring timbers were of softwood, which, if imported or if locally grown would probably be the Scots pine, *Pinus sylvestris*, only minimally trimmed to log form using an axe. The largest diameter sprag was 1.9 m long and 0.21 m diameter, which had approximately 50 growth rings. A diameter of 0.16 m was more typical, down to a minimum diameter of 0.11 m, a rare deciduous example. Lengths were between 3 m down to 0.85 m. Rule-of-thumb dimensions for pit props in the mid 20th century were one twelfth the length down to three inches (75 mm), which suggests these were comparatively stout timbers.

The fast-grown rather rough form suggests they may have been locally grown, especially since transport of such low-value material, second and third quality, only better than firewood, would have been expensive. Scots pine, known in England as fir, was unusual in south-west England in the 18th century and these trees were at least 50 years old at the time of felling. They may thus be partly from the extensive planting of firs by Allen in the 1730s onwards (hence the name The Firs). The wood used in the Phase III and IV quarries included the smaller diameter, more knotty logs derived from the tops of trees, and even the largest log was clearly not the highest value 'butt log' (first log cut above the stump).

Two of the smaller sprags were deciduous, and it was perhaps surprising that more use was not made of such offcuts from local timber yards. One sprag, 1.09 m long and 0.085 m in diameter was slightly crooked and made of elm. A further example of a sprag was distinctive in being manually pit-sawn, 0.94 long and 0.11 m thick. It may be a product of an estate saw pit.

Worked and architectural stone by Ruth Shaffrey

Worked stone

There is a varied and extensive assemblage of worked stone, which has been divided into miners' equipment and worked stone intended for removal to the surface (Shaffrey 2009). It is described by artefact class and, where appropriate, in relation to individual quarry areas and phases. Quarrymens' equipment comprises storage boxes, candle holders, hones and 'graphite' markers which had been found throughout the quarries. Worked stone mainly comprises architectural stonework, internal and external domestic items such as sinks, water containers and staddle stones, or ashlar blocks.

Boxes

A total of 18 stone boxes were recovered, from all phases except Phase VI, ranging in size from small unfinished boxes, to more standard candle boxes and a single large box made of slabs (Fig. 11.23).

Most are fairly crudely worked, with evidence for the use of saws and scappling axes. No two boxes have identical dimensions and, although their overall range is limited, with the majority measuring between 300 and 400 mm in length and 200 and 300 mm in width, the variability suggests there was no template for their manufacture. They may have been made on an ad hoc basis. The majority of the boxes are comparable to some found in the Whittington quarries near Cheltenham. These were probably candle boxes (Price 2007, 91), used to keep candles and the means to light them (flint, tinder etc) dry and away from vermin. The majority of the Combe Down examples are of an appropriate date, predating the invention of the match in 1829, after which dry storage for tinder gradually became irrelevant (Price 2007, 86). They would only have been effective, however, if they were sealed with lids, of which only one was recovered, suggesting that the primary purpose of the boxes was not to keep items dry. An alternative explanation is that they were intended to store small tools or a quarryman's or mason's belongings.

One box merits particular attention. Unlike the others, which were each fashioned from a single piece of stone, it comprised six individual slabs positioned together to form a single box. The two end pieces are the full width of the base and the longer side pieces fit in between. This was clearly constructed to provide a storage place and the presence of a lid, albeit only covering half the box, suggests it was intended to protect the contents, possibly from rats, if food were stored inside it.

Candle holders

Two stone candle holders were recovered from Phase III/IV areas in Far East Firs, Quarry Area 2330. One has a portable paddle-shaped holder. This has a socket in the centre of the paddle, presumably to secure the candle, but the entire surface of the holder was covered with a thick wax deposit, which has been left *in situ* (Fig. 11.23). The other is of squat cylindrical shape, offering no protection to the hands from the hot wax and was thus less practical for carrying around. It is likely to have been positioned in a semi-permanent location on a ledge. It is reminiscent of the single stone candle holder found in the Whittington quarries (Price 2007, fig. 66: 7.207).

Markers

Three 'graphite' markers were recovered. All demonstrate wear consistent with mark-making or writing and were probably used for tallying and similar purposes in the mine. One is from a Phase IV/V location, the others are unphased. David Pollard suggests these were made of a soft coal found in local mines (pers. comm.).

Hones

Two sharpening stones found in the mines would have been used to sharpen tools. One is an unmodified Pennant Sandstone or Old Red Sandstone slab with a deep groove from sharpening. Other slabs of the same sandstone were recovered but had not been used for sharpening, although it seems likely that they were intended for use in this way. A single whetstone is a smaller item, smoothed on its faces from sharpening.

Comment

With the exception of the boxes, which survive in good numbers, the assemblage of miners' equipment is relatively small given the overall size of the quarry complex. However, as published examples of most categories are rare, these are significant finds. The number of storage boxes suggests these were an important everyday item in the quarries, although their size would have made them easier to find than the smaller items.

The general lack of sharpening stones suggests that sharpening of tools was not generally carried out underground. However, the evidence indicates that saws were sharpened underground using a large stone with a longitudinal cut to hold the saw. Many saw-sharpening files and three saw-setting tools were found. It is likely that these may represent some relatively ad hoc sharpening, although as a rule the saws were systematically sharpened each week, and time for this process was allowed, for instance on a Saturday, for this task. The small numbers of stone candleholders suggests they were not particularly common.

Architectural stonework

The following summary report indicates the range of stonework found (Fig 11.24) and the phasing (where identified) of quarrying involved, with the conclusions which can be drawn from it.

Moulded stonework

A total of 24 pieces of decorative stonework were recovered, including pieces that may have functioned as elements of architraves, cornices or string courses. This type of stonework was found in all phases of quarrying activity. A single console support with cyma recta moulding was also identified. The mouldings survive in various stages of completion, including some where two pieces are still joined together in mirror image of one another. The profile of all the mouldings is complete and it is generally other sections of the stone that have not been finished; for example, on those that are rebated (see below) this section was removed last. The quality of workmanship varies significantly between pieces, with some demonstrating very few flaws and completed but abandoned, presumably because they were subsequently broken. Other pieces are unfinished but with clear lack of accuracy in the detail, which may have rendered them of insufficient quality for use.

There are four distinct groups of profiles within the architectural mouldings. The first group of six pieces have the distinctive S-shaped bolection profile; five of these were found in Phases III to V areas while a single slim piece with a partial S-shape moulding was the only piece found in an area attributable to Phase I. Some have a deep square rebate on one side, a common feature of sash window embrasures in Bath, its function being to secure the weights, with the other half of the box being constructed by the joiner. The mouldings with the rebate were probably intended for use as window surrounds, while those without may have been architrave round doorways or possibly for an internal feature such as a fireplace. Although the bolection profile was used on some prominent buildings up to about 1720 it would have been considered old-fashioned well before Ralph Allen controlled the underground workings. Thus with one exception all the bolection profile mouldings were found in much later phases of the quarries than their usual period of use.

The second group has a more delicate and complex profile forming an approximate sequence from the top down: cyma recta; fillet; cyma recta; corona; small overhang; fillet; cyma recta. Such profiles have been recorded in use as cornices and also for architraves above doorways and round windows. One item is from a Phase II or III area of the quarry (913) and two are unphased, but this profile superseded those of bolection type. Early examples in Bath date to the early part of the 18th century, for example at Queen Square, built between 1728 and 1736; much later examples are also known – for example, Green Park Station, built in 1868-9.

A third group has four pieces of simple shape with concave moulding in one corner and with the opposing edge either flat or chamfered. These could have been used either as a simple string course or the cornice at the top of a wall such as that observed on Barton Buildings, Bath, built in the 1760s. Two of these are from Phase IV or V tipping in Far East Firs, one is from a Phase I context in south Byfield and the other is un-phased.

The fourth group comprises items with a simple sequence including either cyma or ovolo-shaped moulding and all rebated on the adjacent side, suggesting these were intended for use as sash window surrounds. All three were recovered from Phase II workings in Far East Firs. A small number of other pieces are associated with window architecture, including two window mullions and a small number of other shaped pieces of stone that may be sills and window surrounds. Of the two probable window mullions, one has ovolo moulding and one has chamfered moulding. The mullion with ovolo moulding was found in Central Firs Quarry 2347, Phase I, and is of a generally early type. Although mullioned windows were gradually superseded by sash windows, they continued to be used in the rear or side elevations of dwellings in Bath and do occur on late 17th-century buildings in Bath. A single pillar from Phase IV Quarry 504 may be a rough-out for multiple mullions having been made into a chamfered mullion shape and then abandoned. A number of stepped blocks were probably intended for use as sash window sills or simple door frames.

Columns/pillars

Three column capitals were recovered from the mine. The two larger pieces were found together in Quarry Area 2347 belonging to Phase I. They do not present a known profile. In addition, three partly worked pillars or slim column shafts were also found. Each had been roughly fashioned into a cylindrical shape with a scappling axe. One is plain, the other has two sections of differing diameters and the third has two parallel grooves partially cut into the stone.

A single large pillar has a square cross-section with a rectangular socket on each of four faces, each measuring approximately 6 x 3 inches (180 x 90 mm). These may represent recesses for timber joists, although it is not clear in what sort of structural situation a pillar such as this might have been employed.

Balusters

A total of 42 baluster segments were recovered, the vast majority (37) from pre-Allen Phase I quarry areas. Of these, 27 were found in south Byfield, Quarry 910. A further four are from Phase II areas of the far southern tip of the East Firs Quarry (2348). The remaining baluster sections were found scattered throughout the mine. All that remains of the vast majority of these are the end pieces, most with the lathe socket remaining in the base. Some survive only as the square end block with a short section of profile. On others only the circular section and part of the shaft survives. In one example the central portion survives without the socket and one example is fully perforated.

Ridge stones

One of the most common components of the worked stone assemblage are the varied lengths of angled ridge stones and their wasters that were used to cover the ridges of roofs. Many of the waste pieces reveal consistent angles, as would be expected on long terraces with uniform roof lines. However, there are examples with shallower angles, which show that ridge stones were put to use on different types of roofs. Ridge stones are also represented by marked-out blocks, not yet separated into individual pieces, which provide a good view of the manufacturing process followed in making the ridge stones. Of two marked-out blocks, one has had the ends of each coping stone individually shaped prior to separation from the main block and the other still has vertical edges although the lines of each block are marked out.

Silt traps

A total of eight large items with vertical sides and angled ends were recovered from Phases I to V contexts. These silt traps were used in the sewerage system in Bath from the 1720s onwards. With the exception of one extremely large example, all the silt traps are of very uniform dimensions, varying from 350-400 mm length, 190-240 mm depth and 180-320 mm width. Silt traps were found in Byfield and all the Firs areas, and a group of them in various stages of completion was found in Far East Firs. All are in their final exterior form but they are at various stages of manufacture, with one interior not excavated, two partially excavated and the fourth fully excavated and crudely finished. It is not clear why these silt traps were abandoned partway through manufacture, as there are no apparent faults in either the objects themselves or in the stone.

Miscellaneous stonework

These include internal and external fixtures as well as items of indeterminate function. Half a sink in two surviving fragments was recovered from Quarry 2001 although it was probably associated with Central Firs Quarry 2368. It is a shallow sink with internally sloping sides, one square and one rounded corner and a base that slopes down towards the missing end. Another trough is flat-bottomed internally and thus not a sink but may have been intended as a water container.

Other items include a ball, possibly for use as a marble and half a circular vessel measuring about 230 mm diameter by 200 mm high. It is quite crudely made and, although it has a lump on one side suggesting a rib and thus could function as a mortar, it would have been too soft for that purpose. It may have been intended for some use in the quarry rather than as an export, or perhaps as a decorative vessel. Three large circular stones were recovered, measuring approximately 600 mm diameter with one flat face and one slightly convex face leading up to a flat surface. The most likely interpretation is that these are staddle stone caps, as they have no obvious architectural function and are of an inappropriate material for millstones.

Ashlar blocks

Over 80 ashlar blocks were recovered from the mine for analysis. These vary in size and finish and, although summarised here, were recorded in full on the archive database. A single block retained a possible Lewis slot (SF 681) (see Chapter 9). The ashlar blocks show no uniformity of size, varying from pieces of 'small ashlar' to much larger blocks, presumably intended to be cut down further. Some blocks retain evidence of this intention in the form of a scored line along the centre, marking out the intended division into two smaller blocks. Others have some marking-out for further development, and one is scored ready for cutting.

The ashlar has been worked and finished to varying degrees with evidence for a combination of sawing, scappling and combing. A total of 45 blocks preserve saw marks on some or (rarely) all faces, as the principal method for producing ashlar. However, 33 sawn blocks also demonstrate the use of a scappling axe on one or more faces. Although very few surfaces survive sufficiently for the width of the scappling axe to be determined – blade widths of 2, 2.5, 3 and 4 inches could be identified. Fourteen blocks were also finished on one or more faces with a comb (mason's drag), of which numerous examples were recovered.

Among the ashlar pieces are a number of oddities. These include three small ashlar blocks with curved tops that serve no obvious structural function. There are also triangular-shaped blocks and an arch-shaped stone.

Conclusion

The assemblage of worked and architectural stone from the Combe Down underground quarries provides a valuable insight into quarry operations, particularly to what degree the stone was worked underground and how the process was organised. The variety and quality of the finish of the worked stone indicate that a whole range of stone working occurred underground including sawing, scappling, combing and the production of more detailed mouldings. The most intricately detailed stone is found in the form of architrave, and these date to most phases of the quarries' use. Although the underground quarrying areas were operated in different ways while they were open, these pieces are a clear indication that some stone was worked to a near-finished state underground.

In addition to revealing the extent of work that was carried out underground, the evidence also helps our understanding of how it was organised. Some areas of the underground quarries produced notable concentrations of worked stone providing the solid evidence that masons worked underground. The recovery of broad ranges of worked stone from single working areas indicates that where masons worked underground, they did so on a multitude of things and that the manufacture of different items was not segregated. Individual categories of worked stone were not generally concentrated in any particular quarry areas, with the exception of the balusters, which demonstrated a significant concentration, a large group of coping stones and wasters and small collections of other items such as the four silt traps found together. Analysis of the petrology of the utilised stone indicated very little variation and it does not appear that selections of stone types were made for different functions (Palmer 2009).

All the stones in this assemblage are either waste products or items that were either discarded or abandoned and in some cases, the reason for this is obvious. The stone of a few pieces was found to be flawed during manufacture. Some of the more detailed mouldings were badly made, with the long lines being poorly aligned for example. Other pieces

Worked stone

39

45

Architectural stone

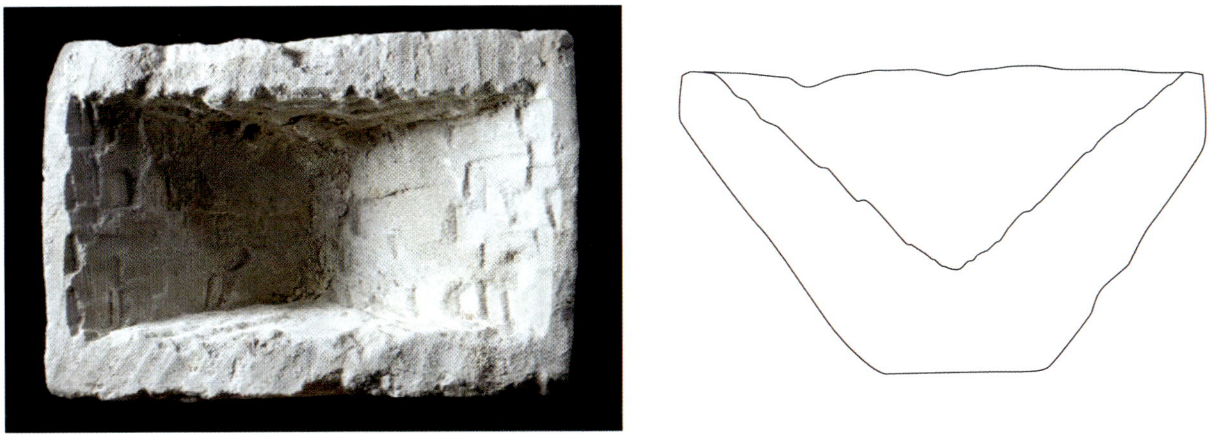

36

8

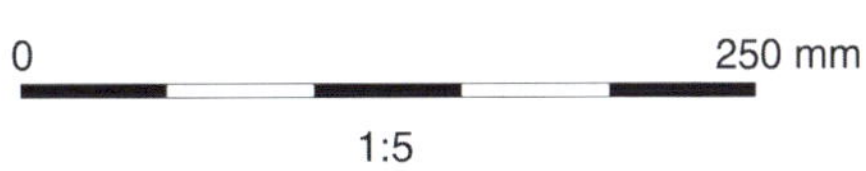

Fig. 11.23 Worked stone and architectural stone

Architectural stone

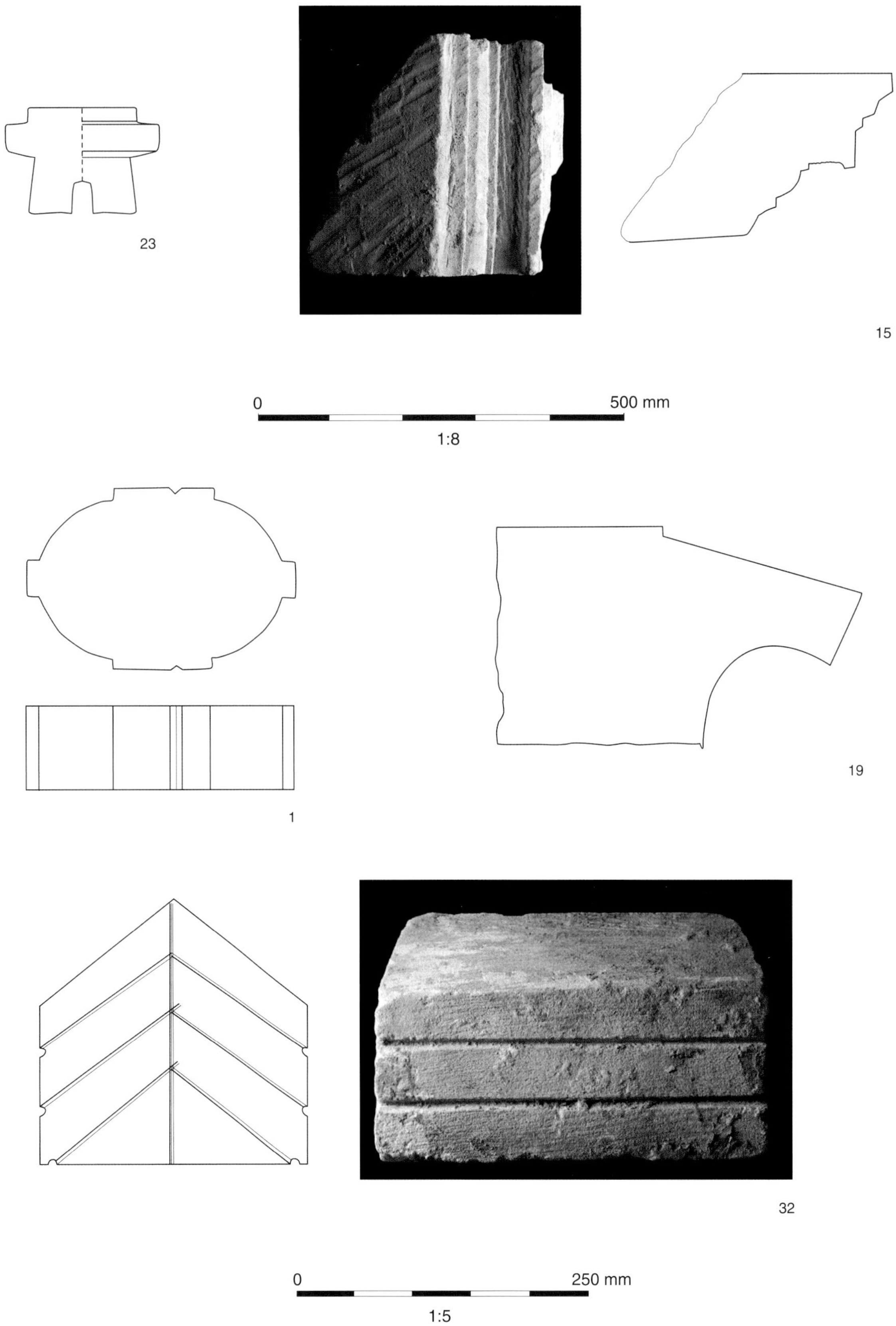

Fig. 11.24 Architectural stone

were probably broken or damaged during manufacture and then discarded. Some items, notably the silt traps, show no obvious reason for discard. This discarded stonework encompasses a broad range of types from simple ashlar blocks to complex architectural profiles. The vast majority of the stone is architectural as would be expected; however a full compliment of architectural stone is not represented. Simple pieces such as voussoirs are not present nor are any examples of distinctive ecclesiastical architecture. There are no examples of detailed carving such as friezes, scrolls and generally very little evidence for any sections of columns. It may be that more complex architectural features and carvings were carved on the building site rather than at the quarry. Certainly the finishing of pieces at the quarry meant there was more chance of them being damaged in transit, a risk no doubt increased for more complex pieces. In addition, higher profile, non-speculative projects may have had other controlling measures in place leading to stone being worked on site. The more skilled masons who were responsible for carving intricate pieces may not have been connected with the underground working at all, and employed directly by individual projects to work on site.

Catalogue of illustrated items (with archive catalogue numbering)

Figure 11.23: Worked stone and architectural stone

39. **Candle Box.** Long slim box, crudely formed. Wall thickness
15–30 mm. Measures 432 x 178 x 152 mm. SF 606. Phase 1 late 17th early 18th.

45. **Candle holder.** Shovel-shaped candle holder with oblong shaped handle and flat upper surface covered in thick layer of wax which still has indentation from where candle was positioned roughly in the centre. Measures 160 x 72 mm maximum width x 40 mm maximum thickness. SF 608. Phase 3 or 4 1768-1800 or 1800-1833. Quarry 2330, Far East Firs

36. **Silt trap.** Fully excavated with internal dimensions of 300 x 155 x 120 mm deep. Fashioned with the use of a scappling axe or hack except on the top face where it had been sawn. Externally it measures 360 x 240 x 210 mm deep. SF 662. Phase 4 or 5 1800-1833 or 1833-1867.

8. **Probable sash window surround.** Bolection moulding with rebate. Ashlar finish. Measures 265 x 254 x 150 mm. SF 35.Phase 3 1768-1800

Figure 11.24: Architectural stonework

23. **Baluster.** Partially worked. Square base with section of baluster above. The square corners are chamfered up to the circle. Flat on top with roughly finished edges to circle. The socket measures 24 x 24 mm. Measures 210 mm diameter across shaft and base x 160 mm high. SF 965. Phase 4 or 5 1800-1833 or 1833-1867. Quarry 2219

15. **Architrave/decorative moulding.** Roughly shaped but with complex profile commonly visible on standing buildings in Bath. One side, base and top neat, the rest damaged. Measures 340 x 220 mm. SF 774. Phase 2 or 3 1729-1768 or 1765-1800. Quarry area 913

1. **Window mullion.** Central piece with ovolo moulding. Measures 227 x 158 x 65 mm. SF 508. Phase 1 late17th early 18th. Quarry 2347, Central Firs

19. **Cornice/string course.** Moulded stone with concave moulding. Sawn on each of the six faces and finished with a mason's drag. Measures 240 x 180 x 140 mm. SF 419.

32. **Ridge stone roughout.** Long block with marking out on the end and along the sides – quite deeply cut in. The waster is still present. Measures 300 long x 230 mm high to peak. SF 272. Phase 4 or 5 1800-1833 or 1833-1867

Clay tobacco pipes by Marek Lewcun

Introduction

The assemblage of clay tobacco pipes comprise not just those pipes recovered from the underground quarries during the Stabilisation Scheme between 2001 and 2008 but also a large quantity of pipes collected by the author in 1985 and deposited in the Roman Baths Museum. During 1985 a total of 355 bowls and over 1150 stems were excavated in the underground quarries, while 84 bowls and 132 stems were recovered during the Stabilisation Scheme. Together, this is a total of 439 bowls and over 1400 stems.

The majority of the pipes in 1985 came from the excavation by the author of silt in the area under and near the Carriage Inn on The Avenue (Quarry 2347); the others were recovered from the foot of steps below a shaft near Tyning Place (Quarry 2360). Those recovered during the Stabilisation Scheme were from throughout the stone mines. Whereas the 19th-century pipes are a fairly standard group of the period, the 18th-century pipes found between 1985 and 2008 are the largest group of complete bowls for that period found in the city and represent an important reference collection for Bath and the surrounding area.

Date range

The earliest two pipes in the assemblage are reliably datable to 1700-1710, based on the maker's date of death, while there are three pipes dating from after 1694 but no later than 1739, again based on the maker's date of death. The vast majority of the remaining pipes date to the period 1750-1790, and were excavated from Quarry 2347 in 1985; a smaller number dated to 1790-1810. Only a very small number of pipes from this latter period were found during the Stabilisation Scheme work. The other pipes found in 1985 came from the foot of the shaft steps at Quarry 2360 and date to the middle and second half of the 19th century, as also do most of the pipes found during the Stabilisation Scheme. A small number of pipes found between 2001 and 2008 might possibly date to the early 20th century.

Manufacturing sources

The two earliest pipes from the mines were made in Norton St Philip, while those of the 1694-1739 period were made in Bath. The vast majority of the pipes from the second half of the 18th century were probably made in Bath, but there is one stem from Bristol (stamped WILL/NICH/LAS) dating *c* 1750-1776. Eight pipes from the period 1790-1810 with makers' marks attributable to Bath makers were found in Quarry 2347. The pipes from the first half of the 19th century were again most probably made in Bath, with just one bowl which can be positively attributed to Bristol. From around 1850 onwards there are a number of pipes which can be sourced to Bath by the initials 'JS' on the spur, while a small number might come from Bristol or further afield. Makers' marks and details of the makers are shown on the website.

Four cigar holders (part of finds reference context 5300, which represents a collection of pipes recovered by the stabilisation mining team from Quarry 2394, were identical to nine others found at the foot of the steps in Quarry 2360 and the only examples recorded in Bath to date. They might have been made in Bath, but could equally have been made in Bristol or elsewhere. Four porcelain pipes are of the Berliner Stummel type, normally mounted with a horn mouthpiece, and were made in Germany in the late 19th and early 20th centuries. These are the first known examples to have been found in Bath, and a fifth example, also from the quarries, is in the possession of a local resident. These would have been far more expensive than the average pipe, and with their inherent risks the quarries seem an unlikely place for their breakage to have been risked.

Decorated pipes

A number of the pipes from the mines are decorated, ranging from no more than leaves along the mould seams on the front and back of the bowl to more elaborate designs such as the Bacchanalian scene on the back of some 18th-century bowls. The Bacchus pipes are unparalleled in Britain and twelve examples were found in 1985, with only a small number found elsewhere in the city. Ornamentation on pipes in the second half of the 18th century is uncommon, and decoration does not make a general appearance until the early years of the 19th century.

From the period 1810-1840 two bowls found during the Stabilisation Scheme are decorated. One has plain or stylised leaves along the mould seams, while a bowl from a pillar ledge has more detailed decoration of a form more commonly produced in Bristol. The latter bowl is one of three identical examples found in Bath, where it was more likely to have been made in this case. From around 1840 onwards decorated pipes become more common, and nine pipes from the underground quarries have better-defined leaves along the seams of the mould, and there are nine with ribs on the sides of the bowl and one of a panelled style, all being types frequently found in Bath.

Slightly more elaborate styles of decoration appear on three bowls found during the Stabilisation Scheme. Moulds were being machine-cut during the later years of the 19th century and many are almost identical, and thus it is difficult to say whether these pipes were produced in Bath, Bristol or perhaps Broseley, where the Southorn family produced pipes which were sold throughout Britain, including Bath, as testified by advertisements in a number of editions of the *Bath Chronicle* newspaper. One bowl with hatched panels intersected by lines of beading is of a type described as a 'French Common Cutty' in a Manchester company's catalogue of *c* 1906. The bowl with the initials RAOB and buffalo horns of the Royal Antediluvian Order of Buffaloes is a common type in the later 19th and early 20th centuries. The Buffaloes, initially organised by actors and stage technicians in Drury Lane, London, were in existence by at least 1822, becoming a philanthropic and charitable body as time progressed. Pipes with the RAOB design were being produced by at least 1880, and probably would have been on open sale to anyone in the streets who had a liking for the design. The 'football' bowl is of a common form of which there are numerous variations dating from around 1880 onwards and into the 20th century; it would have been slightly more expensive to buy than an ordinary pipe.

Catalogue of illustrated pipes *(with archive catalogue numbering)*

1. Stem, stamped RICH/GREEN/LAND, and remains of bowl *c*1700-1710. Made by Richard Greenland of Norton St. Philip, near Bath. SF61
2 Bowl with stem stamped RO/CARP/ENTER/BATH. 1700-1739. SF725
4. Bowl with Bacchanalian scene on the back and initials IS on side of spur. Made by Jeremiah Smith of Bath between 1762 and his death in 1780, but were more probably made in the 1770s. Context 6101
12. Decorated bowl with ribbed sides and chain below lip. 1810-1840 SF815
15. Ribbed or fluted bowl, spur embossed J/S, with remains of long stem. Made by Joseph Sants of Bath, *c*1850-1880. SF30
20. 'Football Pipe', *c* 1889-1920 SF194
21. Porcelain pipe of the Berliner Stummel type, made in Germany *c* 1875-1900. Upper half of bowl white, lower half dark brown. Complete. SF47
22. Porcelain pipe, probably made in Germany *c* 1875-1900, depicting church tower flanked by foliage and with path in foreground below. Remains of copper lid survive on rim of bowl. Complete. SF494

Glass by Ian Scott

In total there were 184 sherds representing some 158 vessels or objects. The majority of the glass comprised 145 bottles and jars representing 19 wine bottles, and 113 other bottles, the base of a large jar

Fig. 11.25 Clay pipes

or bottle and 12 jars. The remaining glass comprised 9 stoppers or lids, fragments of 3 household vessels and 1 printed glass advertising sign.

The majority of the complete or near-complete bottles and jars are of either 19th- or 20th-century date and most of these are either unphased or from post-quarrying deposits. Much of the glass comes from just five contexts, the result of dumping of domestic or retail establishment waste. Since much of the glass assemblage is not securely stratified the bottle and jar assemblage (described in the discussion and catalogue) is treated as a single group for the purposes of this report and is summarised below.

The glass that can be assigned a dated provenance during the working life of the stone mines is limited in quantity, but includes, from Phase 1, a single early free-blown flask or bottle, and from Phase 2, two wine bottles, one of which is an early 18th-century type and the other an early to mid 18th-century type. Undated free-blown bottles include three bottles that date to the early 18th-century, five further bottles of early to mid 18th-century date, three bottles of mid 18th-century date and six bottles of late 18th- or early 19th-century date. Bottles and other glass of a later date are unsurprisingly much more common, including identifiable items from both local producers and further afield.

Embossed beer bottles are all from local breweries or local suppliers, including the Bath Brewery, Cater, Stoffel and Fortt, and Sainsbury Brothers, or from Bristol breweries such as the Bristol United Brewery and the Ashton Gate Brewery. A bottle of the Cheltenham Original Brewery is made of colourless glass and probably a soda or mineral water bottle rather than a beer bottle. Another bottle is embossed on the base 'POWELL | & CO | BRISTOL' and was made in the 3-piece mould patented by Henry Ricketts of Bristol in 1821 (Weeden 1982, 89 and fig. 6). The bottle is probably a sherry bottle and was made in the 1850s. Another bottle made in the mid 19th century is a case gin bottle made in a dip mould with a hand-applied tooled finish. There are two bottles embossed for Frederick William Fuller (Cat. Nos 26-27), who was a Bath based wine and spirit merchant.

There are a number of soda or mineral water bottles embossed with makers' names, most of them local Bath or Bristol businesses. There are three codd-neck bottles, one embossed for Luke Bristow of Bath, another for Hansford & Co of Bristol and the third for Carters Ltd of Bristol. Luke Bristow traded briefly as a ginger beer manufacturer and mineral water manufacturer at the very end of the 19th century. The codd-neck bottle embossed with his name must date to that brief period. The codd-neck bottle embossed for Hansford & Co dates to the very end of the 19th or early 20th century. The third codd-neck bottle is probably embossed for H W Carter & Co, of Wilder Street, 'mineral water & soluble essence manufacturers' (*Kelly's Directory of Bristol 1902*, 242). Again the bottle dates to the late 19th or early 20th century. There are two bottles of Wigmore & Son, Bristol who were mineral water manufacturers based in Bristol although with premises in Bath (*PO Bath Directory 1911*, 433, 457). The bottles both have internal screw threads and threaded rubber stoppers and date to the late 19th or early to mid 20th century. A torpedo or 'Hamilton' bottle embossed for Thomas Ekin of Grantham is an early form of soda water bottle and would fit well with a mid to late 19th-century date. Ekin was a 'manufacturing & dispensing chemist and mineral & soda water manufacturer' with premises in the Market Place, Grantham.

There are 23 pharmaceutical bottles from post-quarrying dumps. This is a large proportion of the total of 31 pharmaceutical bottles from the whole project. They include the base of a thin-walled free-blown cylindrical bottle of 18th-century date (context 717), and 10 small unlabelled pharmaceutical bottles of mid to late 19th- or early 20th-century date. Amongst them is a single patent medicine bottle. This a square section bottle with a short neck made in a 2-piece mould, with a hand tooled finish and cork closure. The bottle is embossed 'GUY'S TONIC'. Guy's Tonic was analysed by the British Medical Association and the results published in 1912. At that time the tonic sold for 1s 1d for a 6 fl oz bottle and was described as a 'rich cordial prepared from Vegetable Tincture and other curative agents found in the British Pharmacopeia . . .'. Although the advertising stated that it did not claim to be a 'cure-all' it was nonetheless recommended for 'Disorders of the Digestive System', 'Functional Derangements of the Liver', 'Disorders of the Blood', 'Nervous maladies', 'General Diseases' and 'Convalescence from Acute Diseases'. The BMA analysis suggested that the ingredients for a single bottle probably cost one halfpenny.

A wide range of glass jars and bottles illustrate the progressively wider distribution of branded items from the mid 19th to the mid 20th centuries, with manufacturers that are still household names such as Burdall, Boots and Bovril. The discussion in the full report includes details of the development of these and many lesser manufacturers.

The lack of secure stratigraphy means that, as is the case with the pottery assemblage, the glass is more important for what it can reveal about the developments in retail history and the social history of the Combe Down settlement than for its direct relevance to the Combe Down Quarries. The finds of blown bottles from the early and mid 18th century may have links with the glassworks which operated then at what is today nearby Odd Down, although confirmatory evidence is lacking.

Catalogue of illustrated glass items (with archive catalogue numbering)

19 **Beer bottle** embossed 'THE BATH BREWERY LIMD' Moulded bottle with hand tooled finish; and cork closure. Embossed 'P & R B' for the makers Powell & Ricketts Bristol. Complete. Dark green

Fig. 11.26 Glass vessels

glass, moulded. Ht: 188 mm. Stub S3, Context 263. [ID 87]

20 **Beer bottle** embossed 'CATER STOFFELL & FORTT LD BATH'. Most of neck missing otherwise complete. Dark green glass. Made in three-part mould. Ht extant: 160 mm. Context 263. [ID 83]

21 Beer bottle embossed 'SAINSBURY BROS' 'BATH'. Complete. Screw cork closure. Dark brown glass, moulded. Ht: 257 mm. Post quarrying Context 189. [ID 23]

22 **Flagon** embossed: 'BRISTOL UNITED', nearly complete. Internally threaded closure for rubber screw cork. Dark green glass, moulded. Ht; c 295 mm. SF 1004, Foxhill, Mine area 2383, Rd 1.25 end. [ID 153]

23 **Flagon** embossed: ' ASHTON GATE / SUNRISE BRAND / BREWERY / Co Ld / BRISTOL'. Complete. Rubber screw cork in situ. Dark brown glass, moulded. Ht: 294 mm. SF 1004, Foxhill, Mine area 2383, Rd 1.25 end. [ID 154]

26 **Bottle** embossed 'F.W.FULLER, BATH' and on shoulders 'IMPERIAL' and 'HALF PINT'. Complete. Cork closure. Pale blue/green glass, moulded, with hand applied tooled finish. Ht: 186 mm. Post-quarrying, Context 189. [ID 21]

27 **Spirit bottle** embossed 'F.W.FULLER / BATH'. Complete. Cork closure. Pale blue green glass, moulded with hand finished rim. Ht: 180 mm. Context 68. [ID 62]

29 **Cylindrical bottle** embossed on base: 'POWELL I & CO I BRISTOL'. Neck missing. Dark green glass. Made in three-piece mould, Probably a sherry bottle rather than a wine bottle. Ht extant: 152 mm. SF 690, Mine area 2344 or post-quarrying, Stub I7. [ID 86]

30 **Codd bottle** embossed 'L. BRISTOW I PETER STREET I BATH', and embossed 'L.B.' on base. Complete. Pale green glass. Moulded bottle, hand tooled finish. Soda water bottle. Ht: 230 mm. Context 263. [ID 84]

32 **Codd bottle** embossed 'CARTERS LTD' and ''BRISTOL', and embossed on base 'DAN RYLANDS I BARNSLEY' '4' and 'SOLE MAKER'. Body only. Pale green glass. Moulded. Soda water bottle. Ht extant: 148 mm. Context 68. [ID 66]

33 **Soda water bottle** embossed: 'CHELTENHAM ORIGINAL BREWERY Co Ltd' around shoulder. Screw cork in place. The maker's name embossed on the base is 'DALE BROWN & Co SWINTON'. Cork embossed: 'CHELTENHAM ORIGINAL BREWERY'. Complete. Colourless glass, moulded. Ht: 264 mm. SF 1015, Foxhill Mine area 2382, Rd 3F set (44). [ID 150]

35 **Soda bottle** embossed 'WIGMORE I & SON I BRISTOL'. Complete. Pale blue green glass. Screw cork closure. Ht: 261 mm. Mine area 518 tipping, Context 342. [ID 75]

36 **Cylindrical bottle**, with rubber screw cork. Has wrythen decoration moulded on shoulders. Embossed: 'BROOKE & PRUDENCIO' around bottom of body and embossed: 'CTC (B & P) 1026' on base. The bottle has a screw cork embossed 'WAR GRADE'. Complete. Colourless glass, moulded. Soda water bottle. Ht: 295 mm. Post-quarrying, Context 189. [ID 13]

52 **Medicine or tonic bottle** embossed 'COMMANS & WILSON I BATH'. Complete. Very pale blue glass, moulded with hand tooled finish. Cork closure. Ht: 167 mm. SF 310, Mine area 2372, Context 5023. [ID 59]

Pottery by John Cotter

Some 655 sherds of pottery were examined (Cotter 2009), with a date range from around 1680 or 1700, to a dated saucer fragment of 1942 with a few sherds likely to be post-1950.

The sequence of wares or pottery types recovered, for the period in question, was much as would be expected for a site in this part of the country and is not particularly different from many other post-medieval sites in England (Fig. 11.27). Commencing, as it does, around 1700, there is the usual scenario where local and regional earthenwares or coarsewares are gradually replaced by the rise of the Staffordshire potteries as the 18th century progressed. Initially, local glazed red earthenwares predominated – simple utilitarian forms such as storage jars and possibly a few bowls and jugs. These were supplemented by a small number of regional products including North Devon gravel-tempered wares, which generally came in the form of large robust storage jars and bowls.

Table 11.1 Breakdown of pottery types from the underground quarries at Combe Down

Pottery type	*Date*	*No. sherds*
Post-medieval red earthenwares (local)	*c* 1675-1925	65
Unglazed red earthenware (flowerpot etc)	*c* 1700-1925	2
North Devon gravel-tempered ware;	*c* 1675-1850	8
Donyatt slipware (Somerset)	*c* 1675-1900	8
Verwood-type ware (Dorset/Hants.)	*c* 1675-1925	1
English tin-glazed earthenware (Bristol/London)	*c* 1700-1840	13
French tin-glazed earthenware (Sarreguemines?)	*c* 1850-1925	1
Staffordshire/Bristol-type slipware	*c* 1675-1900	13
Midlands/Bristol iron-streaked earthenware	*c* 1690-1800	2
Refined red earthenware (Staffs.)	*c* 1740-1800	1
English brown stoneware (Bristol/Staffs./London)	*c* 1675-1850	23
Staffordshire white salt-glazed stoneware	*c* 1720-1780	15
Nottingham stoneware	*c* 1690-1800	2
Chinese porcelainc	*c* 1650-1900	9
English porcelain	*c* 1745-1925	18
Central European porcelain	*c* 1850-1925	4
Later Creamware ("Queensware")	*c* 1850-1925	40
Pearlware (Staffs, etc)	*c* 1780-1830	37
Refined white earthenwares (Staffs, etc)	*c* 1825-1950	342
Yellow ware (Staffs, etc)	*c* 1800-1900	3
Wedgwood-type dipped Blue Jasper ware	*c* 1800-1900	1
Wedgwood-type glazed black Basaltes ware	*c* 1825-1900	6
Brown teapot-type ware	*c* 1825-1925	4
Misc. coloured earthenwares	*c* 1825-1925	1
Modern English stoneware	*c* 1830-1940	33
Coloured stonewares	*c* 1825-1900	1
Pipeclay objects	19-20th C	2
TOTAL		655

A small number of decorative and utilitarian slipware forms were also acquired from the Donyatt potteries in south Somerset. A single jug probably from the Verwood potteries in Dorset completes the list of unequivocal regional wares – but this last piece is difficult to date closely and could be as late as the 19th century. Staffordshire products are present from *c* 1700, mainly combed slipwares, brown salt-glazed stoneware and later on in the century some Creamwares and Pearlwares. All of these however are difficult to distinguish from identical products copied by the Bristol potteries and the assemblage here probably comprises a mixture of both. Bristol, and neighbouring potteries at Brislington and Wincanton (Somerset), also produced their own tin-glazed earthenwares during the 17th and 18th centuries and these cannot always be distinguished from London tin-glazed wares. After *c* 1830 the flood of refined white earthenwares, many with transfer-printed decoration, can mostly be attributed to the Staffordshire potteries and other industrialised potteries of the Midlands and the North. Many whiteware preserve jars of the late 19th and early 20th century bear the mark of 'MALING, NEWCASTLE' and were made at the pottery there. Modern English stonewares, including those with glassy 'Bristol' glazes made after *c* 1835, were made at numerous centres in Britain but many of those from here bear the marks of London potteries while none is definitely from Bristol. The many ' WP HARTLEY' stoneware jam jars present were made at potteries in Lancashire and Glasgow in the late 19th and early 20th centuries – they may have been potted with jam at the factories in Lancashire or London before entering the retail chain and ending up in the shops of Bath, or, in fact, anywhere in the world.

Ordinary domestic pottery types, such as tablewares and storage jars, predominate although the presence of a pharmacist's pestle, a stoneware bottle possibly used for acid or mercury storage, as well as a fairly high number of small ointment jars, might suggest that part of the assemblage may have come from a chemist's shop or similar premises.

A total of 63% of the assemblage (including a high proportion of whole items), were mass-produced modern wares. Very little of this is likely to have been used in the quarries. The great bulk represented domestic rubbish dumping in abandoned shafts and other entrances.

From the viewpoint of dating the quarrying, the potentially most useful finds were from Quarry 2347 in the E4 stub. These were from a drain, from which a collection had been made in the 1980s, which seemed to establish the probability of pre-1750 quarrying there (Pollard 1994). The drain was associated with the road outside the former Carriage Inn on The Avenue. The deposit below the drain contained a number of Westerwald stoneware drinking mugs (see Pollard 1994, fig. 11.21), rarely imported after 1750. In the event, a pre-Allen, Phase I dating for the quarry working was confirmed from other data, but the drain assemblage proved to also have had items of mid- to late 18th century date within the assemblage, making a deposition date of *c* 1770-1780 more likely. However the possibility remains that the deposit accumulated over several decades, leaving an early date of quarry-working still possible.

Quarry 2346, which was on the North side of the Old Rank / de Montalt Place (specifically under the rear of 89 Church Road as it is today) had one of the larger pottery assemblages. It was known to have been partially infilled from the adjacent surface quarry, which had closed by *c* 1800 when the quarry was probably used as gardens. Most of the pottery was dated 1720-50 but with some salt-glazed stoneware of 1720-80, and some possibly intrusive material from the late 19th century. Thus the suspected late 18th-century date for the working of the underground and surface quarries is empirically confirmed.

The largest assemblage of 193 sherds came from below a shaft cone in Quarry 2201, but was associated with later Quarry 2370, where a late 19th-century working of a remaining substantial pillar cut through a probable Allen Phase II cartway. This was the only shaft cone debris that was excavated and recorded under archaeological conditions. It was similar to a typical Victorian bottle dump with mainly refined white earthenwares, modern stoneware, and a single fragment of Chinese porcelain. Two lower fills produced sherds datable to 1830-1850. The only earlier vessel is considered an heirloom piece. The assemblage was thus typical of a late 19th-century work, with the shaft left open for a shortish time to allow domestic refuse to be dumped in from nearby properties. The probable much earlier quarry cartway phase there was thus not represented, and the pottery assemblage gives an approximate date for the construction of the shaft capping at sometime after the mid 19th century.

Similarly, other pottery was useful in providing broad date ranges confirming the general age of individual workings, and, in case of dumps within shafts some indicative evidence of the date of their closure. Overall, however, for dating purposes the assemblages mainly reflect what was discovered or deduced from other data.

The perspective that the pottery gives into the social and economic history of Combe Down is considerable. There was a wide range of domestic wares, some of which have considerable exhibition potential in a local museum context, in emphasising the growth of factory-prepared foodstuffs and other household goods in a national and, indeed international context. Of particularly interest is the saucer fragment dated on its underside to 1942, discovered at a location where shelter was taken from the Baedecker bombing raids. There was also a pair of 'mystery' objects in the form of two pipeclay items of which only a few other instances have been

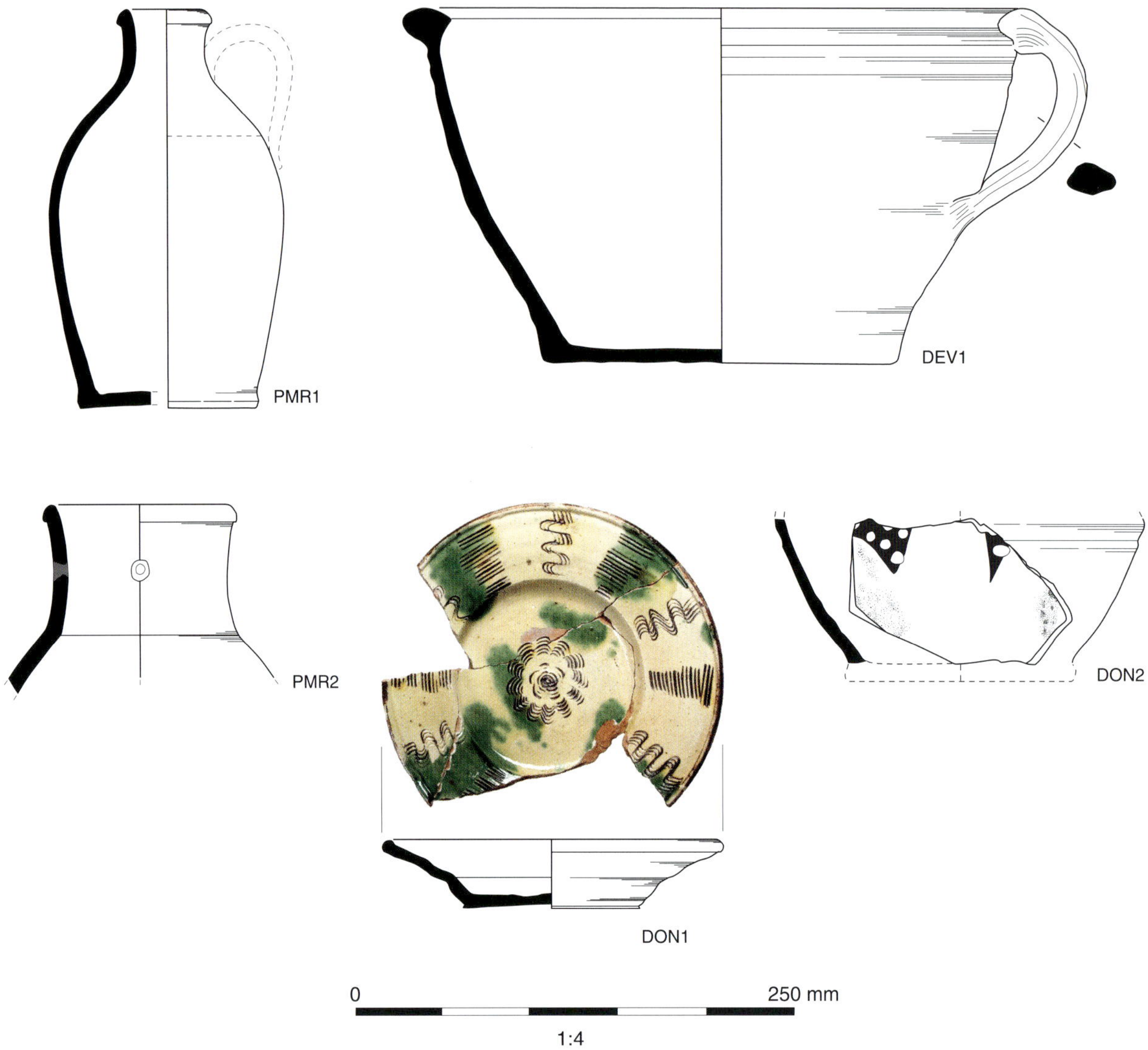

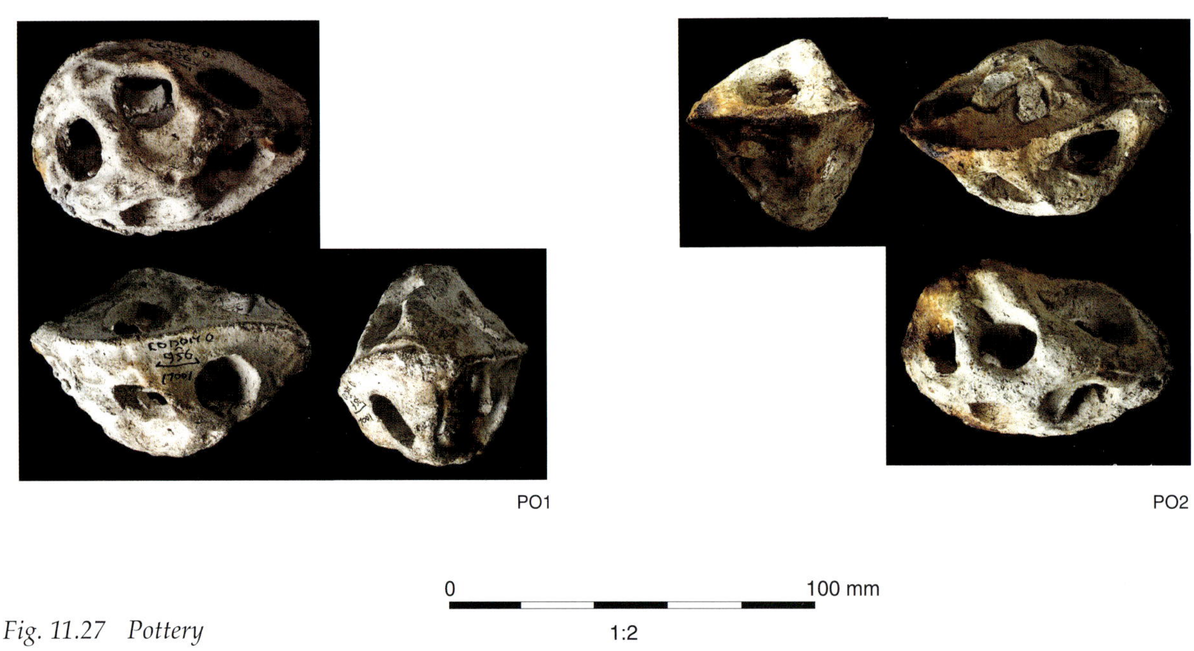

Fig. 11.27 Pottery

found in England and which have created some speculation as to their possible use (Fig. 11.27). The discovery of these in a context of dumped fire-hearth ashes at the Springfield Quarry (2374) may add support to one suggestion that they are a form of firelighter – porous enough to absorb paraffin and thus – when inserted among the kindling – providing a portable and reusable accelerant to encourage a newly-lit fire.

Catalogue of illustrated pottery

Pottery

PMR1	Post-med red earthenware. Bottle-like jug with handle scar. Light orange fabric with clear glaze allover int and on top ext. Rim diam 53 mm. L18-E19C
PMR2	Post-med red earthenware. Jug/jar rim with hole bored through neck. Light orange fabric with clear glaze all over interior and in a band under rim exterior. Possible trace of handle? Possibly a 'bird pot'? Rim diam 110 mm. L18-E19C
DEV1	North Devon gravel tempered ware. Deep bowl with handle. Rim diam *c* 360 mm. 18C
DON1	Donyatt slipware. Dish profile. Rim diam 195 mm.18C
DON2	Donyatt slipware. Deep bowl body sherds. 18C

Pipeclay objects

PO1	Pipeclay object or polyhedron. Max length 71 mm. Springfield Quarry. 19-E20C. SF956
PO2	Pipeclay object or polyhedron. Max length 69 mm. Springfield Quarry. 19-E20C. SF957

Animal bone *by Lena Strid*

Most of the small assemblage of faunal remains retrieved from the underground quarries at Combe Down were derived from waste stone layers and tipping contexts, but some intrusive bones were recovered from deposits originating from roof collapses or from entry via shafts. Generally, the on-site policy was not to retrieve animal bones which were known to have originated from surface deposits but this could sometimes be difficult to ascertain at the time of discovery. There was also potential for some finds to be post-quarrying intrusions.

Sheep and goat were identified to species where possible but otherwise classified as 'sheep/goat'. Ribs and vertebrae, with the exception of atlas and axis, were classified by size: 'large mammal' representing cattle, horse and deer; 'medium mammal' representing sheep/goat, pig and large dog; and 'small mammal' representing small dog, cat and hare. The minimum number of individuals (MNI) was calculated on the most frequently appearing bones for each species.

The assemblage comprised 83 bones, of which 95% could be identified to species. Apart from a single hare bone, all bones derived from domestic species. The faunal remains give the impression of a mix of kitchen waste and dumped material. The bones were in a very good condition, over 85 % being grade 0 or 1. Rodent gnaw marks occurred on a hare tibia, but there was no evidence of gnawing by dogs. No bones were burnt.

Two articulated skeletons were recorded: one juvenile pig from Central Byfield (Quarry 916) and one cat in Central Byfield (Quarry 510). The pig was mostly complete, but lacked the metapodials and phalanges as well as the left femur. While the phalanges are small and easy to miss during excavation (especially in the conditions at Combe Down), the metapodials are usually of a sufficient size to be collected. This opens the possibility that the feet were removed from the carcass before deposition or that body parts were scavenged afterwards, although the lack of gnaw marks argues against the latter. The first permanent molars were fully erupted, but unworn, indicating an age at death of 4-8 months. The lack of butchering marks, in combination with its young age, suggest that the pig died shortly after birth, and was dumped in the mine.

The cat skeleton consists mainly of the upper front half of the body. Cut marks were absent, and it is likely that the bones represent a natural mortality. Apart from the distal humerus, all epiphyses were unfused, suggesting it died at *c* 8.5 months of age.

Cattle are represented by an atlas and three fragmented hind limb bones. Butchery marks were recorded on the atlas, which was split axially, and the femur, which had transverse cut-marks from filleting mid-shaft. Axial splitting of the carcass was a standard butchery procedure from the 17th to the 19th centuries. Epiphyseal fusing was recorded on the two tibiae, from animals of over 2.5 years old and less than 3.5 to 4 years of age respectively. Judging from surface structure, the other two bones were from sub-adults or adults.

The sheep/goat remains consist of a horn core fragment from a male sheep and a neurocranium fragment potentially from the same individual, as well as a radius fragment.

The disarticulated pig remains comprise a humerus fragment and a mandible from a boar. The third molar had erupted but only the enamel was worn, which indicates that the pig was 1.5 to 2 years old when it died.

The horse bones were found in three separate contexts and are therefore not likely to be from a single individual. With the exception of a skull, all elements present derive from the front limb. The bones were all fused, indicating an age at death of more than 3.5 to 4 years. Measurements taken on limb bones confirm that the bones are from horse, rather than donkey or mule. The calculated withers' heights suggest that one horse was a pony and one a small horse. The articulated phalanges in on one example show exostoses laterally and medially: the first phalanx on the posterior side of the bone, and the second phalanx on the anterior side. These are likely to be ossification of muscle attachments and indicate muscle strains. The lateral metapodials were fused to the metacarpal shaft. These pathologies are commonly seen among work horses and are linked to muscular stress in connection with traction. Some lipping was recorded on the coracoid process of the scapula on one example, suggesting wear on the joint. Although only parts of the horses were recovered, there was no evidence of butchery on those bones that were present.

A single hare tibia was collected. The absence of other bones suggests it represents the remains of a meal. After discarding, the bone had lain on the ground for a while, as evidenced by several rodent gnaw marks.

The faunal remains from Combe Down stone mine comprise food remains as well as dumped materials. The food remains, such as leg of hare, parts of long bones from cattle, sheep and pig, and medium mammal ribs, would have been left there by the quarrymen. The lack of gnaw marks, apart from rodents, is consistent with the nature of the site. Dogs would not have roamed loose there. The dumped materials consist of a juvenile pig, a subadult cat and at least three adult horses. The pig might have died from a disease; it would not be considered suitable for eating and subsequently was dumped in the mine.

The horse bones, on the other hand, may feasibly have been remains of old ponies worked in the quarries. Although one of these may have been derived from the surface through a roof collapse, the location of the two other contexts containing horse bones were well away from access routes to the outside. While horses may have worked there underground, one would not expect their bones to remain there. Horse carcasses were generally extensively used in the 18th and 19th centuries: hides for leather, mane and tails for furniture stuffing, meat for pet food, fat for oil rendering and bone for buttons and glue making. The dumping of these bodies in the mine is therefore rather perplexing, since there would probably have been enough monetary value attached to a horse carcass to make it worthwhile transporting it out of the mine. Perhaps these horses were unofficially butchered and dumped in abandoned parts of the mine complex. Whether the cat was deliberately dumped underground or died a natural death there is unknown.

Chapter 12

The Combe Down complex – chronology, development and case studies

Geography of the workings

The geographical and chronological relationships of the various parts of the underground quarries were revealed during the archaeological investigation in the Stabilisation Scheme and from documentary research, especially by quarry leases. The overall time relationships are summarised on the plan inside the rear cover; the various quarries determined by the archaeological survey in conjunction with any documentary evidence available, are displayed in colour-coded form based on the phase in which they were worked. The CD accompanying this volume contains a tabular listing of the features found in each quarry and quarry area on which the chronology was largely based.

Neither features nor stratigraphical relationships were available or sufficient in many cases on their own for dating, but could be used in association with information derived from documentary sources. Stratigraphic relationships in the sense used on a conventional archaeological site were not always available for study, as no actual archaeological excavation was normally allowed. However, micro-detail, of methods of disposing of spoil for example, was often revealed by the activities of the Stabilisation Scheme teams and their mechanical excavators. The most useful stratigraphic relationships were found at the spoil dumps and the spoil barrow-ways as, obviously, dumping of spoil from one quarry area in another, or within a single quarry, illuminates the activities of different dates.

This chapter therefore takes the historical and archaeological observations chronicled in the previous chapters and discusses the overall story that the marriage of these techniques presents. The development of the quarries is discussed by chronological phase and one, or usually two, case studies is presented for each phase. The location of these case studies is shown on Figure 12.1.

Determining chronological relationships

With no important known exceptions, the underground quarries began operations either from the southern margins, at Shepherds Walk and its Beechwood Road extension, or from within the outcrop, for instance north of Summer Lane or near The Brow (which partially may be fault scarps repeating the sequence). Alternatively, underground operations were begun a little further north over the outcrop, from surface declines, in what were probably existing surface quarries such as Sheeps Pasture Quarry (below), dug out to a depth where the freestone beds could be accessed. These locations sometimes provide approximate starting dates with a good degree of confidence, notably from where Allen's cartways were driven around 1730.

The progress of quarry developments can to some extent be assessed from extant leases that, with care, provide us with fixed positions in time. In this we are particularly fortunate in that leases can be used for *c* 1800 right across the site from West to East Byfield and part of Far East Firs, allowing us once again to have a check on the progress of working with a fairly high level of confidence. It is possible also to date the end of working of some of these leases with some precision, because later workings often had rather conspicuous changes in their quarrying methods as new systems were introduced.

In Far East Firs, east of the Long Drung, the starting of works is fairly clear, but the late 18th- and 19th-century quarrying in particular is much more difficult to date because of the more diverse and smaller scale of specific workings found there compared to elsewhere. Relationships with the sale of plots and building above do, however, provide some assessable information and it is likely future local historians will be able to develop this further than time allowed us. Small later-phase workings fortunately tended to attract graffiti, which are potentially revealing about owners and workmen in locations not easily accessible except during the working phase.

For further refinement of a quarry's chronology, it is necessary to rely on technological changes reflected in the archaeology of the individual areas. The conclusions at this level of chronological analysis are less precise in terms of exact dates that can be put on the specific changes, though these conclusions can be tested by comparison with similar activities seen in different areas of the complex. In some cases working hypotheses about technology relating to chronology were tested as opportunity arose to approach other previously unseen areas. To a limited degree external data such as the demand for stone can roughly be related to the volume extracted at the supposed same phase.

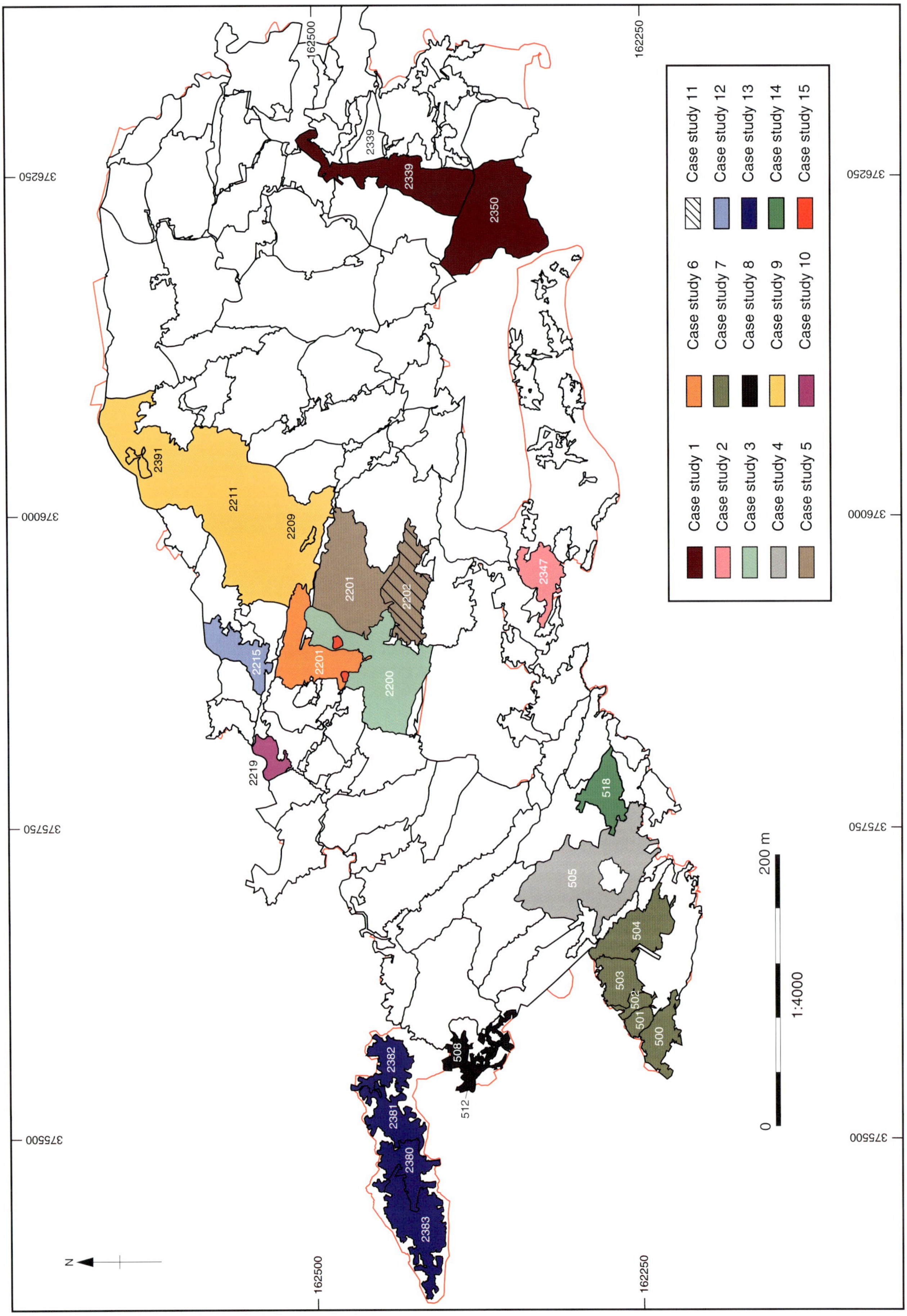

Fig. 12.1 Map of the Quarry Areas involved in the case studies

Although lacking the same level of precision possible with documentary sources, the repetition of associations of features in different areas supports the conclusions we have drawn below.

In addition, in a number of cases we can be reasonably confident about graffiti and artefacts used for dating purposes, even though these are not usually in a clear physical stratigraphical relationship in the normal archaeological sense. A clay tobacco pipe, or a date inscribed in an inaccessible position at the top of a high pillar can reasonably be assumed to have been placed contemporaneously with the original working. Similarly dates on graffiti with a contemporary flavour to drawings can cautiously be used to demonstrate they are contemporary with, or later than, the working date of the area where they are found.

Surface quarries were relatively few in the 18th century, though some map evidence survives, but in the 19th century it is often possible to use maps in conjunction with leases to not only date the starting of a quarry (within a few years), but its progress over time and even, sometimes, the type of technology in use.

The variation of technological change

Since the basis for a major part of our chronology is the technology of the features we have observed, it is important to be aware of its limitations. A key factor is the geology. The type of ground, for instance, would have been a major determinant of pillar shape – for instance the ellipsoidal plan and the hade (angle of inclination from the vertical of a fault) in the sailed pillars in north Central Firs were found where a zone of dislocation, probably a fault, ran through. Local variation in joint spacing and orientation was likely to have been reflected in roof spans and the pillar form and in blocks of unworked stone. The workman often had to tailor methods of working to the character of the rock rather than to what he had done nearby. The reverse could also apply; the apophygate shape of pillars seen at the southern boundary was at first taken by the engineers and others to be solely the result of frost shatter, rather than as a result of a deliberate pillar design, which became obvious when pillars further underground were examined.

Another factor is inertia. Once the infrastructure for the development of a quarry was established, for instance a cartway of certain dimensions or a barrow-way route made through narrow or low openings, then the method originally used would tend to remain the same until the utility of the system had ceased. To this could be added the inertia of both the master, in a reluctance to expend further capital or depart from a system he understood, and of the workmen unwilling to change tools or discard the experience of a lifetime. This could carry an outdated system long past the date when alternative methods were introduced elsewhere. Equally an innovative master might force new methods to be used for efficiency or marketing needs, introducing new men if necessary. Such technological leads and lags can be seen in the West Byfield quarries, for example, in which a single lease had sections sub-let with the result that at least three systems of working were used within a comparatively small area, virtually simultaneously. These factors can distort our retrospective view of the likely date of actual working, although this does not negate the usefulness of analysing long-term trends.

The third factor is re-working. An existing but worked-out quarry might, to a later quarryman's eye, have substantial reserves left on pillars which then become diminished by the sawing-off of faces or even their total removal. The change may be observable by archaeological observation, but in circumstances where this is not possible, the newer faces and spoil dumped over the floors can make totally or almost totally invisible any evidence of the earlier work. This was a particularly difficult problem in many mines and underground quarries and affects parts of Combe Down.

Phasing of the workings

Phasing of the underground quarries

At Combe Down, known starting dates, termination dates (by death of an owner for example) and lease dates make it convenient to divide the two centuries involved into early, middle, and late, give or take a few years.

Phase I: 1700 – c *1728*

Pre-Allen workings mainly of the 17th and early 18th century. Still earlier workings are likely to have been obliterated, or at least, masked, by the later intensive working over most of the area under investigation. They are likely only to be revealed through associated datable artefacts.

Phase II: c *1728 – 1764*

Allen-period workings from around 1728-30 when he entered into the 'Stone Business' to the end of 1764 when the estate leased his quarries to independent freemasons after his death. This phase can be divided into two sub-phases based on changes in pillar form and method of working, probably around 1750. The change may also be associated with the decision to work surface quarries.

Phase III: 1764 – c *1803*

Allen Estate workings (1764 – *c* 1803) with quarries leased by the estate and still almost entirely under its control, until around 1803 when the estate was broken up.

Phase IV: c *1800–1833*

New leases were made for all the remaining ground in the core area from West Byfield to the Long Drung and just beyond in the north corner. Quarries beyond the Long Drung generally were sub-divided by sales of small plots primarily intended for housing. The end date refers to the last of the large underground quarries ceasing work, by which time most of the other *c*1803 leases had recently ceased operations also.

Phase V: c *1833–1867*

The transition period to high, open quarrying areas with use of saws becoming common in primary extraction. Far East Firs had many small quarries, some only extracting a very few metres depth of rock.

Phase VI: 1867 – c *1938*

The period is divisible into two sub-phases, the second commencing around the beginning of the 1914-18 War. Cranes and rail systems became normal with some mechanical power common in the second sub-phase. The second sub-phase only concerned the smaller sites and not the main Firs / Byfield quarries.

Phasing of the surface quarries

Surface quarries have been considered separately – their dating is mainly based on historical data except for Phases I and II which have direct relationships with underground quarries. It seems likely the first extensive surface quarrying began in the later part of Phase II under Allen himself, continuing in the Combe Down core area and beginning in the wider area in Allen Estate times. Surface quarrying began to dominate in Phase IV when a specific type of derrick-type crane eased handling and certainly dominated in Phase V onwards. In Phase VI steam power for cranes and lorries was introduced, though by then the output had become very limited.

Phase I: Pre-Allen quarrying

Demand for stone before 1715

It is clear that despite Combe Down/Greendown/Quarr Down's relatively isolated position from a significant market, there was quarrying long before the early 18th century and Ralph Allen's dramatic entry to the stone business. The Roman villa found just below Belmont had a substantial number of stone coffins found nearby and more were found at the east edge of the Foxhill Estate. But since Belmont is on the edge of the freestone outcrop and was probably also quarried later for the building of Belmont itself and the ground in front dumped over, it is unlikely any direct evidence of Roman period quarrying will be located there. There has been a suggestion that the villa stone came from Vinegar Down Quarry, which is below it. Since this quarry is geologically distinctive compared with the main Combe Down stone, this might be tested palaeontologically, but this was outside the present project's scope. The Foxhill coffins were found nearer the north rim of the freestone outcrop where there has been much less quarrying in recent centuries so it is possible that any ancient quarry there may still be undisturbed. It is not known, of course, to what extent Roman coffins were transported and it is quite possible that other than the closest suitable sources might have been used.

In the case of medieval Bath and for buildings en route into the city, quarrying up to the early 18th century seems to have been concentrated mainly around the old Fosse Way entry to the city, from around and below Odd Down where transport was relatively easy. Lansdown has also been suggested as a source. There are substantial remains of quarrying below Odd Down but these and Lansdown are both outside the present study area.

Closer to Combe Down, buildings and walls on the Prior's Park and Bishop's Park which were the precursors to Prior Park would have required some stone, though it is likely that it was derived from the abundant reserves close by. Small extant quarries adjacent to Prior Park, near the later house site or in adjacent Claverton might be considered but again there has been much subsequent alteration and it is not easy to distinguish early from later small-scale surface quarrying methods.

The main parent parish of Combe Down was Monkton Combe. However this village was more likely to have derived its considerable quantities of stone from the edge of the freestone above the village, near where Shaft Road begins to steeply descend to the houses. There are small roadside surface quarries there but these, though noted in the initial assessment made of the wider area, were not investigated in detail. An easier gradient to Monkton Combe was from around the Kingham Quarry, under which Summer Lane slants across the hillside, but it is substantially further away. The main Combe Down quarried outcrop is even further away. Kingham would have been the logical source for stone needed at the Tuckingmill which lies directly down-slope of it, perhaps using a horse and sledge for transport. The fulling or tucking mill had been long disused by 1700 and it is not even certain that freestone was used as there was substantial rebuilding by William Smith about 1800.

An obvious example of possible early quarrying in this context is Horsecombe Vale where a quarry is mentioned around 1663 though this might better be associated with the ways into Bath or the Cross Keys Inn which are both just outside the present remit. However, on the 1742 Thorpe Map, there is shown a farm or similar building, Horsecombe House, just under the freestone rim. There is a small quarry nearby worked at surface along the cliff face, and underground. The underground workings were infilled during the Stabilisation Scheme, and were partially archaeologically surveyed but the surface quarry remains intact. Horsecombe Quarry may well have been the source for building the early part of Horsecombe House and the very few other buildings in the area. Most of the observed quarry-face surfaces at Horsecombe Quarry, however, seem to date from the later 18th or early 19th century (Phases III or IV), probably reflecting use for other local houses. It is the nature of a quarry face that later work destroys most of the evidence for earlier work.

The pre-Allen quarries up to about 1730

It is possible that the quarrymasters at work before Allen took over already-developed surface and underground quarry sites. Before Allen, there are known to have been about a half-dozen quarrymasters – freemasons and established building masons – working on the Monkton Combe portion of the Greendown/Combe Down area. Presumably each had his own small quarry. As described in Chapter 5, one or two of the quarries may have been east of the Long Drung (that is, outside Allen's later quarry-designated area), operated by Milo Smith briefly and Thomas Greenway possibly, with successors, for the remaining part of the century. Underground quarries of early date which may have been operated by one or both of them may still exist and, given the apparent longevity of the Greenway connection, the substantial quarry (2350) in Far East Firs may well be one of them, with its long cartway extending northwards from what is today The Brow, into Phase II and even later workings (see Case Study 1 below).

West of the Long Drung to Combe Road

West of the Long Drung, across to Combe Road, there are more possible early quarries – certainly enough to accommodate the documented quarrymasters, John Pitcher the Elder, and his son John Pitcher (to the first of whom Richard Jones was apprenticed), William Biggs and, possibly Joseph Biggs and John Pearce (or Pierce). A deed (LL. A91/18/5/16) of some seventy years later mentions John Allen as well, though this may be an error as in other early 19th-century documents he is not mentioned, though he is named in documents as a quarryman at a later date.

The most positive indications of this early quarrying are datable artefacts, but also include inscribed or graffiti dates. Most such positively identified works have a system of roof support involving apophygate pillars and, sometimes, apophygate arches. Where these are isolated from the Allen cartways and rather small there is a good probability of them being Phase 1. However, apophygate pillars were in use substantially after 1728-30 so, in many cases, it was not possible to identify the transition to the Phase II period of working. However, workings penetrating deep inside the hill are likely to belong to Phase II. Conversely, it is possible that at the Allen workings the cartways passed through earlier workings. This seems possible, for example, with the very small arches and apophygate spacing at the eastern Central Byfield entry, though at Central Firs there is some indication in the maturity of the pillar design and spacing that he may have avoided older (nearby) entries and positioned his entries to go in beyond earlier workings.

In the area between the Long Drung and Combe Road the position is fairly clear. Allen seems to have taken over all the quarries there. East of the Long Drung, however, the evidence indicates that Allen did not quarry there, but quarrying continued, run by other quarrymasters.

In Lyncombe and Widcombe parish, west of Combe Road, Allen did not take ownership until 1744, so that pre-existing quarrying (Quarry 910) presumably continued in operation, though perhaps ceasing after his takeover. It is possible that the competition from Allen's railway-based transport had the most severe effect on quarries here, sited slightly even further away from Bath. It may thus be that production here declined during the general expansion of the industry.

The situation close to the Long Drung in south East Firs is complicated by later surface quarrying behind Hopecote, which has obliterated remains of early working comparable to that a few metres to the east, while the Masons' Crane House Quarry has done the same on the southern margin of Central and West Firs. However there is a possibility of extant Phase I (grading probably into Phase II) working at Quarry 2204 at the rear of the quarry behind the Old Rank of Allen's workmens' cottages. Under the end of The Avenue at the junction with Church Road, Quarry 2347, which is largely east of the The Avenue, has dated artefacts and an inscribed date of 1725, despite not using apophygate pillars, making it a certain Phase I quarry (see Case Study 2 below). Quarry 2341 nearby has apophygate pillars, may have been part of Quarry 2204, and probably was accessed from the the north-western edge of Burgess Quarry. Quarry 2338 may have been similarly accessed but, though this is far from clear-cut, there are indications that it may have had an entry from the north side, possibly from the Sheeps House Quarry.

Sheeps House Quarry has early entries predominantly on its southern side, where, either surface declines were made through the Twinhoe Beds to the freestone beds below, or the quarrymen took advantage of a pre-existing quarry. Given that outcrop entries, very close by, would have needed less excavation, then pre-existing surface quarrying seems at least possible. The positions may also have been advantageous in giving access to less-weathered rock, and being further back from the outcrop it may also have avoided quarrying in the unstable cambered strata there, which later caused William Smith such grief at his Kingham Quarry (see Chapter 6).

It is possible that Quarry 2341 was entered from Burgess Quarry on the western side, and it is likely that Quarry 2200 on the north side of Sheeps House Quarry began in Phase 1 also, though its relatively large size suggests a long and intensive working period which would probably have continued into Phase II.

With one exception, Quarry 505 – which may have continued into Phase II – the early quarries in the Byfield Mine area are associated with the southern Oolite outcrop. Those in Monkton Combe, Quarries 911 and 916, are both isolated from any of Allen's cartways and were subsequently approached from the north in Allen Estate (Phase III) times, suggesting that quarrying was given up there when Allen took control of the area and then reopened later. Parts of Quarry 505, which does have Allen cartways (entering at Ralph Allen Yard today) may have pre-Allen working, but if so, it is not at all apparent. There is evidence of working underground south of this area, in Byfield Place (in Lyncombe and Widcombe) which, though the underground quarries were destroyed here in recent decades as part of housing development, they may well have been early, as were those others adjacent to the west in Lyncombe and Widcombe. If so Allen may have leapfrogged them to get access to more secure and whole ground beyond them.

Lyncombe and Widcombe

In this section of Greendown or Quarr Down west of Combe Road, described to differentiate it as 'late Collibees', Allen did not obtain ownership until about 1744. The boundary is nowadays inscribed on a stone on the former malthouse building at the east side of Rockhall Lane but prior to about 1800 when some changes had to be made in agreements, the exact location (affecting our judgement of which entries belonged to which parish) does not seem to have been clearly determined. Underground the Combe Road Pillar probably marks the boundary as understood in the 18th century.

Two entries (additional to those made by Allen on the east of the Combe Road Pillar in Monkton Combe) penetrated westwards from Ralph Allen Yard (the former Jones Quarry), into Lyncombe and Widcombe. Both passed through apophygate pillars into Lyncombe and Widcombe. Quarry 505 was very small, but formed the access to later workings, Quarry 501. To its south was a substantial underground quarry (910) which was, by 1805, part of the apparent surface-quarrying area of Jonathan (and later Sophia) Rudman (see Chapter 4), which has every appearance of being pre-Allen, probably continuing to be worked contemporaneously until at least when Allen took ownership. If the workings destroyed at the time of building the western part of Byfield Place were also early, then there was a considerable area of pre-Allen quarrying in this area. This is not apparent in documentary sources, but would help account for known quarrymen mentioned in documents not associated with Allen.

The phasing plan (inside the rear cover) shows that the west side of Byfield Place fills the gap between Quarry 910 in Lyncombe and Widcombe and Quarries 911 and 516 in Monkton Combe, suggesting an almost complete band of quarries probably originating in Phase I developed close to the southern outcrop within the Byfield Mine area. A further series of Phase I quarries (519, 914 and 2349) appear to be associated with Sheeps House Quarry. Quarries 2347 and 2351 may be associated with the southern outcrop, with Quarry 2351 located in the area east of the Long Drung near The Brow.

Case Study 1: Quarries 2339/2350 – east of the Long Drung. Phase I and II (Fig. 12.2)

This quarry was possibly associated with the Greenway family. However, since it has two, or possibly three cartways within the area shown on the map which are not clearly associated with each other, it may also have been the site of Milo Smith's fairly short-lived venture from circa 1729 when he bought Greendown Combe Down, to 1734 when he sold his quarrying and some other rights to Allen. It is bounded on the east by a long 'boundary pillar' which has been pierced to enter (and perhaps help ventilate) the adjacent quarry, and on the west by the land ownership boundary of the Long Drung which also is followed by a pillar underground for part of its course. To the south is the outcrop and, formerly, a cartway entrance. Northwards the cartway continues through a boundary pillar into workings of a later quarrying phase.

The longer cartway (Figs. 9.11 and 9.12) which dominated the transport system appears to have extended from near The Brow, where building of new houses in the grounds caused destruction of some underground workings some years ago. From there it extends north for nearly 200 m, to the later workings near East Tyning. The cartway varies from those driven by Allen further west in fairly subtle ways. Most significant is that in the observable short sections seen it appears to have been driven two beds lower in the sequence, with the roof under rather than in the picking beds. The worked freestone on both sides was full height, the picking bed having been removed, so the side passages (over spoil dumped in the worked-out side rooms) sloped down into the cartway. This probably gave it a better roof (Allen's cartways suffered badly from falls). It was somewhat wider than Allen's early cartways and was lined with rubble roof-supporting packs visible at the sides of most of the observed sections of the route, with recessed pillars seen only occasionally. A second cartway, whose entry was not traceable, runs some

Fig. 12.2 (facing page) Case Study 1: Phase I and II Quarries 2339 and 2350 – east of the Long Drung

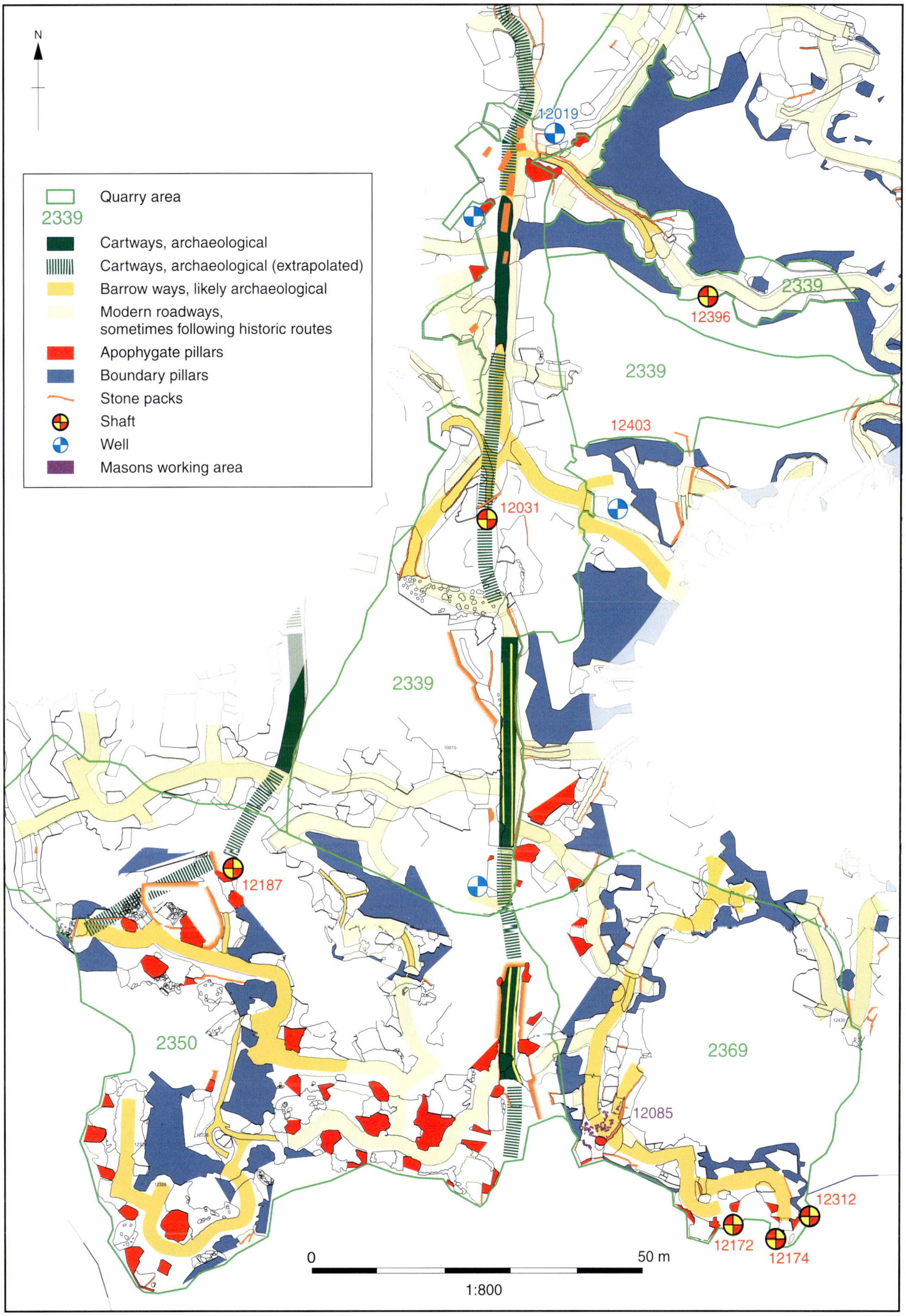
N
Quarry area
2339
Cartways, archaeological
Cartways, archaeological (extrapolated)
Barrow ways, likely archaeological
Modern roadways,
sometimes following historic routes
Apophygate pillars
Boundary pillars
Stone packs
Shaft
Well
Masons working area
12019
12396
2339
2339
12403
12031
2339
12187
2350
2369
12085
12312
12172
12174
0
50 m
1:800

25 m to the west to under Sydenham Place. This is a similar but not regular distance apart as compared with those in Allen's Firs entries (below). A probable third cartway lies close to the eastern boundary pillar, but could not be confirmed due to access limitations.

Many of the pillars in the quarry were apophygate, but often were less well formed than in Allen's quarries, the apophygation often almost vestigial and in some cases looking almost like a poorly defined corbelled pillar. Much of the area seen was a substantial way from the entrance, and this may thus be a similar transition in pillar morphology chronologically comparable to that seen in Allen's quarries. Such corbelled pillars probably mark the second sub-stage in Phase II workings which all suggests rapid adoption of new ideas, though which way these flowed is unknown. The corbelled pillar form involved stepping-out underhanging beds under the roof, probably leaving the upper picking bed (s) *in situ* and thus removing less poor-grade rock than the picking bed removal associated with the apophygate pillar.

Forming the corbel was done by use of jad slots, horizontal slots in the top of the freestone or in the picking bed, which had been cut using a one-inch (25 mm) hack (bladed form of pick) in a rather irregular way (regularity was a feature in most other areas). There were also chamfered jad slots and chamfered jads (v-shaped cuts with the lower surface horizontal) made, often back to a joint, to free the top of a bed below so that it could be barred out. Two examples were found of sawn faces of pillars (most faces are natural-joint defined) which appear to have been original rather than later pillar robbing.

Rooms between pillars were filled with spoil comprising rubble and fines, thus obscuring views in most of the quarry area. Two well-preserved barrow-ways were found, typically about a metre wide, others having been buried under spoil. Both had been used to transport spoil to tipping-fronts at their southern ends, and had also been used as stone barrow-ways to bring out blocks or ashlar. The longest was 46 m, and was lined either side with roof-supporting rubble-stone packs.

Sawn material was found in the spoil in many places and it was clear that banker mason activity had taken place underground. This was confirmed by the discovery of a 'workshop' in the form of a substantial chamber at the southern margin of the areas seen. It had two apophygate arches, with supporting stone packs, probably close to a surface entry, with, in front of the arches, some forty individual sawn and scappled ashlar blocks adjacent to a working area (Fig. 12.3). In a side pack a small sawn stone had been marked *'Francis Oliver 1730'* together with a list of stone required or produced.

The graffiti and the apophygate pillars suggest this area was contemporaneous with Allen's workings or possibly pre-dated them. The workings appear to have continued through much of the century and the penetration of the cartway some 200 m is very similar to that reached by cartways in other areas of the quarries west of the Long Drung by the end of the 18th century. As well as the subtle differences to pillars which differentiate it from quarries further west, the production of substantial amounts of sawn ashlar underground (resulting in the observed waste and the assemblage of semi-finished stone) was something not generally seen in Allen's quarries. Documentary evidence asserts his quarries were used for blockstone, with banker mason work taking place at Allen's two known surface stoneyards. If Thomas Greenway was the mason involved in this quarry, it is perhaps more than a fanciable notion that it was the origin of stone for Beau Nash's house, built by Greenway in 1720.

Fig. 12.3 Apophygate arches and supporting packs with masons working area in the foreground

Case Study 2: Quarry 2347, under the south end of The Avenue. Phase I (Fig. 12.4)

This quarry was somewhat unexpectedly found to be of early date, and though it was subsequently classed as High Grade, thus requiring a high level of care and recording, unfortunately, this was after it had been significantly impacted as part of the E4 Stub, by the Stabilisation team. However, it did as a consequence receive more time to record detail than generally was the situation. The quarry was located at the south end of the Avenue, its southern margin more or less under the north pavement of Church Road. Initially it was anticipated that it was originally accessed from the south, from perhaps where Quarrybottom alongside Summer Lane is now. However, the only evidence of an access route came from a surviving 12 m-length of cartway on the north side, partially infilled at the southern end and fully infilled at the north. The spoil was derived from later workings to the north forming tipping fronts, and the rubble was stacked in packs either side, forming a barrow-way. This would indicate an original entry from the south east corner of Sheeps House Quarry (Rock Lane). It is one of several early underground quarries on the southern margin of this surface quarry which have forced a re-assessment of the latter's age, which thus appears to be definitely pre-Allen.

The quarry, in Hawkins Sector E, had been flagged by previous work (see Pollard 1994) as likely to be relatively early. He had reported a well associated with clay dumped in the workings which had clearly been drawn from it, with pottery sherds including early 18th-century ware. On further investigation (this volume) the date range unfortunately extended to later than the *c* 1750 date he indicated, but two further discoveries, of a clay pipe *c* 1720-1725, and an inscription of 1725 made pre-Allen working extremely likely.

The first notable feature observed as the supported roadway entered the area, was the well (see Figure 11.2). This is or was located in the back yards, more or less under the dividing wall of The Bank and Lansdown House, on the east side of The Avenue between the Carriage Inn buildings, and the eleven houses of the Old Rank or de Montalt Row. Originally the well probably served the Old Rank and the Carriage Inn and, perhaps, also a horse trough, since it was adjacent to Allen's railway in an area with scarce surface water resources.

Close to the well a substantial quantity of once rather fluid clay had flowed from where it had been dumped a metre or two away, over the surface of rubble spoil. This may have been when the pumps were installed a century or more ago, or possibly it was original, derived when the well was sunk (if so it implies the workings were already open when the well was sunk, which indeed seems likely anyway). Another deposit of clay had been investigated by Philip Wooster (amongst others) who explored the quarries in the 1980s. They had found broken household waste items in the clay which seemed to have derived from a square hole near the well seemingly used for dumping by the Carriage Inn, though perhaps constructed to remove water from the front of the inn. The square hole was found infilled with a brownish clay with pottery sherds. The items found included numerous blue-grey fragments of some 80 Westerwald drinking mugs of quart, pint and half-pint sizes, two or more Westerwald jugs and dozens of Bath-made pipes. There were also mallet-type wine bottles and apothecary bottles and much early china including two nearly complete Staffordshire slip-glazed baking dishes. There were two silver teaspoons with insignia on the handles, dating btween 1720 and 1750 (Paul de Ath, cited in Pollard 1994, 57-59). Pottery sherds examined as part of the present study from the latest deposit within the drain (see Chapter 11) suggest 1780 as a more suitable date for the cessation of dumping. Despite this obvious weakness in the evidence of the Wooster Collection its origin in an area away from Phase II development offers at least a hint of the possibility of Phase I working,

This was significantly strengthened by the recovery by Marek Lewcun (Chapter 11) of a clay pipe from a pack on the north side of the cartway, dated about 1720-25. Again this is not conclusive of pre-Allen working, but the present investigation also revealed a rubble stone pack built across the southern passage which had two carefully placed stones on either side at chest height, both bearing the neat inscription '*RJ*' (the 'J' a crossed 'I'). Against the wall another stone in the same pack was inscribed (little more than neatly scratched) '*XXIX*', which might indicate a date, and more positively, just beyond the pack to the north was the inscription, less neat, of '*JM 1725*' (Fig. 12.5).

This is therefore a pre-Allen quarry in a condition only slightly affected by later entrants who were perhaps intent only on robbing pillars in two or three places. The intriguing possibility is that the initials *RJ* are those of Richard Jones himself, who was definitely literate, and an apprentice here from 1715 to 1722 or thereabouts. He was then bound to John Pitcher the Elder, mason of Bath and known to have been already at work at Combe Down years before Allen took over the quarrying area *c* 1730. Pitcher and others were first referred to in a deed of 1723 (see Chapter 5).

The cartway lies on the north side, running roughly N-S, nearly in line with the well and about 3 m wide and approximately 4 m high at the deepest position (Fig. 12.6). The floor is low with respect to the backfilled floors of the adjacent workings which gave access to it from working positions and which would have facilitated loading. The floor showed no cart ruts, but any of these would probably have been overlain by later barrowing and dumping.

South of the well there is a corner room where there was what appeared to have been a banker

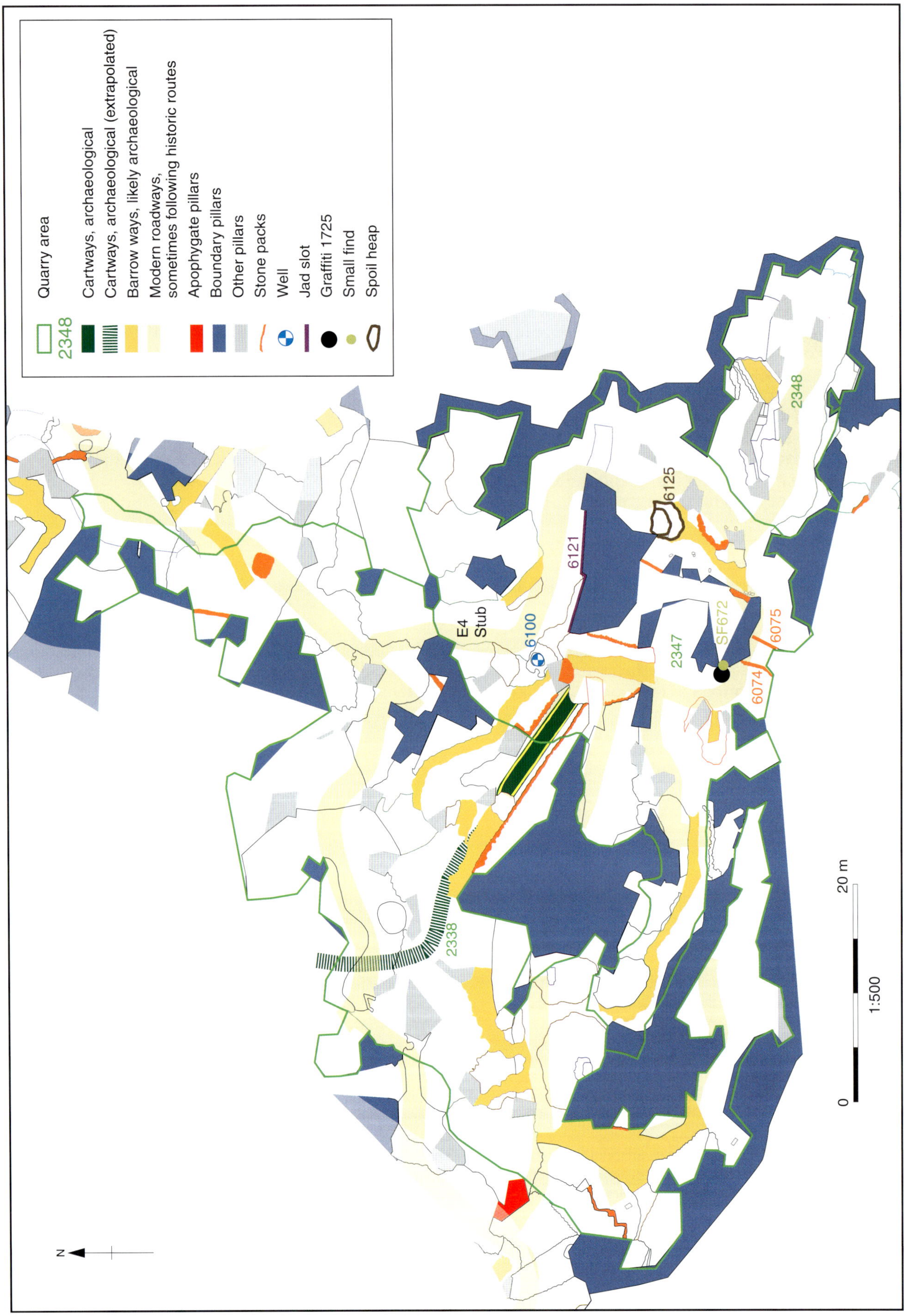

Fig. 12.4 Case Study 2: Phase I Quarry 2347, under the south end of The Avenue

masons' workplace (Fig. 12.7). The site had heaps of fine spoil derived from trimming blocks, which form small ridges running perpendicular from the wall. One or two working positions may suggest the masons had been seated next to the wall or along a plank parallel to the ridges. Several small pieces of stonework of good quality were found a few metres

Fig. 12.5 Graffiti from the E4 stub, indicating early 18th-century workings

to the south in an area of later adjacent quarrying (2348), including a small classical pedestal (unfinished) and what may have been the early roughing out of a medallion. Several small tools were recovered including a comb (for surface finishing) and marking-out tools.

Near this was probably the northern extent of the quarry, which was indicated by the distribution of jad slots and the outermost of the corbelled pillars. The corbels here and elsewhere in the workings were substantial and long, a metre or so projecting under the roof each side along the length and above the pillars (Fig. 12.8). This is substantially longer than seen anywhere else in the Combe Down workings. They also did not seem to show signs of contemporary or post-working stress in the form of a crack in line with the pillar, so were a successful design for the local geological conditions, unlike very many in other places where they had failed. Most but not all pillars had them, with fewer at the east end.

The quarry has two or three parallel Long Rooms, each 3-4 m wide, linked by openings in the long pillars between (Fig. 12.9). The dominant feature, supporting the early date supposed, is that the pillars, including long pillars, are corbelled, using very regular and long jad slots about four inches (100 mm) high which seem to have penetrated to substantial depths. The length of the corbel, at its base, was not always back to a joint, indicating the jad was developed to a position it would break the rock, not the joint. It was possibly this which made them more successful. They were developed in either the second or third (downwards) of the usual three picking beds. A one inch (25 mm) hack had been used, perhaps in conjunction with a jadding bar (notably used to work the underside of the top of the slot to keep it parallel), for this purpose.

Long lengths of jads were found on the bounding walls of the quarry (Fig. 12.10). These were used to free the beds being worked below them. Two were up to 3 m long, though others were less. The jad was normally a much later feature in quarrying at Combe Down. Its use in this working was coupled by both a neatness in working, and in the condition of the workings and floors, which we associated with those areas in which banker masons, rather than just freemasons, worked underground. There were a few other such places. The implication is that the pace of work was slower and the care necessary higher, as a characteristic of the banker mason; perhaps only enough stone was taken at a time as was immediately required.

Parallel to the cartway were two longish rooms (not the type of Long Room described for Phase III working in Chapter 8), only some three metres wide and again with the well-formed jad slots. Eastwards around a corner the southernmost passage was formed on infilled spoil, leaving a height of just under two metres, continued to the rubble pack with the *RJ* initials, with beyond this a short length

Fig. 12.6 The partially infilled cartway suggesting an entry from the north-west side (in Sheeps House Quarry)

Fig. 12.7 Banker mason's workplace indicated by spoil heaps

Fig. 12.8 Long corbels on a pillar in the E4 Stub

Fig. 12.9 Long but narrow room in the E4 Stub

of passage and then a return north to the well. Since access was easy via the area near the well to the far side of the rubble pack, the pack may have been installed simply to guide the air, though the definite placing of the initials would suggest a boundary. Alternatively, perhaps being an apprentice, it reflects nothing more than pride in his work.

At the western end beyond the junction corner, was the adjacent Quarry Area 2348. The workings were again formed of two long and fairly narrow rooms. That on the south side had small rooms between partly formed pillars against the boundary wall with spoil placed on them with a barrow-way in front. Two carved stone capitals were found on the floor in this section (Fig. 12.11). The southern passage ended in a rubble decline with a 4m-high vertical slot formed in the 1.5 m-wide face. The rock here was closely jointed vertically and it is possible work was abandoned as the stone was too broken, though they may have been using the jointed central portion as an easy way to advance. The jad slots continued to the top of the rubble slope but were not used at the face.

Parallel to this, on the north side, ranging from the junction, was another narrow room, up to four metres high, but with an uneven floor, in which a small sawn face was found at a low level, and triangular wedge pits had been used fairly frequently to break rock from faces or doorways in what were incipient pillars. The wedge pits were individual, ie

Fig. 12.10 Development of jads in the E4 Stub

not used in pairs or threes, to break off a large face. At the junction, a vertically well-jointed pillar had wedge pits where the face had been formed by splitting. There were also several small faces that had been cut by vertically picked jads (more often a 19th-century technique). Two faces were noted which had been cut back by a four-inch (100 mm) hack, to allow space to extract the block below. However most faces were derived from natural joints and most extraction of blocks probably only needed a bar.

Fig. 12.11 Capitals for columns. It seems likely these were produced to a pattern, not turned on a lathe (Quarry Area 2348)

This quarry (2348) may have been part of Quarry Area 2347 to the north, but the difference in the extraction techniques and the number of disparities between the toolmarkings in both quarrying areas suggests that they were slightly later than 2347. Both areas and especially 2347 were of considerable importance in a number of ways. Its possible association with Pitcher and Jones has considerable human interest since it is very rare for this time to place people so closely in their workplace, and helps confirm the early date of the workings. The use of jad slots, wedge pits, and corbelled pillars at this early date is a salutary reminder not to take matters and methods for granted – these are simple techniques, at least to a skilled mason, and were probably long known and were also probably one of the most efficient ways of splitting and breaking bedded rocks. Almost the whole gamut of hand-quarrying methods were used. In a parallel way it was interesting to see the contrast in modern removal of stone by our own consultant mason (Gerry Melksham), and that of one of the best modern miners (who shall be un-named) using a pneumatic tool to cut out a slot. Neatness and speed and economy of effort were all to the great advantage of the mason.

Phase II: Workings mainly associated with Ralph Allen 1728-30 to 1764

Working in this phase by Allen took place east of the Combe Road boundary with Lyncombe and Widcombe at Rockhall Lane (Ralph Allen Yard today), from north of Sheeps House Quarry and from the later surface quarries north of the Old Rank (later de Montalt) cottages and Hopecote Lodge (see plan inside rear cover). Other quarries were also active in this phase, east of the Long Drung in the works possibly attributable to Thomas Greenway and Milo Smith Case Study 1 above) and also probably south-west within the Lyncombe and Widcombe workings.

Lack of access prevents certainty, but the largest area of suspected Phase II workings (perhaps with some Phase I) appears to be in East Firs, north of the quarries behind the Old Rank and Hopecote Lodge. These were much affected by roof collapse. The second most important area was certainly in Central Byfield which (for its time) had a large quarry with very wide pillar spacing, though collapses were more modest in scale. Central Firs was somewhat smaller and, in terms of internal layout, apparently slightly more organised, which may suggest it was slightly later to be developed.

The most characteristic feature of Allen's quarrying was its organisation. Much of this has been dealt with in Chapter 5, and archaeologically the features extant at surface are still clear – in the village houses at the Old Rank and the Carriage Inn buildings, in the layout of the roads in the modern village and the former railway route down what is now Ralph Allen Drive past the masons' cottages near the river. There are also the quarries at surface, which will be considered below, and the inclined surface accesses these gave to the underground levels. This high level of organisation was continued underground, in the regular development of pairs of levels or cartways with workings advancing between them and at the immediate sides and less obviously in the provision for both spoil-dumping and stone barrow-ways. Shafts appear to have been sunk in conjunction with the use of levels, again in a regular pattern and the working areas are substantial and completely worked out. The same pattern of development can be seen in each of the probable four sets of entries, to the extent, as fieldwork and assessment progressed, that it was possible to forecast the main features in areas not yet seen.

What seems to have been the traditional working method seen in most of the earlier workings was maintained for between 15 and 25 years, in that apophygate pillar and room working remained the norm. Whether the lower sections of pillars were worked in the same way as previously is, however, unknown, since we have no examples of these because of spoil backfilling in all areas. Except along cartways and barrow-ways, generally there was little space left above the spoil and with the exception of a major underground quarry in Central Byfield, views of more than 10 m ahead were rare and limited.

Access to the areas between The Avenue and the Long Drung (East Firs) was limited, in a large part by the existence of high hazard areas identified by Hawkins (1994) which were at their most common in this area. It was possible to access the cartways within them from further inside. In all there were four or five cartways fitting the Allen blueprint, three at the western end, two at the east. In addition there was a branch cartway from Allen's Sheeps House Quarry, which cut across at least the westernmost of the East Firs cartways. It cut across the back of the collapsed zone and may, with use of shafts, have replaced the earlier cartways. Features including cranes on the first part of this cartway may suggest it was of Phase III construction and it will be considered under that section.

During the archaeological examination, the Stabilisation Scheme access to the southern area of East Firs provided a transit across the Phase II workings there, as represented by the pillar type, showing them to follow a similar sequence to elsewhere: across corbelled into apophygate and across corbelled again, with a further similar sequence to the east. Each set of cartways was associated with two wide shafts placed to the side of the levels. It was possible to view up the 5-6 m of largely unlined shafts to a very-flattened-arch or domed underside of a cap in semi-dressed stone blocks. They were very similar to each other and of architecturally high quality, which might be described as 'mason-like' in contrast to the more common lower quality of other arched caps seen in the workings. It is not clear if the two areas had direct connecting transport links or, in effect acted as separate quarries, as a number of boundary pillars in the areas of East Firs seem to suggest.

Phase II was sub-divided as the significance of the pillar systems became more apparent. It was a significant moment when it was realised there was sufficient information about the workings that a line could be drawn around the apophygate pillars in Central Byfield and that they were surrounded by the second sub-phase of corbelled pillars, with some pillars actually being apophygate on one side and corbelled on the other. Entabled pillars were found in both Central Byfield and East Firs at the junction of the two sub-phases and reflects the taking of all three picking beds for the apophygate working, but only one or two (if any) with the corbelled pillar area. The result was that there was a great reduction in spoil created and a great deal less work needed to open up the beds below for extraction.

The change was thus an important technological advance. The apophygate pillar method was clearly the established way, but if Richard Jones did leave his initials in the early quarry in Sector E, then he was clearly aware of an alternative, the use of jads and corbelled pillars. Apophygate and arched pillars were a common feature in other underground quarries too, such as at Beer in Dorset and in Browns Folly at Monkton Farleigh, in Wiltshire. As practised at Combe Down, the original arch, a very good roof-support structure, was 'stretched-out', leaving just the small apophygation to support a wider spread of roof above, but placing the area between into tension. Given the frequent and often loose joints due to cambering of strata, and the ever-wider spacing, which seems to have been adopted, this was a fundamental weakness. There were substantial roof-falls, largely from this cause in the Central Firs cartways, and even worse in East Firs where the worst areas of unstable ground were located. There were falls too in the Byfield cartways.

It is not clear if it was apparent at the time the pillars and rooms were being worked, but the apophygates had a tendency to spall badly, small thin fragments of rock splitting from the pillar sides. This was caused by the almost imperceptible sag in the roof between pillars causing the span load to bear on the weak edges of the apophygation and, eventually as the pillar thins, on the slightly curved sides of the pillar. It was perhaps these collapses and signs of weakness in pillars which persuaded Jones, or allowed Jones to persuade Allen, to change to the corbelled pillar supports. John Wood (1765) commented on the dangers Allen became aware of in underground mining and the accidents and deaths it caused, and commented that Allen began quarrying from the surface to avoid it. This was probably a reference to surface quarrying, but it might just be a reference to use of shafts as an alternative to the long and sometimes dangerous treks down the cartways. Fairly narrow shafts had been used within the early sub-phase, probably for ventilation only, but wide winding shafts became used later. At least one was probably sunk within the early sub-phase but others as part of the second. This transition between phases, judging by the relative areas of exploitation, probably took place about 1745-55, not necessarily simultaneously in the different quarries.

The enormous amounts of spoil backfilled into the apophygate workings, and the limited heights of the cartways prevent collection of data about the quarrying techniques on the lower parts of the pillars, while the top section of many pillars was further damaged by spalling, and – near entries only – by some possible frost damage. The apophygate style of the pillars appears to be confined to the softer, poor weathering stone of the three picking beds which allowed picks and, probably, the small hacks to form the curved shape. This did, however also open-up the roof to the so-called Bastard Bed which, though much harder, has strong and often fairly open jointing which makes the bed very vulnerable to failure. The consequences were seen especially in the cartways, partly since these were more open to view, but also possibly because lateral support of the pillars from spoil was less. Much of the actual failure would have been delayed until post-working stresses took over and failure progressed, perhaps in a domino effect. The situation was thus less unsafe during the initial working phase, becoming more and more dangerous as time went on. This was possibly still within the time the cartways were still developing in Phase II, but certainly was apparent by the early and mid 19th century when quarrying operations included repairing the damage.

Of tool marks observed there is evidence of the wider blade hack marks on the face of some of the pillars and perhaps smaller picks were used near the roof. There is no evidence visible of sawn work, and use of wedge pits is rare. In corbelled pillars jad slots or chamfered jads were used but long sections of jad are rare, and as with an example in East Firs, probably belong to the next phase.

The transport of stone was clearly well-organised where this can be observed, with wheel ruts in cartways suggesting wagons of about four feet wide, but dumping of spoil from later periods presumably has obscured the systems away from the cartways. There was no obvious trace of two-wheeled carts or wheelbarrows, though rare single-wheel ruts show the latter were used.

The use of wide shafts, sometimes considered as light shafts (erroneously – they only light a very restricted area below), can only have been justified for winding to surface to avoid long distances in the cartways, though they would also have assisted with ventilation. The wide shaft would have allowed long or substantial blocks to be lifted horizontally off a vehicle below, and lowered on to a cart or wagon at surface without the hazard of turning the stone, but there is no actual record of this. It is likely these shafts were mainly introduced towards the end of Phase II. With one exception, in Central Byfield where it is surrounded by apophygate pillars, the four or five wide shafts likely to have been sunk in the Phase II period were placed just to the side of cartways, some tens of metres within the corbelled pillar areas. Their use was probably one of the major reasons, together with likely consolidation of production, why the post-1764 Allen Estate map only showed one crane installed on the underground level entries, that at Sheeps House Quarry.

Of the four wide shafts it was possible to investigate, each of which was just off the cartway, there was found evidence for associated cable haulage, probably over only a short distance. This was represented by cable grooves usually worn in the rock of pillars, and examples were found also near the level entries at Sheeps House Quarry and at the Central Byfield easternmost cartway entry. The shafts examined were the second wide shaft in Central Byfield, the so-called Firs Shaft in Central Firs and two shafts with near-identical caps in East Firs. It is not clear whether the cables were used to drag blocks to the shaft bottom, or whether they pulled wagons or carts along the level. Almost certainly each used a cable (chain or rope) passed down the shaft from the winding gear above, through a pulley block and along the cartway. It was a very simple and, given the presence of a cable down the shaft, an intuitive technology used also in some coal mines in the mid 18th century, which could relieve much labour. That it was common to all four shafts and quarrying areas suggests it was an imposed business-wide technology, perhaps at the very late stage of Allen's exploitation.

This repetition and commonality in development methods also offers the possibility that the change from Phase II to III can be observed in the diversification in the methods of working which followed the leasing of quarrying to individuals after Allen's death, with, of course due deference to leads and lags in developments.

Areas being worked at the end of Phase II

The leases which controlled working in the Central Byfield area after 1803 extended south from the Turnpike (North Road) to include the area around the second wide shaft in Central Byfield. Only one of all the leases (for further east in Byfield, adjacent to West Firs), mentions the existence of substantial old workings within the lease. This and the documentary evidence in the Allen Estate Map of the removal of the crane over the Central Byfield entries, suggests that work shut down at the Central Byfield site at or prior to Allen's death. This may be a reflection of surface quarrying production at the Masons Crane House Quarry whose crane was still in position in 1764, and centralisation of production based on the Central and East Firs quarries.

In Central Firs, Phase II work extended between and alongside the two cartways as far as the Firs Shaft and the immediate vicinity. There may have been working to the west near the two entries, but to an undetermined extent. In East Firs a broad band of workings extended across the south front of the whole area, penetrating northwards as far as or a little beyond the two similarly-capped shafts. Work seems to have continued in all the Firs areas through into Phase III.

Because of the limitations of access in the East Firs area, the two case studies which follow will concentrate on the workings in Central Firs and Central Byfield.

Case Study 3. Phase II: Allen underground quarry 2200, entered from Sheeps House Quarry (Fig. 12.12)

The entries were at the foot of the surface quarry face in Sheeps House Quarry at what later became known as Davidges Bottom and now Rock Lane. There were two belonging to the initial phase which penetrate, not quite in parallel, directly northwards into the freestone beds, almost in a straight line. Above the east side on the cliff was one of the Padmore cranes and it is likely the floor outside the two entries was levelled to provide handling and stacking space reachable by the crane. The floor level was probably somewhat lower than today so as to be at the level of the cartway floors. There were cable grooves on the west wall of the easternmost cartway just inside the entry (Fig. 9.32). Both cartways were initially driven in solid rock formed as apophygate arches. After a few metres the area, some 15-20 m wide between the two cartways, had been worked out, leaving apophygate pillars along the inner side sides and usually, two more in the area between the cartways, the whole backfilled almost to the roof with spoil. The roofs of both cartways had suffered roof falls and the sides had been reinforced with fairly heavy pack walls as necessary: these may well in part have belonged to later phases of use. Prior to the falls the cartways would have been about 2-2.4 m high above the backfilled floor, but falls have increased this. Traces of cart wheel grooves were found in the floor but for the most part this had a rubble surface.

In the eastern cartway, about 55 m in, a shaft about 1 m wide had been sunk from the surface to the centre of the level with a rubble cone below it, although the shaft was found capped at the top. It belled out above the cartway, though the cone shape did not reflect this, suggesting the shaft was sunk to the gate (and not the gate driven to it), and that the rubble tumbled in later before capping. The shaft may well have been at the end of the very first section of driving, providing ventilation to both cartways and adjacent areas and the need for it suggests air conditions had deteriorated. The cartway widened just before the shaft at an intersection with a cartway driven, angled slightly back, to meet the other cartway, the area still within apophygate pillars. On the opposite, eastern side was a branch cartway (Fig. 12.13), driven between apophygate pillars but after a few metres either bounded by an un-pierced wall of rock on the north side, or corbelled pillars on the south. This branch will be considered in the Phase III section below.

Apophygate pillars continued to just above the junction(s) for one or two rooms, but then both cartways entered areas of corbelled pillars. The apophygate pillars extended well to the west but not to the east. They marked the first phase of work in Quarry 2200, which continued in a second sub-phase to include part of the area with corbelled pillars, probably as far north as the Firs Shaft (Fig. 12.14).

Because of access restrictions the detail of the eastern cartway north of the junction was not easily visible, but at about 110 m in from the entry, the cartway intersected with a partially backfilled cart or barrow-way ranging to the west from a wide shaft placed just to the east of the cartway (the 4 m-wide Firs Shaft used in the Stabilisation Scheme). Cable grooves were seen cutting into the pillars immediately supporting this shaft (there was also a number of Lewis slots, but these seemed likely to belong to the next phase in which cranes began to be used underground). The pillars were mainly corbelled with some direct, and substantial rooms were left clear of heaps of waste especially around the pillars and cartway near the shaft, where there was approximately 2 m or more of headroom, presumably as an area for handling and storing stone. What other areas alongside the cartway and between the cartways could be seen had mainly

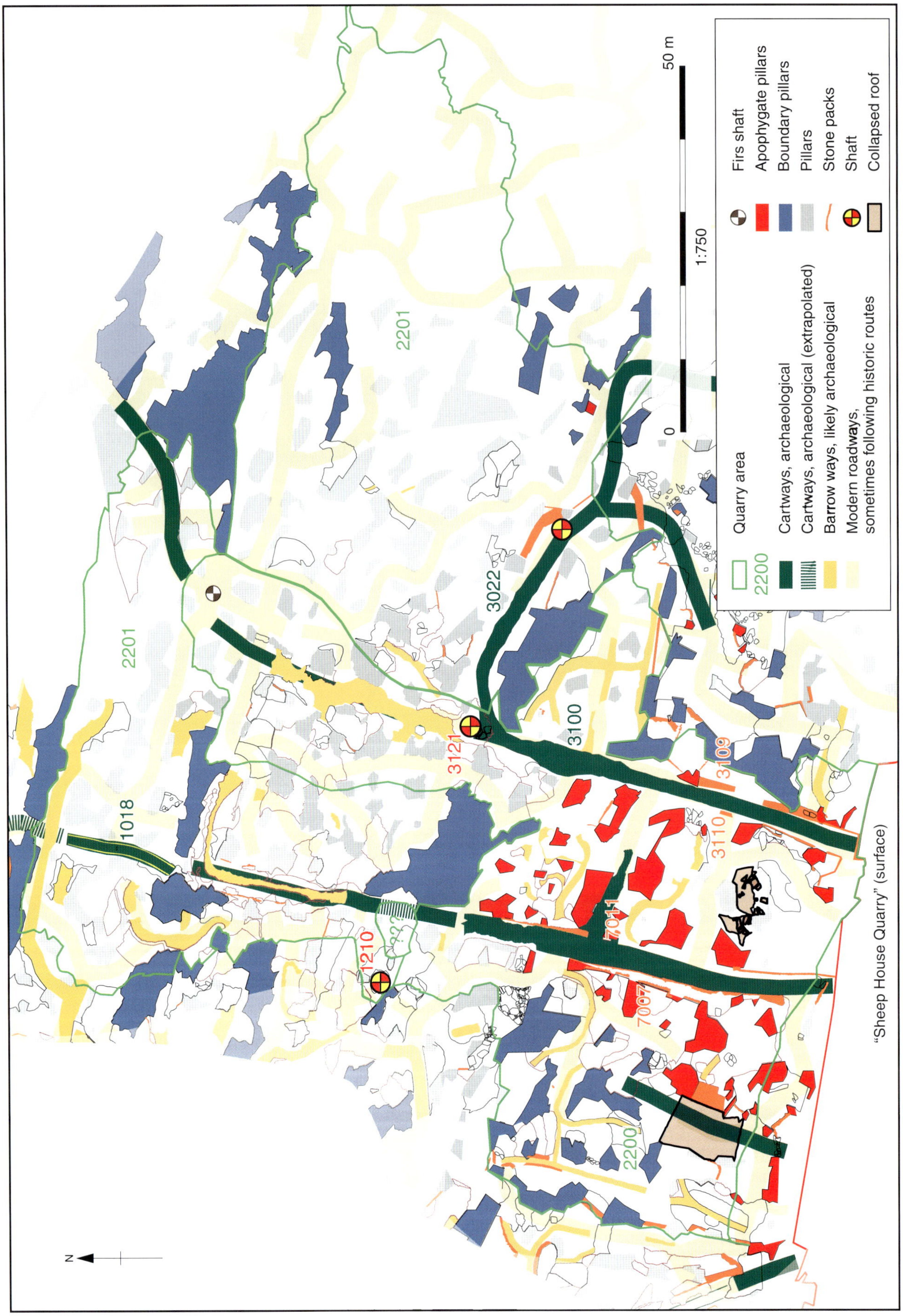

Fig. 12.12 Case Study 3: Phase II: Allen underground quarry 2200, entered from Sheeps House Quarry

Fig. 12.13 View along the eastern Allen cartway at the junction with the branch cartway

Fig. 12.14 View down the eastern cartway near the Firs Shaft

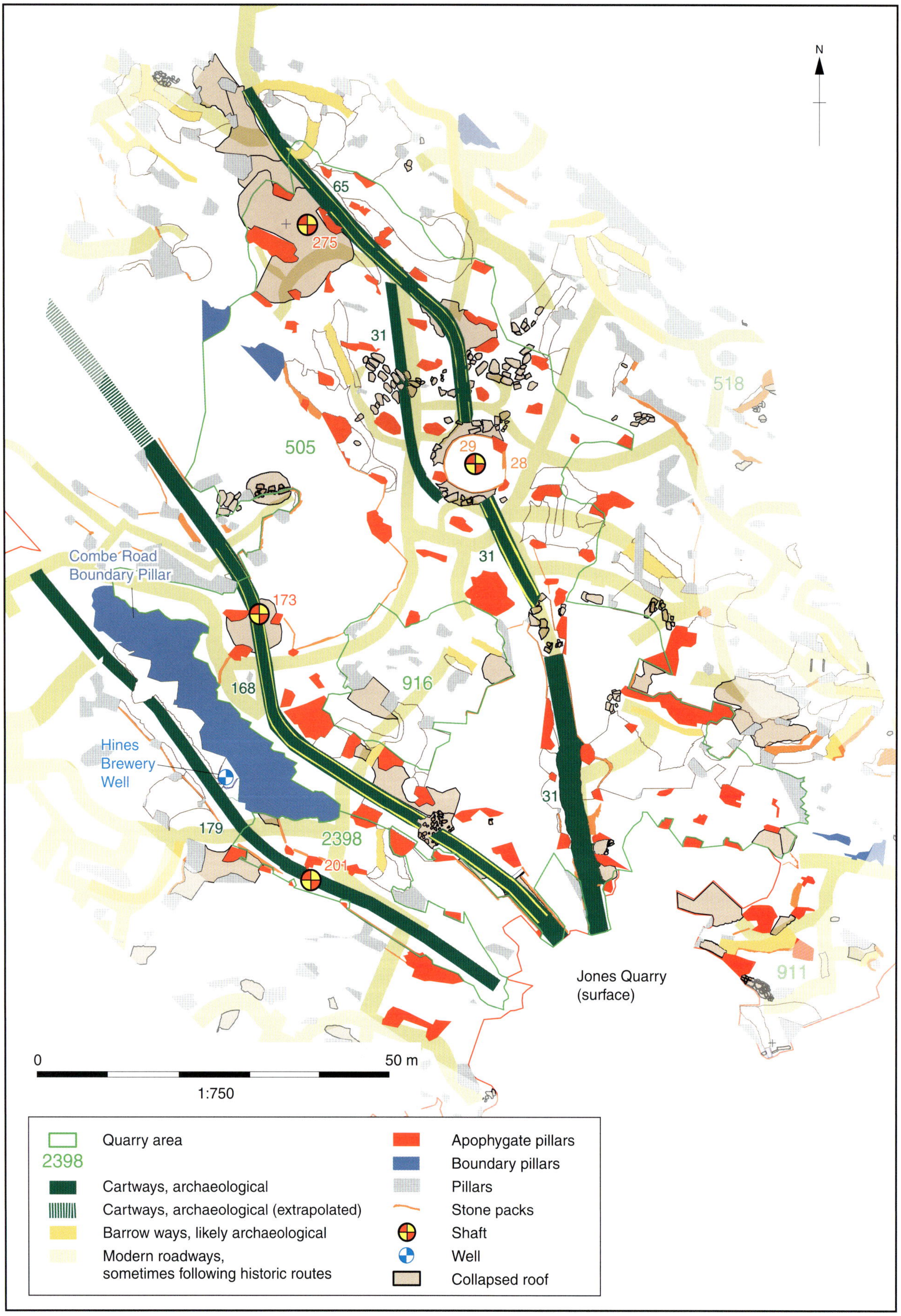
N
65
275
31
505
29
28
518
Combe Road
Boundary Pillar
173
31
916
168
Hines
Brewery
Well
31
179
2398
201
Jones Quarry
(surface)
911
0
50 m
1:750
Quarry area
2398
Cartways, archaeological
Cartways, archaeological (extrapolated)
Barrow ways, likely archaeological
Modern roadways,
sometimes following historic routes
Apophygate pillars
Boundary pillars
Pillars
Stone packs
Shaft
Well
Collapsed roof

large amounts of spoil in heaps and banks. The Firs Shaft may mark the limit of development in the second sub-phase, though some work fairly close by almost certainly belonged to Phase III.

The western cartway continued past the junction to just pass, on the west side, a wide shaft about 70 m in. In this area a late 19th-century quarry (2344) had been developed both on and at the side of the cartway to a greater depth than the Phase II workings and will be considered later, below. Beyond it the cartway continued (but re-used as a later barrow-way associated with Quarry Areas 2344 and 2368), between corbelled and direct pillars, until it widened out into workings with openings around 2 m high, intersecting with the cartway/ barrow-way ranging west from the Firs Shaft. At this point it was certainly part of the Phase III workings and these will also be considered later.

Access limitations make it difficult to determine the extent of the second sub-phase, but it included most of those between the cartways and probably to the west before the later quarry (2344) where there was an unexploited area of stone left. It is also not clear that the shaft in Quarry 2344 was Phase II or was sunk for the much later working. It would fit the pattern of wide shafts established in other Phase II areas, but, given Firs Shaft's accessibility, it was not totally necessary until the later phase.

Case Study 4: Phase II workings in Central Byfield Quarry 505 (Fig. 12.15)

These workings were within the Monkton Combe Greendown, east of the western boundary at Combe Road (today). The parish boundary was reflected underground by the near-continuous Combe Road boundary pillar.

The workings were again entered by two cartways, which diverged from a starting point next to each other in what is now Ralph Allen Yard. This was possibly a small existing quarry with older workings opening off it, but it seems most likely the entries were made directly into freestone so as to avoid old work. Cable groove marks just inside, and deductions about the course of the surface railway make it clear a crane was positioned above the entries.

The western side cartway followed up the east side of the Combe Road boundary pillar, the latter left entire except for a (later?) doorway and short passage through to the Lyncombe and Widcombe workings some 30 m from the entry. The pillars were apophygate, though later falls of rock meant lost definition in their form. On the east the pillars enclosed a triangular area between there and the pillars lining the eastern cartway. The western cartway had well-defined ruts, and banks at each side were built-up of light-coloured mud, apparently cleaned from the cartway floor. After some 50 m the cartway appeared to pass under a shaft, but this was entirely hidden by rubble and rubble packs which were placed, probably in the early 19th century, by adjacent quarries. The northern side of the shaft was ultimately to be accessed from the other direction. Rubble and packs again obliterated the shaft itself, but cart ruts were just visible beyond, extending to the second pillar beyond the shaft. There appeared to be apophygation on a remnant of pillar close to the shaft. The cartway changed to a barrow-way beyond, and was driven further north possibly in this same phase, until it came under the present church, a former telephone exchange, in Combe Road, where the roof was in a dangerous condition, only a metre below surface. Rubble stone roof support packs had been built in this section against the wall, each vertical portion advancing a very short distance and completed to roof level before the next section was started, to cut down risk. The original way forwards was abandoned at this point and other working in the area was probably done from the north-east.

The eastern cartway, which had cable grooves in the east wall near the entry, was eventually continued almost to the North Road, the Phase II section perhaps as far as the second wide shaft (after passing, according to documents, an unseen well shaft in the first section). The pillars were apophygate alongside both the two cartways (but internally between them were direct, reflecting a probable later working).

The level entry to this eastern cartway had been arched over, followed by a rubble pack at the east side as a buttress for a short distance. There were then apophygate pillars on both the west and east for a short distance with very small rooms between – about three metres across and filled with spoil. An opening on the east led into corbelled pillar and room working but otherwise the sides had either rock walls or rubble packs. Just before the cartway reached the first of the wide shafts noted above, and beyond it, there were faint traces of cart wheel ruts (about 1.3 m internal width and 1.5 m external) and a substantial fall of rock.

Near the first wide shaft corbelled pillars appeared to take over (some were possibly transitional) in one or two rooms back from the cartway. This suggests the cartway was first driven through with a room or two either side, then the extraction area was later widened, presumably in the second sub-phase. The first wide shaft was surrounded by apophygate pillars, suggesting a Phase I date, but was walled around and presumably filled, so no detail was available of the shaft bottom and it is not clear whether the shaft was sunk at the time of driving, or later, though its position suggests either alternative was of the first sub-phase date. It may have been sunk specifically to serve the large quarrying area to the north east (see below).

North of this the quarried area was much more open and may have served to store stone, since there

Fig. 12.15 (facing page) Case Study 4: Phase II workings in Central Byfield (Quarry 505)

were two or three substantial squared, scappled blocks piled near to each other, and, nearby, a stone with a groove to allow it to hold saws for sharpening (which may suggest later activity than Phase II). The roof height around the shaft was typically about two metres, but on the east side there had been a very substantial roof-fall, with large and small rubble and boulders under a large cavity in the roof. This and perhaps falls associated with the shaft had blocked the cartway which ran under the shaft, and a diversionary cartway had been made round to the west of the shaft. This resumed the original course to the north, heading for a second wide shaft. Apophygate pillars terminated just short of this shaft. Later wide shafts here and elsewhere, with the exception of the shaft on the west-side cartway, seem all to have been set at the side of the cartways. The position of the shafts on each of the cartways here, amidst the apophygate pillars, may indicate they were the earliest of their type and the lesson of the relation with cartways was possibly learnt through experience here.

North-east of the small area around the first shaft, alongside the original cartway was a very substantial quarried area with very wide spacings between two lines of apophygate pillars (Fig. 12.16). Many of these were very substantial, almost direct pillars, with only vestigial apophygate charcateristics. It is not clear why this huge area remained nearly free of collapse when many other smaller areas between pillars were so damaged. Perhaps the massive pillars used helped, and, as it was at the centre of the freestone outcrop, it is possible that joints were either less in number or less open.

Within the quarry area the floor was dumped over with heaps of rubble and, on the southern side, there was a long boundary pillar that had been occasionally pierced. Through the area to its south the pillars were mainly vertical and in the area were long barrow-ways mainly running eastwards with lateral dumps of rubble indicating the (last) spoil-disposal method used there. The rubble spoil heaps were to about half a metre to a metre under the roof, the barrow-ways up to two metres. South-west of the shaft was another 'organised' dumping area, with barrow-ways with lateral dumps and distributary ends, some dumps with arcuate tipping fronts. These appear to contrast with the dumping of spoil generally and suggest this activity was becoming more systematic, freeing space in the producing or working areas. It may, however, simply reflect later working in the peripheral areas to the quarry under present consideration.

On the south-eastern side of this large quarry most of the pillars were mainly corbelled with some direct, and at the meeting point with apophygate pillars again there was the 'entabled pillar' noted earlier. The corbelled pillars continued to the east until again apophygates were found around what was called Hills Shaft (Quarry Area 519), an opening formed through a probable Phase I small quarry at the west side of Sheeps House Quarry. To the south east the apophygate and corbelled pillars extended into an area known as H Pillar (Quarry Area 517), a much later quarrying area operated by James Riddle after 1900.

Fig. 12.16 View of the 'Great Quarry' in Central Byfield (505). This was by far the largest area open to view in the Allen workings

As in the last case study, the amount of tool-working which could be seen was minimal, except for jad slots under the corbelled pillars. There was one example of a pillar side under a jad slot being worked for a large block which had been partly scappled *in situ*. It seems likely however, this may have been an example of later intrusion for pillar robbing, for which the saw-sharpening bench stone was further evidence, although there is the possibility that it was a relic of the Allen Phase II.

This Central Byfield quarry was thus a very complex area, which clearly provided a substantial part of Allen's production. On the post-Allen estate map (Figure 5.1), the crane above the Central Byfield entries was shown as gone, though the location was clear from a crescent-shaped area marked over the entries. This suggests the present quarries had either shut down or were being served by one of the wide shafts which were not marked on the map, anywhere. The second wide shaft had cable grooves a few metres to the south down the cartway and the marks of where a wagon wheel had backed into a pillar. These marks may have resulted from winding at this shaft, though later activity there is likely too. Otherwise the only cranes known from documents to be apparently operating at the end of Phase II were at the Masons Crane House Quarry and at the Central Firs entries at Sheeps House Quarry.

Phase III: The Allen Estate leased quarry workings 1765 – *c* 1803

Documentary sources show that by December 1764 at the latest Allen's wooden railway had gone and that leasing the quarries had replaced the earlier direct estate management. The cranes were also sold, though by then only the two at Masons Crane House and Sheeps House Quarry had remained at the quarries. There is reasonable evidence to suppose that the quarrying at Byfield had ceased by the time of Allen's death. Underground quarrying appears to have become concentrated on the quarries under the area we now know as Firs, west and east of The Avenue. Quarrying at surface was being done at Masons Crane House Quarry and at the quarry immediately to the north of the Old Rank of workmen's cottages. There may have been some surface quarrying also at the Sheeps House Quarry, at the east side under the old crane position, and at the west side where new entries were made in this phase. At Byfield west of Combe Road, the 1799 Charlton map (BRO copy) at the Byfield location shows no quarrying despite it being clearly shown around Entry Hill and Crossways. Other evidence suggests quarrying, with the exception of one underground area immediately adjacent to the Combe Road pillar, started (or restarted) at West Byfield in Phase IV. This is contrary to a suggestion by Irving (2005, 29-30) that the leases there were to existing occupants, though some had undoubtedly held leases elsewhere.

It can also reasonably be supposed that the most likely quarrymaster-freemasons after 1764 were those who had held responsibility under Ralph Allen and Richard Jones. Jones, in alliance with his son, appears to be one such independent contractor, possibly taking a lease at Byfield west of Combe Road. This suggestion is based upon the somewhat flimsy evidence that his surname seems to have survived there. In other respects we have little documentary material until towards the end of the century, when a number of freemasons are listed with their production of the years 1795-98 (see Chapter 6).

Most of these freemasons listed in the 1795-98 accounts took out new leases from about 1803, at the beginning of Phase IV. At about that date Benjamin Wingrove and Henry Salmon had purchased the whole quarrying area west of the Avenue across into Lyncombe and Widcombe after which it was sold on or leased out (see Chapter 4). The extant leases and the included plans have information about the areas which the new leases applied. This is most useful where the plans provide evidence of the extent of previously-driven cartways – that is. in the earlier phase(s) – and, also where new cartways should be driven.

This line of reasoning has obvious dangers. If the same lessee who had worked the old lease simply took a new lease for the same area, then neither he nor the landlord would require further detail in the new lease, and it might have been considered immaterial to either as payments were on the area worked since the last payment, with no time limit imposed. Previous working was stated in one of the leases, but this may have been because the lessee was new, or indeed that the owner (Nathaniel Hadley) was new. In this respect the portrayal of actual cartways on lease plans is superior as indicating the true (minimum) limit of the working.

It is accepted that there will also have been technological leads and lags, which will mean that the exact boundaries between Phase II and III, and Phases III and IV, may have considerable overlap. Accepting there will be likely inaccuracy in detail, we can, however, assess the general limit of working of Phase III in certain areas, and by assuming that similar technology elsewhere is of roughly the same phase, apply this to other areas.

The differentiation of Phase II from Phase III is difficult and in a large part has to be done by elimination. It is considered likely to be about half the area quarried up to 1803. Not only did the cartways penetrate further north, but we can also expect areas to be worked east and west of the original Allen workings along his cartways near the southern margins. These should all be distinguishable by the technology used. It is also possible to rule out some areas that have demonstrably later phases of operation.

As noted above, this phase may be indicated by new methods and diversity in their use; that it happened is certainly the case. There were two new

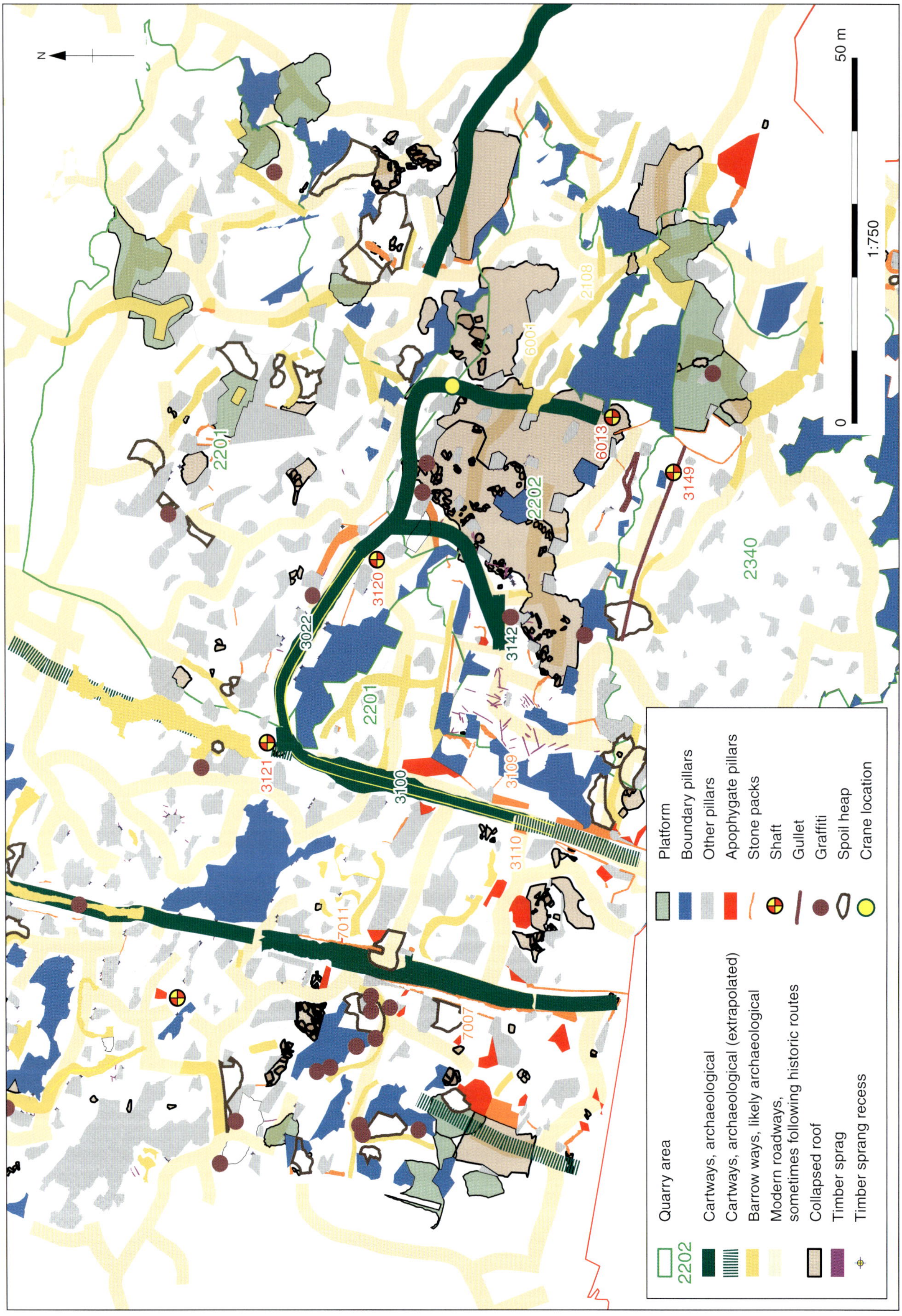

Fig. 12.17 Case Study 5: Phase III. Workings east of the Allen cartway in Central and East Firs

methods of working, Long Room, and Long Wall with gullet and pillars, the first of which may have been adopted for reasons of stability and was not generally used, while the second came into general use over wide areas. There was also more systematic use of jad slots in some places, perhaps associated with a single controlling individual, and the common but not particularly systematic use of wedge pits and the wedge-and-chip for breaking stone. Most dramatic, however, was a trial of the use of cranes underground, though in a small area, apparently only in conjunction with Long Rooms, and then abandoned.

Phase III working largely took place in two main areas, the largest probably west of the Allen Phase II workings in Central and West Firs, including a substantial portion of the spoil-backfilled area (Quarry Area 2397) on the western Firs boundary which is a blank on Hawkins survey (1994). Second, there was a large area on the east side of the east Allen cartway and largely north of the branch cartway in East Firs. The more eastern areas of the latter were probably served by shafts. At Byfield, there were underground workings just west of the Combe Road Boundary, reached through much older apophygate pillar workings (910) on the west side of Ralph Allen Yard, ranging under the William the IVth Public House. In the far east, beyond the Long Drung, it is likely the area north of The Brow and alongside the Long Drung continued to be worked, possibly, by the Greenway family who were still associated with it in 1800, reaching, by the end of the phase within some 120 m or so of the Turnpike (North Road). However, the lack of access to the workings there make this difficult to show. This was a very substantial output area to help satisfy the demand of the building booms in Bath, but in addition to this were the outputs from the surface quarries in the centre of Combe Down north and south of the Old Rank cottages and also at Sheeps House from under the former crane position and maybe also near the west side entries. Also substantially before the end of this phase, in direct competition and with possibly a small transport advantage, four or perhaps five quarries, one of which was underground, were in operation at the western limit of the Combe Down area, at Entry Hill and Crossways. There may also have been other output from east of the Long Drung; Mount Pleasant and other quarries were either already in operation or were being planned.

Case Study 5: Phase III. Quarries 2201/2202 Workings east of the Allen cartway in Central and East Firs (Fig. 12.17)

The branch cartway runs eastward from the junction on the eastern of the two Allen cartways, from under the first metre-wide air shaft about 40 m from the entry. The shaft and the first two rooms lies within the apophygate pillar area and the branch to the east extends about 20 m to a further 1 m-wide shaft from surface. Here the pillars are both corbelled and direct and either side, beyond the apophygate pillars, had distinctive jad slots, common but not systematic use of wedge slots and Long Rooms nearly entirely filled with spoil. The height of the cartway was about 4 m, although roof collapse in sections had left a greater height exposed.

There were very substantial rubble roof-support packs installed on the north side of the cartway following the substantial roof falls which involved large blocks, as is apparent from the gaps in the roof. At a later phase, to which these substantial packs belong, the cartway branched and was extended further to serve quarries to the south but, initially, it was driven eastwards and probably passed just south of a boundary pillar, working the rooms immediately adjacent. Beyond this the cartway was covered with a large fall of rock, but could be accessed again at the other side, continuing eastwards and linking with at least one, and possibly two of the cartways in the east of the Long Drung, then ranging north and connecting to the area just south of the Grand Canyon south boundary pillar. Long Rooms were found in much of this area, and also earlier workings as the cartway cut across them.

The most unusual features were the remains of cranes. These seemed, from remaining Lewis slots used as anchors and from cable grooves (Fig. 9.32), to have been worked from immediately after the apophygate pillars were passed. They appear to have comprised a form of mast crane, with stays or cables of chain or iron rods holding the top of the mast in position – probably from four or five points. Locations of two cranes could be deduced. One was on the north side of the cartway opposite the 1 metre-wide shaft. This position was behind the massive but later pack lining the cartway, and would (presumably) have loaded on to wagons on the cartway. At least two of its chains passed over the cartway (one leaving cable grooves where it passed around a pillar) and others were located close to it. The second crane was on the north side of the roof fall and used wedge slots, Lewis slots and a cable around a pillar as anchors. It may have also dragged stone to what seemed to be a low loading bay a few metres away, and in line with the crane and the workings. Other slots along the cartway just before the fall suggest a crane or cranes were moved as work progressed. Conditions locally appeared very confined, though the full height was, of course, not available to see, and it is for this reason, perhaps, that crane use seems to have been discontinued, although later it became a normal feature of stone quarrying underground, at Combe Down and elsewhere.

It is possible that this branch cartway was driven across to this area, either to avoid unstable areas, which abound to its south, or perhaps to centralise handling at the Sheeps House Quarry, or both. Together with shafts, of which there are several, it enabled the working of the northern half of the area between the older Phase II workings and the later Phase IV workings by Burgess in the north. The abandonment of old cartways passing under it may also have facilitated the surface working of the quarry behind the Old Rank cottages.

Case Study 6: Quarry Area 2201. The long-room and gullet workings west of the Firs Shaft

These workings were at the upper end of the western cartway at Sheeps House Quarry and a further cartway to the west was also started in this phase, penetrating perhaps 35 m by the end of the 18th century. These may have divided further in to serve this area, though the actual divisions were not accessible to us. Of the short sections seen, they were at least 4 m high, wider than the earlier Allen cartways, and had traces of cartway ruts. Jad slots had been used for exposing the beds below, and again wedge pits were common but not systematic. There is a regularity seen in the pillar plan made by Hawkins (1994) in this northern half of the supposed Phase III working which is not apparent further south, where often irregular but more substantial pillars have been left. These latter were possibly of the earlier phase and may partially be due to ground conditions which were often not good. The pillar plan (Figure 1.9) suggests smaller pillars and a higher proportion of stone removed than is the case further south, which was presumably the result of the method of working.

At the north end of the Allen Phase II workings (2200) a former cartway or barrow-way, discernable under systematically dumped spoil, extended from the Firs Shaft to the west and was followed by one of the modern supported roadways, which was thus unfortunately but necessarily destructive of substantial detail between archaeological visits. This cartway followed the south side of a long boundary pillar and later experience suggests its partially rubble-filled hollow against the pillar was, in fact the final stage of a gullet following the earlier working face. It first crossed the west Allen cartway, which had passed the boundary pillar and which was associated with Long Room working. After some 20 m it crossed a further cart or barrow-way into Quarry 2224. This was the site at which the diagram for the Long Room and pillar method was sketched out (Fig. 8.12), though the area was destroyed before full recording was possible.

On the east side of the cartway, to the north the freestone had been removed and the area backfilled to a walking height below the roof. The infilled area had clearly been served by the cartway, since the latter's deep trench had been blocked by a substantial rubble pack at its south-west corner. West of this rubble pack the cartway curved and opened out into a gullet running along the working face with behind it a row of corbelled pillars. The gullet was about 4 m deep (but with substantial rubble fallen in) below the roof and about 15 m long. Two pillars were beginning to be formed against the long backwall pillar and a wrist stone had been removed to allow the face to be advanced. At the back of the gullet, more or less in line with the fully formed pillars, was a low and irregular rubble wall from which a bank of spoil rose to the upper floor level at a normal angle of rest. Thus the cartway had accessed the centre of the quarry working area, the east side had been worked-out and its final gullet filled, and the new gullet progressed forward. The partially formed pillars suggest the intention had been to advance further.

This simple system, which apart from the pillars resembles surface working methods in many pre-crane stone quarries, probably occurred over wide areas, notably in the workings in East Byfield (of the later Phase IV) where the pillar plan also shows regularity and where small traces of similar gullets could occasionally be seen.

Phase IV: Workings between *c* 1803 and *c* 1838

The main areas of working were laid out by Salmon for west of Combe Road and through Wingrove and Layton eastward to The Avenue (see Chapter 6). Most of the land west of the Combe Road was surface-quarried, but underground William Hulonce took the area under the later William IV Public House extending to what is now a grilled bat entrance. There was also a fairly small area in the north-east corner, inside the Combe Road / North Road junction worked underground by John Scrace and close to it another very small quarry of extraordinary neatness, suggesting again a mason at work. Both these areas are considered in the case studies below.

Any attempt at describing the underground landscape in the northern parts of either central and East Byfield or West Firs is dogged by the limited availability of viewpoints during the Stabilisation Scheme and the apparently chaotic distributions of spoil and often low barrow-ways used to transport it from working areas. On closer examination this was less chaotic than first appeared, the spoil from one area usually passing southwards to be dumped in the adjacent worked-out area, which thus became almost totally hidden below spoil. The area between the notional Byfield and Firs quarries, the southern part of which was worked in Phase III became almost totally filled with spoil. The most coherent way to approach a general description is by means of the cartways of which some are shown, together with shafts mentioned on leases, or otherwise planned during the archaeological survey. The planned cartways were frequently used in conjunction with shafts and are shown in Figure 9.2.

The first leases extended east to a boundary running south to Rock Lane from Stonehouse Lane (a boundary probably established by the original Combe Lane track). This included unworked ground to the east, beyond Allen's workings) and to the south, within the curve of Combe Road / Church Road / Rock Lane.

In Central and East Byfield four 40-perch-long quarries were measured out west-east from Combe Road to the Stonehouse Lane-Rock Lane boundary. The Phase II cartway (originally Allen's east cartway) to the second wide shaft was extended northwards into Quarry 2213 (that on the west was abandoned as it was in bad ground), to serve the quarry on the west (2342 and 2343) belonging to Samuel and John Pearce, as well as that of Richard Lankasheer who leased the adjacent quarry with the cartway at its centre (2220). As the cartway progressed northwards a further two wide shafts were sunk on it, regularly spaced so that in all four shafts served the workings to the northern boundary (North Road), almost in a straight line (only three are shown on the early Ordnance Survey maps – presumably by then the southern shaft had been filled in, perhaps because of instability).

The two remaining quarries on the eastern half of Byfield were originally leased to Samuel Pearce and Ann Godwyn (Quarries 2221 and/or 514 and 520) . They also were to share an entry, and the cost of making it, against the value of the stone got from it, was to be shared between them. It seems to have been made on the west side of Sheeps House Quarry, probably utilising an old apophygate pillar working (Quarry 519). This area became known as Hills Shaft in the Stabilisation Scheme – a misnomer for a partially backfilled level entry). The actual cartway followed a more direct route than indicated on the lease plan to a denoted shaft in virtually the centre of Godwin's workings, Quarry 514. This cartway in its northern section ranged north to a small loading bay for wagons which were pulled back southwards to the shaft (hidden by a fall) by cable which has left traces of wheel-ruts and cable grooves and abrasions (curved faces just above floor level) on pillars on its slightly winding course back to the shaft (Fig. 9.13). The loading bay area had graffiti including the date 1816 on a sawn face (Fig.11.8), indicating saw-use at that fairly early date for quarrying rather than banker mason work.

A second entry was shown on Job Salter's 1816 lease, just south of the above Hills Shaft entry. This was driven into John Davidge's working (East Byfield,Quarry 520), with from here a further cartway ranging south, taking a slightly curving course through a fairly narrow worked area. The cartway was in a deep trench, the pillars were direct and corbelled, and there were distinctive jad slots and both single and rows of wedge pits (Fig. 12.18). The way was flanked by interconnecting, more or less spoil-filled rooms. The presence of many Lewis slots showed that mast cranes had been used for handling stone blocks throughout the length of the cartway. It was not clear how transported blocks were hoisted out, possibly via an incline either just inside or at surface.

A major (negative) feature of the Hawkins survey had been an unexamined area between the Firs and Byfield Mines. During the Stabilisation Scheme it was shown to be almost entirely infilled with spoil. It took in halves of two quarries, one the eastern half of a moiety of Samuel Pearce's quarry, then owned (1816) by John Davidge (Quarry 2397), and the other part of the western side of Godwyn's Quarry then

Fig. 12.18 John Davidge's working, to the south of Hills Shaft in East Byfield (Quarry 520)

owned by Job Salter, Quarry 520. The original cartway came to its eastern margin, but no trace was found of the arrangements in Davidge's half. It is possible that the arrangements were shared.

A further tranche of leases affected land east of this which remained west and north of ground worked in Phase IV, largely in the north of West and Central Firs. Extending east of this, John Greenway had a lease in partially worked and unworked ground, and east of Firs Shaft took a large area of unworked ground in a triangular area with its apex opposite the Prior Park carriageway gates (under the Hadley Arms today). This area, when leased to John Burgess, was described as of three acres and this gave its name for our identification purposes to this substantial quarry. This group of quarries, all let from around the same time showed both rapid changes of ownership and a considerable diversity of methods used (emphasising how this might be valid as a phase change criteria between Phases III and IV). The methods used were usually very traditional, but there does appear a willingness to retry methods used but not generally applied earlier, maybe stimulated by the ability of the Kennet and Avon canal to bring in competing supplies of stone if they failed to gain efficiencies to protect their local market. By comparison, however, as will be seen below, the organisation of the quarrying methods was far less sophisticated than taken up by John Burgess further east. However, it is possible that the lease-areas granted were too small to allow the larger-scale methods used by Burgess.

The cartways shown in leases show substantial working took place between the above leases and the Three Acre Quarry, in West and Central Firs. The quarrymasters here were Samuel and John Pearce who seem to have worked mainly from the cartway in the north-west of Sheeps House Quarry (Fig. 12.19). Southern parts of this area had been worked in Phase III, and at least one cartway, west of the Allen cartways, was shown extending about 36 m from its entry in 1803, but this would have left substantial amounts of stone in the northern and north-western parts of Firs for the Pearces. Probably two or three of the cartways in the north-west of Sheeps House were used if not initiated by them. They too followed a fairly conventional mode of quarrying.

The plan (Fig. 4.2) deduced from the lease plans and underground observations, shows the likely grant, which excludes an area in the four acres known to already have been worked previously and another part of the four acres not worked until significantly later in the century. The remaining part, known as the Three Acres Quarry and within it, what seems to be a late, even last, stage of working in Phase IV became known as the 'Grand Canyon' because of its spectacular near full-height pillars and gorge-like topography. Two contiguous small areas of $^{1}/_{2}$ acre each at the top of Combe Park and the Tyning were also worked by Burgess. This is the subject of one of the case studies below.

Fig. 12.19 Late 18th- to 19th-century cartway with cart ruts in West Firs (Quarry 2342) which belonged to Samuel and John Pearce

Case Study 7: Phase IV. Quarries 500/501/502/ 503/504 west of Combe Road – Hulonce's Quarry (Fig. 12.20)

William Hulonce was one of the two Lyncombe and Widcombe quarrymasters who took leaseholds from Harry Salmon *c* 1805 (see Chapter 6) and developed underground quarries. He almost immediately subleased the western half of his ground to Patrick Byrne and he may have also subleased another part to Henry Street when Hulonce possibly gave up quarrying about 1816, although the evidence is somewhat confused (Irving 2005, 41-2). This is reflected in the diversity within the workings.

The underground area stretched from Ralph Allen Yard, under the William IV Public House and yard ranging for about 100 m west to the present grilled bat entrance. A large part of the underground area remains open today for a bat sanctuary, but has been almost completely impacted by concrete support to the original pillars, masking archaeological features.

Hulonce took a lease which included workings of earlier phases. The eastern end workings (and south of his leased area roughly under Tor House) are both pre-Allen and Allen period, though unlikely to have been worked by Allen (Phases I and II), and display mostly apophygate pillars (Quarry 910). The entry envisaged in the lease has a fine stone arch, as seen from inside, and a cartway of probable Phase III age (Quarry 501) passed through the apophygate pillars and through a Long Room-form of working alongside the Combe Road pillar, at the far end of which a wide shaft had been sunk. These Phase III workings, possibly those of Richard Jones and Son, may have curved round the north side of the quarry a short distance to another wide shaft. Hulonce's decision, alone amongst his immediate neighbours, to quarry underground, may have been influenced by the existing underground access. Even so, only the eastern half of his lease was eventually quarried underground.

Despite the existing access, a new entry was made on the south side, behind (the later) Tor House, in the form of a decline (Irvings Incline – see Fig. 7.5). It has a wide arch with cut stone including some sawn pieces and, from its size, was clearly intended to be the main access to the centre of the Hulonce workings leaving the extant pillar between them and the older work. Its exit seems to have turned sharply east behind Tor House, following the low cliff and wall at the end of the pub yard. Unfortunately the decline hit bad ground at its lower end, caused by a large gull or open joint, and a heavy rubble support pack had to be built across its path. This left a 1 m-wide opening with the main entry turned to the west, then down another, smaller, declined ramp into the central area. To the east it links via a low doorway with both sawn and picked (jad cut) sides opposing each other to workings and a passage leading through the Combe Down pillar, perhaps for ventilation needs. The main Stabilisation Scheme entry to the Byfield workings was at the north-west end and may have been formed (or enlarged) by adjacent surface quarrying and was almost certainly at a late stage in development there.

At the north west corner of the central area (Quarry 501) adjacent to what was known during the Stabilisation Scheme as Bat Alley – now a bat incubator, along the north wall and adjacent to the Phase III workings near the northeast corner, the existence of a long high bank of dumped material suggested this was the first work done, probably from the Ralph Allen Yard entry. The direct pillars in the corners have jad slots used to excavate the picking beds, with 0.9 m high cavities left at the top of the fairly substantial pillars. Similar picking bed removal follows along the northern side, at the northern boundary pillar. It seems likely that a way was formed round the central area in a U-shape as the initial work in the area, possibly using the Phase III cartway entry, though the cartway has definite features distinctive from it. It was an idiosyncratic form of working by that time, though somewhat similar features can be seen in northern parts of Central and East Byfield.

The main part of the central area served by Irvings Incline had high pillars, with approximately 5 m of space above the floor (Fig. 12.21). These workings were also of an unusual form for Combe Down, some with jad slots in the picking beds and either natural vertical faces or vertical jad cut faces distinguished by sweeping pick marks from roof to floor, or with wedge-cut faces. Subsequently the picking bed-portion of workings at the north side was used as a dump for large quantities of spoil composed of fines with some rubble, apparently largely derived from the vertical jad cuts. This formed a wide bank, to within a metre of so of the roof, which was partially held back by the first line of the jad cut pillars, but otherwise sloped steeply to the lower floor. It was not possible to determine the full height of the working, as no gullet area remained open, possibly because of some re-use of the workings for storage at a later date.

The western end, between the Bat Alley and the western entrance had direct pillars made using jad slots of which some two metres remained above floor level. The floor had probably been cleared to allow a secondary use including, apparently, storage of mushroom boxes (Quarries 500 and 501). At the eastern side was a line of rubble stone pillars supporting ground with a sewer pipe running above (the Horsecombe Vale sewer).

The Hulonce quarry, particularly at the western end, had many sawn faces. On the whole these appear to be later than the general work, but some may be contemporary with the main working phase. Such faces appealed to the literary sense of many who came in, including those who worked

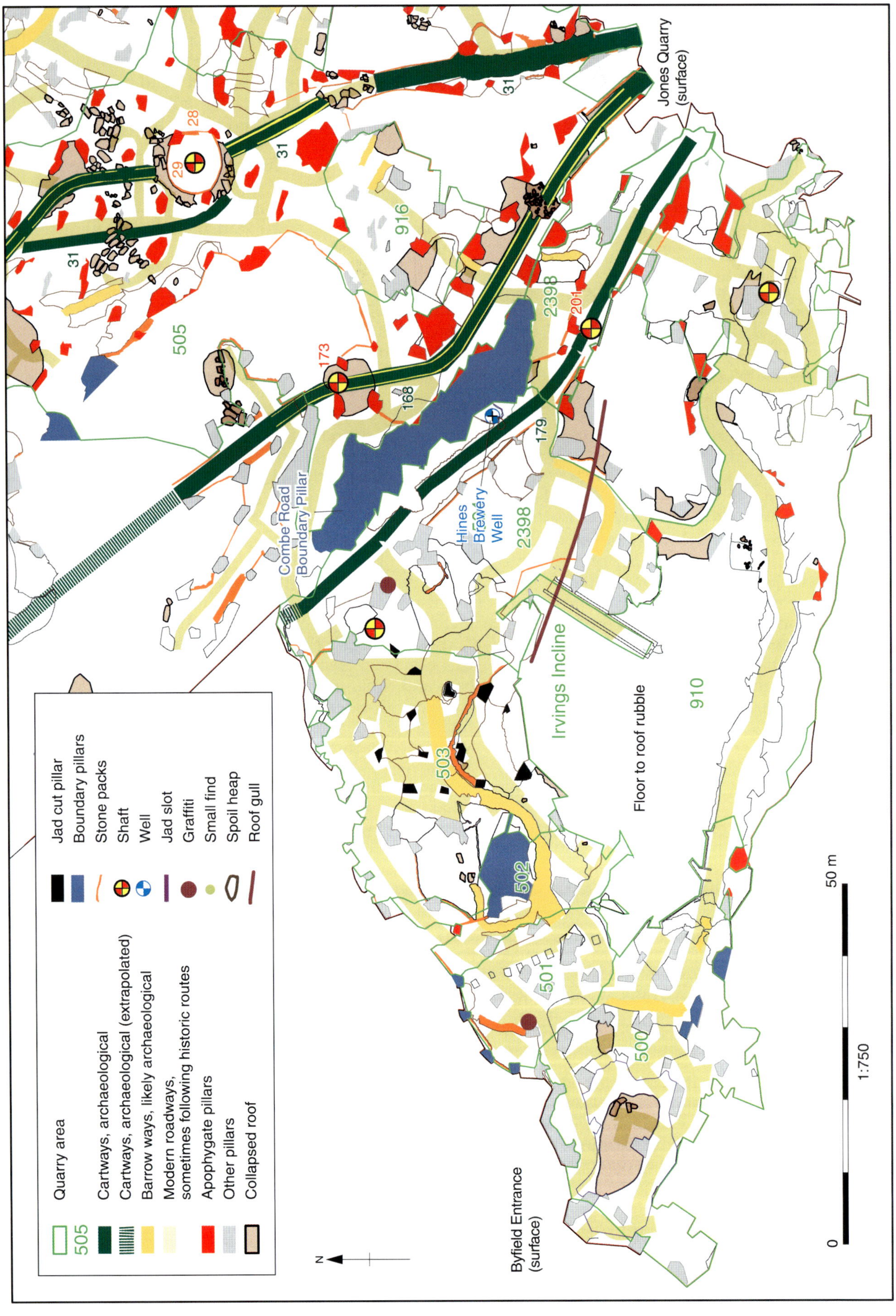

Fig. 12.20 Case Study 7: Phase IV Quarry west of Combe Road – Hulonce's Quarry

Fig. 12.21 High pillars in William Hulonce's Quarry (West Byfield)

there. Of particular value is one close to the entry displaying a graffito listing names of workmen and (probably) the production which can be attributed to them. This particular graffito (Fig. 11.18) was removed as a sawn slab, and so remains available, but it, and others await detailed analysis outside the scope of this report.

The picking bed system seemed decidedly old-fashioned to have been used in Phase IV. The jad slot used at the top of the high pillars in the central area is much more typical of its time, but the very few examples of vertical jad cuts are known in all the quarry areas in Combe Down. They are however, a very common feature indeed in the Wiltshire quarries such as at Box, with literally miles of jad cut faces recorded. Whether the technique was derived from Wiltshire, or whether the technology flow was the other way is currently unknown. The picking bed work was clearly earlier, as demonstrated by the dumping over its locations, so it may be this was done by Hulonce, and since the Irvings Incline would have opened on to the edge of the earlier area, then it too probably belongs to Hulonce's time. Whether the work at the west end is by Hulonce, Byrne or Street is unknown, but, as it is later and Street eventually succeeded to all the area, it was most likely his.

Of other features within the quarried area, the most notable is the Hines Brewery well (see Fig. 11.1). This had been sunk by 1813 when access to it was granted by Henry Street by Hulance, then presumably for a private house (Irving 2005). The original brewery was started in 1824 by Job Salter. The date of walling around is not known, but walling seems to have been a normal practice when the area was originally built-up. The Hines Brewery was built soon after 1840 (Bone 2000, 112) and subsequently became the William IV Public House.

Case Study 8: Phase IV. Quarries 508/512, John Scrace's two-level quarry and the mason's quarry (Fig. 12.22)

Scrace's very small but unusual quarry lies at the south-west corner of the North Road and Combe Road junction. It was a small part of the leasehold property, running north to south lengthwise, granted to him by Harry Salmon in *c* 1805 – Quarries 508 (upper) and 512 (lower). The working was bedevilled by a bed of freestone, between about 2 to 3 m below the usual picking bed roof, which contained 'cockles'. These are small cavities within the bed, also called vugs, lined with calcitic, 'dogtooth' crystals, which ruined the masons' saws and the smooth appearance of the built stone. They were not commonly found elsewhere and were usually either discarded or the bed made into coursed rubble for backs of buildings or interior dividing walls. Scrace's access is not clear since the workings on the opposite side of Combe Road did not approach there until much later. He may have had a shaft, and his workings area on the south side has a well that might originally once have served as a shaft. Unless he worked at surface on the south of his plot from 1805, rather than later, an access from it is also difficult to conceive. An incline or level entry was probably possible, and would have been more suitable, from close to or in his neighbour Abraham Sumsion's quarry with whom he probably shared a crane and two winches (as revealed by details of the sale of Scrace's property). The upper working, apart from a stooping-height

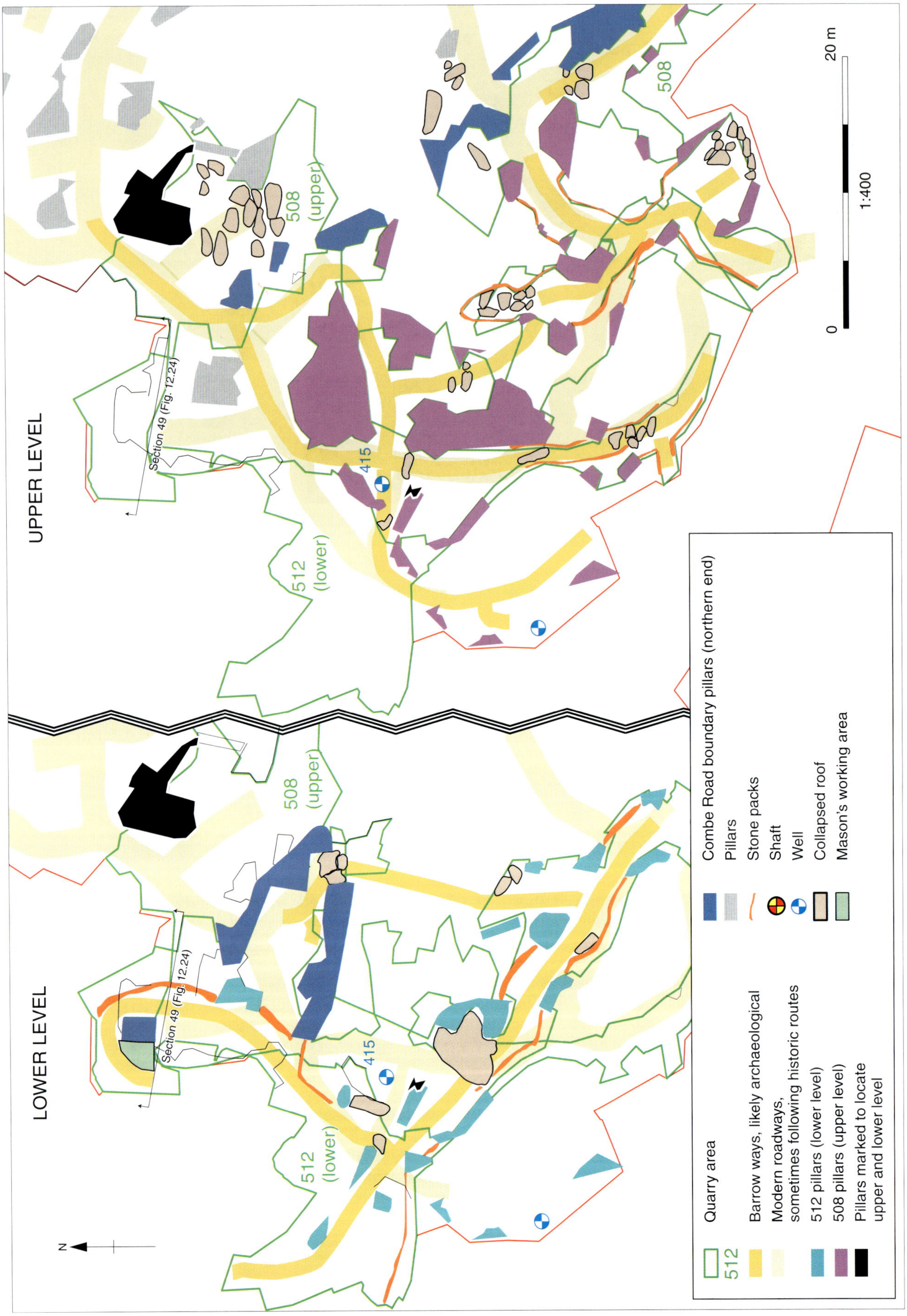

Fig. 12.22 Case Study 8: Phase IV, John Scrace's two-level quarry and the masons' quarry

Fig. 12.23 The unstable lower working in Scrace's quarry (Quarry Area 512, Byfield)

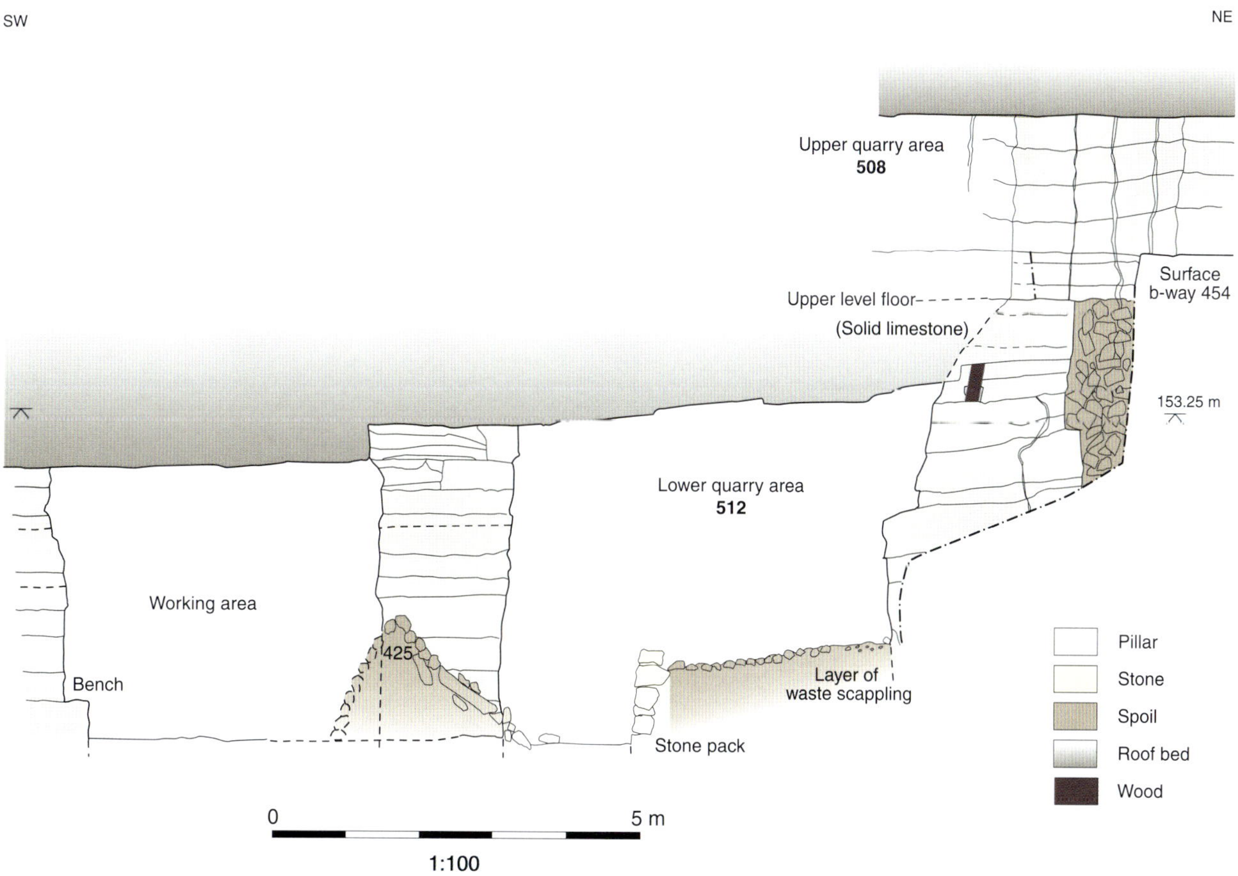

Fig. 12.24 Section of the masons quarry, showing step-down from John Scrace's Upper Quarry (508) on the right

barrow-way was found almost completely filled with rubble dumped behind packs, perhaps because of the instability more apparent in the lower working. This led at the south end to the small quarry working area (422) where it was necessary to step down a metre as the roof was lower with no upper working directly above.

The northern part of the lower working was a long chamber with a blocked side (Fig. 12.23), showing signs of extreme instability and many already fallen blocks of stone. At the north end was an opening in the roof, giving access to the upper level. The instability to the north was probably caused by the upper working pillars not being directly superimposed on those of the lower. Scrace went bankrupt, and most of his plot was subsequently worked by surface means.

The masons' quarry working area was reached by a 1 m-step down from the upper workings (Fig. 12.24). It was remarkable for its neatness, both in its working faces and the careful stacking of spoil on a low bench at the side. The floor appeared almost to have been swept before abandonment. The north-south barrow-way, used for both stone and waste, entered around a rectangular direct pillar some 3 m high, with neat jad slots and regular broken faces, each bed emphasised (Fig. 12.25). At the back of the working, a row of triangular wedge pits was found low down. Behind the pillar several cubic or rectangular, scappled blocks, with minimum dimensions of 300 mm were scattered around. Two lengths of timber lay on the floor. It is difficult to believe this was by the same workmen as seen elsewhere in Scrace's lower working. Possibly it was that of a mason and labourer, perhaps as in some Far East Firs workings, exploiting the stone below to build the house above.

A further small quarry working area within the same phase was located some 8 m to the north of the workings described above and was linked to it by a barrow-way. It was fairly similar in appearance but had been substantially filled by spoil and was also next to a roof collapse. The well, mentioned earlier, was 3-4 m away in a short branch on the west side

Fig. 12.25 The masons' quarry: note the clear floor, spoil neatly stacked and the clear-cut faces on the pillar

of this further working area. What appears to have been clay from the Fullers Earth Formation below was dumped over the floor of the connecting branch, suggesting the bottom of the shaft, if not the upper section, may have been sunk after quarrying there had ceased. At surface the well is now under the entrance road to the recent Byfield Mine access just off Combe Road. A supposed former bakery was demolished some years ago at the well site and houses were built there about by about 1830, by when Scrace's and the mason's working was presumably finished.

Case Study 9: Phase IV. Quarry 2211, The Three Acre Quarry and the Grand Canyon (2209) c *1809-1838*

The Three Acre Quarry occupies a roughly triangular area with its apex under the road junction at the then Prior Park Gates (Hadley Arms today), extending south-west to a nearly continuous boundary pillar. This seems to have terminated the Phase III workings; on the west it came against Phase II and Phase III workings in Central Firs near the cartway to Firs Shaft and the cartway to its north. The north boundary was the Turnpike Road (North Road today), perhaps extending to half way across under the Turnpike (Fig. 12.1). There seem to have been several small underground workings alongside the Turnpike, some possibly preceding the Three Acre, and at least one after (Quarry 2212) possibly worked by William Stennard from an entry under the North Road in his surface quarry near what is today Selway House. John Burgess acquired the Three Acres via John Greenway on a lease from Nathaniel Hadley in 1809, along with two further $^1/_2$ acre plots either side of the north end of the Long Drung. Burgess presumably worked them together, perhaps starting at the north-east end and proceeding westward (see below). He had previously operated the surface quarry on the north side of Hopecote Lodge, off Church Street and there are strong indications that by 1810 it was nearly worked-out by the time building on the south edge of the site began.

Working the Three Acres was clearly carefully planned from the start. In the two small plots at the north end of the Long Drung, two wide shafts had been sunk, one near the apex of the properties, the other under what is today the floor of the motor car repair garage on The Avenue. In the Three Acre Quarry itself, a cartway was driven centrally through the area from a shaft just south of the present day Hadley Arms, south-west to a point close to the Firs Shaft. This may have been done from the shafts simultaneously or progressively from the first one sunk, but stone extraction would have proceeded immediately it was feasible. The shaft near the Hadley Arms was not seen but it and the next two shafts were regularly spaced at about 40 m apart, but the nearest of the new shafts to Firs Shaft was some 60 m away. The haulage distances along the cartway are thus very short to one or another shaft.

Firs Shaft probably dated to the late stage in Allen's time. The second shaft ('Chestnut Shaft') was located close to the War Memorial on the northern side, and under the chestnut tree on Firs Field. It was tunnelled under during stabilisation and was found with much rubbish at its bottom. The third, on the east edge of the field was used recently as the materials shaft and was called by the survey team the Arched Shaft. The fourth shaft was not accessible, and was encircled by floor-to-roof stone packs to retain surface-derived materials. The Firs Shaft was certainly used by the quarrymen for winding stone out from a depth of about 11 m to the cartway floor, since wheel-ruts led from the nearest working (the Grand Canyon) towards it but were not seen in the cartway beyond (Fig. 9.14). There is evidence in cable grooves at the shaft and a worn curved surface on the first north side pillar on the cartway that a cable was used to drag wagons back to the Firs Shaft. There were also Lewis slots in the shaft itself centring on a post about 1.5 m from the south wall. This may have been a mast crane situated in the shaft hauling from the cartway. There can be little doubt similar technology was used elsewhere within the quarries. It is unsure whether the Chestnut Shaft and that near the Hadley Arms were ever important for winding, since the facilities at the bottom of the Chestnut Shaft were minimal, and possibly their function was mainly ventilation, but the other, recent materials shaft was clearly important and was probably the main shaft used.

On its west side the Arched Shaft had a spectacular stone-arched opening into which the cartway level directly led (Fig. 9.15). It and the shaft whose actual opening lies approximately 2 m beyond, were reinforced by a very large rubble stone pack, at least 5 m high, facing west but continuing around the north side. The round arch was about 2 m high and 3.5 m wide, springing off the sides of two pillars about three metres from the cartway floor. It and the walls were about 0.9 m thick. The upper part of a second arch was subsequently found during heavy machine works to convert the shaft to the recent use, on the south side, the top of which was roughly at the level of the upper arch springing. It probably led into another, concealed cartway, perhaps at a slightly lower level.

From beneath the west-facing arch, the shaft was seen to bell-out at the bottom, so the rock underside sloped back some four metres to what was the base of the Twinhoe Beds. Thus the west-side top of the shaft was some 4-5 m east of the inside of the arch. It was heavily buttressed on the south side, and above the arch there was rubble and clay fill surmounted by a rubble pack to support the natural roof. From the inside, the rubble pack on the north wall was seen to line the shaft and was probably additional to that seen outside the shaft, rising some 9-10 m to within the Twinhoe Beds some two metres from surface.

On the east side the lower part of the shaft similarly belled out, and though the shaft had a huge clay shaft cone within it (mainly not touching the sides) it was clear the east side cartway had been filled in and was disused. This supports a hypothesis of quarry development from east to west.

The belling-out gave a large area for manoeuvring loads at the bottom of the shaft, but at the expense of stability and a large investment in reinforcing stonework (Fig. 12.26). The shaft was much more massive in all respects than any of the other shafts observed and would have probably have needed a substantial hoisting arrangement above it. On the east side of the shaft the wooden shaft of a (probable) two-wheeled cart was found. It was approximately 150 mm square in section, and reinforced by iron bands which ran down the sides and joined over the end: presumably this was to withstand colliding with the roof when it was moved into a vertical position during loading. It had a stout crosspiece inserted near the end to allow one or two men to handle the load, but the actual carriage and platform were missing.

The cartway route was visible from the west side of the Arched Shaft back to the Firs Shaft and had been impacted by fallen rubble from the sides, so only in the section from near the Grand Canyon to the Firs Shaft was the floor visible, again emphasising the east-to-west progression of working. On the north side of the cartway, continuing from the main shaft to just east of the Grand Canyon entry, at a similar floor level, the adjacent ground had been kept clear for between 3 – 5 m width. Beyond this, a rubble pack extending to near roof level held back spoil. This may have been either a working area for banker masons or a storage area. On the southern side a rubble pack rose from the side of the cartway, again up to or near the roof, supporting spoil. In either case there was no visibility beyond the packs, illustrating how thorough and organised spoil disposal had been, a feature noted elsewhere in the work done under Burgess.

The Grand Canyon junction was only about 15 m from the Firs Shaft, and by it the cartway passed through long pillars on either side. This appears to be the long boundary pillar dividing the Canyon from earlier work. Between there and the Firs Shaft the cartway cuts obliquely through two Long Rooms from the Phase III working north and north-east of Firs Shaft.

In the area eastwards beyond the Arched Shaft the cartway was not visible, but towards the later Hadley Arms, the workings were generally found fairly open between the pillars, with level floors consisting of waste stone and fines. The tall vertical pillars had some sawn faces in the majority of the workings. The pillars were generally some five metres high above the floor with spoil presumably stacked some distance away on the south side, beyond view. The pillars, and one strong continuous pillar were aligned so as to cross the predominant regional joint system as was the boundary pillar at the Grand Canyon. Both these boundary pillars had successive sections offset, presumably to prevent a linear weakness in the roof developing through the system. This can clearly be seen on the

Fig. 12.26 Supporting stone packs at the Arched Shaft

pillar plans and demonstrates knowledge of how to exploit weakness for working purposes without succumbing to it – a fine example of underground quarrying. This area also has several survey marks in the form of an incised cross, usually on pillar walls, but with one example on a boulder.

The working of the freestone seems to have gone out under the Turnpike or North Road and one section has a serious collapse which may have penetrated to surface and been filled. East from this, a very substantial rubble-pack wall had been built, following the edge of the road, then curving under the Hadley Arms walls. This was probably following the original roadside opposite the Prior Park carriageway gates. It is not known when this was done but it may have been at the time of quarrying. Alternatively it may have been done at the time of road widening for the electric tramway, *c* 1900.

It is probable that the unsupported, partially collapsed areas along the road west of the strong supporting pack or wall (near to the surveyor's marks) were not carried out by Burgess, and they have been shown as separate quarry areas in the survey based on their archaeology. Their existence may add credence to the North Road Wall being built under Burgess.

The North Road Wall was pierced in two places near the Hadley Arms, first at the west of the Hadley Arms yard for the Hadley Steps, an entry with a corbelled arch which predated the wall and the 1845 building of the Public House. Another corbelled arch was found a short distance away, leading to the Hadley Arms well.

Opposite the former Prior Park gates, underground, was a further similarly corbelled-arched entry. This, however, had a Roman-arch branch within, leading under the road at the junction. A passage and chamber under the road has been driven from the Prior Park side to it. A surface quarry was worked at the south-west corner of the Park in the early-and-mid 19th century. Crownholing (upward development of a roof collapse to surface) and subsidence under the tramway took place around 1920 (Hawkins 1994, 13), presumably under an unsupported section.

The Grand Canyon – a complete working area

It was unusual anywhere in the quarries to observe complete working places because of their subsequent use for backfilling. The so-called Grand Canyon was the most spectacular, and canyon-like feature of all, and the only one observed where a complete system of major working remained unbackfilled (see Fig. 12.27). It was not possible to physically enter the deep trench of the gullet due to access restrictions, so views and features seen were generally from above from a short length of spoil platform on the western side and nearly the whole length along the dumped spoil on the east side all from just under the roof, and from a slightly lower viewpoint at the southern end. The west end was obscured by a corner in the cartway which ran for a few metres before disappearing into the gullet out of sight. Use of cameras extended up to five metres on arms or slides, or running on a cable across the width allowed more detail to be recorded but the lower sections have, at best, only been viewed from a distance, though some detail was recovered of these by laser-scan survey.

The dominating feature in the Grand Canyon was a row of direct pillars up to 8.5 m high exposing almost the whole of the worked freestone sequence with a narrow gullet between them and in the front of a worked face. This had a barrow-way running along it on the south side along the long boundary pillar (Fig. 12.28). There was also a subsidiary working face in the north east corner connecting with working rooms behind on the north side. Between some of the line of pillars, tall rubble-stone packs had been constructed retaining rubble spoil (Fig. 12.29). The widest part was about 19 m and it was 70 m long overall with an incline leading east from it as a barrow-way to transport spoil into largely worked-out areas. At its west end the gullet and solid pillars alongside turned at a slight angle into the Firs-Hadley Arms cartway; this ran back to the Firs Shaft and displayed wheel ruts showing the Grand Canyon operation to be the last in the immediate area.

It is the most western of the lines of workings, many presumably with similar features but buried under spoil. Its survival was presumably caused (allowing for some secondary working) by it being the last part of the Burgess Quarries and the last major working within the general vicinity. Working of the wider area began about 1809 by John Burgess and ended, at the lease termination, under his son Harry in 1838. Stratified clay tobacco pipes have been recovered from the western end of the canyon and have indicated dates from the mid to late 1830s suggesting that the workings were indeed worked towards the end of the Three Acre Quarry's operations.

The pillars were developed along the joints, especially the main NW-SE direction but apparently to achieve a regularity of dimension, joints aligned some 40° from the main set were largely ignored. To do this the other rock faces were broken by splitting, sometimes by use of wedge pits and wedge-and-chip. Most pillars were thus rectangular in form, vertical, mainly direct but some with a corbel (Fig. 12.30). However viewed from the south-east side one was split into a V-shaped plan, presumably due to a joint direction affecting it. A photograph, taken remotely from a position just under the roof and directly along the gullet showed the pillars to be well-jointed with about five vertical joints in the width. These, probably under post-working stresses were opening, suggesting that stability today is poor, though no doubt it was adequate when worked. Typically the pillars were about 1.5 m wide and some 2 m long with gaps between around 4 to 5 m. Several pillars had sawn faces, and at least one

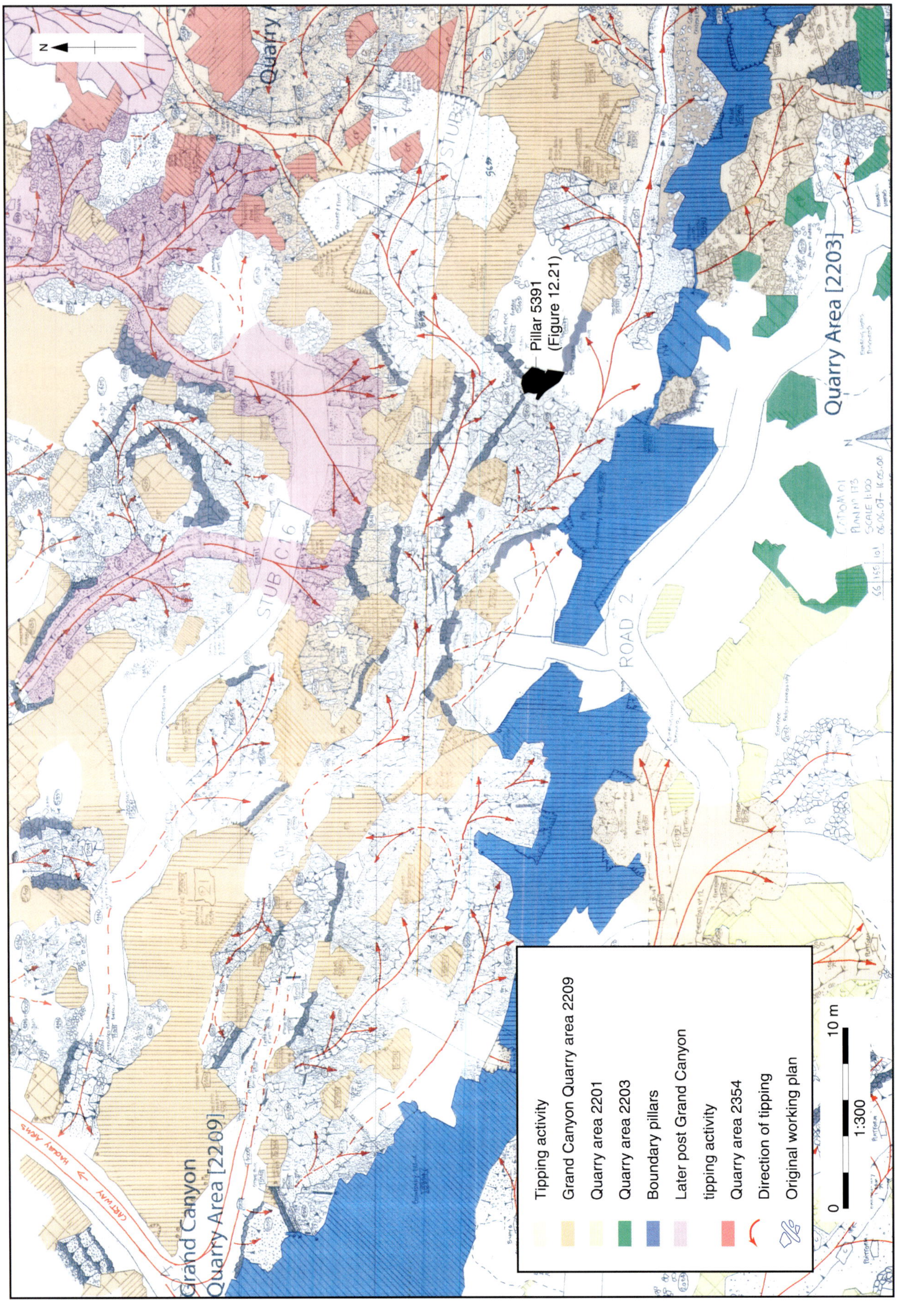

Fig. 12.27 Plan of the Grand Canyon area (2209) of the Three Acre Quarry

Fig. 12.28 View towards the north-west from just under the roof of the Grand Canyon, directly over the central gullet, showing the full 8.5 m maximum height of the direct pillars and the barrow-way below, with over 5 m-high rubble stone packs at the side

Fig. 12.29 Direct pillars seen from the southern end of the Grand Canyon

such face was so positioned that this must have been part of the original working, but was certainly not the usual method of working. Other sawn faces might reflect a final aspect of the working process – taking the last morsel, as it were, or might have been an entirely separate secondary pillar robbing operation.

Pillar 5391 was located at the east end of the Canyon and had an exposed face surviving to a height of about five meters with a significant amount of evidence for tool-markings. These included the initial chamfered jad slot beneath the corbel at the top of the pillar (Fig. 12.30). Below, on several beds are a series of wedge holes with associated chip impressions, showing their use to extract blocks of stone in conjunction with a vertical jad cut on the pillar face; wedges were also driven in along the bedding plane to prize the blocks from their horizontal beds. Numerous $3^1/_2$ inch and 4 inch-wide hack marks (88 to 100 mm) or scappling axe marks were noted on the face. These were used to trim the face of the pillar to prepare an access to the bed beneath. A pillar located on the northern edge of the Canyon had similar tool-marking features,

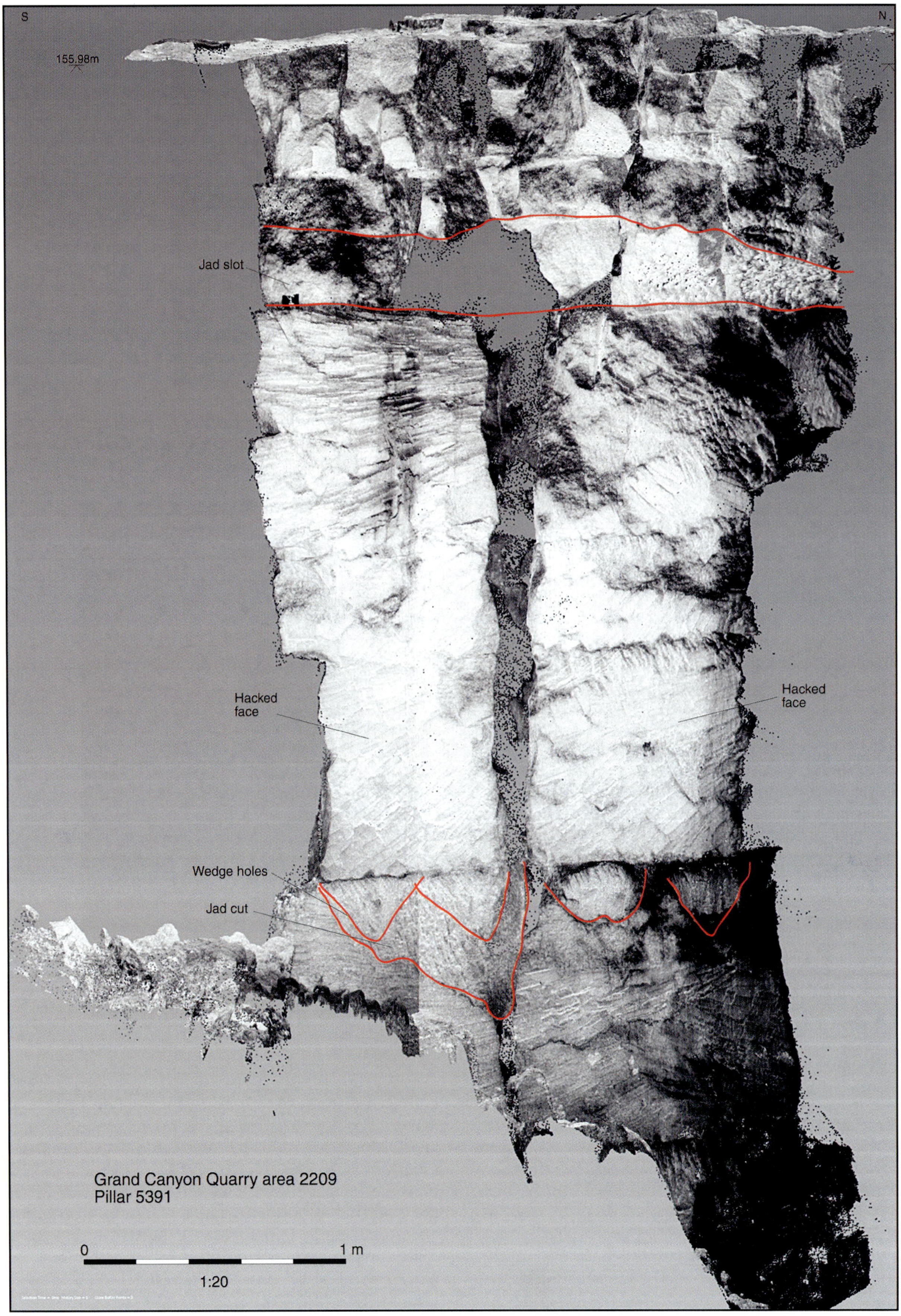

Fig. 12.30 Laser scan image of the east face of pillar 5391 (within the Grand Canyon), showing working features

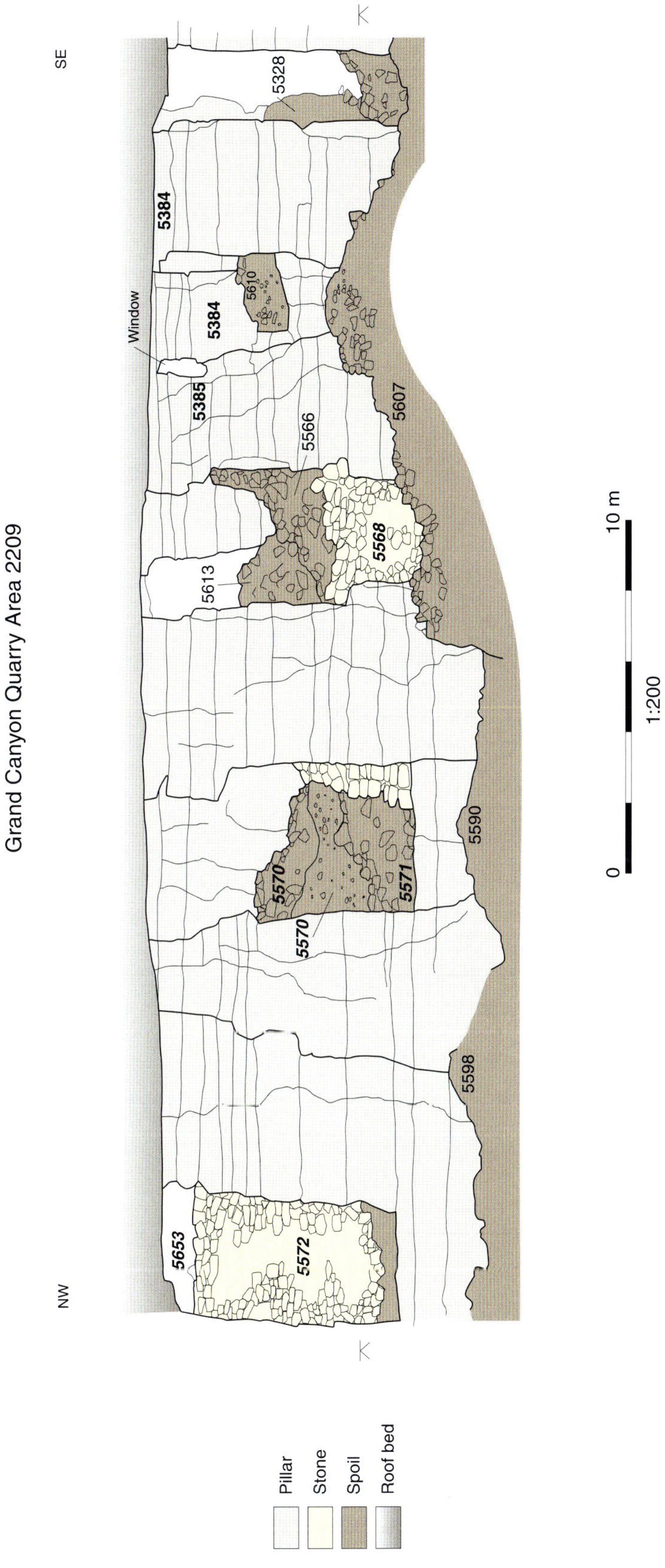

Fig. 12.31 Long elevation of the Grand Canyon produced by laser scanning, viewed to the north east of the central row of pillars, intervening spoil-supporting packs and the north-east wall between. Note the over-deepening of the floor shown by the higher level of the base of the rubble packs

especially the vertical jad cuts and the wedge holes. The two pillars are fairly representative of the type of extraction techniques used in the Grand Canyon, and exploited the secondary joints, perpendicular to the primary NW-SE joints. Evidence for the use of crowbars to lever blocks of stone from the quarry face was always difficult to determine archaeologically, the occurrence of such tools (elsewhere) indicates they were almost certainly used in conjunction with the wedges and jad cuts.

The south-west working face against the long boundary pillar and the opposing north-east faces were not very instructive, as these – like the pillars – were aligned with the main joint direction and had long sections of face unmarked by tools. More could be seen in corners at an angle to this, where it was clear jad cuts had been used to free the beds below the picking bed for 1.4 to 1.6 m, with then three or four thinner beds broken out – as a wrist stone to allow extraction to proceed regularly sideways and down. Frequently ledges had been left.

The method of working is a variant of the common Long Wall, gullet and pillar system, in which the advance is along a working face in front of a trench or gullet which provides access along it, and with spoil being dumped behind. However, the unusual height here, at least $1^1/_2$ times that suspected in some earlier areas of working, in the absence of cranes required an adaptation of the system. The elevation (see Fig. 12.31) shows secondary deepening along the gullet floor in the central part of the Grand Canyon, perhaps done on the final retreat. The rubble packs are shown in the laser scan to have their footings at least 2 m above the final floor depth, resulting in the exceptional overall height of 8.5 m. It was normal for the face to extend approximately 2 m below the cartway floor levels which originally here would probably have been only about 5 m below the roof). Unfortunately, although the use of ramp or step could be deduced from considerations of roof heights at either side, the actual position was obscured by the corner in the Grand Canyon length, leaving the problem unsolved of how blocks from below the cartway level were handled.

The south-east corner of the Grand Canyon offered a useful insight into the method of working in these conditions. For approximately 10 m this end appears not to have been initially worked, with only a sufficiently wide opening made to allow spoil to be transported out of the way on a spoil-incline (see below). To work this end a narrow vertical opening was opened on a cross-joint alongside the packs and dumps behind the gullet, through the long joint face forming the east-side

Fig. 12.32 Reconstruction drawing to show how stone may have been worked at the south east end of the Grand Canyon

wall, communicating with other near contemporaneous workings of a similar type behind.

From a level barrow-way at the bottom of the gullet at the above opening, leaving pillars as the advance was made, a spoil-covered ramp was developed, possibly accessed from the adjacent spoil barrow-way-incline to some three metres below the roof, so extraction could begin in the normal way at the top. Thus the main joint was at the left or eastern wall, and the last breaking-out of rock occurred at the top. Presumably as a block was extracted, it was dragged down the spoil-covered ramp, and any spoil, surplus to the ramp-needs, was sent out via the inclined barrow-way (Fig. 12.32). In other words the end, at least, was worked along the length of the room between the joint-faced north wall and the opposing row of pillars. It is unsure whether a form of bedded 'staircase' was made under the rubble or the face kept near-vertical and a rubble incline was built to a vertical face. The spoil would have been removed in either case as the lower beds were revealed.

To what extent this was a typical symptom of working in the Grand Canyon Area we cannot tell, though it could be adapted in several ways, such as turning into the eastern wall (leaving pillars as appropriate), for example, at the top, and then working the upper part forward in the conventional way. In another part of the quarries, heavy and long planks were located and clearly, as well as being used for transporting spoil across exposed heights to dumping areas, these could be used as temporary ramps to manoeuvre blocks and some form of temporary shear-legs would also sort many problems with blocks too heavy to manhandle. Raising from below the gullet floor in other areas within Firs was certainly facilitated by ramps. Allowing blocks to fall from such heights as here was unlikely due to the risk of fracturing. Sawing or splitting *in situ*, which was certainly done in surface quarries, was an option underground though sawing as a normal method of working new faces was still a few years off at the time of the Grand Canyon. As seen there the method of working can be thought of as transitional, the problem being finally solved in Combe Down underground stone quarrying by the mid and late 19th-century use of underground cranes.

Spoil disposal was always a major problem and here it was a particularly complex operation. The south-eastern end of the area was well-defined by the boundary pillar, and only a narrow gap end gave access in this direction, but at the eastern side the workings had impacted and truncated the western limits of Quarry 2354. The western and south-western side has broken into former quarries (2201, 2203, 2206 and 2355), and there had been both tipping from the present working into these and tipping from elsewhere into it (see Fig. 12.27).

There were clear phases of tipping fronts and barrow-way activity within the Grand Canyon area. The upper beds of the quarry, extracted close to the roof and forming the first areas to be extracted, produced significant waste material. Side piercings in the south-western boundary pillar forming windows, had been used for access for successive barrow-way routes, including driving southwards into the former Phase III (Allen Estate) quarries 2201 and 2203 delivering material to substantially-sized tipping fronts.

There were also significant amounts of waste material transported to the south-eastern edge of the quarry, into an earlier Quarry (2353), on long lengths of barrow-ways with dumps with arcuate tipping fronts (Fig. 12.33) at the intermediate quarry heights, at 3 – 6 m below the roof. The waste stone was deposited laterally into adjacent long-abandoned Long Rooms at the different levels.

Fig. 12.33 Arcuate tipping front in Quarry 2353 from the Grand Canyon

Waste at the base of the Grand Canyon had been tipped eastwards on level barrow-way routes, but the most significant feature was the inclined barrow-way delivering material southwards into Quarry 2334. The inclined barrow-way had an overall length of 50 m with several spurs or distributaries and associated tipping fronts at the easternmost ends, and had an incline of about 1 in 6 (Fig. 12.34). Substantial amounts of small to large rubble had been deposited laterally along the inclined barrow-way edges and some had also been stacked into low stone packs where space was available. Exterior to the Grand Canyon itself, successive tipping ways could be seen to be one above another, and at inclines with similar gradients suggesting that the gullet end has retreated away from the dumping area to maintain the angle and/or to access higher levels.

On the southwestern side of the Grand Canyon a platform of vertical stone packs had been constructed to a height of 4.5 m from the floor, and had probably been formed from both an inclined barrow-way on the western edge and partly from stone transported at a higher level in its final construction. It was possibly accessed from routes of barrow planks, placed at height above the floor, and delivered from the northern edge of the Grand Canyon from other contemporary barrow-way routes. Another barrow-way transported materials from the extraction cartway or barrow-way at the western end of the Grand Canyon, situated about 6 m below the quarry roof level, to about 1.5 m below the roof located at the tipping front at the eastern end. Several successive barrow-way surfaces consisting of mixed fines and small rubble formed the feature.

Though the Grand Canyon at the southernmost working of the Three Acre Quarry was initially considered to have represented the end of that extraction period, there have been workings on what were probably residual blocks close by. From the north two main barrow-way routes had been used to deposit significant quantities of waste stone on the northern edge of the canyon, so must have remained active while the Grand Canyon was being exploited, or very soon after.

Fig. 12.34 Inclined barrow-way at the south-east end of the Grand Canyon

Phase V: Small-scale working in the mid-19th century 1838-68

With the opening of the railway in 1840 from London through Bath to Bristol, the competition already felt at Combe Down from the Wiltshire quarries by the building of the Kennet and Avon Canal intensified, and it was not to be long before the Combe Down quarries were eclipsed by their competitors. However, there was still a preference, justified or not, for Combe Down stone over Bath Stone in Bath, and quarrying continued, though probably under pressure from reduced real prices, especially in more distant markets. It is this that probably motivated changes in working methods, some, for instance, probably forced on the local producers by the example of the much larger quarries developing at Corsham and Box. Several small underground quarries are described below, which illustrate how they reflected changes taking place in the wider industry.

Large parts of the central area of Combe Down by 1838 were nearly worked out. Henry Street continued until he retired about 1851 at the surface quarries west of Combe Road, though some production seems to have continued, but (with the exception of secondary pillar robbing) only at surface. Pillar robbing seems to have been the only activity in most of the Monkton Combe Byfield underground quarries, though John Davidge may still have been at work in the southern part of East Byfield workings, where even after 1900 a block of unworked ground remained for James Riddle to quarry (518).

North of the Turnpike or North Road, a large area of surface quarrying continued, leased from the Thomas Trustees, with probably five separate operations at, or soon after the time of the Tithe

Award *c* 1851-2; West of Stonehouse Lane was worked by Isaac and Giles Sumsion (Addison 1998, 45), William Stennard (Henry Stennard after 1870) east of the Lane; next was James Stone up to Farrs Lane, then Richard Lankasheer and James (Edwin after 1872) Love at the east, coming up against the Blind Lane or Popes Walk as it now is known. By then the quarries were approximately 100 m back from the road towards the north rim of the Down. This later reduced to three, as Lankasheer and Stone ceased operation. Their surface quarries gave opportunities for easy access under buildings or under the Turnpike and south of the road.

Two blocks of ground near the Turnpike were worked underground. At the northwestern end of the long boundary pillar bounding the Three Acre Quarry, (just east of Farrs Lane today), an area north (Quarries 2216 and 2218) and south of the Turnpike (2214 and 2215) possibly began work in the previous phase. The last of these to be worked (2215), seems to have been worked out by soon after 1880 so probably began in the present phase under William Stennard, continuing into Phase VI (see Case Study 12). Another small area of ground, this time north of the Turnpike under the oldest buildings in Combe Down at the (present) entrance to Stonehouse Lane could not thus be surface-worked, and so was tackled underground by means of entries, probably from the Isaac and Giles Sumsions's surface quarry, around or soon after the mid 19th century.

In the southern part of Central Firs a block of ground, Quarry 2202, remained unworked on the east side of the Allen entries in Sheeps House Quarry, and to its south under Gay Staithe. It is not known who worked these, and though the name Davidge's Bottom was applied to the area, this was probably because of his brewhouse having earlier been built there. It is quite possible it was another quarrymaster, since Greenway had earlier acquired rights to the underground entry areas, and he or a successor would certainly be seeking a return on this long-ago investment. Another small block was left half-way along the western Allen entry, Quarry 2344, perhaps marking the Phase II cessation of work there, but its working seems to have been in the late century.

In Combe Park, now considerably built over, the Hadley Estate leased the underground to Richard Lankasheer in 1856; graffiti suggests an earlier start than 1856, but many parts were not datable. His royalty, expressed in the quantity of stone extracted, rather than the area worked, suggests that though new areas may have been worked, this was essentially a secondary working operation. This may explain the substantial amount of pillar robbing that went on there, not all of it commendably carried out.

This left the larger number of underground quarries of this period, though probably far from the larger production, in the area east of the Long Drung. Here quarrying, some on a very small scale, continued until the century end. These quarries (denoted on the Phase Plan as Phase V or VI), all apparently accessed from shafts on the small leasehold properties let out since about 1830, were very diverse and suited either the needs of a single mason, or for building the house above.

Probably a greater amount of stone was being, by the mid century, produced in the outlying quarries which came within the bounds of the new Ecclesiastical Parish of Combe Down in 1854. They included those around Entry Hill; at the top above Horsecombe, and at Foxhill on the west, and at Lodge Style eastwards to near Combe Manor on the east. Probably only the Shaft Mine area had underground quarrying before the latter part of the century. Mid-century surface quarrying may have continued east of the built area of Combe Down and, possibly north of North Road in Prior Park and the Rainbow Wood areas. There was also a small quarry north of Brassknocker Hill which may be mid-century also.

Case Study 10. Phase V, Stonehouse Lane Quarry (2219)

There were two houses above the quarry, possibly with outbuildings shown on the 1742 Thorpe Map of Bath and Five Miles Around. One house remains, on the west corner of the Lane with North Road. The other was at the eastern end of the plot shown by Thorpe, and was demolished in about 1950. The underground quarry extends south of the road, but the southern workings are dissimilar from those to the north side, possibly suggesting a separate operation (with whom and when unknown). The north side would almost certainly have been worked by Sumsions who owned the surface quarry which gave access to it.

At the east end there was a boundary pillar and, under the road a series of rooms of which those at the east appeared to have dumps derived from the west. Close to the eastern end there was a well, which had been partially infilled from above, forming a shaft-foot cone at the freestone working level. Remains of a rod and pipe suggest it had a pump, but detail was not observable. The well would have been at the rear of the east side house.

Further to the west behind pillars with jad slots and wedge slots, was a clear floor (Fig. 12.35), with nearby scappled blocks or slabs leaning against the wall. More slabs were found north of the pillars along the front of the clear area. The area narrowed as it went north and at the west side were openings walled up with packs or with spoil of a type we had learned to associate with surface quarries. The multiple entries from the surface quarry suggest the underground, after the initial working, was probably used as a convenient shelter, either for working stone (possibly in the winter) or for storing it. As found today the open workings were largely just west of the

Fig. 12.35 Cleared floor at Stonehouse Lane Quarry, with scappled slabs

modern lane, but it seems likely they actually followed under the original quarry road, which maps show curved to the north-east.

Case Study 11. Phase V, Quarry 2202 – east of Sheeps House Quarry

This was accessed via the east Allen cartway and the branch cartway from its east side. The cartway and branch, much of it through apophygate or corbelled pillars, had clearly severely deteriorated by the mid 19th century, and, in the first section, large blocks, probably of bastard stone, had collapsed from the roof, or had been barred down. The substantial rubble packs lining the route were probably derived from them. The somewhat wider and higher branch cartway appears to have been in an even worse state, with large collapses over the whole of the roof, and massive rubble packs were built at the sides (in front of the old crane position) virtually all the way to the entry into the quarry (2202). The quarry too was also prone to falls – a massive example lay just inside, which was obviously post-working and was used as a test site by Hawkins (1994) in his survey. This was to determine whether such stone falls were cumulative or catastrophic, though this particular example obstinately seems to have remained stable since.

The quarry was worked Open Room (with wide cleared floors between and round pillars rather than a narrow gullet – see the Grand Canyon, above). It was L-shaped around and up against older workings and cut into spoil dumped in them. The use of mast cranes was probable from numerous anchor points left, mainly Lewis holes, although inability to access the area prevented possible location of actual crane sites. Some form of winch or pulley system was probably also used, held back to a pillar by spigots in holes hand-drilled either side. The line-up suggested it was used to drag stone up a ramp at the foot of (the final) vertical working face at the north-west side of the quarry on to the wide flat working floor of the quarry (Fig. 12.36). The ramp was steep, vertically rising about two metres, so the overall stone quarried was over seven metres high, with nearly six metres height left above the working floor. Another pillar had two spigots high on the pillar with a chain stretched between them around a small portion of the circumference. The purpose was not obvious as a substantial pull on the centre of the chain would probably have dislodged it.

Pillar faces, when not natural joints, were usually split by wedges or wedge and chip, but some were sawn, and at the north-west end of the quarry a right-angled projection from a boundary pillar was sawn roof to floor on the three exposed faces. The outstanding differentiating feature from any other workings at Combe Down was the use of large numbers of timber sprags or scorters (Fig. 12.37). Some were still in position, but the holes chipped into the rock for them were very numerous indeed, in both the individual pillars and the boundary wall and in the roof. They were used to support roof blocks, probably by pressing the adjacent blocks together so they held by friction. Clearly they gave the quarrymen considerable confidence and apart from the fall near the entrance, their confidence seems to have been justified.

There were large amounts of spoil, but some would have been derived from the workings broken into which spilled out on to the floor. A long

Fig. 12.36 *Open room working floors in Quarry 2202*

Fig. 12.37 *Pillar with scorter recesses*

barrow-way incline was used to transport the spoil to a level high enough to be dumped in old workings (2340).

On a high ledge, where they must have been thrown or left before the rock was cut down, were a pair of leather shoes, definitely too soft and delicate a quality for working in a quarry. There were also several graffiti items, including dates of 1856 and 1886, and an argument over the respective merits of the beer and its price ($3^1/_2$d a quart) at the Hadley Arms (built about 1845) and Coxes (perhaps the Foresters Arms) (see Figure 11.13). Actual working dates are not known – several other sawn faces had the name 'James Morris' and the date '1863', which are thought to be associated with secondary pillar robbing activity, but the graffiti dates here were probably close to the working period of the mid 19th century.

South of this working, extending nearly to under Gay Staithe, a further block of stone was left (Quarry 2340), possible worked contemporaneously or slightly earlier than the above. The principal feature was a high final working face, worked from a narrow gullet, perhaps as high as eight metres, Spoil had been dumped into it preventing any detailed assessment of how the quarry had been worked. The origin of the spoil was unclear but it obviously represents later working, probably in the same area and possibly small-scale reworking, It appears to have completed the working of substantial blocks in the southern area of Central Firs, though why such conveniently available stone to Sheeps House Quarry and under the core area of the village should be left is hard to discern.

Phase VI: Working from the late 19th century to the early 20th century

By about 1868, it is likely that working had just begun or at least resumed on the underground Stennards Quarry (2215) served from his surface quarry north of North Road. The working (2202) near Gay Staithe (above) was perhaps completed in the early part of this phase, while two small reworking Quarries (2367 and 2368) and the small but deep working at the shaft on Allen's western cartway (2370) extended through it. Quarry 2370 exploited a very small block at the junction of boundary pillars and is the only known example of re-excavation below the old (Phase II) cartway floor into the freestone beds remaining below, to depths which had become normal by the second half of the 19th century. The quarryman clearly recognised that the shaft, located to the west of the cartway, allowed easy handling using a direct surface hoisting arrangement which would compensate for the costs of moving existing deposits of spoil. Another Quarry (2344) located 30 m to the north, is also thought to have been worked during a similar period, possibly was part of the same late scavenging operation, also used the same shaft for handling stone. The quarry had survey marks delimiting the area, so was probably preceded by a professional assessment of this area of Combe Down. Graffiti dates continue to suggest that pillar robbing, though on a smaller scale, remained a frequent activity, and small-scale working and banker mason activity is also evidenced from the Far East Firs area.

There was substantial activity continuing on the outer fringes of the Combe Down parish, at the Springfield Quarry especially at Entry Hill and in the surface and underground quarries between Combe Manor and the Lawns, including the Lodge Style or Shaft Road underground workings and St Winifred's (cited as the only quarry underground still working in 1895 in *The Builder*) and at Mount Pleasant. Nearer to, but outside the older quarrying area, four small underground quarries, including Coxes Vertical Shaft Mine and Tankfield Quarry operated under what is today the Foxhill Ministry of Defence establishment back towards the Combe Road junction. All these were active from the 1870s, closing by the early 20th century. In the main body of underground mines, James Riddle's small quarry (518) completed all underground work there about 1911, though officially it remained active until 1914.

The major business development of the period was the formation of the Bath Stone Firms Ltd, a combination of some of both Combe Down and Wiltshire quarries which aimed, with reasonable success, to regulate the Bath Stone business. This took place in 1887 and involved the Combe Down quarries of Henry Stone and family (adjacent to the Prior Park carriageway) and Isaac Sumsion (Lodge Hill or Shaft Road underground Quarries and the adjacent Combe Down Quarries, and the Mount Pleasant Quarries) and Randell and Saunders (Springfield Quarry and later, under Foxhill). The initial chairman was CJ Pictor (Pictor and Sons of the Corsham quarries), but soon after Isaac Sumsion succeeded him, with his son taking over about 10 years later. The name of the group was changed in 1911 to the Bath and Portland Stone Firms Ltd and the stone business remained effectively within their control into the 1930s. Early on they rationalised some minor quarries and remained active until the around 1938-39 when demand fell away because of the imminence of war. The underground quarries on Shaft Road: Lodge Hill and Mount Pleasant both closed about then, as did St Winifreds.

Independent operators still continued to operate, including the Hancock family at Lodge Style (the Lawns) which still operates today, and James Riddle who was the last to operate underground in the old, traditional area of underground quarries, at Allotment Shaft (518).

Case Study 12 – William Stennard's underground quarry (2215)

The quarry was worked from Stennard's surface quarry on the north side of North Road, but was originally accessed by the survey from the extension of the Allen cartway (313) where it turned north-east. This is just through the long boundary pillar and at a point of a probable fault that had resulted in the sailed pillars and poor stone which led to abandonment in Allen Estate period (Phase III) since the pillars appear to have been worked in the Long Room way, similarly to the workings to the south. This seems to have led to the abandonment of ground to the north which was not worked until the mid and late 19th century. Here the undated workings around Quarry 2215 to the north and north-east were accessed from north of the road, with a cartway entering into Quarry 2217. These were older than Stennard's quarry 2215, probably belonging to the previous phase since the surface quarries almost certainly had to be there to service them. Details of their relationship with both Samuel and John Pearce, the adjacent quarrymasters and the Hadley Estate, the landowner, are unknown.

Stennard's Quarry, which probably dates from around 1870 was worked using cranes. In effect it was a substantial flat-bottomed hole with dumped material on all sides, some from the hole, some remaining in place from adjacent quarries. This was a form of wide room, still with pillars but with spoil removed from the immediate area allowing the space for cranes to operate efficiently, much as they would in a surface quarry. These cranes were probably very similar to the surface cranes, though possibly slightly smaller, with a central mast supported by cables (iron wire or chains) running back to anchors, and a swinging jib which could be used to haul blocks towards it. One cable anchor on the west side was a large Lewis (Fig. 9.33) which was recovered. Other Lewis slots were noted and chains were probably also wrapped around pillars. No crane base was found.

The pillars here were nearly all sawn, with a wide jad slot just under the roof to let the razzer in, followed by the frig bob, completed bed by bed to the bottom. Most pillars had their faces inclined inwards, so that the base of the pillars could be very thin, dangerously so in places, but fortunately the area was not large and thus stresses were less. As a working method sawing probably produced much less spoil. The spoil generated by the working of the pillars, and from the collapsing in of older waste, was stored on wide flat-topped or bench-like tips at the sides, reached via small inclines. This was a more systematic method and was used in several of the later workings, possibly facilitated by a reduction in the spoil produced.

The quarry probably closed about 1880, so would have begun late in the present phase. The 1st edition OS 25 inch (1882-3) showed it as accessed by a wide hole left behind from the current quarry face, located to the side of the later Selway Court; it probably used a crane to hoist stone to surface level. The second edition OS map showed Selway Court built over the area.

Case Study 13: Phase VI. Foxhill Underground Quarries: Wilks Quarry, Coxes Vertical Shaft Mine and Tankfield Quarry (2380, 2381, 2382 and 2383) (Fig. 12.38)

This is a group of three or four (two may belong to the same stage of working) small, connected underground quarry workings extending from the north-west corner of Combe Road across the front of Bradford Road under the houses and west under the MOD establishment which occupies the former Foxhill Farm. It was probably either here or at one of the Sumsion quarries that rail and post crane techniques were introduced underground to Combe Down, almost certainly from Wiltshire where Randall and Saunders of the Corsham Quarries pioneered them (Pollard 1994, 43).

The site was originally part of Widcombe Field. There was a surface quarry north of the junction of Combe Road, known by Addison (1998) as Wilks Quarry, and a quarry is shown here in its early development on the 1882-3 OS 25 inch map. There is very little to indicate the other three quarries on the maps. The water tank which gave the mine its name was shown on the Ordnance Survey map for 1882-3 onwards and to its west is a small square of ground which appears to be a quarry, possibly with a shaft or depression which may be at the shaft site between Quarries 2390 and 2393, and a well (shaft?) is marked north-west of the Wilks Quarry. On the 1902 map the Wilks site had its Bradford Road frontage built but and no indication of quarrying activity at all. Indeed a house has also been built in the south-west corner of the field. The 1912 map does show possible activity just north of the water tank where there are two small buildings or enclosures which may relate to the Coxes Vertical Shaft being used by the Tankfield Quarry, and the well. By 1930 houses had extended along the frontage and the west side of the field, except, perhaps significantly where the Coxes Vertical / Tank Shaft had been. The area was by then known as Ralph Allen Park. Taken together, this seems to indicate a phase of activity in the late 1870s and 80s, and (with the other evidence below) a renewal of activity around 1912, with termination early in the 1920s allowing housing development to commence.

Wilks Quarry underground (2382)

The underground workings 2382 appear to be associated with this quarry as the entries all emerge on the north or north-west side. It is unlikely that the Stonehouse Quarry to the east of Combe Road end developed the underground workings since it was over 50 m away. Wilks Quarry was shown on the 1882-3 OS 25 inch map as a small area of worked ground at the south end, with a crane, and it is probably the 'new quarry near the Turnpike' (almost certainly a surface quarry from absence of reference to shaft or underground) mentioned in connection with the death of 70-year-old James

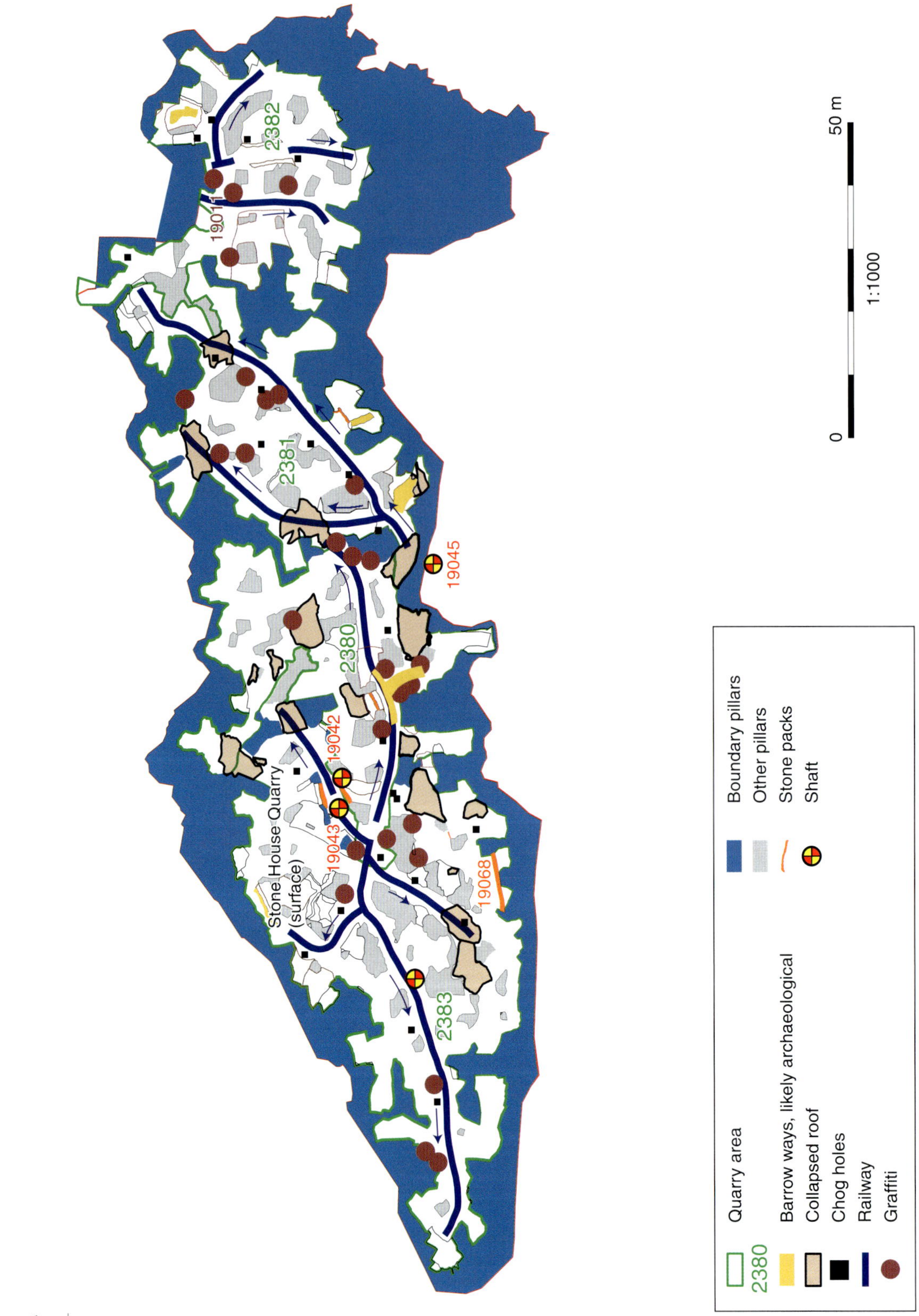

Fig. 12.38 Case Study 13: Phase VI. Foxhill Underground Quarries: Wilks Quarry, Cox's Vertical Shaft Mine and Tankfield Quarry (2380, 2381, 2382 and 2383)

Hillyar who was struck by the windlass handle he was operating (*Bath Express* 11 September 1875). It or the Coxes Shaft may also have been the Randell and Saunders quarry where Edward Filo was killed by a falling four-ton block of stone two years later (Keenes Bath Journal 28 July 1877). According to Addison (1998) the houses were built alongside Bradford Road by 1891, strongly suggesting that nearby quarrying activity had ceased by then, and certainly by 1902. This seems to date the Wilks Quarry and its underground section as operating between 1875 and the late 1880s.

The Wilks (2382) underground section appears to be functionally separate from the others and the only connection was a 'window-sized' opening on the west side. It is possible that initially, *c* 1880, the south side of the surface quarry was used for stacking the 'ridding' – the extracted material – from behind, where the crane stood. Rather than remove the spoil dump, entries were made under it and the stone brought out while the crane was in a suitable operating position to handle stone from both surface and underground.

The archaeological survey found that four entries had been made, all from the north or north-west side. The two north-west entries may have been of small enough penetration to allow the direct dragging-out of blocks of stone. The two southern entries had sleeper voids indicating use of rails with the eastern-most track dividing into two. The tracks, possibly two to two-and-a-half feet (750 mm) gauge, ran down long Open Rooms between substantial pillars. On the main track, two post crane positions, marked by chog holes about $12^1/_2$ inches (313 mm) square, were placed either side of the track at about 9 m intervals. These marked the advance of the working face that was about 12 m ahead of the southernmost crane. From roof to railway was approximately 3.7 m, suggesting that about 5 – 6 m of stone was removed. Three other chog holes were noted close to the eastern and western quarry entries. All had an associated Lewis hole, used for pulling the post into position. An *in situ* Lewis was recovered from the southernmost crane position, perhaps resulting from its last use.

There was graffiti, most of it near the entries, some of which may have been contemporary with working. Unfortunately no dates obviously related to the working period one example, 'P Long 1818' was clearly much too early for the type of working and other dates all appear to relate to later entry, possibly via the window (unless the well was a shaft remaining open into these workings). However, a list of tools was much more likely to be contemporaneous: '7 saws, Bars – 3 big and 1 small – 2 pair of Lewises – 1 snatch block – 8 planks – 1 crane – 1 jack – 1 shovel'. These are mostly obvious in their use, but the snatch block (an openable pulley block) was probably used for redirecting the crane cable to the required pulling direction, and the jack was probably necessary for getting loaded wagons back on rail, or for manoeuvring a block of stone.

Of the later graffiti, 'BEC 30 Dec 1944' indicates that the group of four workings was still open at that time (at least at the western end). This may reflect its wartime usage as a shelter.

Coxes Vertical Shaft Mine (2381)

Addison also refers to Coxes Vertical Shaft Mine or Tankfield Quarry (the tank was present on the 1882-3 map and was demolished in 1973). The Tankfield is now thought to be identified as Quarry 2383, through archaeological evidence. The quarry has its shaft near the MOD gates and the tank. This shaft, re-utilised and effectively destroyed by the stabilisation programme before we gained access seems to originally to have served the adjacent quarry to Wilks. This leaves two further quarries adjacent to the west with a further shaft, probably the later Tankfield Quarry (see below). There is an obvious relationship with the name Coxes Quarry for that south of Bradford Road, opposite, and, probably a name of a Public House seen on graffiti in Firs Mine.

The quarry was apparently operated by Randall and Saunders between about 1875 and 1885 and there was a graffito date 1882 at the shaft bottom which seems to confirm the dating. This is very similar to that ascribed to the Wilks Quarry, but this may reflect a landlord issuing leases simultaneously. The quarry seems to have been developed using a railway from the shaft bottom to the NE corner, where it came against the Wilks Quarry boundary pillar, there fairly thin. A second railway may have been developed against the eastern boundary pillar. A length of rail was recovered but the gauge could not be determined and the track had been dumped over.

There were four chog holes in the roof at about 8 m, then 10 m and 12 m intervals and a fifth possible but obscured by a roof fall. They ranged along the north side of the railway, the first $9^1/_2$ inches (238 mm), the rest each 10 inches (250 mm) square with the usual Lewis slot. This suggests the first crane may have remained *in situ*, a second being moved along as work progressed. There are two lines of pillars in the central area and another two in the south east – a lack of chog holes suggest stone from there was pulled directly back to the shaft.

Graffiti may indicate names of workmen, though this is always uncertain, including Edwin Riddle 1889 (cf James Riddle, later a quarrymaster in East Byfield) and J.E. Dowling and W. Hole 1882. The Dowling family left their names very widely through all the Combe Down workings (and on the Palladian Bridge in Prior Park).

Coxes Vertical Shaft Mine may also have been worked, in sequence, with the 'Unknown' Phase working below, which may be supported by graffiti dating between 1898 and 1911.

Unknown and Tankfield Quarry, (2380 and 2383)

The early or first stage quarry (2380) cannot be firmly dated (except by accepting graffiti dates between 1898 and 1911). It has traces of a railway with $9^1/_2$ inch (238 mm) chog holes above, at eight metre intervals, crossing roughly east-west, and several Lewis holes. This does not serve the whole quarried area well and it may thus be that the cranes pulled from substantial distances and possibly the old Coxes Vertical Shaft was used for the parts near to it. One of the chog holes near the shaft seems to indicate a crane position was re-used in the second stage of operation

The second stage, westernmost, quarry had a notably more complex railway following the long Open Rooms within the pillars with a notably central open area in the southern part. The main length was developed south-west from near the shaft, but is substantially overlain by dumping. Four $12^1/_2$ inch (320 mm) chog holes were somewhat irregularly placed along its length, located 8 and 12 m apart on its north side. Two more chog holes were found to the north and two to the south, while two collapses may mask others. A length of wire cable was found below one chog hole, and a single chog remained *in situ*. Two Lewis holes were found at one chog hole.

The name seems to be derived from the water tank placed near the Coxes Vertical Shaft Mine and may suggest this shaft was used as part of the present quarry development. The Coxes shaft is placed at the boundary between the older and newer quarries, but a second winding shaft was installed between the two phases of the present workings with a nearby smaller diameter shaft possibly remaining from the first phase, no doubt for ventilation or statutory escape purposes.

The Combe Down (Bath) Freestone Co. Ltd was set up in July 1909 with £2000 capital to acquire a lease of 21 years for the Bath Stone Quarry in Tank Field, owners James Tucker, Frederick Armstrong and Frederick Box, with equal numbers of shares, the property involved extending to an acre. It continued to be registered until 1924, though by 1921 the registered address was Ralph Allen Park, suggesting interest had by then turned to property development. It employed up to sixteen workers, in 1912, but seven or eight were more usual up to 1914. During the War, only one or two were employed until 1917, then none until in 1924 nine workers were engaged.

The main working shaft, 3.5 m wide, and found backfilled, appears to have been that between the two phases, placed centrally, which are separated by a pierced boundary pillar. Dumping via barrow-way from the western working into the east establishes the phasing, with the eastern clearly completed first so that the railway system is almost completely sealed. Because the Coxes Vertical Shaft is also placed centrally between phases (Quarries 2381 and 2380), in absence of other firm dating of Phase I it may belong to either Coxes Vertical or to Tankfield Quarry, or, of course to both. Certainly Tankfield had control of the earlier phase in order to carry out dumping.

These workings had many artefacts left within them, including two saws at the base of the shaft, sharpening files used as coat hooks, containers for water (drip cans for sawing, or bottles) and items associated with the railways – bars, wedges and oil-lamps.

Comment

The choice of underground working at these quarries is puzzling. Wilkes was partially worked underground, and a rational reason can be advanced for this, but the other three linked locations were only worked underground. It points to a landlord insisting on it and this could only reasonably be because housing was conceived as a future use and undermining was less damaging. If so the explanation for Wilks Quarry having some workings in the same range almost certainly has to have the same underlying reason, though storing spoil on it temporarily would be a sensible option still.

The two earlier workings appear to be contemporaneous, and it is possible the third was wholly or partly worked in sequence after them as part of the Coxes Vertical Shaft Mine, but probably after Wilks had closed. The delay before Tankfield Quarry operated is possibly an indication that other quarrying methods were economically preferable.

Case Study 14: Phase VI – James Riddle's Quarry (518) 1905-14

This was probably the last area of the main Byfield/ Firs complex to be worked (Fig. 12.39), and was known as Allotments Quarry (Addison 1997, 43). It was at a very small remnant block of stone surrounded by older worked-out areas. It was the only one known to use an underground railway in conjunction with post cranes: the former left sleeper troughs in the floor and the latter a series of chog holes in the roof, and one instance of a crane base stone in the floor. The two areas were at one end separated at what became known as 'H Pillar', a significant landscape feature of unidentifiable age which had been extremely overworked and left in an unstable condition (Fig. 2.4). Close by H Pillar the railway, which ran back to shaft around a corner cut through an older spoil bank, came to a loading point served by a crane. The crane, indicated by a base stone and chog hole directly above in the roof stood next to a pillar (Fig. 9.34).

The first area to be worked seems to have been on the south side where pillars are of two forms, vertical with deep jad slots and sawn, split and natural faces and a northern area where the majority of faces are sawn vertically (see Fig. 8.30). Pillars on the south side also had cable grooves and it is possible a mast crane was also in use at some time, though part of the area had several post crane chog holes too.

Fig. 12.39 Pillars in James Riddle's Quarry

Both areas have been worked Open Room, an obvious advantage for where cranes are in use, with spoil from the earlier area apparently dumped in old Phase I and III workings, which Riddle's workings abut, (516 and 915 respectively). Spoil from the northern area had been dumped on flat-topped banks on the southern area. The floors in the area are built up above the bottom of the worked stone and kept flat and clear, with stone dragged by the crane(s) to the rail-served loading point.

The northern area (and part of the southern) with totally sawn pillars is arranged in an avenue about 15 m long and about 7 m wide. There were two sets of rooms extending for approximately 8-10 m on the west side. A chog hole in the roof shows the position of a crane inside the first of these, possibly situated there to drag stone back to where the railway once extended, or alternatively, to where the crane noted above stood next to the railway.

The eastern line of pillars had been worked in a straight line along a single joint which was open to some 50-100 mm width: the western line was less obviously weakened. Timber sprags or scorters had been placed just below the tops of the pillars to help support the roof and the fact it had not collapsed perhaps justifies their optimism.

Case Study 15, Phase VI – The deep quarry on Allen's west cartway in central Firs – Quarry Areas 2368 and 2370

This study examines the reuse of an Allen western cartway (see Chapter 6) in the late 19th century. The southern extent of this cartway had passed under a narrow shaft 70 m in where there was no substantial evidence of cart-ruts. A 3 m diameter shaft was located near the interface between the apophygate pillars and the slightly later corbelled phase of the Allen workings (Phase II), though several aphophygate pillars were noted to the north of the shaft, abutted by later 19th-century stone packs. The shaft was probably used during the 18th century for winding stone out and was certainly used in the 19th century for localised pillar robbing activity associated with Quarry Areas 2368, 2370, and also 2344. The northern extent beyond the shaft was sealed by later barrow-way activity and a loading platform associated with activity from Quarry Area 2368. pillar robbing from Quarry Area 2370 beside the shaft had cut through the cartway section and the loading platform of Quarry 2368, fortuitously exposing a stepped face cross-section through the Allen cartway route, extending down to the natural freestone.

A large shaft cone of surface-derived material, including pottery sherds, had formed at the base of the shaft. This had accumulated prior to capping of the shaft with a slightly domed stone-block construction during the mid-late 19th century.

The upper stepped face in the sequence through the cartway route (see Fig. 9.26 – Section 191) was investigated across the whole width of the former cartway and mainly recorded in section the later barrow-way activity, from the earliest sequence (1443) to the latest (1226), from the pillar robbing from Quarry Area 2368. Earlier barrow-way surfaces below this had sealed the latest extant cart-ruts (1018). The later barrow-way activity was probably mid 19th century, as the deposits appeared to be contemporary with cartway (3022), dated on other evidence to the 1850s or 1860s. Graffiti within Quarry Area 2368 was dated to 1863 with the initials

'J M', which were probably written by the quarryman James Morris, who had left other examples of dated graffiti in the year 1863 besides his full name during other episodes of pillar robbing activity noted in Quarries 518 and 2201. The recorded barrow-way activity included several later barrow-way surfaces and associated lateral tips above the cartway. These were mainly associated with dumps of waste stone produced during quarrying of the working faces, while those used within the construction the cartway were layers of finer limestone spoil deposited well above the base of the quarry working level, noted in section 192 at 5.55 m below the roof. There was little deposition depth in the later thin layers above the ruts, which were about 4.1 m below the roof. This suggests that at least part of the cartway sequence was of 19th-century date. Several phases or layers of small rubble and limestone fines were noted below the uppermost surviving of ruts in plan (1018). There was an absence of the concave features in the centre of the cartways, usually located between the ruts, and typical of many of the observable cartway sequences seen in plan within other Combe Down complex quarries. The characteristic low convex heap of fines that formed in the centre of the cart-ruts were partly indicated within the fines/cartway surface layer 1457. There was also an absence of clearly defined cart-ruts below the uppermost 19th-century cartway. The potential absolute absence of ruts, ie use of wheeled carts in the early period, and the absence of the distinctive concave feature were also noted below the contemporary cartway (Ctx 3100) located to the east. The lack of evidence for the use of carts in both cartway sequences may indicate that a change of methods occurred in the nineteenth century with the introduction of the carts underground within the quarries and not just employed at the surface for the moving of the stone. This may suggest the use of rollers or sledges for the removal of the stone from the underground quarries in the earlier 18th century.

The lower part of the stepped section (Figure 9.26 Section 191 & 192), records the lower sequence of cartway activity, with the uppermost cartway surface (1018), being the contemporary height of the cartway floor leading from this location southwards back to the surface quarry. It is unclear how much further this cartway extended northwards beyond the section. However, it may have been extended during the Allen period as far as the northern boundary pillar (1161), and the line of the route was well-defined by the adjacent wider pillar alignment. A small section of cart-ruts were preserved close to the boundary pillar, and a corresponding breach in the pillar suggests that the route may have extended through this quarry boundary. The period of this breech could not be established archaeologically but Lynn Willies had noted the cartway alignment, covered with debris, extending through the pillar face. By around 1800 leases show the cartway extending beyond the boundary pillar (LL. 91/18/5/30) for about 30 m to the north-east. A shaft which may be the present day Firs Shaft is shown 20 m to the south of the north-east extension.

The easternmost cartway route was over 100 m long and 3.60 m wide between the apophygate pillars. The southern stretch of the route was re-used by a later cartway (3022) branching towards the east, and later Quarry 2202. A shaft thought to be contemporary with the early development of the cartway, located about 70 m from the entry in an analogous position to that on the western cartway, was *c* 1.40 m in diameter. It was one of the few examples of a probable ventilation shaft and was of a similar diameter to another located 30 m to the east above the branch cartway. The latter may have been contemporary with the later part of Allen Phase II or, more likely, was an early Estate, Phase III cartway, as it was associated with Long Room working and crane usage.

Beyond the 70 m distance from the entry shaft the cartway was extended northwards to the Firs Shaft and probably just beyond during the late Allen period of working. This change was noted in the development of pillar type from apophygate to corbelled along the route. The shaft, though much modified by modern recent use, had Lewis holes positioned on one side so as to suggest a crane post support, and cable grooves were found on the pillars. Though no date could be confidently ascribed to either, the adjacent areas had Long Rooms similar to those further south-east associated with crane use, considered as either late Allen or early Allen Estate (Phases II and III).

Chapter 13
Quarrying at Combe Down – consideration of the evidence

Introduction

In the foregoing chronological analysis and illustrative case studies, historical sources and archaeologically-derived information have, where possible, been deduced to support and test each other. This section seeks to synthesise them to provide an overall account of developments within the area of the ecclesiastical parish of Combe Down after 1854, and in the context of the City of Bath after 1912 so far as the links with quarrying justify.

Although localised quarrying must have existed previously in the region, notably in Roman times, three centuries ago the area known as Combe Down or Greendown was still largely devoted to sheep grazing, occupying an elevated plateau and circumscribed by a precipitous rim. Modern Combe Down coalesced from the amalgamated distant parts of Monkton Combe, Lyncombe and Widcombe, and Claverton parishes. The new settlement formed around the core quarrying settlement set up by Ralph Allen which began to expand, comparatively rapidly after his residual estate was finally sold around 1803 onwards. Underlying this development was the excellence of the Combe Down oolitic limestone for building, and the proximity of a market, the City of Bath, which then, as now, depended on the stone for its buildings. Although the central area saw decline in its quarries after *c* 1840 and the underground quarries almost ceased after *c* 1860, quarrying at surface in the surrounding area of Combe Down continued to be a substantial part of the village economy well into the 20th century. By then Combe Down had come under the administration of the City of Bath, its suburbanisation emphasised by the electric tramway link along North Road.

Freestone quarrying

Prior to the 18th century available information is fragmentary and hard to ascribe to specific localities. It is clear that the larger quarrying area supplying Bath with its building stone was further west at what is today Odd Down where communications with Bath were more direct down the old Fosse route. Although small-scale quarrying was certainly taking place in Monkton Combe, Lincombe and Widcombe's Combe Down or Greendown, the documentary evidence for quarrying before Ralph Allen's time is minimal. However the published history about subsequent quarrying within the context of village history was usefully summarised by Addison (1998) little more than a decade ago and the present study, effectively begun by Pollard in 1994, with his preliminary assessment of the underground quarries can add substantially to their accounts, not just for the quarries, but also for the origins of and relations of quarrying with the modern settlement.

The phasing of underground workings map (inside rear cover) based on the archaeology allows some very approximate levels of underground production to be estimated, which, invites caution in regard to approximations occasionally drawn from historical sources alone. This is summarised in the following table:

Table 13.1: Estimated output by phase for the core area underground workings of Combe Down

Phase	*area worked (sq. m)*	*area worked (adjusted) (sq. m)*	*output (tons)*
I	9610	5189	64863
II	24930	13462	168275
II and III	11136	6013	75162
III	15651	12677	158463
III and IV	11711	6324	79050
IV	54359	29354	366925
IV and V	14364	7757	96963
V	6737	3638	45475
V and VI	33494	18087	226088
VI	6818	3682	46025
Total	18.881 ha.	10618 ha.	1,327,289

Notes

The area worked is based on a computer calculation of the area of the individual quarry areas (see inside rear cover) attributed by the archaeologists to each phase. This has been reduced (column 2) by 10% for the pillars and a further 40% for the proportion left behind as spoil to allow calculation of useful output. The tonnage calculated is based on five metre depth of average working, and a specific gravity of the freestone of 2.5. Considerable caution is needed in interpreting these figures. For comparison purposes below, half the buffer zone figure of the inter-phases (above and below) is attributed to the phase between.

How much stone?

In the table the most reliable figures are the overall area worked underground in the central area of Combe Down, about 18.9 hectares, for which the main Byfield and Firs quarries were about 18 hectares, with the remainder mainly in outliers, notably Foxhill. The total production is far more of an estimate, as we do not accurately know the size of each pillar (only a few have been seen full-height), nor the depth worked. We know, for example that the Three Acre Quarry as seen generally, exceeded five metres, and that areas in the Far East Firs – mainly small scale ventures – were often well below this. But the overall production of saleable stone was probably around 1.33 million tons, about 7500 tons a year average over the two centuries involved. Below the possible tonnages are considered under each phase.

Early quarrying: Phase 1 – Pre 1728-30

Around 1725 there were some half dozen quarry-masters or freemasons at work under the Monkton Combe portion of Greendown or Combe Down while, as the archaeology suggests, there were possibly more operating in the adjacent area of Lyncombe and Widcombe, at (Greendown or Quarr Down). With underground workings, frequently there is a mismatch between what seems reliable documentary evidence and archaeological findings. Such was certainly the case for Richard Irving, (2005 and pers. comm.) who was disconcerted to find, a year after his publication on the Byfield Quarry, that some of the early underground quarrying in Lyncombe and Widcombe extended completely under his house. It also raised the probability that Allen's paired accesses were placed either next to pre-existing underground access and associated ways, or actually drove though them.

It was similarly surprising for the present writers, on analysis of the distribution of early workings, to realise that early working had taken place on three sides of the Sheeps House Quarry at Rock Lane, including the possible entry for the John Pitcher working. Thus quarrying was most likely taking place here too before Allen drove his two almost parallel cartways north from it. There he probably avoided the bulk of earlier workings, but still utilised existing surface facilities for his access. It will not be possible to prove the same thing happened east of The Avenue, since surface quarrying has removed all evidence of the original frontage, and it was not possible to approach the front internally during stabilisation, but, given the evidence of two of his other sites, it must be a strong possibility. The archaeology thus suggests that pre-Allen, Phase I underground and possibly associated surface-working on a small scale was more widespread than previously considered, both at the southern scarp margin and 50-100 m back from it.

Phase 1, prior to Allen, based on the areas on the Phasing Map, appears to have produced some 65,000 tons overall. But this includes workings since possibly the end years of the 17th century, with some probably continuing for a time after Allen started, which we were unable to differentiate. This is a level of only a few hundred tons per year, divided amongst several freemasons, enough for just a few houses, a whole order of production below that produced by Allen.

Allen's quarrying: Phase II – 1728-30 to 1764

The basic story of Ralph Allen is well known. He became Bath's premier citizen in the mid 18th century, either the town's own 'Squire Allworthy' (Boyce 1967, 183) or 'the one headed corporation' (Peach 1895b, 180), depending on preference, with a rise from humble beginning to a vast fortune within a few years. Both his biographers cited above seem to consider his stone business a major success. The only doubt seems to have been a near-contemporary comment by Philip Thickness, that whereas people considered he had earned his fortune from 'picking up stones' underground, it really came from his postal business (Peach 1895b, 98). Any contemporary or biographical lack of appreciation of this was and is understandable, since the two businesses grew together after 1727 (when the post business was probably still unprofitable). But while the post business was dispersed over a large part of the country and run from a small office in Bath, the stone business was intensely local. The stone was conveyed in large quantities and was probably only too visible on the roads and construction sites within the city. However, analysis of the price structure for his stone business shows that a true profit including cost of capital was likely to be modest when the full picture is examined and that, given his consumption pattern, it is probable that the postal profits were indeed very substantial, as Thickness implied. It was probably convenient for Allen to allow the misconception, if it really existed, to continue.

Even if the stone business was not especially successful in financial terms, it remains true that it was both very innovatory in terms of its approach to production and was technically successful. Ralph Allen and his clerk Richard Jones, who was responsible for the whole of his stone business, provided an unprecedented combination of capital provision and industrial organisation in setting up an integrated quarrying, masonry and stone building and export business. Innovations included the ways quarries were operated, in the way they moved stone, and in the way Allen advertised it to what he saw as his major markets, whether in Bath or potentially, in London.

The business used the new Avon Navigation connection to Bath for both long- and short-haul movement and provided a wooden gravity railway a mile and a half long to the quarries from it. There were cranes for loading outside the quarries and a crane for loading on board ships on the Navigation and a roll-on-roll-off barge system to convey stone across the river and on to building sites in the City. He provided houses for his key underground staff and his stoneyard masons, and regular work and shelter for the latter in his yards. By his improved transport he reduced the price of stone in Bath by a full quarter. Wood the architect and friend of Allen complained that Allen's competitors then reduced it a further sixpence. Wood was more receptive to the reduction in masons' wages by 10%, which followed reducing the masons' loss of time (and their wages) in scurrying between their quarries and their building activities in the city by concentrating their work in his stoneyards.

Allen used advertising techniques for selling stone, which could well have been an exemplar for the later methods used by Wedgwood and Bentley for marketing quality pottery. This included the building of his Prior Park house for entertaining the noble and the fashionable, where, even Thickness agreed, he maintained a generous table. In combination with John Wood the architect (and later Wood's son John), he played a major role in the building of 18th-century Bath. In contrast, he found his export of blockstone had difficulties due, substantially, to the disruption caused by the war with France and this certainly disappointed him.

While Pollard's study (1994) had identified the essential features of some of Allen's underground working, the extent of Allen's working and many details were still a puzzle. The archaeology showed, however, that Allen's phase (II) of quarrying could be distinguished. Two of his sets of paired entries were well established from the estate plan, and underground his work was distinguishable, largely by the type of pillars used and by the organisation of the workings. The use of what we termed apophygate pillars was entirely conventional and can be seen in a number of other underground quarries both locally and further afield, but the use of pairs of, rather than single, cartways penetrating directly into the freestone beds under the Down was unusual. The disposition of these apophygate pillars and a later corbelled form about the cartways at the two known entrances led to the realisation that there were probably two other sets of entries with similar characteristics dating to this phase of quarrying. This understanding was reinforced by a realisation that documentary sources pointed to two further cranes to equip them and that the limits of his work may well be at shafts in all four instances, each having similar relationships with the cartways and pillars. This regular use of similar techniques both at surface and underground within the workings with the same adaptation of new methods seems to us to be a product of systematic management ensuring the best methods were introduced as fast as possible. The final and most convincing discovery, of grooves from cables running down via each of the four shafts, to haul stone or wagons to the shaft bottom, seemed to be both the last piece of the jigsaw and the innovatory hallmark of his phase of working. It allows the likely extent of his phase of working to be fairly reliably placed on the survey map, but with a still substantial proportion listed in the 'buffer zone', that is classified as Phase II or III..

Simple as this all was in principle, at both surface and underground it was substantially more capital-intensive than usual and clearly designed to enable a substantially larger scale of production by greatly increasing the number of working faces in what was still a young quarry, and by multiplying this output by having four separate units in operation (five if the Masons Crane House working is included).

Allen also operated surface quarries, probably mainly in the latter half of his time in the business, possibly since he had established a market for some of the Twinhoe ridding beds which continued to remain untouched underground. This included the surface working at Masons Crane House Quarry, in front of the Old Rank cottages, which was well established by the time of Robins drawing of 1759, extending through modern Quarryvale, and though its eastward extent is not exactly known, from near the present Ralph Allen Yard to somewhere east of the later church is likely. It also seems likely that the quarries behind (north of) the Old Rank also began being worked. In both cases this quarrying certainly continued after his death, but both sites were worked out by *c* 1800 or soon after.

He does not appear to have worked in Lincombe or Widcombe (except, perhaps, for a minor transgression underground due to poor identification of the boundary), nor in the area east beyond the Long Drung. From the Byfield entries at what is now Ralph Allen's Yard on Rock Hall Lane he worked north alongside his cartways to some 150-200 m from the North Road, at what probably became his largest single quarry. In Firs the distance of his cartways from the North Road was barely 50 m, though it was still largely undeveloped at the inner end. In the area east of The Avenue to the Long Drung, although more difficult to survey, he had probably penetrated some 100 m, but across a wider front with two separate quarry operations. These areas are shown on the plan of working phases (see inside rear cover.

In Chapter Five, historical data, based mainly on demand, was used to suggest possible outputs in Ralph Allen's time (Phase II, *c* 1730-64). The possible maximum level at peak periods may have amounted to 20,000 tons per annum, with a minimum of about 5000 tons and a possible average of about 12,500 tons. As well as underground production, Allen also had surface quarries, especially after the mid century, but even so, given the areas known (but not then calculated as above) it was considered this level was too high. As shown in the table, with the added 50% of the buffer zone, it seems likely his average underground production was only a little above 6000 tons per year. A very large proportion of this must have been the roughly worked coursed rubble stone used in partition walls and backs, rather than the more expensive blocks intended fro the production of ashlar. It also means that the export, claimed by Richard Jones, of 1800 tons of blockstone in what must only have been good years, is perhaps more impressive than has been implied.

This suggests several things. First that we may have underestimated the area attributed to Allen's phase of working, but inclusion of the entire buffer zone production would not substantially increase the amount. It may be that he had surface quarries working for much of his time whose outputs are not included here. It may be too that others also were producing, possibly from the Odd Down area, thus satisfying the overall demand. It could also be that in the latter part of his time Allen was producing at a much lower level than in the earlier, wooden-railway-monopoly years. Evidence for this is the contraction in working areas, back to the Central Firs workings only, as evidenced by the *c* 1764 Estate Plan sites. The figures show that the level of profitability of his quarrying enterprise was probably even worse than suggested in Chapter 5, and it is easy to see why Gertrude Warburton had no hesitation in lifting the track and putting the quarries out on lease after Allen's death in 1764.

However, his was not, as sometimes claimed, the only quarry on the Downs. The detailed archaeological study also allowed for the identification of underground working that were not attributable to Allen. He certainly had a monopoly in the area he set aside for quarrying, easily seen on his estate plan since he planted the surface with fir trees and rides through them from his house. It was a roughly triangular area – now the core of the Combe Down village. At least one other freemason, Thomas Greenway, was still working after 1730 beyond Allen's eastern boundary (a footpath known as the Long Drung), despite the land being in Allen's possession by 1737. It is possible from the evidence of three separate cartways that there was another quarry being operated, possibly by Milo Smith. Greenway was famous in Bath as an ornamental sculptor for the highly decorated house where 'Beau' Nash lived for a time. It was possibly his high and rare level of skill, and thus status, which caused Allen to tolerate him. However, even taking into account possible production by others at Odd Down and Entry Hill there is little doubt that Allen was indeed pre-eminent in the supply of stone in the mid years of the 18th century building of Bath.

Probably the very low marginal cost of moving his stone downhill to Bath was sufficient on its own to secure demand for his stone, and the ability to supply the vast amounts of stone needed for buildings like the Parades would certainly have established his position. Had he been able to establish a large London market too, in the way that was possible using long distance railway network between quarries and markets a century later, then it is possible this cause of his lack of substantial profit could have been overcome or, at least, reduced. The original cost of his wooden railway had to be repaid (or lost) by the sole producer and user, unlike the Turnpike roads used by his competitors and, indeed the later public railway system, and this was the most likely reason for his likely poor financial performance.

Moreover, whereas he might have been expected to profit from his downstream activities, exporting and speculative building, his forced reductions in the cost of stone and the cost of masons in Bath were probably equally helpful to his competitors too. Whatever the cause, in his later years, after the mid 1740s, his attention seems to have turned towards his interest in social and political matters and management of his estate, all heavily subsidised from his postal business, as is evident in the cash-flow problems after his death, when the postal business contract ceased.

Assessing the success of Ralph Allen

This story of Ralph Allen's stone business was not an unusual one in the 18th century and has many echoes in both contemporary and slightly later development in diverse industries such as mining, metallurgy, textiles, and pottery. In all these the scale of working and use of paid labour increased, the applied capital grew, and innovation in transport and advertising or other means of capturing markets played their parts in Britain's industrial revolution. However, unlike the achievements of Arkwright or Wedgwood, Allen's business remained stubbornly local. The reason for this, even disregarding the dislocation caused by wars with France, was largely lack of suitable inland water transport between Bath and London. Locally however, his success in lowering costs of the stone and of masons' work, and his not disinterested support for the vision of the Woods, played a fundamental part in keeping out alternative materials, such as brick and render, and in establishing the building of Bath with results so important to the image of the city today.

Quarrying in the Allen Estate: Phase III – 1764 to c 1803

Documentary and literary evidence is much more scarce for the Phase III, Allen Estate period of working. After Allen's death, the quarries, surface and underground, appear all to have been leased to independent quarrymasters or freemasons. Details of who they were and how much they produced are not available until near the end of the century, at which time the statistics, because of intense economic depression are unlikely to be representative. They did not have the use of the wooden railway, which was demolished in late 1764, but did a few years later gain much better access to Bath by the extension of the Turnpike across Combe Down. Despite competition developing at Crossways and

Entry Hill, and east of the Lawns, their cumulative output cannot have been much inferior to that of Allen, since the building industry in Bath remained fairly prosperous until just before the end of the 18th century, with some massive building projects under the younger John Wood.

In the underground workings, there is evidence of diversity in approach in this period. Two new methods of working seem to have been introduced, the Long Room System and the Long Wall, gullet and pillar, the latter eventually becoming dominant and widespread in this phase and later. The pillar plan (Fig. 1.9) shows the system in West Firs was applied with a regularity in the form and offset layout of pillars not seen before which suggest one quarrymaster at least had developed a controlled and systematic approach. The Long Room method was only applied over a limited area, but in part of it a system of hauling and loading stone was introduced using mast cranes, though this probably proved premature, requiring more space than was available with this method.

The extent of working in this phase can be fairly well deduced from information in a series of leases granted at the beginning of the following phase, when the Allen Estate sold its properties to new owners, who were as much involved with land speculation as quarrying. The Allen Estate freemasons or quarrymasters appear not to have worked in the surveyed areas of Byfield, except for a possible fairly small working operated by Richard Jones and son, and located just west of the Combe Road boundary and pillar (Quarry Area 2398). In Central Firs, the working was completed to some 25 m from the North Road (2200), and to the west of the Allen entries, two or three new entries were made, one of some 30-40 m length, to exploit the western portion of Firs. In East Firs, up to the Long Drung and immediately to the east of it (2339), Phase III working penetrated to within about 120 m from the apex near the Prior Park Gates. It is probable that at least some of these quarries continued into the next phase, under the same lessees.

Based on the Phasing map, Phase III appears to have achieved a production level somewhat higher on average than Allen's, at around 7000 tons. There was rising demand as Bath expanded, and the freemasons who took underground leases would not in general have had the same option as Allen had, that is to switch to surface production. Given the dreadful years at the end of the century when underground production sank back to a few hundred tons, the general level may have been even higher than suggested. There was a substantial rise in competition in these years from other freemasons in the wider area.

The early 19th century: Phase IV – c 1803 to 1833

Quarrying in the new century saw both a continuation of the old quarries together with old scale working, and the introduction of many new quarries, notably at surface, but also a larger-scale quarry underground. Sale of the area in parcels to several land speculators, led to them in turn laying out quarries in regular areas typically of two to three acres each in total, to up to a dozen freemasons.

West of Combe Road in Lyncombe and Widcombe, Harry Salmon divided the area today known as Coxes Quarry and Byfield Mine into some six or seven blocks, with half a dozen quarry owners. Only two blocks had any archaeological evidence for workings underground. What is now the Bat Sanctuary was probably worked by three different owners and the other, John Scrace's quarry in the north-west corner (512), was probably difficult to work economically because of a bed of poor stone, and he went bankrupt. The surface quarries seem to have been much more successful, the southernmost owned by Samuel Nowell (no.1 on Figure 4.2), most likely flourishing because of his son Phillip's renowned abilities as a mason which probably led to orders for all the quarries. There were other family relationships there, notably between Street and the Nowells, which must have helped spread this prosperity, and Irving (2005), in his description of the area points to several examples of co-operation between freemasons which must have been mutually beneficial in what sometimes were particularly hard times. The area was substantially quarried over by around 1830, but it is possible that reworking at greater depth over part of the area took place for several more decades.

The area north and east of Allen's working in Central Byfield (505) was also let to several freemasons, with changes of lessee suggesting they found times no less hard. Their leases commenced 40 perches south from North Road (almost exactly 200 m), although this was probably nominal, interacting slightly with the older Allen-period workings. The tone of the leases suggest that shafts were initially not welcomed by the new landlord, Edward Layton, and a little later by his heir Nathaniel Hadley, even though at least two and possibly three had been made previously. Three more were indeed sunk, all in a single row and aligned to a length of cartway below. Shafts, cartway and a cable hauled system were here developed as an integrated system of quarrying which also, as seen in slightly earlier workings in Firs, may have involved the regular offset use of pillars for support. In West Firs previous working had been more extensive and the potential working life there and in the western part of Central Firs north of the Allen working was limited, even in 1803.

The largest underground working of this phase was however in the eastern side of Central Firs, overspilling in the north into the top of Combe Park and into the north-west corner of the area just east of the Long Drung. Here John Greenway seems to have put together a somewhat larger quarrying area than had his competitors, some three acres in Central Firs (2211) and just under two acres immediately adjacent. This area was actually partially worked by John Burgess and son in a regular and apparently economic manner from around 1809 to 1838, including the Grand Canyon Quarrying Area 2209. It again had a central cartway running beneath wide shafts mainly under the north-east of Firs Field, some use of cable haulage and in particular higher faces of up to 8 m, more than seen in any other contemporary or earlier workings. Essentially this was the last substantial block of ground left near the village, and subsequent working in the core quarrying area was predominantly tidying-up.

The production area for Phase IV, for the first third of the 19th century shows it was was indeed the golden age. Because the Three Acre quarry was the main source of Phase IV and V output, probably terminating in 1838, the actual level is probably higher than the figures would suggest. The 12,500 tons calculated by using the usual calculation may have been, in actuality, as much as 14,000 -15,000 tons.

The mid-century years : Phase V – 1833 to 1867

By 1860, according to Tyte (1897 cited in Hemmings 1983), there was no underground quarrying in Combe Down. If so it was only a temporary period of inactivity, but it does indicate the marked reduction by then, and despite a little recovery after the turn of the century, it was generally avoided in favour of surface methods.

The production in this phase calculated from the table would suggest some 7600 tons output annually. However as the largest annual amounts were probably prior to closure of the Three Acre Quarry in 1838, the actual output after this probably fell below around 6000 tons, possibly well below that of Allen's period. It probably was well below this by 1860, when, by general agreement the central area of Combe Down was effectively worked out, leaving only small blocks for the future.

The unknown area is principally the far eastern area of Firs, under Gladstone Road and east to the boundary. Much of this area was not examined because of floor to roof spoil dumps (themselves suggesting that lower heights were worked) and the few areas where the roof to floor height was seen were only some 3 m high. The large area there probably exaggerates the production attributable to Phase V (and even more to Phase VI).

The few small parcels of unworked ground left after Burgess finished in 1838 were worked mostly by use of winches or cranes and by an increasing use of saws. To use these methods effectively, the Open Room method of working evolved. It is not entirely clear if this was copied from fast-growing areas around nearby Monkton Farley, and Corsham and Box in Wiltshire or whether it was a response to the fierce competition set up by them, though the former is probably more likely. However, surface quarrying had still a considerable advantage in the closeness of Bath and the perceived better quality of Combe Down stone. In more distant markets Combe Down could not produce the quantity of stone, nor had the direct rail access needed effectively to compete, The Wiltshire stone, available in thick massive beds, was more easily quarried on a huge scale and the term Bath Stone, by this period, meant it was Oolite from anywhere but Combe Down, though it is unlikely most customers usually either knew or cared.

This did not mean the end of substantial quarrying. North of North Road, John Ovens Thomas had let out quarrying leases between Combe Road and what is now Popes Walk, which continued to benefit his descendants after his death *c* 1830, until the 1870s and after. This area was almost entirely worked at the surface except for a few small areas under the buildings at Stonehouse Lane (2219), and some small workings south of North Road in Central Firs, reached from the surface quarries to the north, including Quarry 2212. These underground sections continued to about 1880. Similarly surface workings at the Lawns (Lodge Style) and Shaft Road (Lodge Hill or Combe Quarries) on a very extensive scale continued operating well into the 20th century. Here and around Entry Hill and Crossways the growing activity was no doubt helped by the easy accesses to Bath by road, and after about 1810, to wider afield with access to either the Kennet and Avon or its branch, the Somersetshire Coal Canal.

The late 19th and early 20th centuries: Phase VI – 1867 to 1938

Though with a production of only some 1700 tons average annually in the central area of Combe Down, which as explained above may still be a little high, at last Allen's dream of Bath Stone being predominant in the market was achieved. *The Builder* (20 April 1895) wrote that it was used in immense quantities. Unfortunately it coupled this with the remark that it was also the 'gerry-builders stone par excellence', a statement not refuted by a comment from one supplier that they supplied whatever the customer needed, even calling it Portland. It is possible that Combe Down freestone, with diminishing supply and mainly used locally was somewhat freer of these problems, though John Wood, in his and Allen's time also decried such subterfuge.

By the end of the century, however, at surface, north of North Road, quarrying was retreating to the north limit of the Down, though a small narrow strip between Popes Walk and the Prior Park Drive (Ralph Allen Drive today) had finally been given over to the quarrymen, and the end near Combe Road was still quite close to the main road. Most of the effort was expended at and near Combe Quarry at Shaft Road, including Mount Pleasant. In 1887 several quarries were amalgamated under the Bath Stone Firms Ltd and it became easier to match output to fluctuating demand by temporarily closing and re-opening quarries. The former seems to have happened about 1895 when *The Builder* reported that only Combe Quarry (east of Shaft Road) was active at the time.

Around 1895, underground quarries became active again, working where surface working would have adverse impacts. The underground workings at Shaft Road and at what was recently termed Freylings Shaft, but originally was probably part of Mount Pleasant Quarry, both seem to be active about this time, with another 'slope entry' nearby and with St Winifreds developing somewhat later. The Shaft Road workings exploited the area under that road, and successfully took stone from as little as a metre below the surface! At Foxhill, west of the extended Combe Road at what is now the Rugby Football Ground and the MOD establishment, a small late 19th- and early 20th-century quarry worked at surface at the former almost opposite the older Stonehouse Lane Quarry. Addison (1998) calls this Wilks Quarry and refers to the traditional use of barrows conveying the ridding on planks high across the gullet or working trench. Apparently blocked by the recent housing at the south side next to North Road, they took their working underground, undermining by an entry from the surface quarry. Adjacent to this, over the first 20 years or so of the new century, three more small workings were opened, including what Addison called Coxes Vertical Shaft Mine, near the MOD entrance. All these turn-of-century underground mines used a similar technology. Headings were rail-served and the rooms off either used winches and a snatch block for dragging stone blocks, or this was done by hand-operated post cranes, their original presence revealed archaeologically by chog holes in the roof. Almost identical technology (allowing for the fractured nature of the Combe Down freestone) can be seen in the Wiltshire Quarries though there the scale was usually very much greater. Hand saws were widely used, but large blocks were still more likely to be broken out by wedges.

One last blow was struck in the old core area, when James Riddle, operating using a petrol winch from the Allotment Shaft at the western side of Sheeps House Quarry in about 1905, began a similar operation on a small area left unworked there. By 1914 however, in that small area too, it was all over. In the other areas, working was reduced in scale. The Bath and Portland Stone Firms, as it became known eventually, ended business about 1937. Since then the single company operating has been Hancocks' Quarry at Lodge Style, who are still in business today.

Growth of the Combe Down settlement and the relationship with quarrying

The growth of the Combe Down settlement was not organic in the sense that it grew only to serve the quarrying community. What would nowadays be called a brownfield site, a mix of disused quarries and potential quarries was not in itself immediately attractive for housing, either because the intrusion of housing makes quarrying activity more difficult, or because living in the immediate proximity of industrial-scale working quarries is not obviously desirable, except perhaps for the convenience of the workmen there. That the landscape was less damaged than it might have been was a consequence of the ownership of wealthy Ralph Allen whose house was nearby and who developed a tree-planted landscape designed to preserve and enhance its amenities. His successors in ownership however, saw its proximity to Bath and its fine outlooks and salubrious climate as a speculative building development opportunity.

Although the great house of Prior Park was omitted from the new parish of Combe Down, perhaps as it was seen as more important to Bath than to an outlying parish, it was the early ownership of the Down by the Allen Estate which defined the boundaries of modern Combe Down, and the core of its original quarrying activity defined the core of the modern village and ward. Allen had probably not foreseen this – his village or township was limited to some dozen houses for his key workers and a public house (he built a brewery too, but nearer the city). Others must have walked in to work from surrounding communities. Until just after 1800, the estate assembled by Allen remained substantially intact, connected by family links through his niece and heir Gertrude Warburton and, more tenuously, through the marriage of another niece, Mary Bennet, to Cornwallis Maude. He was soon to become the Earl de Montalt, was later created Lord Hawarden, and purchased the Allen properties stage by stage from Gertrude. Almost certainly undercapitalised and perhaps damaged by the depression of the mid 1790s, Hawarden sought the aid of two of Bath's land speculators, Harry Salmon and Benjamin Wingrove. Together they seem to have laid out the principle roads for modern Combe Down: The Avenue, Combe Road, Rockhall Lane, Church Road, Beechwood Road and even the path running south of the Church to Belmont Road. The most conspicuous development by de Montalt was Isabella Place, upgraded with the current polite three-story front built on to an older row of cottages. It was named after his wife, and complemented the changes also made to the Old Rank, which then became de Montalt Row. The construction of cottages down Rock Lane may also have been begun under him.

Lord Hawarden, as he had become by the end of the century, clearly saw a new type of settlement replacing the quarries in the central area, whose activities were coming to an end. It seems likely he saw his new buildings as largely for the middle class, but generally this *embourgeoisement*, today likely to affect even the smallest cottage in the old core area, came somewhat later. Hawarden's death in 1803 simply hastened what was already happening and the land was transferred to the two speculators who quickly sold it on, both to individuals in small parcels and, as an estate to Edward Layton. Following his death, it became the Hadley Estate. This saw the first houses built on Combe Park, accessed from The Avenue but, more rapidly, the workmen's cottages on Summer Lane (Quarrybottom) were developed. The Hadley Estate probably either built a dozen houses, or sold land leasehold for building on in the next decade of so: among these was John Davidge's brewhouse down Rock Lane, giving his name to the bottom.

What land did not go to Salmon and Wingrove remained with the Prior Park Estate, but this too was sold on the death of the second Lord Hawarden in 1807, to another businessman, a Quaker, John Ovens Thomas. He came to live at Prior Park, but was no slower to see his opportunity and leased the areas close on the north of North Road beyond Pope's Walk for quarrying, which lasted over the whole of the 19th century. After his death in 1830 a family trust benefited. In conjunction with another speculator and surveyor, Henry Cotterell, he had the area east of the Long Drung as far as the boundary with the Lawns parcelled out for housing. The remaining unsold area was acquired by Cotterell after Thomas's death. This is the area known as the Tyning and Gladstone Road today.

The land here was sold in substantial parcels and some certainly attracted masons and small speculators who built both individual and rows of houses which, in the mid 19th century were mostly occupied by wage earners in quarrying or linked trades. This proportion of occupiers linked to quarrying appears to have diminished after 1860. The Tyning and Gladstone Road did not, however, have the crowded appearance seen today (which is largely modern infilling) and large areas seem to have been gardens or allotments, though there must also have been at least ten or so shafts to the workings below and possibly, small masons or builders yards associated with them still at work until near the end of the century.

Generally, building of houses, as in the core of the village, followed quarrying. The old Masons' Crane House Quarry soon housed the Quarry Vale cottages alongside Summer Lane. Claremont, the Unitarian Chapel and Hopecote Lodge developed after Burgess moved his operations underground, and finally the Church was built along what was then The South Parade and is now Church Road. The Vicarage and school, and larger houses such as Belmont (Mrs Cruikshank's house – heir of John Thomas) were built on Belmont Road as dumping of spoil from the Burgess' quarry above had also ended. This was also so, for instance, at Farrs Lane before and around 1850, where, as the quarries retreated back away from North Road, houses began to develop one by one near the roadside on the spoil heaps. Something similar happened on the south side of North Road beyond Combe Road, as far west as the Foresters Arms, and also down Combe Road.

More houses and thirsty workmen led naturally to Public Houses being built. John Davidge had his brewhouse down Rock Hall lane, the William IV arose from Job Salter's brewery, later Hines Brewery, and the Hadley Arms (*c* 1845) may also have made its own beer. The Carriage Inn was probably still in business, another Public House was located down Rock Hall Lane and at least two others were built, one either end of the village on North and Bradford Roads.

The end of quarrying was delayed at Combe Park when Richard Lankasheer took a lease to quarry under it in 1856 but housing there was by then well under way. In the Tyning and Gladstone Road, Cotterell devised the ultimate selling point for house-plots, with plots sold with the stone underneath suitable for building them. Others found themselves in a similar situation accidentally – the back areas of Gay Staithe for example, had unexploited blocks of stone left underneath which was taken around or after the mid-century.

Though surface quarrying was clearly more favoured in the 19th century, this was not necessarily so to landlords or house owners, who gained the revenue from stone below their properties by letting it out for undermining while also collecting rents for the surface. Not all buyers of land under the Tyning and Gladstone roads were quarrymen. Others must have let out their rights to account for the very widespread mid and late-century undermining of stone which took place there, presumably to mutual advantage. In a few examples, rubble stone packs were built to support the roof underground beneath houses above, and other areas filled with spoil to the roof may also have had this in mind. Under the early 18th-century house at Stonehouse Lane, the adjacent quarry (2219) undermined the houses and the quarry access road some time after 1850. Possibly John Scrace's quarry (512) in the corner of Combe Road and North Road did much the same somewhat earlier. At Foxhill, undermining beginning at the junction with Combe Road continued parallel to the road westwards, while housing, by the 1920s began to extend eastwards from Foxhill Road. Here housing initially dominated, with the grandly named Ralph Allen Park, though in the Second World War this development was eclipsed by War Office and later defence needs.

The significance of the archaeology

Interpreting the workings

The archaeological recording of the underground structures allowed a plausible phasing of workings to be developed and provided a physical context for the historical narrative. Recording and interpretation of the detailed features associated with the quarrying process enabled a deep understanding of the working methods and how they developed, again enhanced by the historical setting.

Art and craft amidst the dumps: the finds

The finds within the underground quarries were significant both for the interpretation of their working and the equipment used, and through the graffiti, of their attitudes and human preoccupations. Less expectedly, the finds also provided clues as to the consumption and disposal patterns of the community above.

Graffiti was useful especially in providing dates, which, with care, helped develop the chronology of the workings, notably the pre-Allen 1725 date from Quarry 2347, and also the approximate time of the introduction of the frig bob saw by dates on sawn faces. Help for dating also came especially from the study of clay pipes carried out by Marek Lewcun, which also provided evidence of the early workings in what became known as the E4 stub under the south end of the Avenue (2347). Pottery found there, deriving presumably from the Carriage Inn, while not proving an early date, certainly allowed for one.

One of the conundrums which faced the investigation was in the use of horses underground. Despite the wooden railway at surface, archaeological evidence suggests that Allen did not extend its use below ground. Probably because the distances were not large, horses were certainly not generally used either by him or later quarrymasters. However, the finds of remains of three horses or ponies, two of them remote from any shaft entry, together with several fragments of horse shoes and a possible cartway with hoof marks does suggest horses were, at least occasionally, used underground. Certainly their use would be very advantageous for bringing out especially large blocks.

A substantial range of stonework, such as for window and door mouldings and other features, was also found underground. It was widespread, and given that many areas of the mine were backfilled at later dates than they were worked, it may have been even more common. It showed, for instance, that banker masons' work was done underground under Allen, despite his having two surface stoneyards, and that smaller quarrying ventures fairly commonly must have done so. The stonework did not seem to include much of the more detailed work, perhaps because it would have been more susceptible to damage, but a fairly wide range was produced, some of which can easily be related to surviving buildings, some which cannot. Actual use of stone equipment within the quarries included candle holders, boxes for candles and perhaps tinder, and probably for protecting food against rats but overall use of worked stone for the quarry (only in regards to the Stabilisation Scheme and in general for the Byfield and Firs areas) and quarrymen was very small.

Artefacts and the community

One of the results initially least expected to be revealed following the decision to record the archaeology of the underground quarries was to be able to reveal some of the social aspects of the settlement. Graffiti, some of which can be seen as a form of vernacular art, has a long survival time compared with the surface, and under Combe Down there were very many examples. Much of it remains enigmatic, though, for instance, initials and signatures may gradually yield their secrets when combined with more research into village historical minutiae. The website contains a table of underground graffiti, in reference to known quarrymen, quarrymasters and stone producers). The table includes information already researched by NRH, David Pollard, Rosemary Simmons of the Combe Down Heritage Group and the acknowledged research by Peter Addison; further research was largely beyond the scope of the present study. What must be drawings of quarrymen and acquaintances were fairly common and sketches such as that of 'Holly' and the 'Naked Lady' yield an insight into what occupied the minds of mid 19th-century young men – much the same as today has to be the conclusion. There is evidence too of a radical frame of mind in two of the drawings, which might well be explored by further historical research.

The range of pipes found has been particularly rewarding, though many of those reported here were recovered in the 1980s. The 18th-century range provides a reference collection for Bath, while the 19th-century samples included a range of Bacchus pipes unparalleled elsewhere in Britain. This may be a sign of comparative prosperity, as may also be a 'football' pipe. Whether this prosperity came from quarrying is difficult to assess.

Overall, however the surface-derived dumping which took place into open shafts and other surface quarry entries revealed mostly what might be expected in any Victorian bottle dump. What was mundane when these were dumped, however, can be a source of great interest and nostalgia today. Many items found will hopefully find a home in exhibitions and displays in the future. Some of the mass-produced pottery, in particular, provides a vivid insight into the growth of domestic food, cosmetic remedies and medical items some of which remain household names today.

Combe Down today

Combe Down has gone, over three centuries, from 'sleight pasture' for sheep through extensive quarrying to a being a dormitory suburb of Bath. It has become obvious that everywhere in the core or old area of Combe Down village, almost every scrap of land has either been quarried from the surface or undermined from below. This is also true for very extensive areas adjacent on the north and east sides, and also for areas around Entry Hill and at Odd Down. This begs the question, why were some areas largely exempted? At least one is explicable: William Smith, the so-called 'Father of Geology', picked a quarry along a fault line which had damaged the stone, though it was probably more his impatience to get on with his study of the national geology which prevented him driving on and eventually succeeding as a local quarrymaster.

Substantial areas remain unworked under the Foxhill Estate and south of it between the Bradford Road and Shepherds Walk. There were indeed a couple of small quarries of probable mid-19th century date, but these seemed to fail to grow like others. Possibly it was the geology, with either fractured ground, more clay layers, or calcitic vugs ruining the stone for monumental purposes. Though there was surface quarrying in the former Claverton part of Combe Down north of North Road it too only took out a restricted area and may have been more shallow-depth working (based on the limited time it seems to have been open) compared with west of Prior Park. We found the quality of at least the upper beds of freestone were deteriorating – thinner, more brashy – in the north-east corner of the underground workings near Gladstone Road, so this may offer an explanation for the nearby Claverton area too.

Conclusions

The physical work of stabilising the underground quarries in The Combe Down Stone Mines project was largely finished by November 2008, after eight years effort. We have been fortunate in benefiting from an unprecedented opportunity to examine over 200 years of underground and surface quarrying archaeology in considerable detail, in conjunction with use of literary and documentary sources. In the first chapter we laid out the original and later modified aims of the archaeological project. There is now a very large body of digitised plans, contexts, photographs, videos and laser scans, with a substantial collection of artefacts which are archived and reported upon. This mass of data and results will go on to the website. This database has been used for this report and will be a resource for future interpretation and displays.

One of the benefits of such a long-term project has been the opportunity to get to know the local area very well, and to have many opportunities to follow up information in local libraries and archives. This is reflected in the considerable portion of this report being based on historical data. In conjunction with the archaeology, this has yielded a sum which is greater than the separate parts and, in particular, it has been possible to assess the role of Ralph Allen and his stone business in a way previously not possible, even though much of the basic information used was previously published. This reassessment of him will be, we hope, an important contribution to the 'Building of Bath'.

We hope it will also be a major contribution to quarrying history. It has certainly been the largest archaeological recording of any quarry or mine in the UK, and is rivalled by only a very few such projects in Europe. It has been in many ways a triumph of co-operation between contract managers, engineering consultants, contractors and archaeologists in overcoming the difficulties in working together in the difficult and potentially very hazardous conditions of the underground quarries. This was in part due to the initiative of the Bath & North East Somerset Council and English Heritage, but much more to the staff of the companies and their personal interest in the remains. Carrying out the recording required considerable innovation to discover the methods most appropriate to the circumstances.

There were obvious limitations to what was possible. The end of any project is likely to leave a sense that more could have been done or could yet be done and this is certainly true here. There is potential for further research notably into the social aspects of Combe Down quarrying; the detailed examination of housing at Combe Down and the relationship with quarrying is still only slightly explored, in particular who lived where and what they were involved in.

Many areas of the underground quarries could not be entered or even distantly seen, either because of safety, with the requirement always to be under modern engineered roof supports, or because they were already infilled with spoil. Possibly this amounted to over half the underground area. On the other hand, the overall area was so large and it was possible to observe repetition in what had been done in different parts of the workings giving confidence that overall we had a sufficient sample from which to draw valid conclusions.

The result has been a narrative which has been especially strong in dealing with the innovatory phase and area of Ralph Allen working, but which has also been able to identify working phases before and up to the early part of the 20th century. There are very extensive areas of underground quarrying in Britain, and some parts of these will be even older, but it is likely that nowhere else will be able to provide such detail for 18th-century quarrying (underground or surface) as it has been possible to assemble at Combe Down. It should, however, be possible to carry out 19th-century studies of equivalent potential in other areas.

Lynn Willies
Neville Redvers-Higgins
Ianto Wain

Bibliography

Addison, P, 1998 *Around Combe Down.* Millstream Books, Bath

Anon. *c* l 750 *New and Universal Dictionary of Arts and Sciences*

Anon. 1953 Roman Britain in 1952 *J of Roman Studies* **43**, 123

Anon. 1963 Boys lost in Combe Down Tunnels *British Caver* **37**, 28

Anon. 1966 Bath and Box Stone Workings. *Mendip Caver* **2** (8), 63

Anon. 1967 Reprint from the Builder, 13 and 20 April 1895 *British Caver* **47**, 32-49

Anon. 1970 *A Guide to the History and Geology of Bath Stone* Kingston Minerals

Anon. 1981 *Brewers Quarry Access* Mendip Nature Research Committee

Anon. (No Date) *The Building of Bath*. The Building of Bath Museum.

Ashbee, J, 1996 *The Quarrying Industry. Step 1 Report* Lancaster University Archaeological Unit for the English Heritage Monuments Protection Programme

Brown, A (ed), 1900 *Bainbridge's Law of Mines and Minerals,* (5th ed), London

Bone, M, 2000 The Rise and Fall of Bath Breweries 1736-1960 *Bath History* **8**, 106-33

Boyce, B, 1967 *The Benevolent Man: A Life of Ralph Allen of Bath* Harvard University Press, Cambridge, Mass

Bristol Master Builders Association 1904 *Historical References to Bath Stone* (documents held by Bath Central Library)

Buchanan, B J, 1986 The Evolution of the Turnpike Trusts: lessons from a case study *Economic History Review* **39:2**, 223-43

Buchanan, B B, 1996 The Avon Navigation and the Inland Port of Bath. *Bath History* **6**, 63-87

Chapman, M, 1996 *A Guide to the Estates of Ralph Allan around Bath, based on his map in the Bath Record Office*. Survey of Old Bath, Bath

Chapman, M, Hawkes, J, and Holland, E, 1998 *The J Charlton Map of Lyncombe and Widcombe, 1799* Survey of Old Bath, Bath

Chitty, G, 1996 *The Coal Industry, Step 4 Report.* English Heritage Monuments Protection Programme

Combe Down Womens' Guild 1963–4 *Combe Down History* (documents held by Bath Central Library)

Cotter, J, 2009 The Pottery in *Combe Down Stone Mines: Specialist Reports* Oxford Archaeology (www.combedownstoneminesproject.com)

Cotterell, T S, 1904 *Paper read at the 61st Congress of the British Archaeological Association held at Bath, August 1904* (document held by Bath Central Library)

Cottingham, L N, 1822 *Plans, Elevations, sections, details and views of the magnificent Chapel of King Henry the Seventh, at Westminster Abbey Church; with the history of the Foundation and an authentic account of the Restoration. Vol. 1* London

Cranstone, D, 1992 *The Lead Industry*. Cranstone Consultancy for the English Heritage Monuments Protection Programme

Cunliffe, B, (ed) 1979 *Excavations in Bath 1950–1975* (Western Archaeological Trust)

Cunliffe, B, 1986 *The City of Bath* Gloucester

Davenport, P, 1988 Bath Abbey *Bath History* 2, l-26

de Labelye, C, 1734 A description of the carriages made use of by Ralph Allen Esq. To carry stone from his quarries, situated on the top of a hill, to the waterside of the River Avon, near the City of Bath, in J T Desaguliers. 1734 A *Course of Experimental Philosophy*, Vol. 1

Desaguliers, J T, 1734 Ralph Allen's Crane at Dolemeads, designed by John Padmore in *A Course of Experimental Philosophy*, Vol. 1

Egan, P, 1819 *Walks Through Bath* Bath

Eglin, J, 2005 *The Imaginary Autocrat: Beau Nash and the Invention of Bath* London

Elton, A, 1962-3 The Prehistory of Railways. *Proc. Somerset Archaeological and Natural History Soc* **107**, 31-59

Flynn, M W, 1984 *The History of the British Coal Industry, Vol. 2: 1700–1800 The Industrial Revolution* Oxford

Foster, C le Neve, 1901 *A Text Book of Ore and Stone Mining* London

Forster, A, Hobbs, P R N, Monkhouse, R A, and Wyatt, R J, 1985 *Environmental Geology Study: Parts of West Wiltshire and South East Avon*

Fox, A, and Fox, C, 1960 Wansdyke Reconsidered. *The Archaeological Journal,* **115**, 1-48

Goodburn, D M, 2009 The Historic Woodwork in *Combe Down Stone Mines: Specialist Reports* Oxford Archaeology (www.combedownstone minesproject.com)

Gregory, W, 1886 *Ralph Allan and Prior Park*

Grenvill, A B, 1841 *Spas of England and Principal Bathing Places* (1971 Reprint) Bath

Hall, C J, 1984 *"Tanky Elms" Bath Stone Quarryman.* Corsham

Hartley, R F, 1994 Tudor Miners of Coleorton, Leicestershire in: T Ford and L Willies (eds) *Mining Before Powder.* Peak District Mines Historical Society, Matlock Bath, Derbyshire, 91-101

Hemmings, P I, 1983 *The Stone Industry in Bath.* Unpublished Dissertation. Typescript Copy in Bath Central Library

Irving, R T A, 2005 *A History of the Byfield Mine, Combe Down* Combe Down Heritage Society, 79

Jennings, P. (No Date) *Materials for a History of Truro, Vol. II, Families.* Mss in Royal Institution of Cornwall, Truro

Jones, Richard *(No date – late 18th Century) Life of Richard Jones*. Mss. in Bath Reference Library, compiled by J. Gregory

Keevil, A J, 1989 The Fosse Way at Bath. *Proc Somerset Archaeol and Nat. Hist Soc.* 133, 75-102

Keevil, A J, 2000 Barrack(s) Farm, Wellsway, Bath. The Estate and its Holders *Bath History* **13**, 28-55

Lewcun, M, 2009 The Tobacco Pipes in: *Combe Down Stone Mines: Specialist Reports.* Oxford Archaeology (www.combedownstonemine sproject.com)

Lewis, M J T, 1970 *Early Wooden Railways* London

Macdonald, D M J, Whitehead, E J, and Butcher, A S, 2000 Data Review leading to a conceptual model of the hydrogeology of the Combe Down Area, Bath. British Geological Survey Report, WD/00/35. Keyworth

Major, A F, and Burrow, E J, 1926 *The Mystery of Wansdyke* London

Manco, J, 1992 Bath and the Great Rebuilding. *Bath History,* 4, 25-51

Mayhew, H, 1851 *London Labour and the London Poor*

Neale, R S, 1981 *Bath 1680-1850: A Social History* London

Oxford Archaeological Unit 2001 Combe Down Mines, Bath Archaeological Assessment. Unpublished report prepared for Parsons Brinkerhof and Bath and North East Somserset Council

Page, W, (ed) 1906 *Victoria County History Somerset* **1**

Palmer, T J, 2009 Petrology of Limestone Artefacts Rescued from the Combe Down Stone Mines, Bath. In: *Combe Down Stone Mines: Specialist Reports* Oxford Archaeology (www.combedown-stoneminesproject.com)

Peach, R E M, 1895a *Bath Old and New* (reprint) Kessinger Publishing, Montana

Peach, R E M. 1895b *The Life and Times of Ralph Allen* (reprint) Kessinger Publishing, Montana

Phillips, J, 1844 *Memoirs of William Smith LL.D.* London

Pitcairn, D L, and Richardson, A, 1924 *Historical Guide to Monkton Combe, Combe Down and Claverton* Bath

Pollard, D, 1982 Bath Stone Quarry Railways 1795-1830 *J Bristol Industrial Archaeol Soc* **15**, 3-19

Pollard, D, 1989 Digging Bath Stone *J Bristol Industrial Archaeol Soc* **21**, 28-36

Pollard, D, 1994 Historical and Archaeological Assessment of the Underground Quarries Known as Firs Mine and Byfield Mine. Unpublished report for Bath and North East Somerset District Council

Ponsford, M, and Jackson, R, 1995 Post-Medieval Britain in 1994, in *Post-Medieval Archaeology* **29,** 153

Price, A J, 2007 *Cheltenham Stone: The Whittington Quarries.* Cotteswold Naturalists Field Club, Cheltenham

Price, L, 1984 *Bath Freestone Workings* The Resurgence Press, Radstock, Bath

Raymond, R W, 1870 *Statistics of Mines and Mining in the States and Territories West of the Rocky Mountains*. Washington

Salter, C, 2009 Metallurgical Report on Samples from eight iron mining tools from Combe Down Stone Mines, in *Combe Down Stone Mines: Specialist Reports* Oxford Archaeology (www.combedownstoneminesproject.com)

Scott, I R, 2010 Glass, in *Combe Down Stone Mines Specialist Reports* Oxford Archaeology (www.combedownstoneminesproject.com)

Scott, I, 2009 Metal Finds in *Combe Down Stone Mines: Specialist Reports*, Oxford Archaeology (www.combedownstoneminesproject.com)

Shaffrey, R, 2009 The Worked Stone in *Combe Down Stone Mines: Specialist Reports*, Oxford Archaeology (www.combedownstonemines project.com)

Silvertant, J, and Mhairtin, J N, 2008 *What Lies Beneath: Report of the 3rd International Symposium on Archaeological Mining History 2008.* Institute Europa Subterranea, Maastricht

Stanier, P, 2000 *Stone Quarry Landscapes* Stroud, Glos

Strassburger, M, 2008 Mining Archaeology in South West Germany. Methods, Results and Protection of Mining Sites in Silvertant and Mhairtin 2008, 39-80

Thorpe, T, 1742 *An Actual Survey of the City of Bath and of Five Miles Around*

Toulmin Smith, L, (ed) 1907 *Leland's Itinerary in England and Wales Vol 1*

Toulmin Smith, L, (ed) 1910 *Leland's Itinerary in England and Wales Vol V*

Tunstall, J, 1847 *Rambles About Bath*

Tyte, W, 1897 *Combe Down in Olden Times*

Vile, N, 1991 *Exploring the Kennet and Avon Canal: Country Walks between Bristol and Reading* Countryside Books, Newbury

Wagner, G A, and Weisgerber, G, 1986 *Silber, Blei und Gold auf Sifnos. Prehistorische und antike Metallproduktion.* Der Anschnitt, Deutches Bergbau-Museum, Bochum

Weisgerber, G, and Goldenberg, G, 2004 *Alpenkupfer – Rame delle Alpi.* Der Anschnitt, Bergbau-Museum, Bochum

Willies, L, 1993 *Underground Metal Mines, Step 1 Report.* Cranstone Consultancy for English Heritage Monuments Protection Programme

Winchester, S, 2001 *The Map that Changed the World*, London

Wood, J, 1765 *Description of Bath* Reprinted (1967) by Kingsmead, Bath

Woodward, H B, 1876 *Geology of East Somerset and the Bristol Coal Fields* Longman's Green and Co. for HMSO

Wooster, F, 1978 The Stone Industry at Bath. *J Bristol Industrial Archaeol Soc* 11, 5-13

Other references consulted

Allsop, N, 1987 *The Kennet and Avon Canal*, Millstream Books, Bath

Allsop, N, 1988 *The Somersetshire Coal Canal Rediscovered.* Millstone Books, Bath

Arlett, M, 1986 *The Somerset and Dorset at Midford.* Millstream Books, Bath

Attwood, B, 1970 The BIAS Paper Mills Survey. *J Bristol Industrial Archaeol Soc* 3, ll-21

Bentley, J B, and Morless, B J, 1987 *Somerset Roads. The legacy of Turnpikes, Phase 2. East Somerset* Somerset Industrial Archaeology Society

Blunden, J, 1975 *The Mineral Resources of Britain* London

Broome, J, 1988 Tuckingmill. *J Bristol Industrial Archaeol Soc* 20, 2-3

Buchanan, A, 1969 Combe Hay Caisson Lock *J Bristol Industrial Archaeol Soc* 2, 27-30

Buchanan, A, and Browning, D, 1968 A Note on the Bath Roads *J Bristol Industrial Archaeol Soc* **1**, 15-16

Buchanan, R, and Cossons, N, 1969 *The Industrial Archaeology of the Bristol Region.* Newton Abbott

Butcher, A, and Mayhew, R, 1971 Combe Down Freestone Mines, *Shepton Mallet Caving Club*, **5:1**, 3-9+ii

Calvert, M, 1963 Bath Stone Mines: Their history and method of operation. *Belfrey Bulletin*, **185** July, 2-5

Chapman, M, 1988 The Somerset Coal Canal: a Cartographic Survey. *J Bristol Industrial Archaeol Soc*, 20, 4-21

Chipchase, N, 1977 Combe Down Quarry Mine. *Cerberus Speleological Society Newsletter* **7** (15), 375-78

Chipchase, N, 1978 More underground ramblings at Combe Down. *Cerberus Speleological Society Newsletter*, **8** (5), 178-9

Clark, G, 1987 *Prior Park: A Compleat Landscape* Bath

Clew, K R, 1968 *The Kennet and Avon Canal* Newton Abbott

Clew, K R, 1970 *The Somersetshire Coal Canal and Railways* Newton Abbott

Cotterell, T S, 1905 Bath Stone. *Antiquity* (New Series) **1**:5, 87-91

Dallimore, K, 1988 *Exploring Combe Down* Millstream Books, Bath

Defoe, D, 1738 *A Tour through the Whole Island of Great Britain*

Donovan, D, 1964 *Exposures in the Great Oolite Series near Bath* Geology Association Guide

Green, G W, and Donovan J, 1969 The Great Oolite Series of the Bath Area. *Bulletin of the Geological Survey* **30**, 1-64

Green, M A, 1904 *The 18th Century Architecture of Bath*

Hardy, S, and Lambert, D, (eds) 1994 *Parks and Gardens of Avon*

Ibbetson, Laporte and Hassell, J, 1793 *Picturesque Bath, Bristol and Hot Wells*

Irving, R T A, 2009 *William Smith and his venture into Stone Manufacture* electronic publication, Combe Down Heritage Society Website, http://combedownheritagesociety.wordpress.com/

Ison, W, 1948 *The Georgian Buildings of Bath 1700-1830* Reprinted, Kingsmead, Bath in 1969

Lonsdale, W, 1832 On the Oolite District of Bath. *Transactions of the Geological Society, Series 2*, **3**, 241

Mehew, R, 1976 Combe Down Stone Mine. *Shepton Mallett Caving Club Journal* 6, l, 3-5

National Trust 1996 *Prior Park Landscape Garden* London

Page, W, (ed) 1911 *Victoria County History Somerset*, **2**

Palmer, T, 2008 Limestone Petrography and Durability in English Jurassic Limestones, in H Doyle, T Hughes, and I Thomas, *England's Heritage in Stone: Proceedings of a conference at Tempest Anderson Hall, York. 15-17 March 2005* English Stone Forum, Folkstone, Kent, 66-78

Perm, I E, Merriman, R J, and Wyatt, R J, 1979 *The Bathonian Strata of the Bath: Frome Area.* IGS. Report No. 78/22. HMSO

Pocock, R, 1888 *Travels Through England* ed. J J Cartwright, Vol. I

Ponsford, M, and Jackson, R, 1996 Post-Medieval Britain in 1995, in *Post Medieval Archaeology* **30**, 282

Ponsford, M, and Jackson, R, 1997 Post-Medieval Britain in 1996, in *Post-Medieval Archaeology* **31**, 296-7

Pope, C, 1765 *The New Bath Guide* (3rd edition)

Post Office 1837 *The Post Office Bath Directory*

Powell, J, 1978 Fullers Earth from Midford. *J Bristol Industrial Archaeol Soc* **11**, 22-25

Price, G, 1979 Combe Down Stone Mines *Cerberus Speleological Society* **9** (2), 51-3

Price, R, and Watts, L, 1981 Rescue excavations at Combe Hay, Somerset, 1968–73 *Proc. Somerset Archaeol and Natur Hist* **124**: 50

Pugh, R B, (ed) 1911 *Victoria County History. Somerset* Vol. **4**

Randells, 1864 *Paper for the British Association for the Advancement of Science* (held by the Bath Central Library)

Reader, W J, 1980 *The MacAdam Family and the Turnpike Roads 1798–1861* London

Reaich, N, 1992 Two Stone Mines near Bath. *Shepton Mallet Caving Club Journal* **9** (3), 33-36

Roche, J S, 1931 *A History of Prior Park College* London

Scarth, H M, 1864 *Aquae Solis, or Notices of Roman Bath* London

Schinkel, K F, 1986 *Reise Nach England, Schottland Und Paris im Jahre 1826* Berlin

Schofield, R B, 2000 *Benjamin Outram: 1764–1805, An Engineering Biography* Cardiff

Seiveking, G de G, 1979 Grimes Graves and Prehistoric European Flint Mining in H Crawford (ed), *Subterranean Britain: Aspects of Underground Archaeology* London, l-43

Shepton Mallet Caving Club (*c* 1998) *Box Freestone Mine Survey, Wilts.* Shepton Mallet Caving Club, Priddy, Bristol

Smith, A, 2006 *The Ringing Grooves of Change: Brunel and the coming of the railway to Bath* Bath

Stanier, P, 1995 *A historic Photographic Record of Quarries of England and Wales* Truro

Stell, F, 2007 *Take my pickaxe mate, I'm late for foreplay class.* Times 2, October 2007

Torrens, H, 1978 *The Evolution of a Family Firm: Stothert and Pitt of Bath* Bath

Trueman, M, 1996 *Monuments Protection Programme: The Lime, Cement and Plaster Industries, Step 1 Report.* Lancaster University Archaeological Unit.

Ure, A, 1849 *Dictionary of Arts, Manufactures and Mines.* Appleton and Co. New York.

Ward-Perkins, J W, Brooks, A T, and Pearce, A E McR, 1979 *Bath Stone: A quarry History.* University College, Cardiff

Watts, M, 1975 John Padmores cranes at Bath and Bristol *J Bristol Industrial Archaeol Soc* 8, l 17-19

Index